*Piet W.N.M. van Leeuwen*
*and John C. Chadwick*

**Homogeneous Catalysts**

## Further reading

Arpe, H.-J.

**Industrial Organic Chemistry**

**Fifth, Completely Revised Edition**

2010
ISBN: 978-3-527-32002-8

Behr, A. / Neubert, P.

**Applied Homogeneous Catalysis**

2012
ISBN: 978-3-527-32641-9 (Hardcover)
ISBN: 978-3-527-32633-4 (Softcover)

Roberts, S. M.

**Catalysts for Fine Chemical Synthesis**

**Volumes 1-5. Set**

2007
ISBN: 978-0-470-51605-8

Platz, M. S., Moss, R. A., Jones, M. (eds.)

**Reviews of Reactive Intermediate Chemistry**

2007
ISBN: 978-0-471-73166-5

Sheldon, R. A., Arends, I., Hanefeld, U.

**Green Chemistry and Catalysis**

2007
ISBN: 978-3-527-30715-9

Dalko, P. I. (ed.)

**Enantioselective Organocatalysis**

**Reactions and Experimental Procedures**

2007
ISBN: 978-3-527-31522-2

*Piet W.N.M. van Leeuwen and John C. Chadwick*

# Homogeneous Catalysts

Activity – Stability – Deactivation

**WILEY-VCH**

WILEY-VCH Verlag GmbH & Co. KGaA

**The Authors**

*Prof. Dr. Piet W.N.M. van Leeuwen*
Institute of Chemical Research
of Catalonia (ICIQ)
Av. Paisos Catalans 16
43007 Tarragona
Spain

*Dr. John C. Chadwick*
University of Eindhoven
Chemical Engineering & Chemistry
P.O. Box 513
5600 MB Eindhoven
The Netherlands

**Library of Congress Card No.:** applied for

**British Library Cataloguing-in-Publication Data**
A catalogue record for this book is available from the British Library.

**Bibliographic information published by the Deutsche Nationalbibliothek**
The Deutsche Nationalbibliothek lists this publication in the Deutsche Nationalbibliografie; detailed bibliographic data are available on the Internet at http://dnb.d-nb.de.

© 2011 Wiley-VCH Verlag & Co. KGaA, Boschstr. 12, 69469 Weinheim, Germany

**Typesetting** Thomson Digital, Noida, India
**Printing and Binding** Fabulous Printers Pte Ltd
**Cover Design** Adam Design, Weinheim

Printed in Singapore
Printed on acid-free paper

**Print ISBN:** 978-3-527-32329-6

# Contents

# Preface

Homogeneous catalysts have played a key role in the production of petrochemicals and coal-derived chemicals since the 1960s. In the last two decades, transition metal catalysts have revolutionized synthetic organic chemistry, both in the laboratory and in industrial production. The use of homogeneous catalysts in polyolefin synthesis started in the 1980s and triggered enormous R&D efforts, leading to hitherto inaccessible polymers and to greatly improved control over polymer structure and properties. The introduction of new processes and catalysts continues in bulk chemical production, as exemplified by new routes that have recently come on stream for the production of 1-octene and methyl methacrylate.

For all catalysts, selectivity and rates of reactions are crucial parameters and in the laboratory even the rate may not concern us that much, as catalyst loadings of 5% or more are often applied. For industrial applications, however, high turnover numbers are required for economic reasons, which may be more complex than simply catalyst costs. For the bulk chemical applications, studies of catalyst activation, activity, stability, deactivation, recycling and regeneration have always formed an integral part of catalysis research. A considerable research effort has been devoted to this, mainly in industry, but explicit publications are scarce, although some stability issues can be deduced from the patent literature. Catalyst stability has been a highly important factor in transforming advances in catalysis into practical applications, notably in the areas of polymer synthesis, cross-coupling chemistry, hydrogenation catalysis, carbonylation reactions and metathesis chemistry.

In heterogeneous catalysis, the study of activation, deactivation, and regeneration of catalysts has always been a major research activity. These topics have been addressed in many articles, books and conferences, and literature searches with these keywords give many relevant results. For homogeneous catalysts this is not the case, with the possible exception of metathesis A wealth of knowledge can be found in a vast number of publications, but this is not easily retrieved. The approach in homogeneous catalysis is entirely different to that of heterogeneous catalysis, especially before industrial applications come into sight; in homogeneous catalysis, the general approach to improving the catalyst performance is variation of one of the catalytic components, without much attention being paid to the question of why other catalyst systems failed.

In this book, we address a number of important homogeneous catalysts, focusing on activity, stability and deactivation, including the important issue of how deactivation pathways can be avoided. Key concepts of activation and deactivation, together with typical catalyst decomposition pathways, are outlined in the first chapter. Chapters 2–6 cover homogeneous catalysts for olefin polymerization and oligomerization, including the effects of catalyst immobilization and polymerization rate limitation as a result of dormant site formation. The following sections of the book, Chapters 7–10, describe catalyst activity and stability in asymmetric hydrogenation, hydroformylation, alkene-CO reactions, methanol carbonylation, metal-catalyzed cross-coupling catalysis and, finally, alkene metathesis.

We hope that the contents of this book will be valuable to many scientists working in the field of homogeneous catalysis and that the inclusion of a broad range of topics, ranging from polymerization catalysis to the synthesis of speciality and bulk chemicals, can lead to useful cross-fertilization of ideas.

We would like to acknowledge useful comments and contributions from Rob Duchateau, Peter Budzelaar and Nick Clément. We thank Marta Moya and María José Gutiérrez for polishing the final draft. We are also indebted to Manfred Kohl from Wiley-VCH, for convincing one of us at the 10th International Symposium on Catalyst Deactivation to embark on this book project. We thank him, Lesley Belfit and their team for the perfect support provided throughout.

February 2011

*Piet W.N.M. van Leeuwen*
*John C. Chadwick*

**Piet van Leeuwen** *(1942) is group leader at the Institute of Chemical Research of Catalonia in Tarragona, Spain, since 2004. After receiving his PhD degree at Leyden University in 1967 he joined Shell Research in 1968. Until 1994 he headed a research group at Shell Research in Amsterdam, studying many aspects of homogeneous catalysis. He was Professor of Homogeneous Catalysis at the University of Amsterdam from 1989 until 2007. He has coauthored 350 publications, 30 patents, and many book chapters, and is author of the book Homogeneous Catalysis: Understanding the Art. He (co)directed 45 PhD theses. In 2005 he was awarded the Holleman Prize for organic chemistry by the Royal Netherlands Academy. In 2009 he received a doctorate honoris causa from the University Rovira I Virgili, Tarragona, and he was awarded a European Research Council Advanced Grant.*

**John Chadwick** *was born in 1950 in Manchester, England and received his B.Sc. and Ph.D. degrees from the University of Bristol, after which he moved to The Netherlands, joining Shell Research in Amsterdam in 1974. He has been involved in polyolefin catalysis since the mid 1980s and in 1995 transferred from Shell to the Montell (later Basell) research center in Ferrara, Italy, where he was involved in fundamental Ziegler–Natta catalyst R&D. From 2001 to 2009, he was at Eindhoven University of Technology on full-time secondment from Basell (now LyondellBasell Industries) to the Dutch Polymer Institute (DPI), becoming DPI Programme Coordinator for Polymer Catalysis and Immobilization. Until his retirement in 2010, his main research interests involved olefin polymerization catalysis, including the immobilization of homogeneous systems, and the relationship between catalyst and polymer structure. He is author or co-author of more than 60 publications and 11 patent applications.*

# 1
# Elementary Steps

## 1.1
## Introduction

Catalyst performance plays a central role in the literature on catalysis and is expressed in terms of selectivity, activity and turnover number. Most often catalyst stability is not addressed directly by studying why catalysts perform poorly, but by varying conditions, ligands, additives, and metal, in order to find a better catalyst. One approach to finding a suitable catalyst concerns the screening of ligands, or libraries of ligands [1] using robotics; especially, supramolecular catalysis [2–4] allows the fast generation of many new catalyst systems. Another approach is to study the decomposition mechanism or the state the catalyst precursor is in and why it is not forming an active species. For several important reactions such studies have been conducted, but they are low in number. As stated in the preface, in homogeneous catalysis there has always been less attention given to catalyst stability [5] than there is in heterogeneous catalysis [6]. We favor a combined approach of understanding and exploration, without claiming that this is more efficient. In the long term this approach may be the winner for a reaction that we have got to know in much detail. For reactions, catalysts, or substrates that are relatively novel a screening approach is much more efficient, as shown by many examples during the last decade; we are not able to study all catalysts in the detail required to arrive at a level at which our knowledge will allow us to make predictions. We can reduce the huge number of potential catalysts (ligands, metals, co-catalysts) for a desired reaction by taking into account what we know about the decomposition reactions of our coordination or organometallic complexes and their ligands. Free phosphines can be easily oxidized and phosphites can be hydrolyzed and thus these simple combinations of ligands and conditions can be excluded from our broad screening program. In addition we can make sophisticated guesses as to what else might happen in the reaction with catalysts that we are about to test and we can reduce our screening effort further. To obtain a better understanding we usually break down the catalytic reaction under study into elementary steps, which we often know in detail from (model) organometallic or organic chemistry. As many books do, we can collect elementary steps and reverse the process and try to design new catalytic cycles. We can do the same for decomposition

*Homogeneous Catalysts: Activity – Stability – Deactivation*, First Edition. Piet W.N.M. van Leeuwen and John C. Chadwick.
© 2011 Wiley-VCH Verlag GmbH & Co. KGaA. Published 2011 by Wiley-VCH Verlag GmbH & Co. KGaA.

processes and first look at their elementary steps [7]; the process may be a single step or more complex, and even autocatalytic. In this chapter we will summarize the elementary reactions leading to the decomposition of the metal complexes and the ligands, limiting ourselves to the catalysis that will be dealt with in the chapters that follow.

## 1.2
## Metal Deposition

Formation of a metallic precipitate is the simplest and most common mechanism for decomposition of a homogeneous catalyst. This is not surprising, since reducing agents such as dihydrogen, metal alkyls, alkenes, and carbon monoxide are the reagents often used. A zerovalent metal may occur as one of the intermediates of the catalytic cycle, which might precipitate as metal unless stabilizing ligands are present. Precipitation of the metal may be preceded by ligand decomposition.

### 1.2.1
### Ligand Loss

A typical example is the loss of carbon monoxide and dihydrogen from a cobalt hydrido carbonyl, the classic hydroformylation catalyst (Scheme 1.1).

$$2\ HCo(CO)_4 \xrightarrow{-H_2} Co_2(CO)_8 \xrightarrow{-CO} Co\ metal \downarrow$$

**Scheme 1.1** Precipitation of cobalt metal.

The resting state of the catalyst is either $HCo(CO)_4$ or $RC(O)Co(CO)_4$, and both must lose one molecule of CO before further reaction can take place. Thus, loss of CO is an intricate part of the catalytic cycle, which includes the danger of complete loss of the ligands giving precipitation of the cobalt metal. Addition of a phosphine ligand stabilizes the cobalt carbonyl species forming $HCo(CO)_3(PR_3)$ and, consequently, higher temperatures and lower pressures are required for this catalyst in the hydroformylation reaction.

A well-known example of metal precipitation in the laboratory is the formation of "palladium black", during cross coupling or carbonylation catalysis with the use of palladium complexes. Usually phosphorus-based ligands are used to stabilize palladium(0) and to prevent this reaction.

### 1.2.2
### Loss of $H^+$, Reductive Elimination of HX

The loss of protons from a cationic metal species, formally a reductive elimination, is a common way to form zerovalent metal species, which, in the absence of stabilizing ligands, will lead to metal deposition. Such reactions have been described for metals such as Ru, Ni, Pd, and Pt (Scheme 1.2). The reverse reaction is a common way to regenerate a metal hydride of the late transition metals and clearly the position of this equilibrium will depend on the acidity of the system.

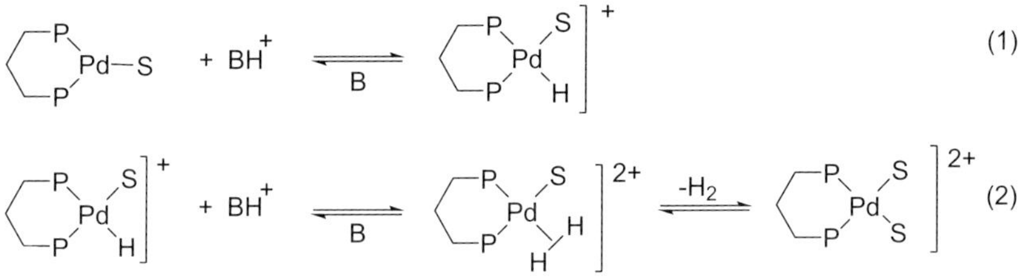

**Scheme 1.2** Reactions involving protons and metal hydrides.

Too strongly acidic media may also lead to decomposition of the active hydride species via formation of dihydrogen and a di-positively charged metal complex (reaction (2), Scheme 1.2). All these reactions are reversible and their course depends on the conditions.

As shown in the reaction schemes for certain alkene hydrogenation reactions and most alkene oligomerization reactions (Schemes 1.3 and 1.4), the metal maintains the divalent state throughout, and the reductive elimination is not an indissoluble part of the reaction sequence.

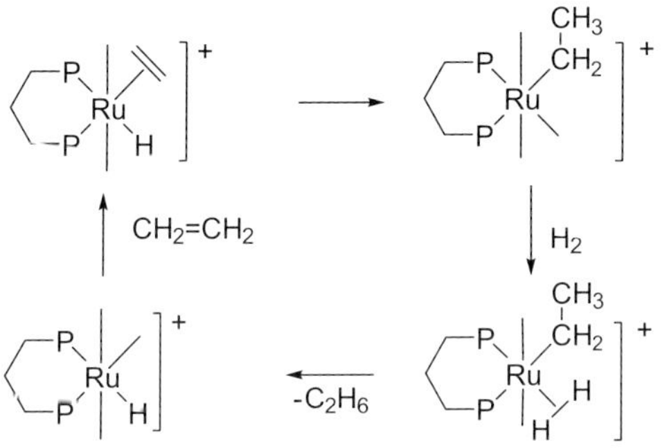

**Scheme 1.3** Simplified scheme for heterolytic hydrogenation.

The species $L_nMH^+$ are stabilized by phosphine donor ligands, as in the Shell Higher Olefins Process (M=Ni) [8] and in palladium-catalyzed carbonylation reactions [9].

We mention two types of reactions for which the equilibrium, shown in Scheme 1.2, between $MH^+$ and $M + H^+$ is part of the reaction sequence, the addition of HX to a double bond and the Wacker reaction. As an example of an HX addition we will take hydrosilylation, as for HCN addition the major decomposition

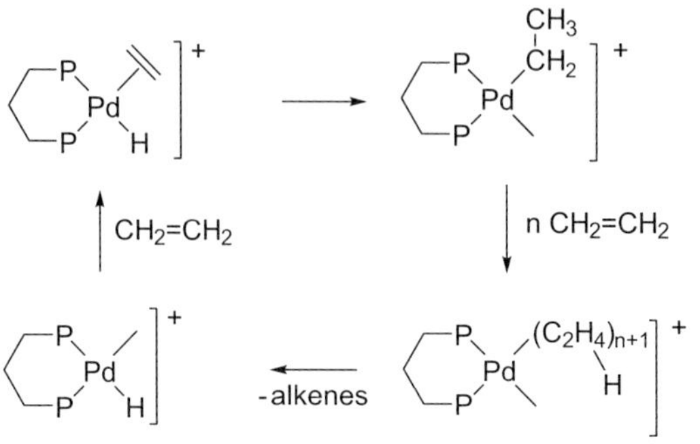

Scheme 1.4 Alkene oligomerization.

reaction is a different one, as we will see later. The hydrosilylation reaction is shown in Scheme 1.5 [10].

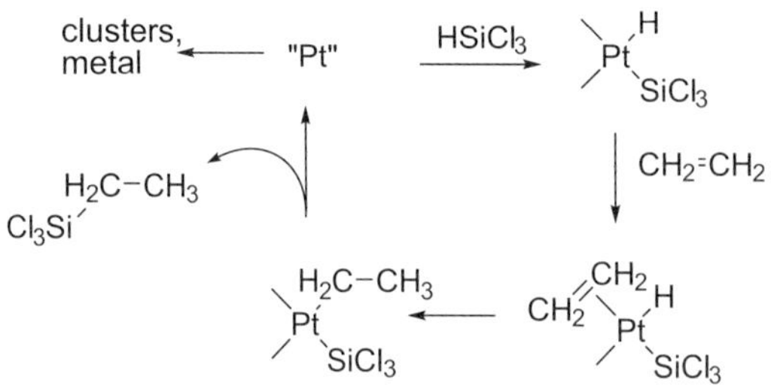

Scheme 1.5 Simplified mechanism for hydrosilylation.

In the Wacker reaction, elimination of HCl from "PdHCl" leads to formation of palladium zero [11] and the precipitation of palladium metal is often observed in the Wacker reaction or related reactions [12]. In the Wacker process palladium(II) oxidizes ethene to ethanal (Scheme 1.6) and, since the re-oxidation of palladium by molecular oxygen is too slow, copper(II) is used as the oxidizing agent. Phosphine ligands cannot be added as stabilizers for palladium zero, because they would be oxidized. In addition, phosphine ligands would make palladium less electrophilic, an important property of palladium in the Wacker reaction.

In the palladium-catalyzed Heck reaction (Scheme 1.7), as in other cross coupling reactions, the palladium zero intermediate should undergo oxidative addition before precipitation of the metal can occur. Alternatively, Pd(0) can be "protected" by ligands present, as in the example of Scheme 1.7, but this requires another dissociation step before oxidative addition can occur. Both effective ligand-free systems [13] and ligand-containing systems have been reported [14]. A polar medium accelerates the oxidative addition. The second approach involves the use of bulky ligands, which give rise to low coordination numbers and hence electronic unsaturation and more reactive species. Turnover numbers of millions have been reported [15].

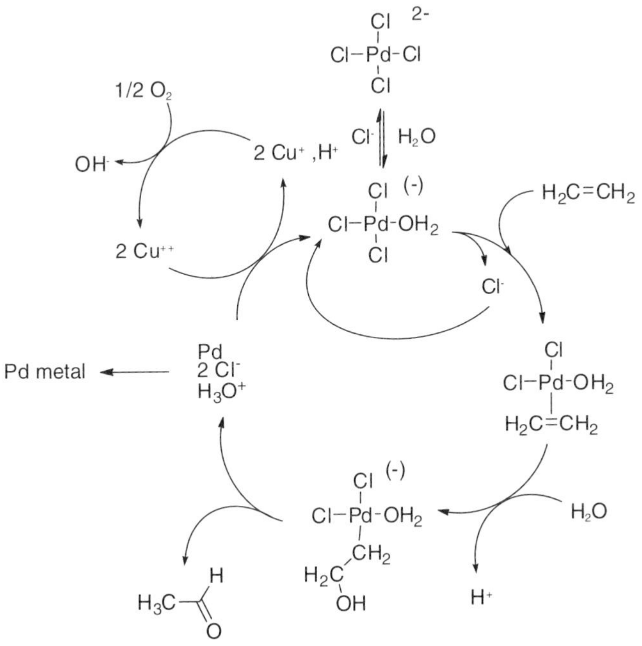

$$\begin{array}{c} \text{Cl} \\ | \\ \text{Cl}-\text{Pd}-\text{Cl} \\ | \\ \text{Cl} \end{array}\bigg]^{2-}$$

1/2 O₂

2 Cu⁺ ,H⁺

OH⁻

2 Cu⁺⁺

$$\begin{array}{c} \text{Cl} \\ | \\ \text{Cl}-\text{Pd}-\text{OH}_2 \\ | \\ \text{Cl} \end{array}\bigg]^{(-)}$$

H₂O

H₂C=CH₂

Cl⁻

Pd metal ← Pd 2 Cl⁻ H₃O⁺

$$\begin{array}{c} \text{Cl} \\ | \\ \text{Cl}-\text{Pd}-\text{OH}_2 \\ | \\ \text{H}_2\text{C}=\text{CH}_2 \end{array}$$

$$\begin{array}{c} \text{Cl} \\ | \\ \text{Cl}-\text{Pd}-\text{OH}_2 \\ | \\ \text{CH}_2 \\ \text{H}_2\text{C} \quad \text{OH} \end{array}\bigg]^{(-)}$$

H₂O

H⁺

H₃C— (O)

H

**Scheme 1.6** Ethanal formation from ethene via a Wacker oxidation reaction.

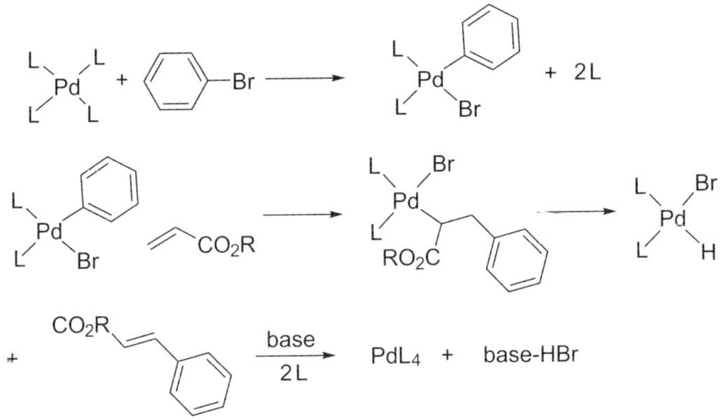

$$\begin{array}{c} \text{L} \quad \text{L} \\ \text{Pd} \\ \text{L} \quad \text{L} \end{array} + \bigcirc-\text{Br} \longrightarrow \begin{array}{c} \text{L} \quad \bigcirc \\ \text{Pd} \\ \text{L} \quad \text{Br} \end{array} + 2\text{L}$$

$$\begin{array}{c} \text{L} \quad \bigcirc \\ \text{Pd} \\ \text{L} \quad \text{Br} \end{array} + \diagup\diagdown\text{CO}_2\text{R} \longrightarrow \begin{array}{c} \text{L} \quad \text{Br} \\ \text{Pd} \\ \text{L} \quad \text{RO}_2\text{C} \end{array} \longrightarrow \begin{array}{c} \text{L} \quad \text{Br} \\ \text{Pd} \\ \text{L} \quad \text{H} \end{array}$$

$$+ \quad \text{CO}_2\text{R}\diagup\diagdown \xrightarrow[\text{2L}]{\text{base}} \text{PdL}_4 + \text{base-HBr}$$

**Scheme 1.7** The mechanism of the Heck reaction using excess phosphine.

## 1.2.3
### Reductive Elimination of C-, N-, O-Donor Fragments

Many cross-coupling reactions have been reported in the last decades. The palladium and nickel-catalyzed formation of C−C, C−N, C−O, and C−P bonds has become an important tool for organic syntheses [16]. The general mechanism for C−C bond formation is depicted in Scheme 1.8. Again, the zerovalent state of the metal is an intrinsic part of the mechanism. As in the Heck reaction, the phosphine ligands must prevent metal deposition and/or oxidative addition of a hydrocarbyl halide should be faster than metal deposition.

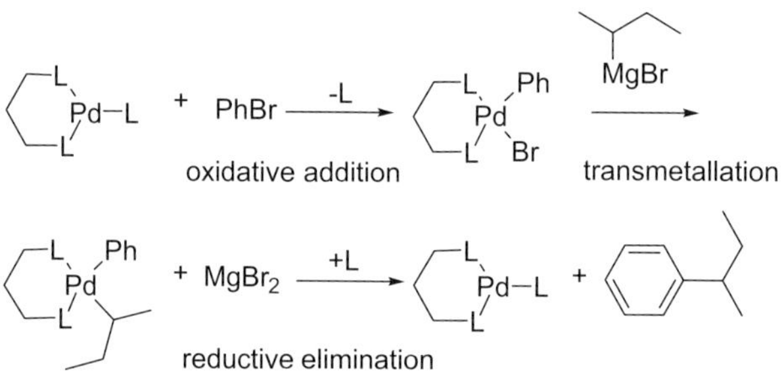

**Scheme 1.8** General mechanism for cross-coupling reactions.

### 1.2.4
### Metallic Nanoparticles

Formation of metal agglomerates probably starts with dimer and trimer formation and occasionally this has been observed via mass spectroscopy or EXAFS; the latter case concerns a palladium diphosphine catalyst forming first dimeric and trimeric species before clusters were observed [17]. On the way from metal complexes to bulk metal particles the system passes undoubtedly through the stage of metal nanoparticles (MNPs). Often this can be deduced from the intermediate yellow, green and blue solutions before a black precipitate is observed. MNPs can also be synthesized on purpose and used as a catalyst [18]. The selective formation of "giant clusters" or MNPs [19] can be regulated by the conditions, metal to ligand ratios, stabilizing agents [20] such as polymers, solid surfaces [21], ionic liquids [22], surfactants [23], and dendrimers [24]. MNPs are soluble, recyclable species, which may present an intermediate between homogeneous and heterogeneous catalysts [25]. Reactions typical of heterogeneous catalysts, such the hydrogenation of alkenes and aromatics, may take place on the surface of the MNPs, and most likely they remain intact [26]. In oxidation catalysis palladium giant clusters (called such at the time) have been known for quite some time [27], but the nature of the actual catalyst is not known. Reactions with PdNP catalysts that strongly resemble homogeneous catalytic processes, such as cross coupling, the Heck reaction, and allylic alkylation have been the subject of much discussion as to whether the PdNP serves as the catalyst or as a sink/precursor for monometallic complexes [28]. Ligand-free palladium "atoms" (solvated, though) are probably very active catalysts in C−C coupling reactions and this may explain why nanoparticles can lead to active catalysts, and even to "efficient" recycling, as only a very small amount of the catalyst precursor is consumed in each cycle. Even asymmetric MNP catalysts have been reported, and examples include Pt-catalyzed hydrogenation of ethyl pyruvate [29], Pd-catalyzed hydrosilylation of styrene [30], and

Pd-catalyzed allylic alkylation of racemic substrates [31]. Modification of surfaces with chiral molecules has been known for several decades to give rise to enantio-selective catalysis [32], but the similarity of ligands used in homogeneous and MNP-based enantioselective catalysis seems suspect. Evidence is growing that the latter reactions are catalyzed by homogeneous complexes [28, 33].

In the pyridine-palladium acetate catalyzed oxidation of alcohols the formation of PdNPs was observed by transmission electron microscopy (TEM) measurements [34], but, by using dendritic pyridine ligands containing a 2,3,4,5-tetraphenylphenyl substituent at the 3-position of the pyridine ring, this was suppressed successfully by Tsuji and coworkers [35].

A key issue for homogeneous catalysis is that MNPs can form in a reversible manner, while the formation of larger metal particles is usually irreversible, both thermodynamically and/or kinetically. MNPs still hold promise for new reactions to be discovered and as precursors for molecular catalysts they have shown advantages, but the control of their size during catalysis seems an intrinsic problem not to be solved easily.

## 1.3
## Ligand Decomposition by Oxidation

### 1.3.1
### General

The main tool for catalyst modification in homogeneous, catalytic processes is modification of the ligands. By changing the ligand properties we try to obtain better selectivity and activity. Decomposition of the ligands and their complexes has a large influence on the performance of the system. Catalysts based on late transition metals often contain phosphites and phosphines as modifying ligands. The decomposition routes of these ligands have received a great deal of attention. They are sensitive to many reactions as we will see. Nitrogen-based ligands, such as amines, imines and pyridines, are much more robust in general, but they are much less effective as ligands for low-valent late transition metals, such as rhodium(I) in rhodium-catalyzed hydroformylation. In ionic complexes though, we have seen an enormous increase in the use of nitrogen donor ligands in catalytic reactions that have become highly efficient and selective.

### 1.3.2
### Oxidation

#### 1.3.2.1 Catalysis Using $O_2$
Homogeneous catalysts for oxidation reactions using $O_2$ do not contain modifying soft ligands, but they are ionic species solvated by water, acetic acid, and the like. Examples include the Wacker process for making ethanal (palladium in water) and the oxidation of *p*-xylene to terephthalic acid (cobalt in acetic acid) [36].

Ligands based on nitrogen donor atoms are the ligands of choice; they stabilize high-valent metal ions and are not as sensitive to oxidation as phosphorus- or sulfur-based ligands. For example, phenanthroline ligands were used for the palladium-catalyzed oxidation of alcohols to ketones or aldehydes [37], and diamines are effective ligands for the copper-catalyzed oxidative coupling of phenols in the synthesis of polyphenylene ether [38]. We are not aware of commercial processes utilizing polydentate nitrogen ligands yet, although many interesting new oxidation catalysts have been reported in recent years [39]. Oxidation of the ligand backbone may be a concern as even porphyrins should be used in halogenated form in order to enhance their stability in oxidation reactions [40].

### 1.3.2.2 Catalysis Using Hydroperoxides
The commercial processes using hydroperoxides (*t*-butyl hydroperoxide and 1-phenylethyl hydroperoxide) for the epoxidation of propene involve "ligand-free" metals such as titanium alkoxides and ligand oxidation is not an issue for these processes [41]. For the asymmetric epoxidation using Sharpless' catalyst, ligand oxidation is also not a major issue [42].

Phosphorus ligands are very prone to oxidation. Therefore, oxygen and hydroperoxides have to be thoroughly removed from our reagents and solvents before starting our catalysis. In spite of this common knowledge, oxidation of phosphine ligands has frequently obscured the catalytic results.

When phosphines are bonded strongly to a transition metal such that no or little dissociation occurs, their oxidation by hydroperoxides will not take place. This is the case for the platinum-catalyzed epoxidation reaction of alkenes by hydrogen peroxide developed by Strukul [43]. The bidentate phosphine ligands "survive" the hydroperoxidic conditions and asymmetric and regioselective epoxidations have been achieved, proving that the chiral ligands remain intact and coordinated to platinum. Typically, turnover numbers are 50 to100, and while the use of hydrogen peroxide is attractive from a green chemistry point of view, these modest numbers have so far not led to industrial applications. Clearly, from a cost point of view, the Sharpless catalyst seems more attractive.

## 1.4
## Phosphines

### 1.4.1
### Introduction

Phosphines and diphosphines are widely used as the ligand component in homogeneous catalysts. Large-scale processes include rhodium-catalyzed hydroformylation for propene, butene, and heptene, ethene oligomerization, cobalt-catalyzed hydroformylation for internal higher alkenes, and butadiene dimerization. Small-scale operations include asymmetric hydrogenation of enamides and substituted acrylic acids, asymmetric isomerization to make menthol, alkoxycarbonylation (ibuprofen), and Group 10 metal-catalyzed C—C bond formation (Heck reaction, Suzuki reaction). Future possibilities may comprise selective trimerization and tetramerization of ethene [44], more alkoxycarbonylations (large-scale methyl methacrylate, more pharmaceuticals), hydroxycarbonylations, a large variety of new C—C [45] and C—N coupling reactions, asymmetric alkene dimerizations, alkene metathesis, and new hydroformylation reactions.

Garrou [46] reviewed the decomposition of phosphorus ligands in relation to homogeneous catalysis many years ago. Many interesting studies on phosphorus ligand decomposition have appeared since, but Garrou's review is still a useful collection of highly relevant reactions. At the time, the formation of phosphido species seemed to be the most common fate of our phosphine-containing catalysts, but in the last two decades many more reaction types have been added, as we reviewed in 2001 [7]. In a recent review Parkins discusses reactions taking place in the ligand in the coordination sphere, a large number of them being examples of phosphines [47]. The reactions in Parkins' review are ordered by metal. Exchange of substituents at phosphorus in metal complexes has been reviewed by Macgregor recently and his review shows that many new reactions have been discovered since the review by Garrou was published [48].

## 1.4.2
## Oxidation of Phosphines

Oxidation of free phosphines was mentioned above (Section 1.3.2.2) as a reaction leading to phosphine loss. Phosphines are used extensively in a large number of organic synthetic reactions in which they usually end up as the phosphine oxide, that is, they are used as reducing agents. Well known examples are the Mitsunobu reaction to generate esters from alcohols and carboxylic acids under very mild conditions, and the Appel reaction to convert alcohols into alkyl halides, and the Wittig reaction. Therefore it is not surprising that, in catalysis, oxidation of phosphines is a common way to deactivate catalytic systems. Common oxidizing agents are dioxygen and hydroperoxides. High-valent metals may also function as the oxidizing agent. Sometimes this reaction is utilized on purpose and the reducing function of phosphines is used to activate the catalyst; for example palladium(II) acetate can be reduced by an excess of phosphine ligand (see Scheme 1.9, third reaction).

$$PR_3 + H_2O \longrightarrow H_2 + O=PR_3$$

$$PR_3 + CO_2 \longrightarrow CO + O=PR_3$$

$$PR_3 + Pd(OAc)_2 + H_2O \longrightarrow Pd(0) + 2\ HOAc + O=PR_3$$

$$PR_3 + 1/2\ O_2 \longrightarrow O=PR_3$$

**Scheme 1.9** Examples for oxidation of phosphines.

For instance, in cross coupling chemistry a palladium(II) precursor is reduced *in situ* by phosphine to generate Pd(0), the "active" species (the oxygen atom is provided by water). Molybdenum(VI), tungsten(VI) and water have been reported as oxidizing agents of phosphines [49]. Rh(III) carbonate oxidizes triphenylphosphine forming $CO_2$ and Rh(I) [50]. Rh(III) in water was found to oxidize tppts yielding tppts-O and Rh(I) [51]. Thermodynamics show that even water and carbon dioxide may oxidize phosphines to the corresponding oxides. These reactions may be catalyzed by the transition metal in the system, for example, Rh for $CO_2$ [52]. Water oxidizes phosphines using palladium as a catalyst and palladium has to be thoroughly removed after its use in a P−C cross coupling synthesis [53]. It should be mentioned, however, that, in view of the many successful applications of water and $sCO_2$ as solvents in homogeneous catalysis, these oxidation reactions are relatively rare.

Oxidation, or partial oxidation of phosphine can also be turned into a useful reaction if an excess of phosphine retards the catalytic reaction. Above we have mentioned that phosphine-free palladium compounds may be very active catalysts for cross coupling reactions, and, thus, intentional or accidental ingress of oxygen may be advantageous for the catalysis. Another example is the oxidation of one of the phosphine molecules in the Grubbs I metathesis catalyst.

Nitro and nitroso compounds are strongly oxidizing agents and, for instance, they have been reported to oxidize $PH_3$ [54]. Thus, nitrobenzene and phosphine give azoxybenzene and phosphorus acids under harsh conditions. In the palladium-catalyzed reductive carbonylation of nitrobenzene it was found that phosphine ligands are not suitable as they are oxidized to phosphine oxides [55]. Nitrosobenzene and isocyanate complexes of zerovalent Group 10 metals will transfer oxygen to triphenylphosphine and also form azoxybenzene [56]. Nitrosobenzene is much more reactive than nitrobenzene towards phosphines as it will oxidize arylphosphines in the absence of metal catalysts, forming azoxybenzenes at ambient temperature in the presence of base [57].

Many sulfur-containing compounds will also oxidize phosphines, and either form phosphine sulfide or, when water is present, phosphine oxides. This reaction has been known since 1935 [58] and, especially with water-soluble tris(2-carboxyethyl) phosphine (TCEP), it is of interest in biochemical systems [59]. It has been studied a couple of times over the years, but only in the last decade has it become extremely popular in biochemistry and molecular biology to reduce protein disulfide bonds (Scheme 1.10), for example in labeling studies, and as a preparatory step for gel electrophoresis [60].

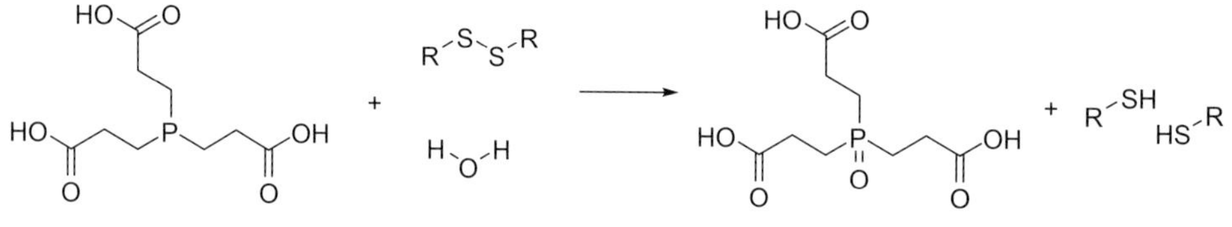

**Scheme 1.10** Reduction of disulfides by TCEP.

TCEP has also been employed for the reduction of sulfoxides, sulfonylchlorides, N-oxides, and azides (Staudinger reaction), thus showing that these compounds also present a potential hazard for phosphines in catalytic systems [61].

### 1.4.3
### Oxidative Addition of a P–C Bond to a Low-Valent Metal

In the next four sections we will discuss four additional ways of phosphine decomposition: oxidative addition of phosphines to low-valent metal complexes, nucleophilic attack on coordinated phosphines, aryl exchange via phosphonium species, and substituent exchange via metallophosphorane formation. Interestingly, in all cases the metal serves as the catalyst for the decomposition reaction!

In his review [46a] Garrou emphasizes the first mechanism, oxidative addition of the phosphorus–carbon bond to low-valent metal complexes (or reductive cleavage of P–C bonds) and formation of phosphido species. In the last two decades experimental support for the other three mechanisms has been reported (Sections 1.4.4–1.4.6). In Scheme 1.11 the four mechanisms are briefly outlined.

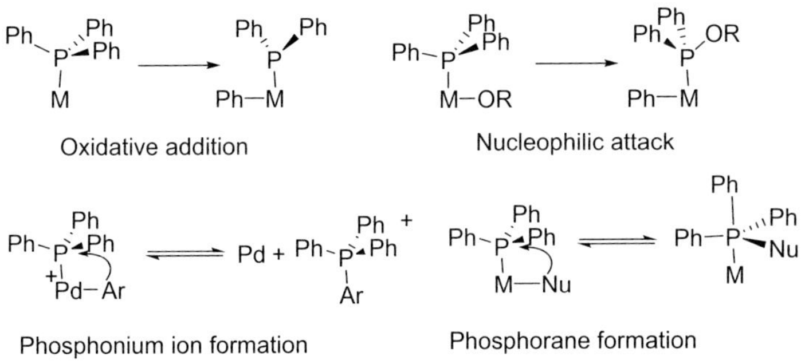

Scheme 1.11 Four mechanisms for phosphine decomposition leading to P–C bond cleavage.

Reductive cleavage of the phosphorus–carbon bond in triaryl- or diarylalkylphosphines is an important tool for making new phosphines [62]. Metals used to this end in the laboratory are lithium, sodium (or sodium naphthalide), and potassium. Cleavage of triphenylphosphine with sodium in liquid ammonia to give $Ph_2P^-$, benzene, and $NaNH_2$ is carried out on an industrial scale for the synthesis of the ligand of the SHOP process, obtained via reaction of sodium diphenylphosphide with o-chlorobenzoic acid [63]. The cleavage reaction works well for phenyl groups and methyl and several methoxy-substituted phenyl groups; most other substitution patterns lead to a Birch reaction or cleavage of the functional group [62b]. It is not surprising, therefore, that low-valent transition metals will also show reductive cleavage of the P–C bond, although mechanistically it involves interaction of the metal with the carbon and phosphorus atoms rather than an initial electron transfer as is the case for the alkali metals. The reaction with transition metals is usually referred to as an oxidative addition of the $R'-PR_2$ molecule to the metal complex.

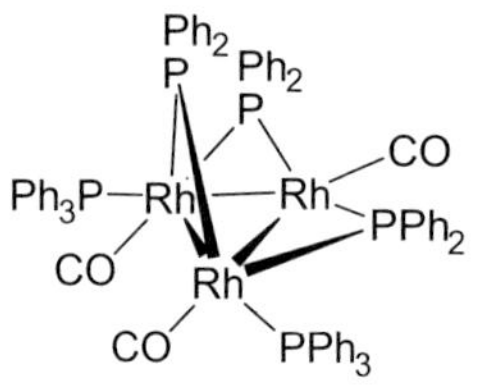

**Figure 1.1** Rhodium cluster that may form in hydroformylation mixture work-up.

Oxidative addition of C−Br or C−Cl bonds is an important reaction in cross-coupling type catalysis, and the reaction of a P−C bond is very similar, although the breaking of carbon–phosphorus bonds is not a useful reaction in homogeneous catalysis. It is an undesirable side-reaction that occurs in systems containing transition metals and phosphine ligands, leading to deactivation of the catalysts. Indeed, the oxidative addition of a phosphine to a low valent transition metal can be most easily understood by comparing the $Ph_2P-$ fragment with a chloro- or bromo-substituent at the phenyl ring; electronically they are very akin, see Hammett parameters and the like. The phosphido anion formed during this reaction will usually lead to bridged bimetallic structures, which are extremely stable. The decomposition of ligands during hydroformylation, which has been reported both for rhodium and cobalt catalysts [64] may serve as an example.

Thermal decomposition of $RhH(CO)(PPh_3)_3$, the well known hydroformylation catalyst, in the absence of $H_2$ and CO leads to a stable cluster, shown in Figure 1.1, containing $\mu_2$-$PPh_2$ fragments, as was studied by Pruett's group at Union Carbide (now Dow Chemical) [65]. It is not known whether P−C cleavage takes place on a cluster or whether it starts with a monometallic species (see the reactions below taking place in clusters).

After heating, the corresponding iridium compound led to the formation of a dimer containing two bridging phosphido bridges. The phenyl group has been eliminated (as benzene or diphenyl), see Scheme 1.12. In view of the short Ir−Ir bond the authors suggested a double bond [66].

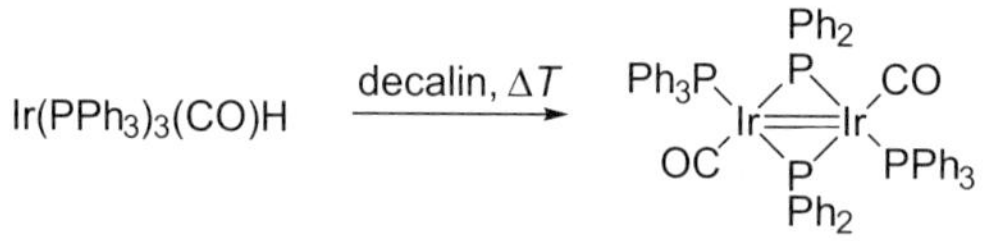

**Scheme 1.12** Iridium dimer that forms in the absence of syn gas.

Several authors have proposed a mechanism involving orthometallation as a first step in the degradation of phosphine ligands, especially in the older literature. Orthometallation does take place, as can be inferred from deuteration studies, but it remains uncertain whether this is a reaction necessarily preceding the oxidative addition (Scheme 1.13).

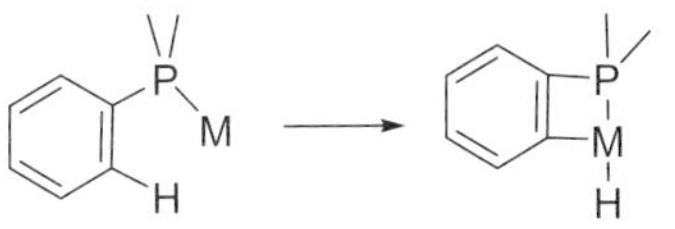

**Scheme 1.13** Orthometallation of a phosphine.

Subsequently the phosphorus–carbon bond is broken and the benzyne interme-diate inserts into the metal hydride bond. Although this mechanism has been popular with many chemists there are many experiments that contradict it. A simple para-substitution of the phenyl group answers the question whether orthometalla-tion was involved, as is shown in Scheme 1.14.

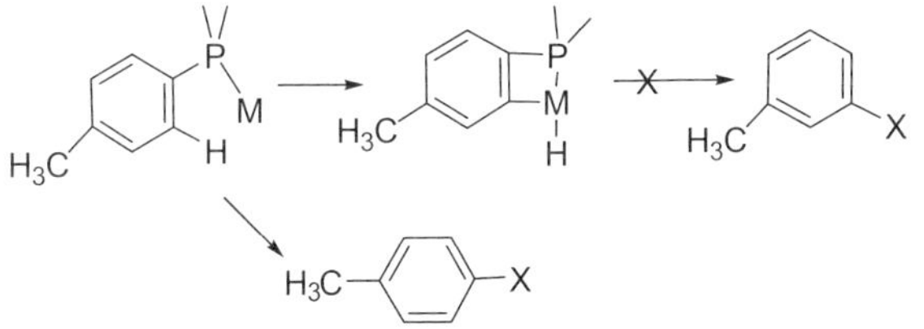

**Scheme 1.14** Disproving of orthometallation as a decomposition pathway.

Decomposition products of p-tolylphosphines should contain methyl substituents in the meta position if the orthometallation mechanism were operative. For palla-dium-catalyzed decomposition of triarylphosphines this was found not to be the case [67]. Using rhodium-containing solutions of tri-o-, tri-m-, and tri-p-tolylpho-sphines Abatjoglou et al. found that only one isomeric tolualdehyde is formed from each phosphine [68]. Thus, the tolualdehydes produced are those resulting from intermediates formed by direct carbon–phosphorus bond cleavage. Likewise Co, and Ru hydroformylation catalysts give aryl derivatives not involving the earlier proposed ortho-metalation mechanism [69].

Several rhodium complexes catalyze the exchange of aryl substituents of triar-ylphosphines at elevated temperatures (130 °C) [68]:

$$R'_3P + R_3P-(Rh) \rightarrow R'R_2P + R'_2RP$$

Abatjoglou et al. proposed as the mechanism for this reaction a reversible oxidative addition of the aryl–phosphido fragments to a low valent rhodium species. A facile aryl exchange has been described for complexes $Pd(PPh_3)_2(C_6H_4CH_3)$ I [70]. These authors also suggested a pathway involving oxidative additions and reductive eliminations. The mechanisms outlined below in the following sections, however, can also explain the results of these two studies.

Phosphido formation has been observed for many transition metal phosphine complexes [43]. Upon prolonged heating, and under an atmosphere of CO and/or $H_2$, palladium and platinum also tend to give stable phosphido-bridged dimers or clusters [71].

A "prototype" of an oxidative addition with concomitant P−C bond cleavage is the reaction of **1** (Scheme 1.15) with $Pd_2(dba)_3$, which gives the addition of an aryl group to palladium and formation of phosphido bridges [72]. The interesting feature of this example is that the aryl group is a pentafluorophenyl group, for which only very few examples of this reaction have been reported. Hydrogen analogs dppe and dppp in their reaction with low valent metals, for example, Pt(0) give metalation instead of P−C bond-cleavage (**2**, Scheme 1.15) [73].

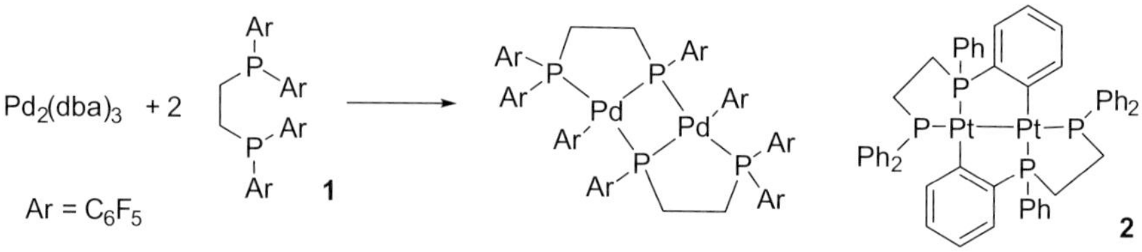

**Scheme 1.15** Reactions of diphosphines with zerovalent palladium and platinum complexes.

Bridging diphosphine metal complexes have been characterized that may be "en route" to P−C cleavage, such as shown in Figure 1.2 [74].

During the studies of the isomerization of butenyl cyanides, relevant to the hydrocyanation of butadiene to give adiponitrile, the intermediate (TRIPHOS)Ni (CN)H complexes were found to decompose to benzene and highly stable μ-phosphido-bridged dimers (only one isomer shown) that deactivate the catalytic process. Oxidative addition as the mechanism would invoke Ni(IV) and therefore one of the mechanisms to be discussed later, nucleophilic attack or phosphorane intermediates, may be operative (Scheme 1.16).

Cluster or bimetallic reactions have also been proposed in addition to monometallic oxidative addition reactions. For instance, trishydridoruthenium dimers will cleave P−C bonds in aryl and alkyl phosphines to give phosphido-bridged hydrides. Alkylphosphines give alkenes as the co-product, but phenylphosphines give benzene. For phenylphosphines the intermediate containing a bridging phenyl group has been observed, thus showing that that the reaction is an oxidative addition of the P−C bond along the ruthenium dimer, and not a nucleophilic attack of a hydride at a phosphorus atom. The alkene products of the alkylphosphines are in accord with this mechanism, as they are formed via β-elimination of the intermediate alkylruthenium groups (Scheme 1.17) [75].

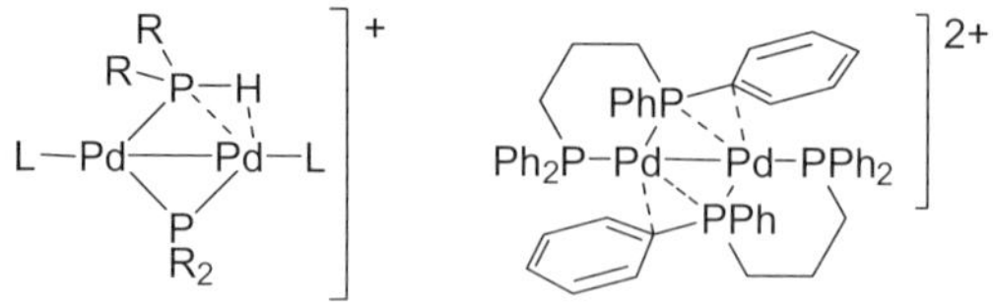

**Figure 1.2** Interactions en route to P−C cleavage.

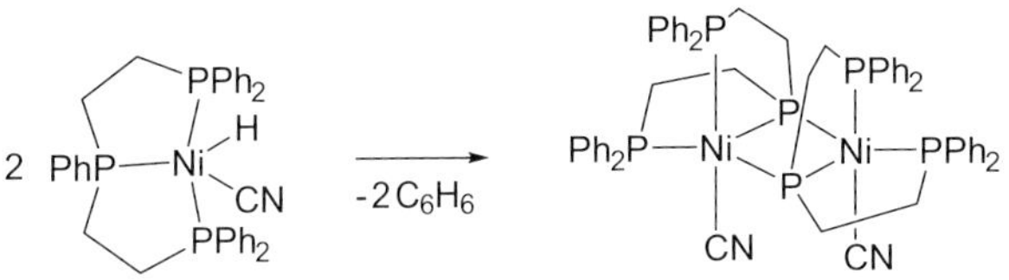

**Scheme 1.16**  P—C cleavage in a hydrocyanation catalyst.

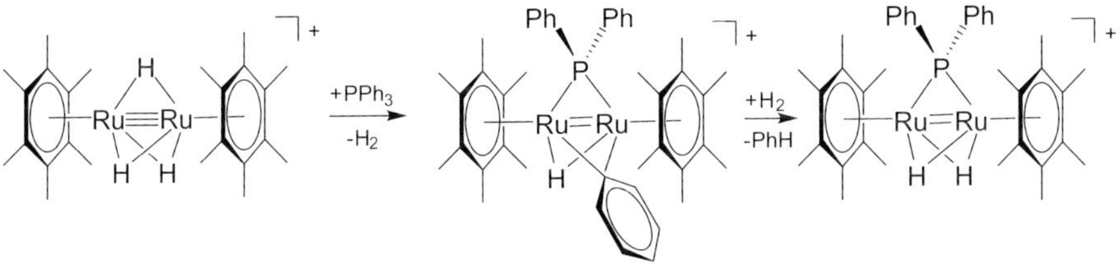

**Scheme 1.17**  P—C cleavage in a ruthenium dimer.

It has been known for a long time that clusters cleave P—C bonds. Nyholm and coworkers showed that $Ru_3(CO)_{12}$ and $Os_3(CO)_{12}$ react with $PPh_3$ to give, for example, $Os_3(CO)_7(\mu_2\text{-}PPh_2)_2(\mu_3\text{-}C_6H_4)$ [76]. Similar reactions have been reported for 1,8-bis(diphenylphosphino)naphthalene and $Ru_3(CO)_{12}$ [77].

As an illustration that rather complicated clusters may initiate the P—C cleavage reactions we mention that iridium-iron-monocarborane clusters have been reported containing $\mu_2\text{-}PPh_2$ groups and phenyl groups resulting from $PPh_3$ cleavage [78].

As in monometallic species, the reverse reaction can also be observed in dimers and clusters. An interesting example is shown in Scheme 1.18, which involves an insertion of ethene into a rhodium hydride bond when, subsequently, the ethyl formed migrates (or reductively eliminates) to the neighboring μ-phosphido group, forming a phosphine under very mild conditions [79].

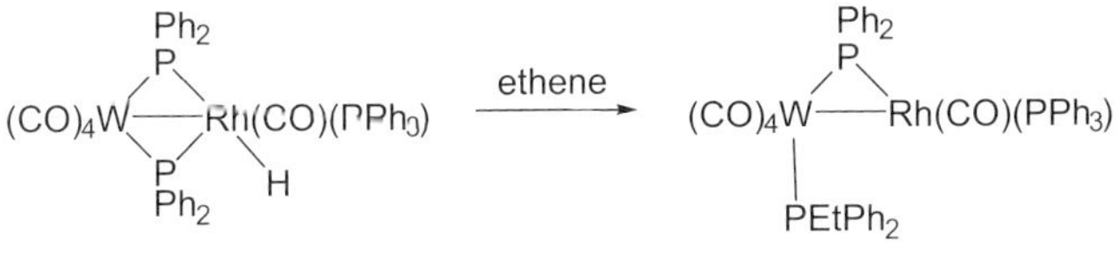

**Scheme 1.18**  P—C bond formation in a heterodimeric complex.

Most studies focus on unsubstituted phenyl groups, but, as in reductive cleavage with Group 1 metals, the substitution of the aryl groups will influence the rate of P—C cleavage and substituted aryls can perhaps be used to our advantage, to avoid transition metal catalyzed cleavage. In summary, a lot more studies are required to understand and suppress undesired cleavage of phosphorus–carbon bonds.

1.4.4
**Nucleophilic Attack at Phosphorus**

Here we are concerned with the metal-catalyzed or metal-aided nucleophilic attack at phosphorus in phosphines, but there are examples of butyllithium and Grignard reagents replacing hydrocarbyl substituents at phosphorus (halide and alkoxide replacement by hydrocarbyl anions is an ubiquitous reaction!). We mention three examples, but surely there are more reports in the literature. Pentafluorophenyl in $(C_6F_5)_3P$ was found to be replaced by ethylmagnesium bromide to give $Et(C_6F_5)_2P$ and $(C_6F_5)Et_2P$ [80]. The phenyl group in **3** (Scheme 1.19) is replaced by a methyl or butyl group upon reaction with methyllithium or butyllithium. Therefore, lithiation of the hydrogens ortho to oxygen should be done with phenyllithium to avoid side-product formation [81].

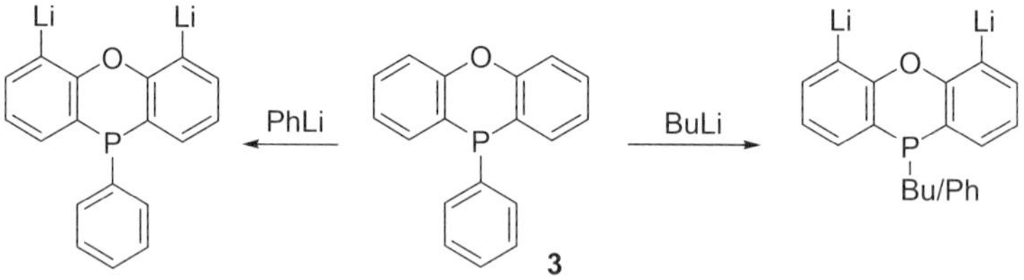

**Scheme 1.19** Nucleophilic displacement at phosphorus.

Hypervalent phosphoranes have been invoked as possible intermediates or transition states for these exchange reactions [82]. Theoretical studies show that lithiophosphoranes may be transition states in these exchange reactions, rather than intermediates [83]. Another example of nucleophilic attack at phosphorus by carbon nucleophiles is the reaction of 2,2'-biphenylyldilithio with $Ph_2PCl$, which gives 9-phenyldibenzophosphole and triphenylphosphine quantitatively, rather than the expected diphosphine [84]. Schlosser proposed a lithiophosphorane intermediate for this reaction (Scheme 1.20) [85].

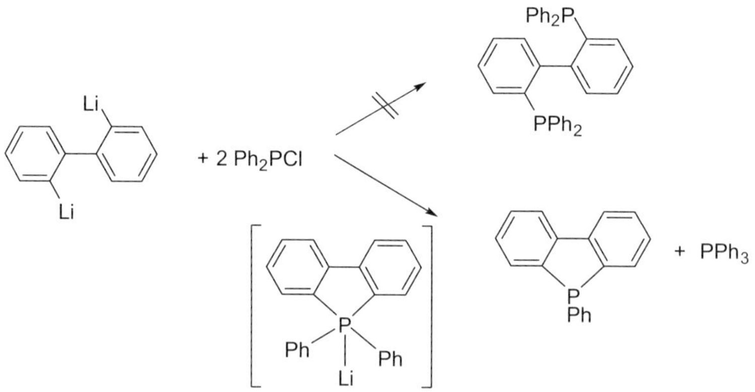

**Scheme 1.20** Nucleophilic displacement leading to dibenzophosphole.

For a long time the literature underestimated the importance of nucleophilic attack as a mechanism for the catalytic decomposition of phosphines coordinated to metals, especially with nucleophiles such as acetate, methoxy, hydroxy and hydride (Scheme 1.21). For examples of nucleophilic attack at coordinated phosphorus see Refs [71, 86]. A very facile decomposition of alkylphosphines and triphenylphosphine (using palladium acetate, one bar of hydrogen and room temperature) has been reported [71a]. Acetate was suggested as the nucleophile, but hydride as the nucleophile cannot be excluded.

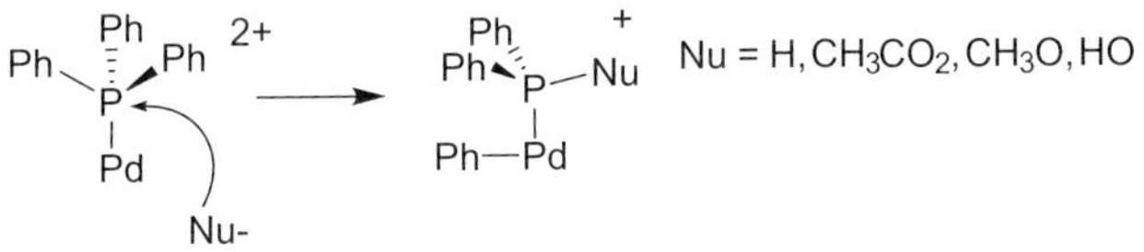

**Scheme 1.21** Phosphine decomposition via nucleophilic attack.

A detailed reaction proving the nucleophilic attack was shown for platinum complexes [86d]. The alkoxide coordinated to platinum attacks phosphorus while the carbon atom coordinated to phosphorus migrates to platinum. Thermodynamically the result seems more favorable, but mechanistically this "shuffle" remains mysterious. See Scheme 1.25. Coordination to platinum makes the phosphorus atom more susceptible to nucleophilic attack, and the harder (P and O) and softer (C and Pt) atoms recombine as one might expect. The same mechanism was proposed by Matsuda [86a] for the decomposition of triphenylphosphine by palladium(II) acetate. In this study the aryl phosphines are used as a source of aryl groups that are converted into stilbenes via a Heck reaction. Even alkyl phosphines underwent P−C bond cleavage via palladium acetate.

It is surprising indeed that phosphines can be effective ligands in cross-coupling reactions under basic conditions, for instance phenoxide will react with triphenylphosphine in palladium chloride complexes at $0\,^{\circ}$C giving $PdCl(Ph)(PPh_3)_2$ together with phenyl phosphites, phosphonites and phosphates [87]. It was concluded that the intermediate $PdCl(OAr)(PPh_3)_2$ is highly labile. The mechanism most likely is a nucleophilic attack of the phenoxide at the phosphorus atom of the coordinated triphenylphosphine.

There are several examples of nucleophilic attack of hydroxide on coordinated dppm (bisdiphenylphosphinomethane) and its methyl analog. The P−C bond cleaved is the one with the methylene unit, which apparently is not very stable. For instance $(dppm)PtCl_2$ will react with NaOH in liquid ammonia to give $Ph_2MeP$ and an SPO (diphenylphosphine oxide) both coordinated to platinum, involved in amide bridges with a second platinum unit (Scheme 1.22) [88]. The intermediate methylene anion formed has been protonated by ammonia.

Dmpm undergoes the same reaction when coordinated to manganese, Scheme 1.23 [89].

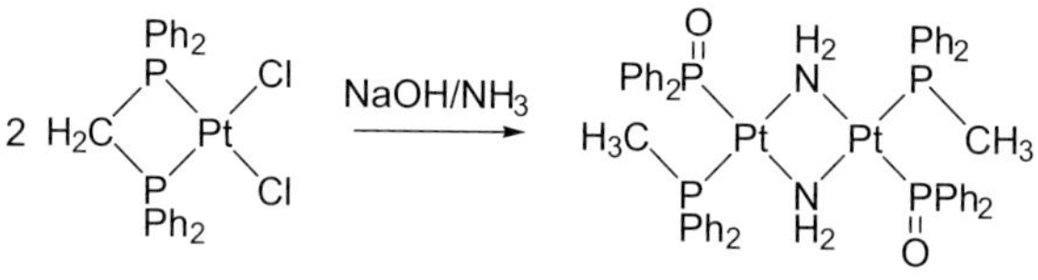

**Scheme 1.22** Dppm decomposition via nucleophilic attack.

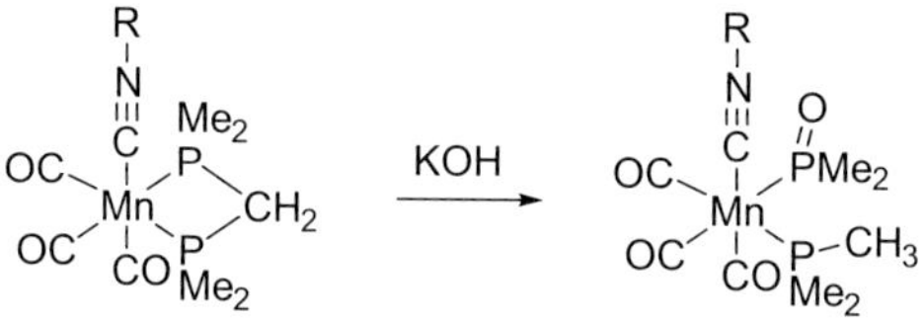

**Scheme 1.23** Dppm decomposition via nucleophilic attack.

The pentadentate ligand with NP4 donor set in Scheme 1.24 undergoes a P–C cleavage reaction with methanol giving a phosphinite complex, while the methyl group formed by protonation of the methylene anion shows an agostic interaction with iron [90]. Surprisingly, when 10 bar of CO is applied, the reaction is reversed and the intact NP4 ligand is regenerated.

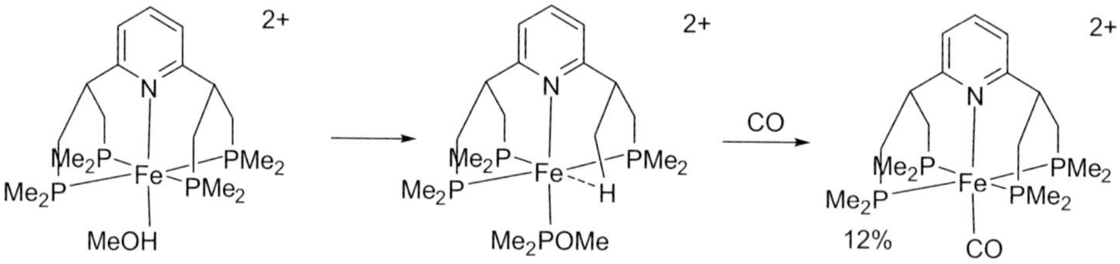

**Scheme 1.24** NP4 decomposition via nucleophilic attack.

Ruthenium aryloxide complexes containing trimethylphosphine also showed an intramolecular nucleophilic attack, leading to methyl–phosphorus bond breaking and formation of a phosphinite ligand [91a]. In Section 1.4.6 we will discuss metallophosphoranes as intermediates in similar exchange reactions and it may well be that transition states or intermediates of this type occur in exchange reactions which we have presented here as nucleophilic mechanisms, or a metal–O/phosphorus–C "shuffle" reaction (Scheme 1.25) [86d].

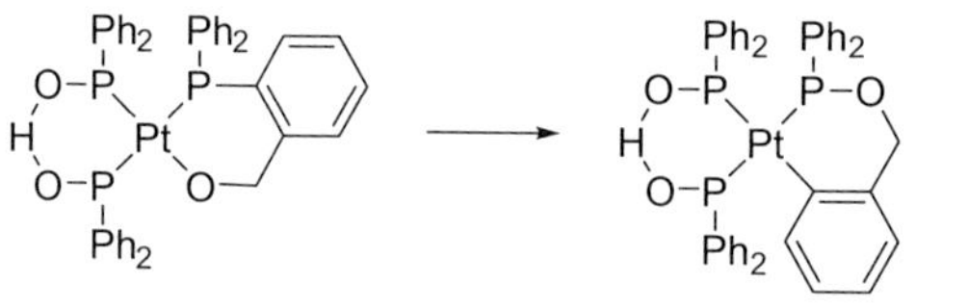

**Scheme 1.25** Metal-O/phosphorus-C "shuffle" reaction via nucleophilic attack.

A catalytic decomposition of triphenylphosphine has been reported [92] in a reaction involving rhodium carbonyls, formaldehyde, water, and carbon monoxide. Several hundreds of moles of phosphine can be decomposed this way per mole of rhodium per hour! The following reactions may be involved (Scheme 1.26).

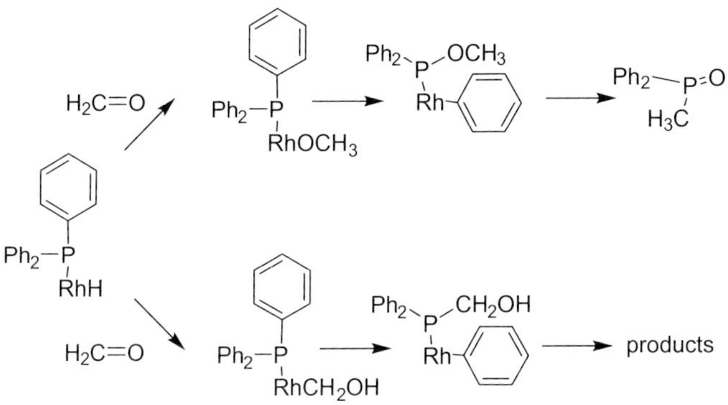

**Scheme 1.26** Catalytic decomposition of triphenylphosphine with formaldehyde and rhodium.

Related to this chemistry is the hydroformylation of formaldehyde to give glycolaldehyde, which would be an attractive route from syn-gas to ethylene glycol. The reaction can indeed be accomplished and is catalyzed by rhodium arylphosphine complexes [93], but clearly phosphine decomposition is one of the major problems to be solved before formaldehyde hydroformylation can be applied commercially.

### 1.4.5
### Aryl Exchange Via Phosphonium Intermediates

Phosphonium salts frequently occur in palladium and nickel chemistry, as reactant, intermediate, or as product. Decomposition of $(PPh_3)_2PdPhI$ gives $Ph_4PI$ as was reported by Yamamoto [94]. Grushin studied the decomposition of a range of these complexes and found that iodides are more reactive than bromides and chlorides, and also that excess of phosphine and halides strongly influences the rate of decomposition (for aryl fluorides see Section 1.4.6) [95]. He suggests that exchange reactions during catalysis (vide infra) can best be avoided by using excesses of phosphines and halides. The latter are present anyway in cross coupling chemistry and there is no choice either, once we have decided on the substrate. Excess of phosphines may retard the reaction.

Aryl (pseudo)halides can be reacted with triphenylphosphine to give phosphonium salts, and metals such as nickel and palladium catalyze this reaction. Nickel works well for only a few substrates and high quantities are required [96], but palladium acetate is an active precursor and shows a wide scope for this reaction [97]. Phosphonium salts can be used as an aryl halide replacement in cross coupling reactions and thus phosphonium salts add oxidatively to palladium(0) producing a triarylphosphine as one of the products [94]. Addition of aryl halide to phosphine and exchange of the aryl groups has been used as a synthetic tool for making phosphines by Chan and coworkers (Scheme 1.27) [98]. As more or less statistical mixtures are obtained, the yields do not exceed 60%, but the method is highly tolerant of functional groups and avoids the use of sensitive and expensive reagents.

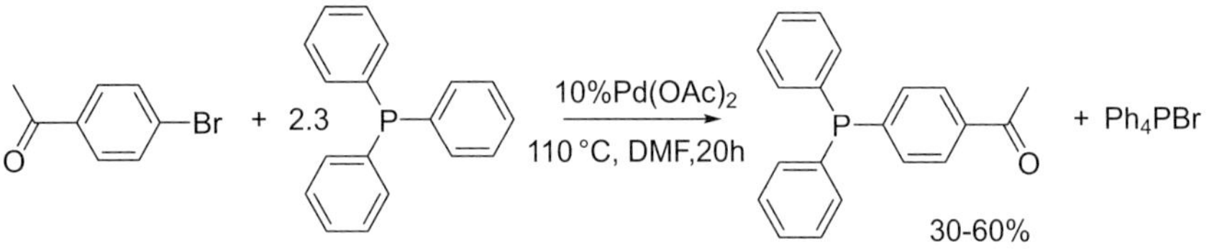

**Scheme 1.27** Catalytic transfer of aryl groups in triphenylphosphine.

The interest in cross coupling reactions in the last decade has led to a large number of reports dealing with the involvement of the phosphine ligands in these reactions [99]. The mechanism has not been elucidated for all cases. The oxidative addition and nucleophilic attack discussed above may explain some of these results but, especially since the work of Novak, phosphonium intermediates have been considered as intermediates [99c]. Formally the mechanism also involves nucleophilic attack of a hydrocarbyl group at coordinated phosphines, but after the nucleophilic attack the phosphorus moiety dissociates from the metal as a phosphonium salt. Effectively, this is a reductive elimination. To obtain a catalytic cycle the phosphonium salt re-adds oxidatively to the zerovalent palladium complex (see Scheme 1.28), in accord with the findings of Yamamoto [94].

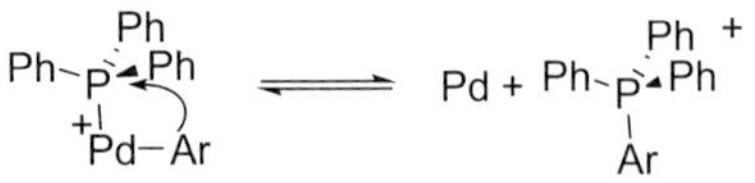

**Scheme 1.28** Substituent exchange via phosphonium intermediates.

For the exchange of a palladium bonded methyl group and a phenyl group at phosphorus, Norton *et al.* [99a] proposed an intramolecular phenyl migration to palladium followed by reductive elimination of $MePh_2P$. They excluded the intermediacy of phosphonium species as deuterated phosphonium salts of the same composition did not participate in the reaction (Scheme 1.29). A metallophosphorane intermediate or transition state (Section 1.4.6) would also nicely explain the

**Scheme 1.29** Intramolecular phenyl migration to palladium [99a].

**Scheme 1.30** π-Allylpalladium phosphine complex giving a phosphonium salt.

intramolecular character, and would not invoke the occurrence of tetravalent palladium intermediates.

Aryl exchange has a deleterious effect on the yield and selectivity of the palladium-mediated coupling of aryl halides with alkenes (Heck reaction) or hydrocarbyl metals (Mg, B, Sn) [99, 100]. Polymerizations toward high molecular weight suffer especially from this side-reaction [101]; arylphosphines will be incorporated in the growing chain and the chain growth may stop there.

Formation of phosphonium species can also lead to catalyst deactivation in other systems. For example, the reductive elimination of allylphosphonium salts from cationic allyl palladium species (Scheme 1.30) may lead to "ligand consumption" and thus reduce (enantio)selectivity or activity of the catalyst.

## 1.4.6
### Aryl Exchange Via Metallophosphoranes

Metallophosphoranes and phosphoranes as intermediates in the reactions studied here are closely related mechanistically with the routes presented in Sections 1.4.4 (nucleophilic attack) and 1.4.5 (phosphonium salt formation). Already as early as 1971 Green and coworkers proposed the intermediacy of metallophosphoranes in the smooth, room-temperature exchange of methyl groups in methyllithium and phenyl groups in PPh$_3$ coordinated to NiCl$_2$ [102]. Among other products the formation of free PMe$_2$Ph and PMePh$_2$ was observed. Full details of the intermediates are not at hand and a schematic mechanism is outlined in Scheme 1.31.

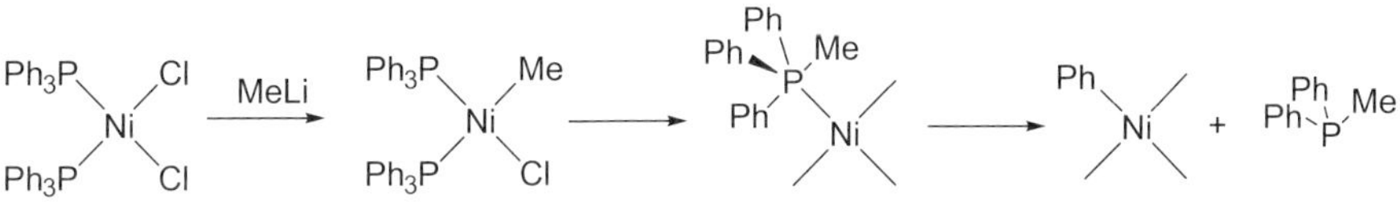

**Scheme 1.31** Hydrocarbyl exchange via a metallophosphorane.

The decomposition of CoMe(PPh$_3$)$_3$ also leads to scrambled methyl/phenylphosphines [103]. In Scheme 1.32 a typical example is shown of a deprotonation of

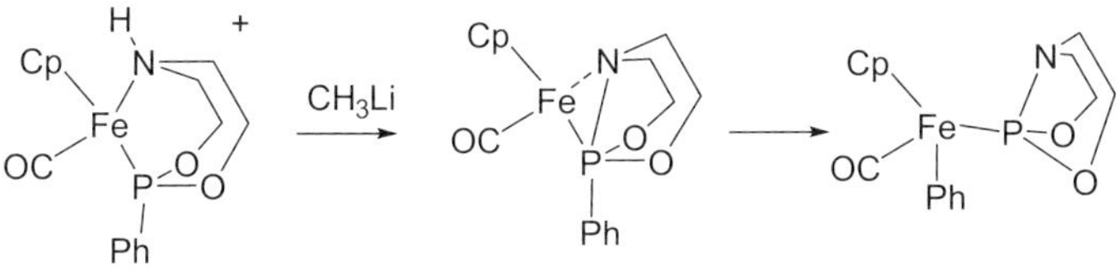

**Scheme 1.32** Exchange of nucleophiles via a metallophosphorane.

a coordinated amine, followed by a nucleophilic attack of the amide at the coordinated phosphonite, forming a metallophosphorane [104]. Upon warming to room temperature the phosphorane rearranges by migration of the phenyl group to iron. Addition of HCl to the latter gives back the starting material, thus showing that nucleophilic attack and its reverse are facile processes in the coordination sphere.

Several nucleophilic attacks described in Section 1.4.4 actually might proceed via a metallophosphorane, for instance the example in Scheme 1.25 would now read as shown in Scheme 1.33. As mentioned in Section 1.4.4 hydrocarbyl exchange in pyridylphosphines seems to involve lithiophosphoranes as transition states rather than as intermediates, as was indicated by DFT calculations [83]. This may also be true for some of the examples shown here.

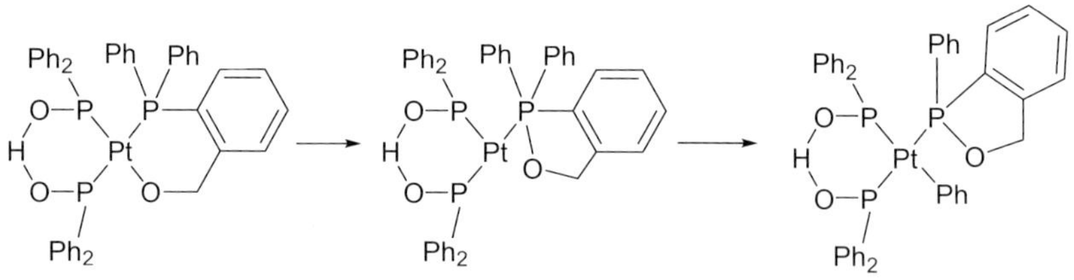

**Scheme 1.33** Exchange of nucleophiles via a metallophosphorane.

Grushin, in addition to the arylphosphines complexes of palladium halides, also studied the fluoride complexes of palladium and rhodium, in particular the aryl exchange reactions and their decomposition [95]. Heating of $RhF(PPh_3)_3$ in benzene led to an exchange of phenyl and fluoride groups to give $RhPh(PFPh_2)(PPh_3)_2$, while in chlorobenzene *trans*-$RhCl(PFPh_2)(PPh_3)_2$ was found as the product, together with biphenyl. DFT calculations showed that the most likely pathway is the formation of a metallofluorophosphorane as an intermediate [105]. In the decomposition of $(PPh_3)_2PdPhF$ Grushin found that $Ph_3PF_2$ was formed, thus, especially with fluorine atoms, the formation of phosphorane compounds is highly likely [95]. BINAP(O) palladium compounds show the same propensity for fluorophosphorane formation [106].

The findings by Pregosin on BINAP, ruthenium and tetrafluoroborate are highly relevant to catalysis. The use of inert anions such as $BF_4^-$ would not arouse much suspicion, but his work shows that fluorodiphenylphosphines may form in such catalysts even at very low temperatures (Scheme 1.34) [107]. Water and acetate may

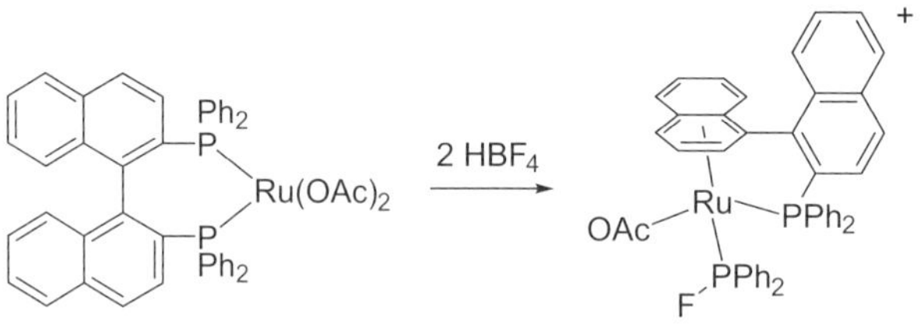

**Scheme 1.34** F—C exchange in BINAP.

also function as the nucleophile. This is mentioned here because it might well involve phosphorane intermediates, as observed in the work reported by Grushin. Metalla-phosphoranes were reviewed by Goodman and Macgregor in 2010 [108].

## 1.5
## Phosphites

Phosphites are easier to synthesize and less prone to oxidation than phosphines. They are much cheaper than most phosphines and a wide variety of structures can be obtained. Disadvantages of the use of phosphites as ligands include several side-reactions: hydrolysis, alcoholysis, trans-esterification, Arbuzov rearrangement, O−C bond cleavage, P−O bond cleavage. Scheme 1.35 gives an overview of these reactions. In hydroformylation systems at least two more reactions may occur, namely nucleophilic attack on aldehydes and oxidative cyclizations with aldehydes. Lewis acids catalyze the Arbuzov reaction at room temperature [109].

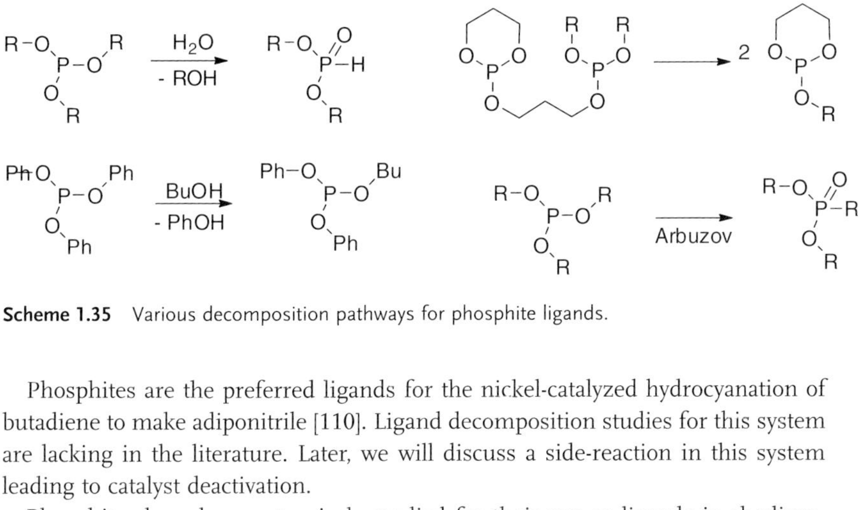

**Scheme 1.35** Various decomposition pathways for phosphite ligands.

Phosphites are the preferred ligands for the nickel-catalyzed hydrocyanation of butadiene to make adiponitrile [110]. Ligand decomposition studies for this system are lacking in the literature. Later, we will discuss a side-reaction in this system leading to catalyst deactivation.

Phosphites have been extensively studied for their use as ligands in rhodium-catalyzed hydroformylation. The first publication on the use of phosphites was from Pruett and Smith at Union Carbide [111]. The first exploitation of bulky *monophosphites* was reported by van Leeuwen and Roobeek [112]. Diphosphites came into focus after the discovery of Bryant and coworkers at Union Carbide Corporation that certain bulky diphosphites lead to high selectivities in the rhodium-catalyzed hydroformylation of terminal and internal alkenes (see Figure 1.3) [113].

It should be noted that all phosphites reported are *aryl* phosphites (sometimes the backbones may be aliphatic) and that the favored ones often contain bulky substituents. One of the reasons that aliphatic phosphites are used only sparingly is that they are susceptible to the Arbuzov rearrangement while the aryl phosphites are not. Acids, carbenium ions, and metals catalyze the Arbusov rearrangement. Many

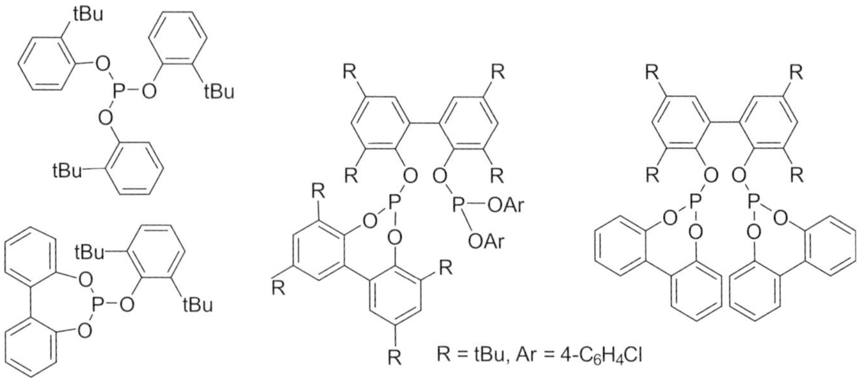

**Figure 1.3** Typical bulky monophosphites and diphosphites.

examples of metal-catalyzed decomposition reactions have been reported (see Scheme 1.36) [114].

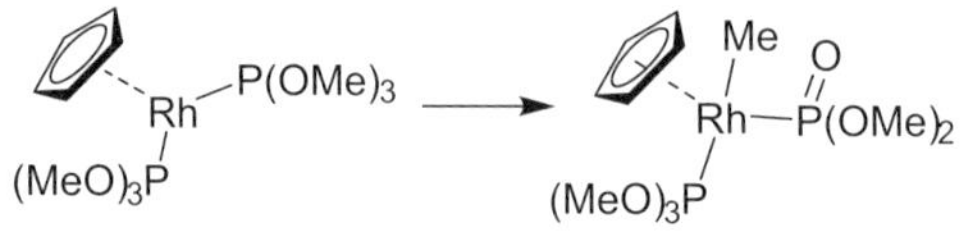

**Scheme 1.36** Metal-catalyzed Arbusov reaction leading to phosphite decomposition.

Thorough exclusion of moisture can easily prevent hydrolysis of phosphites in the laboratory reactor. In a continuous operation under severe conditions traces of water may form via aldol condensation of the aldehyde product. Weak and strong acids and strong bases catalyze the reaction. The reactivity for individual phosphites spans many orders of magnitude. When purifying phosphites over silica columns in the laboratory one usually adds some triethylamine to avoid hydrolysis on the column.

Bryant and coworkers have extensively studied the decomposition of phosphites [115]. Stability involves thermal stability, hydrolysis, alcoholysis, and stability toward aldehydes. The precise structure has an enormous influence on the stability. Surprisingly, it is the reactivity toward aldehydes that received most attention. Older literature mentions [116] several reactions between phosphites and aldehydes of which we show only two in Scheme 1.37.

The addition of a phosphite to an aldehyde to give a phosphonate is the most important reaction [115]. The reaction is catalyzed by acid and, since the product is acidic, the reaction is autocatalytic. Furthermore, acids catalyze hydrolysis and alcoholysis and, therefore, the remedy proposed is continuous removal of the phosphonate over a basic resin (Amberlyst A-21). The examples in the patents illustrate that very stable systems can be obtained when the acidic decomposition products are continuously removed. The thermal decomposition of phosphites with aldehydes is illustrated in Figure 1.4.

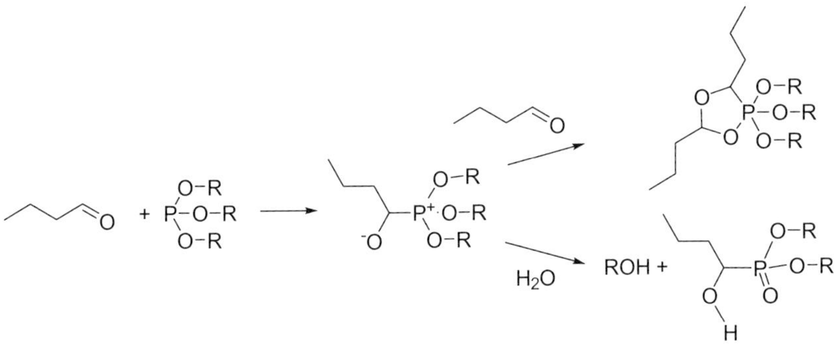

**Scheme 1.37** Reactions of phosphites and aldehydes.

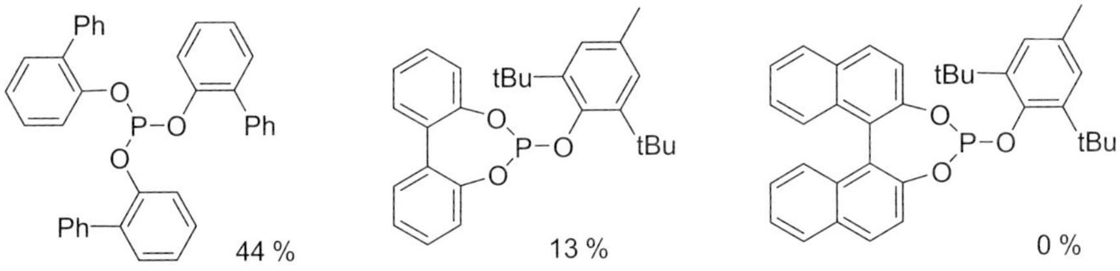

**Figure 1.4** Reactivity of various phosphites toward C5-aldehyde [115]. Percentage decomposition after 23 h at 160 °C.

A decomposition reaction that looks like an Arbuzov reaction but actually is not was reported by Simpson [117]. The decomposition of an iridium triisopropyl phosphite complex involves a metallation of one of the propyl groups before an apparent Arbuzov reaction takes place. It is a nice example of the complexity of the decomposition pathways one can encounter (Scheme 1.38). The final complex contains a π-allyl group and a diisopropyl phosphite ligand.

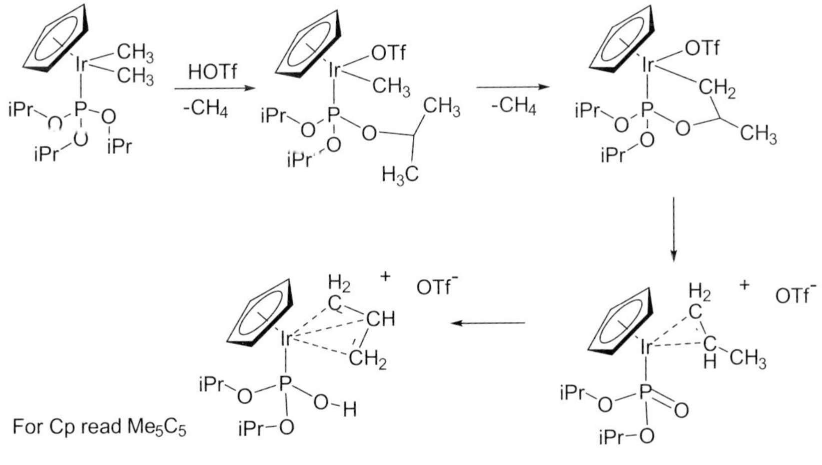

**Scheme 1.38** Phosphite metalation followed by Arbuzov-like reaction.

Dealkylation of trimethyl phosphite in complexes of ruthenium is an acid-catalyzed reaction; the resulting phosphite is $MeOP(OH)_2$ [118]. Metallation of phosphites will be discussed in Section 1.9.3.

## 1.6
## Imines and Pyridines

Intrinsically, pyridines and imines are less susceptible to decomposition reactions than phosphorus ligands. They are less suitable for the stabilization of low-valent metal complexes and, at first sight, they may not be the ligands of choice for cross-coupling type chemistry [119], although even enantioselective Heck reactions have been reported for chiral bisoxazolines [120], and Suzuki–Miyaura reactions are catalyzed by palladium complexes containing nitrogen donor ligands [121], and, while these reactions may proceed without ligand (Sections 1.2.2–1.2.4), the occurrence of enantioselective reactions using nickel proves that the nitrogen bidentate donor remains coordinated during the process [122]. In the last decade nitrogen ligands have received a great deal of attention in copolymerization reactions [123] and oligomerization reactions [124]. No industrial applications have been reported so far, but for ethene oligomerization the activity and selectivity of the new imine-based catalysts is very promising.

The latter development comprises the use of iron- and cobalt-containing pyridinediimine ligands. Extremely fast catalysts were reported simultaneously by the groups of Brookhart and Gibson [124]. Turnover frequencies as high as several millions per hour were recorded! Catalyst activities are very high and a separation of catalyst and product in a two-phase system as in SHOP may not be needed. The molecular weight distribution though will still require the isomerization/metathesis sequence as in the SHOP process. An example of such a catalyst is depicted in Scheme 1.39.

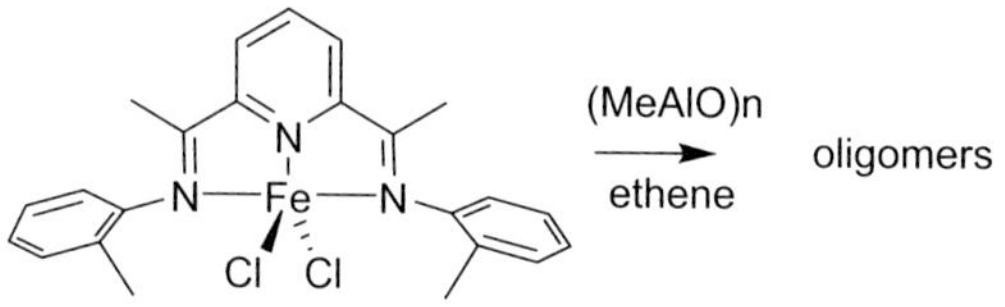

**Scheme 1.39** Pyridine-imine-iron-based ethene oligomerization catalyst.

Activation of the chloride precursor involves alkylation with methyl aluminoxane and thus side-reactions such as those occurring in early-transition-metal catalyzed polymerization may occur (vide infra). No detailed studies about ligand stability have been published and ligand decomposition may not be an issue at all in this instance. A few reactions that may lead to decomposition of imines and pyridines are collected in Scheme 1.40.

We confine ourselves to the following reactions of imines:

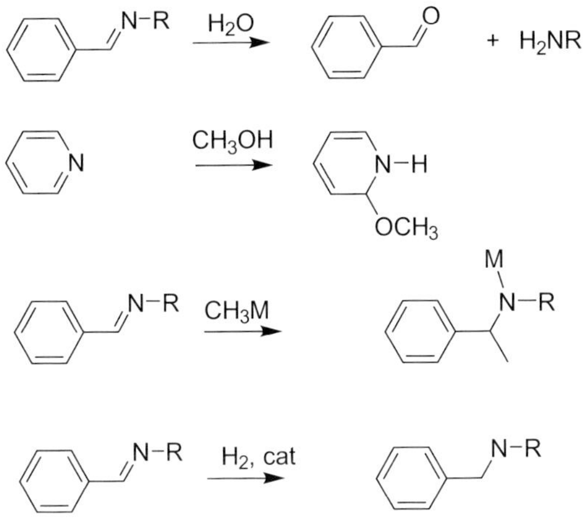

**Scheme 1.40** Decomposition reaction of imines and pyridines.

- Hydrolysis of imines leads to aldehydes and amines, the reverse reaction of the most frequent synthesis.
- Addition of water and methanol is catalyzed by metal ions. Like the former reaction it can be avoided if the catalytic reaction does not involve these reagents.
- Addition of metal alkyls leads to metal amides. The chemistry is complex and several reactions may occur [125].
- Catalytic hydrogenation to give amines may occur if the catalyst is accidentally active for this reaction and when hydrogen is present [126].

## 1.7
## Carbenes

### 1.7.1
### Introduction to NHCs as Ligands

Both the Fischer carbenes and the Schrock carbenes play an important role in catalysis, the former as ligands and the latter as initiator or substrate fragments. The discussion on carbenes as ligands will be limited to N-heterocyclic carbenes, as the older, different hetero-atom-containing Fischer carbenes have hardly been exploited in catalysis. The use of NHCs (N-heterocyclic carbenes) as ligands in complexes and catalysis [127] has enormously increased in the last decade. They are highly effective ligands in metathesis [128], cross coupling reactions [129], butadiene dimerization [130], polymerization [131], and, in this chemistry, results have been spectacular in terms of rate, selectivity, and turn-over numbers. Meanwhile NHCs have been found to be suitable ligands in hydrosilylation [132], carbonylation [133], gold C–C bond formation catalysis [134], hydrogen transfer [135], conjugate addition [136], hydroarylation of alkynes [137], oxidation [138], azide-alkyne click reactions [139], nickel-catalyzed ring closure reactions [140], ethene oligomerization [141], and so on. Their stability and reactivity have been reviewed by Crudden and Allen [142].

NHCs have been known since 1962 and many complexes have been synthesized by the Lappert group since the 70s [143]. The interactions of the orbitals on the nitrogen atoms and the carbon atom make the NHCs highly stable [144]. Especially since the introduction of steric bulk at the nitrogen atoms by Arduengo, which rendered them even higher in stability, they have become very useful and versatile ligands [145]. NHCs are strong σ donors and larger alkyl substituents on the nitrogen atoms increase their basicity further [146]. In many metal complexes NHCs can replace phosphines, bringing about drastic changes in the properties of the complexes, including the catalytic properties. The electronic and steric properties of NHCs have been reviewed by Nolan [147]. The development of the field of NHC ligands has been extremely rapid in the last decade, and one might say that the knowledge already gained in phosphine ligand chemistry has been transferred to carbene chemistry at high speed, giving birth to a plethora of metal complexes containing monodentate NHCs, but also bis-, tris-, tetra-, hexa-dentate and pincer-shaped NHCs [148]. NHCs are highly basic and form a strong σ-bond with the transition metals, which is much stronger than that of phosphines. Thus, while at low catalyst concentrations ($10^{-3}$ M or less) an excess of phosphine is required, especially, for example, when carbon monoxide is a competing ligand (and substrate!), it was thought that NHCs could perhaps be used in stoichiometric amounts. In several instances this seems to be the case, not counting yet possible decomposition routes, which will be discussed below. In Section 1.7.2 the main side-reactions of NHCs will be mentioned and in Section 1.7.3 we will present a few decomposition routes for Schrock carbenes, which feature as substrate fragments in catalysis rather than as modifying ligands.

## 1.7.2
### Reductive Elimination of NHCs

The introduction of NHC ligands in metal complexes is somewhat trickier than the introduction of phosphine and phosphite ligands, or, if one wishes, the chemistry at hand for NHCs is richer, as phosphorus ligands are rarely generated *in situ*! NHCs can be introduced by way of "electron-rich alkenes" (Lappert) [149], free carbenes (Arduengo) [145], as their imidazolium salts and base, by transfer from other metal complexes such as silver, or by *in situ* ring closure on the metal of amine isocyanides (see Scheme 1.41; in line with most literature we have drawn the metal–NHC bond as a single one) [150].

One of the most prominent decomposition reactions of NHC complexes is the reductive elimination discovered by McGuinness and Cavell [151] and we owe most of the information to the Cavell group [152]. Above we have seen that phosphines are susceptible to a number of decomposition reactions in their complexes, the present reaction is akin to the elimination of phosphonium salts from hydrocarbyl metal phosphine complexes (metal = e.g., Ni, Pd). The first reaction discovered was the elimination of trimethylimidazolium tetrafluoroborate from [PdMe(dmiy)(cod)]BF4 (dmiy) 1,3-dimethylimidazolin-2-ylidene), presented schematically in Scheme 1.42.

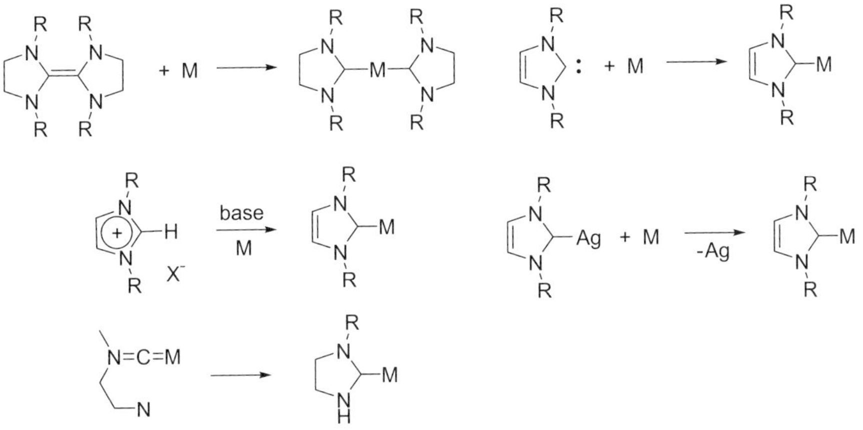

**Scheme 1.41** Synthesis of NHC complexes.

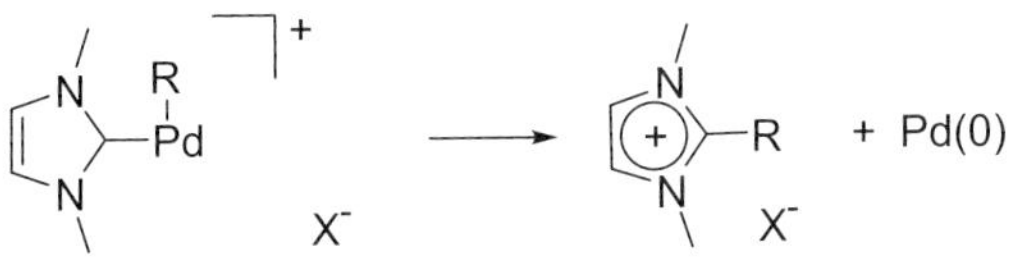

**Scheme 1.42** Reductive elimination of an NHC-alkyl salt.

Cationic complexes react faster than neutral complexes. During the Heck reaction with NHC-modified palladium catalysts reductive elimination products were also found, containing either the aryl group or the phenylethyl group in the imidazolium salt. During carbonylation studies acetylimidazolium elimination products were identified [152c] and 2-alkylsubstituted imidazolium salts were also obtained during the NHC-Ni-catalyzed olefin dimerization reaction [141b, 153]. Subsequent reports have also shown the formation of 2-arylimidazolium salts from NHC-Pd-Aryl complexes [154].

DFT studies show that the mechanism of the reaction is a concerted reductive elimination and not a migration of the hydrocarbyl group to the carbene [155]. Most likely it can be compared with a migratory reductive elimination, as described for cross coupling reactions [156], but a detailed comparison for the geometries of the transition states of the two reactions has not been carried out. Other DFT studies indicated that electron-donating groups at the nitrogen atoms will slow down reductive elimination and, thus, *t*-butyl groups combining steric bulk and strong donor capacity should give the best results [157]. These results also show that the reductive elimination is very similar to the migratory reductive elimination in cross coupling, because in those reactions indeed electron-withdrawing substituents on the aryl group receiving the migrating hydrocarbyl group will enhance reductive elimination [158].

In spite of their facile elimination reactions NHC palladium complexes give highly active Heck catalysts [151], which the authors explain by competing reactions of decomposition and catalysis (alkene insertion and β-elimination); as long as substrate is present the catalytic process is faster than the reductive elimination and catalysis continues (at higher temperatures). Several suggestions were put forward as to how to prevent reductive elimination, such as the use of oligodentate NHCs or bulky NHCs. Indeed, bulky NHCs were found to be the best ligands in Suzuki coupling chemistry [159]. Another effective approach to limit reductive elimination of NHC—R is the use of an excess of imidazolium salts as an ionic liquid solvent, which requires that oxidative addition of the salt to palladium(0) occurs, but this was suggested to be the case. In addition, the ionic medium will stabilize palladium(II) species and may facilitate separation of product and catalyst. Heck reactions under such conditions have been successfully carried out by Seddon and co-workers [160]. Examples of the oxidative addition of imidazolium salts to zerovalent nickel and palladium complexes leading to M(II) hydrides have been published meanwhile [161].

As in the case of the reversible oxidative addition and reductive elimination of phosphonium salts the imidazolium reactions can also be used to our advantage. In this instance the new substituent is introduced as an alkene, which inserts, effectively in the overall scheme, in the C—H bond of the imidazolium salt with the use of a nickel catalyst (Scheme 1.43, spectator phosphines or NHCs omitted on Ni) [162].

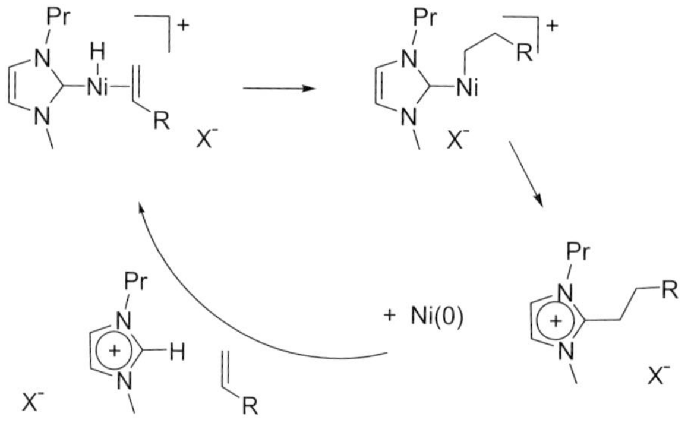

**Scheme 1.43** Synthesis of alkyl imidazolium salts.

Grubbs and coworkers reported on a decomposition reaction of NHCs in nickel complexes that we will mention here in an attempt to collect reactions of potential interest, although its importance for catalysis was not yet known at the time of publication [163]. The reaction is shown in Scheme 1.44 and, most likely in an attempt to make a SHOP-type catalyst based on a phenol-NHC ligand instead of a phosphinophenolate, complex **4** was reacted with base and the phenylnickel precursor shown. We have drawn a likely intermediate that undergoes the reaction shown to yield the reported product (at room temperature in 60% yield). The authors point out the relationship with the NHC alkylation reactions reported by the Cavell group (vide supra), but the relation with phosphine chemistry can be stretched even further, because the reaction shown is a nucleophilic substitution at the carbene

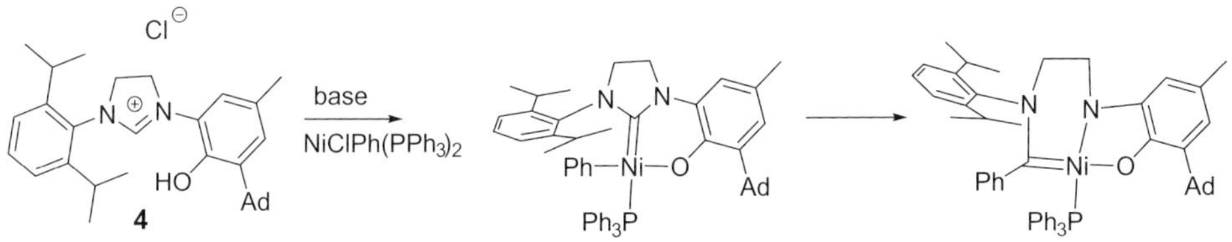

**Scheme 1.44** Nucleophilic attack on NHC followed by ring opening.

central atom. The mechanisms proposed for nucleophilic substitution in phosphorus ligands are presented in Sections 1.4.4–1.4.6 and Schemes 1.19–1.26. The difference with phosphine chemistry is that in phosphorus chemistry the harder anion ends up at the phosphorus atom and the softer carbon atom at platinum or palladium. Another ring expansion of an NHC has been reported, but in this case the reaction most likely proceeds via the alkene dimer of the NHC [163b].

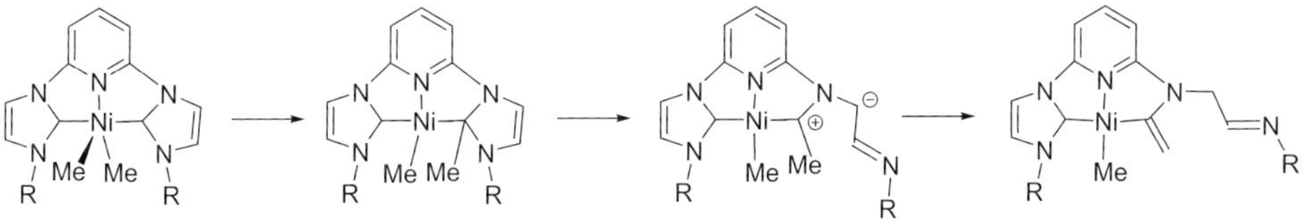

**Scheme 1.45** Nucleophilic attack on NHC followed by ring opening.

An unusual ring opening [164] was reported by Danopoulos and coworkers in a reaction between the TMEDA dimethylnickel complex and a pyridine bisNHC pincer ligand (Scheme 1.45). At first sight this may seem a new reaction, but the mechanism proposed involves a nucleophilic displacement of the nitrogen atom at the carbene donor atom and a proton transfer completes the reaction. The proton transfer may be aided by the TMEDA present. The carbene carbon atom ends up as a vinyl (anion) bound to nickel. For the last compound we are not aware of a phosphorus analog, as this would be a nickelaphosphormethylene complex (a "Wittig" reagent), but would probably rearrange to a phosphinomethylnickel moiety ($R_2PCH_2Ni$).

## 1.7.3
### Carbene Decomposition in Metathesis Catalysts

The first decomposition reaction shown comprises a reaction between the NHC ligand and the carbene fragment involved in the metathesis reaction; the reaction is very specific for the precursor and ligand, but it presents a transition from the previous paragraph to this one! Blechert and coworkers found that the Hoveyda–Grubbs catalyst undergoes a reversible cyclization, as shown in Scheme 1.46, which, after oxidation of the ruthenacycle, renders the compound inactive. The oxidation with oxygen re-establishes the aromatic ring and a carbine–ruthenium bond not suitable for catalysis [165]. The compound was obtained in low yield and most likely the reaction can be prevented by an appropriate substitution pattern on the aryl ring.

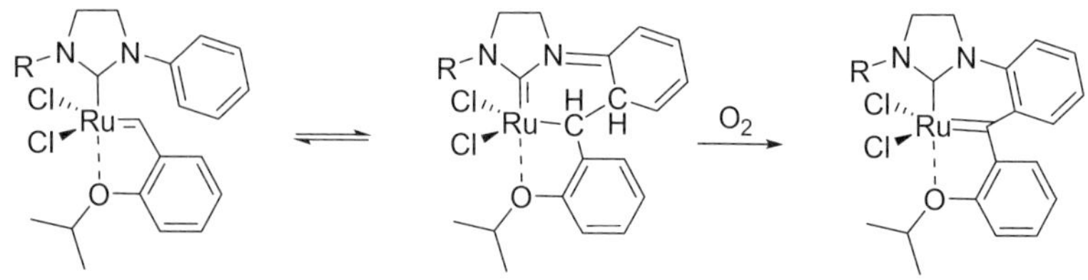

**Scheme 1.46** Reversible cyclization involving ligand and carbene in Hoveyda–Grubbs catalyst.

Another deactivation reaction involving both the NHC ligand and the reactant carbene was discovered by Grubbs for complexes containing non-substituted phenyl groups on the NHC nitrogen atoms [166]. The authors propose that it may involve a double C–H activation and only the first step is shown in Scheme 1.47 while the remaining part will be discussed later. In the absence of polar compounds C–H activation is an important starting point for decomposition reactions. It can be reduced effectively by choosing the proper substitution pattern on the aromatic rings [167].

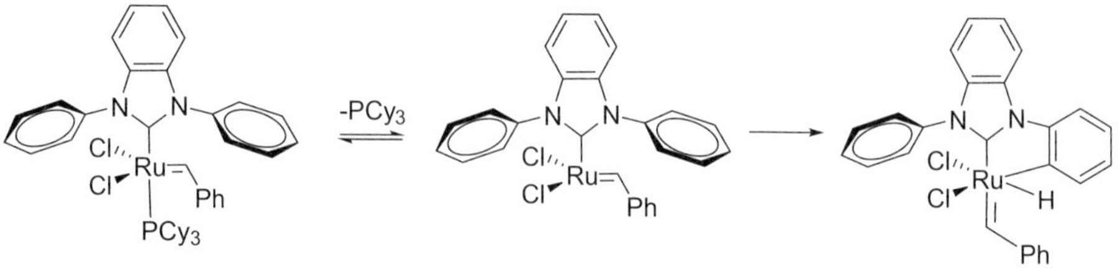

**Scheme 1.47** C–H activation in NHC metathesis catalyst.

In early studies Grubbs has shown that methylidene ruthenium catalysts decompose in a monomolecular reaction, while alkylidene ruthenium catalysts show a bimolecular decomposition pathway involving phosphine dissociation; the inorganic products were not identified at the time [168]. The methylene ruthenium species features in many catalytic reactions and it was concluded that this decomposition reaction is the reason why high catalyst loadings are often needed.

In the same year (1999) Hofmann and coworkers described the attack of electron-rich phosphines on the alkylidene coordinated to a Grubbs 1 catalyst; in this instance the phosphine is a bidentate and the attack may be aided by the intramolecular character [169]. They suggested that this reaction might be of importance in catalyst decomposition, as in many cases dissociation of phosphine from the ruthenium precursor is required to start catalysis. The reaction is shown in Scheme 1.48. After

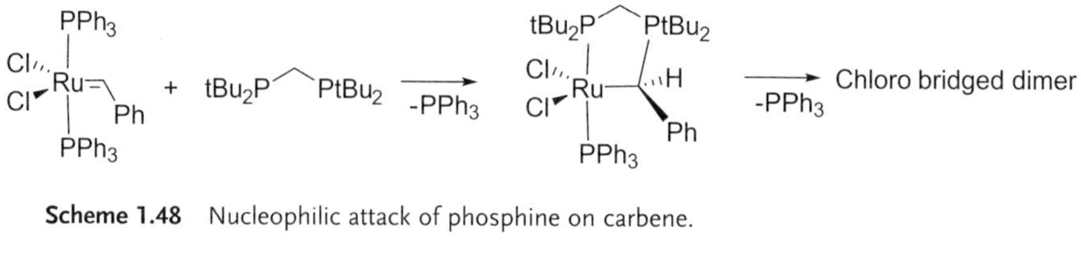

**Scheme 1.48** Nucleophilic attack of phosphine on carbene.

formation of the ylid triphenylphosphine dissociation takes place and the complex dimerizes (not shown).

The decomposition pathways of the most sensitive catalyst intermediate, the ruthenium methylidene species, were further studied by Grubbs and coworkers and they reported that nucleophilic attack of PCy$_3$ on the methylidene fragment was the initiating step of the process [170]. The highly basic phosphorus ylid formed abstracts a proton from another ruthenium methylidene and methyltricyclohexyl-phosphonium chloride, one of the products observed earlier, is formed. The resulting alkylidyne ruthenium complex reacts with the other, coordinatively unsaturated ruthenium species with formation of the dimer shown in Scheme 1.49.

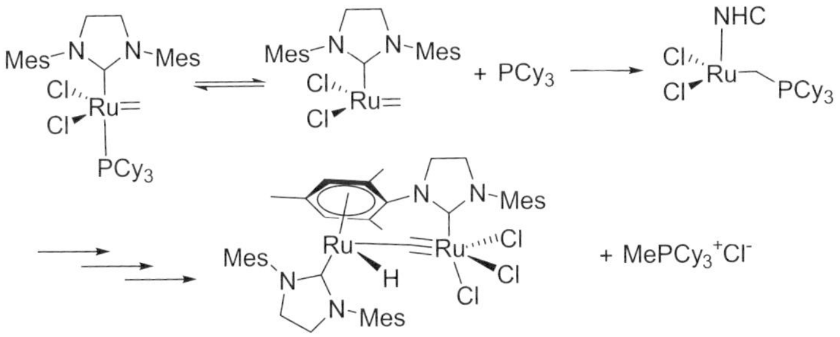

**Scheme 1.49** Grubbs catalyst undergoing nucleophilic attack at carbene by Cy$_3$P.

The sensitivity of metathesis catalysts for oxygenates changes drastically when we move in the periodic table from titanium to ruthenium. This has been noticed in both heterogeneous and homogeneous metathesis catalysis [171] and a simple metathesis reaction of the intermediate carbene metal complexes can be envisaged, and in some cases has been observed, that will convert metallocarbenes to metal oxide and the corresponding organic product. Clearly, reactions with oxygenates occur more frequently with molybdenum catalysts than with ruthenium catalysts. A few schematic reactions have been depicted in Scheme 1.50. It can be imagined that carboxylic

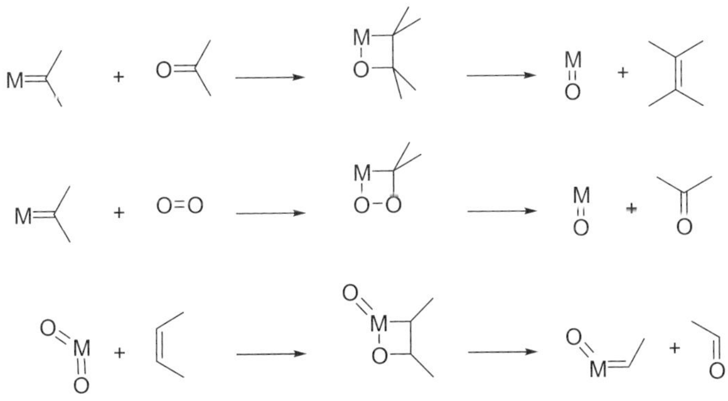

**Scheme 1.50** Reactions of metal-carbenes with oxygenates and oxides.

acids and esters, aldehydes, imines, and so on will lead to similar reactions, in particular where early transition metals are concerned.

A Schrock molybdenum catalyst reacts cleanly with benzaldehyde according to Scheme 1.50 to give the expected products. In this instance ethyl acetate did not react with the catalyst [172]. Carbene–oxo transfer between metals (in this case tantalum and tungsten) has also been reported [173]. The reaction of metal oxides and alkenes gives the alkylidene species (reaction 3, Scheme 1.50) and aldehyde, but by the same token the reverse reaction could lead to catalyst deactivation [174].

Ruthenium catalysts are much more resistant to oxygenates than early transition metal catalysts, as has been known since 1965 [175], but the TONs of ruthenium catalysts for functionalized molecules are also much lower than those for purified, purely hydrocarbon alkenes. In the case of oxygenates 1% of catalyst is often used, while turnover numbers for the alkene substrates may be as high as one million. As in many other catalyzed reactions using alkenes as substrates, the removal of hydroperoxides is important in order to achieve high turnover numbers [176]. Note, however, that $Me_4Sn$-activated tungsten catalysts will also convert up to 500 molecules of oleate substrates [177]!

The degradation of the first-generation Grubbs metathesis catalyst with primary alcohols, water, and oxygen, and the formation and catalytic activity of ruthenium(II) monocarbonyl species has been studied by several groups [178]. For several reactions they reported $RuHCl(CO)L_2$ as the final product. A potential mechanism as proposed by Dinger and Mol [178a] is shown in Scheme 1.51.

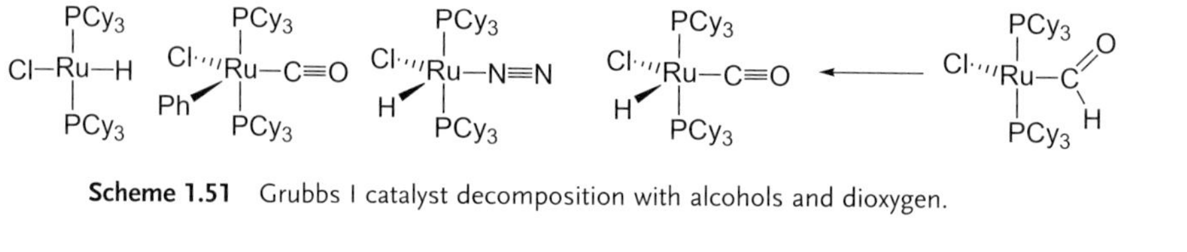

Scheme 1.51   Grubbs I catalyst decomposition with alcohols and dioxygen.

The Grubbs 2 catalyst (Scheme 1.52) reacted faster with methanol under basic conditions than the Grubbs 1 catalyst and gave a mixture of products, including a ligand disproportionation reaction giving the same hydride that results from Grubbs 1; the mechanism was proposed to be the same as that shown in Scheme 1.51 [179].

The hydrides formed are highly active isomerization or hydrogenation catalysts, and thus their formation could drastically influence the selectivity of the metathesis reaction. The metathesis reaction is still faster than the methanolysis and thus

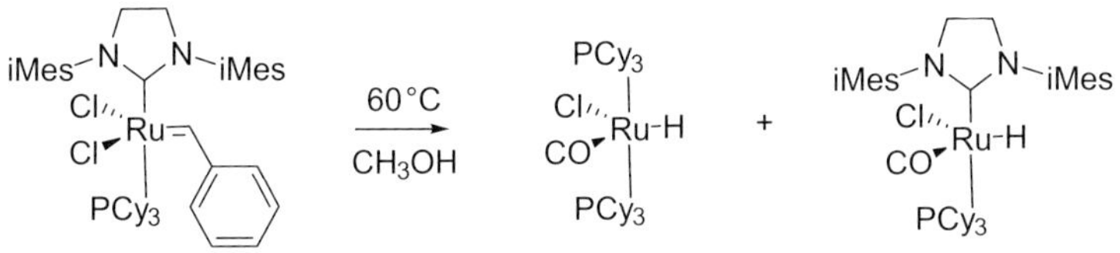

**Scheme 1.52** Grubbs II catalyst decomposition with alcohols and dioxygen.

carrying out the reaction in alcohols has only a small effect on the outcome. Alternatively, the alkylidene catalysts were transformed on purpose into hydride species in order to carry out a tandem metathesis/hydrogenation reaction, as shown by Fogg and coworkers [178d]. Dinger achieved TOFs as high as $160\,000\,\mathrm{m\,m^{-1}\,h^{-1}}$ for octene-1 hydrogenation at $100\,^\circ C$ and 4 bar of hydrogen [179].

The reaction of the Grubbs 2 catalyst converted into a cationic triflate complex with water under slightly acidic conditions and acetonitrile gave benzaldehyde and a solvent ligated NHC-Ru complex [180]. The reaction starts most likely with a nucleophilic attack of water on the carbene carbon atom coordinated to a cationic ruthenium ion. It might be compared with the formation of phosphine oxides from phosphines coordinated to divalent palladium and water (Sections 1.4.2 and 1.4.4).

Decomposition of the intermediate metallacyclobutane complex is a reaction that has been mentioned many times, but for which not all that much solid evidence has been presented [181] (the by-products formed via this route are always present in only tiny amounts, and often the products can be accounted for by metathesis reactions). One such reaction is β-H elimination of the metallacyclobutane complex, especially in heterogeneous catalysis. For immobilized rhenium complexes the reaction has been proven [182] and DFT calculations show that also for ruthenium catalysts this reaction should have a low activation barrier (Scheme 1.53) [183]. The allyl metal hydride species can react in various ways, for example giving a metal alkene complex. The reverse reaction is also mentioned as a way to make the catalyst initiator from a metal complex and alkene [184].

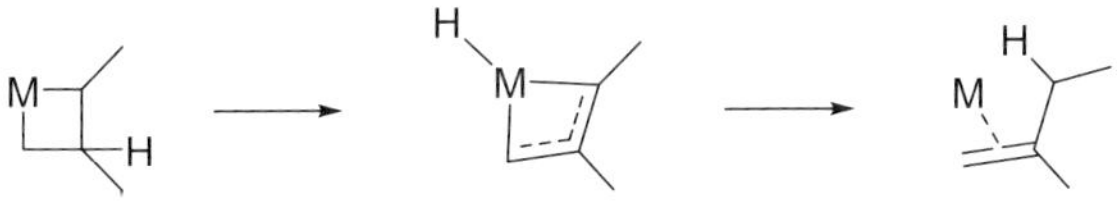

**Scheme 1.53** Metallocyclobutane decomposition to metal alkene complex.

A Schrock catalyst [185] for metathesis of alkynes can also undergo hydrogen loss from the intermediate, in this instance a metallacyclobutadiene generated from terminal alkynes (Scheme 1.54) [186]. Terminal alkynes can react in various ways with metal complexes and metathesis is not a common reaction. Mortreux found as one of the products of the decomposition reaction complex (top right, Scheme 1.54) and *t*-BuOH. This complex is highly active as a polymerization catalyst of alkynes and therefore it could not be identified when an excess of alkyne was present. The other complex, (bottom left, Scheme 1.54), can also be formed directly from

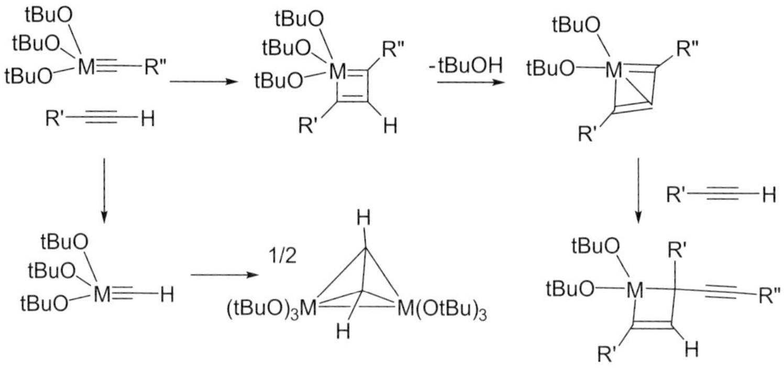

**Scheme 1.54** Schrock alkyne metathesis catalyst decomposition pathways.

metal-alkylidyne complexes and their dimerization has already been mentioned by Schrock as a potential deactivation mechanism [187].

## 1.8
## Reactions of Metal–Carbon and Metal–Hydride Bonds

### 1.8.1
### Reactions with Protic Reagents

The simplest reactions of organometallic compounds are those of the early transition metal alkyls and hydrides with water, acids and alcohols. Oxophilic metal ions such as titanium, zirconium, and vanadium (but also main group metals such as lithium, magnesium, aluminum, zinc, to name a few) react very rapidly with water and if no care is taken to exclude protic reagents thoroughly this presents the most common pathway for their decomposition. Exclusion of such species, including oxygen, is common practice in "Ziegler" catalysis. Usually the aluminum alkylating agent is used in excess so as to remove the last traces of compounds that may decompose the catalysts. Pretreatment of reagents and solvents has been extensively studied and can be found in the appropriate literature [188]. Metal hydrides may undergo similar reactions.

For late transition metal alkyl compounds the situation is more complex and many are resistant to such reactions with water. Perhaps this was somewhat unexpected and it has retarded the development of transition metal chemistry and catalysis in water [189]. Since the 1970s we have learned that many transition metal alkyl species are stable in water and main group metal alkyl species have been known to be stable in water. Examples of main group metals include toxic alkyl heavy metal species such as methylmercury derivatives (a product of aquatic anaerobic organisms!), tetraethyl-lead and/or alkyltin derivatives (that require a copper catalyst for their decomposition in acidic water), and so on. Dimethylpalladium complexes react readily with weak acids, but monoalkyl species of palladium react very slowly, even at $100\,°C$ with strong acids, thus enabling the copolymerization of ethene and carbon monoxide in methanol in the presence of acids [190, 191]. Likewise, the SHOP process [192], involving nickel alkyl species, can be carried out in 1,4-butanediol. The Ruhrchemie–Rhone Poulenc process for the hydroformylation of propene in water involves rhodium hydride and propyl species which are not decomposed by water at neutral pH.

Hydrocyanation of alkenes using nickel phosphite [193] or phosphine [194] catalysts involves nickel hydride intermediates which may react with HCN to give dihydrogen and nickel dicyanides. This reaction is irreversible and leads to catalyst decomposition. In the carbonylation reaction of methanol the Rh(III) methyl intermediate survives the strongly acidic conditions and rhodium methyl or acetyl species react with protons only as a side-reaction. This leads to trivalent rhodium halides, which have to be reduced before they can restart the catalytic cycle as Rh(I) undergoing oxidative addition of methyl iodide.

## 1.8.2
### Reactions of Zirconium and Titanium Alkyl Catalysts

In addition to the obvious reactions of reactive metal alkyls with protic reagents a few reactions have been reported for zirconium and titanium alkyl compounds that may be relevant to the decomposition of these catalysts. Teuben reported that zirconium alkyl compounds react with propene to give an alkane and a π-allyl zirconium species [195]. The latter is inactive in alkene polymerization and thus stops the polymerization reaction (see Scheme 1.55). In ethene polymerization this reaction is less likely as it can only occur with products formed by β-hydride elimination and not with the substrate, ethene. Conceptually this reaction might be compared to the reaction of Pd(II) salts, especially acetates, with alkenes to give π-allyl palladium complexes and acid. Aluminoxanes and methylzirconium compounds lead to methane formation and bridging methylene species, as was reported by Kamisky [196].

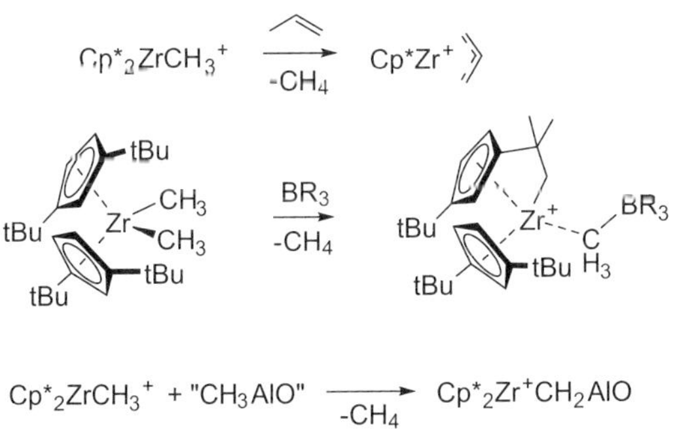

**Scheme 1.55** Deactivation of zirconium catalysts.

Another example of catalyst deactivation via the formation of bridging methylene species, reported by Marks, is illustrated in Scheme 1.56 [197].

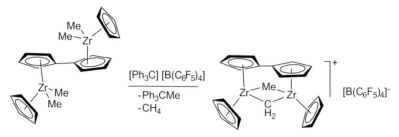

**Scheme 1.56** Catalyst deactivation via μ-CH$_2$ complex formation.

Hydride intermediates formed in chain transfer with hydrogen during olefin polymerization may also lead to deactivation reactions. For example, the presence of ester modifying agents in propene polymerization with Ziegler–Natta catalysts can lead to the formation of inactive titanium alkoxides [198]. The esters can also be reduced to alkoxides by reaction with excess aluminum alkyl [199].

$$\text{Ti-H} + \text{RCOOR}' \rightarrow \text{Ti-OCH(R)OR}'$$

$$\text{PhCOOEt} \cdot \text{AlEt}_3 + \text{AlEt}_3 \rightarrow \text{PhEt}_2\,\text{COAlEt}_2 + \text{AlEt}_2\,\text{OEt}$$

## 1.9
### Reactions Blocking the Active Sites

Impurities in the feed or alternative reactions of substrates can cause the formation of inactive species. For instance when alkenes are the substrate one can imagine that hydroperoxides, dienes, alkynes, enones, and so on can react with the catalyst to give inactive species. Sometimes this may be temporary and the catalyst activity can be restored, in which case one might say that "dormant" sites are formed. These reactions are highly specific and most of them will be treated in the chapters that follow. Dormant sites are well-known in propene polymerization catalysis and in hydroformylation catalysis. In metallocene-catalyzed olefin polymerization, coordination of AlMe$_3$ to [Cp$_2$ZrR]$^+$ can lead to dormant, alkyl-bridged species of type [Cp$_2$Zr(μ-R)(μ-Me)AlMe$_2$]$^+$.

### 1.9.1
### Polar Impurities

Ligands competing with the substrate for the open coordination sites are an obvious cause for the slowing down of a reaction. Undoubtedly, hundreds of examples could be collected. We will mention only one that has been of practical importance, namely in the asymmetric isomerization of the allylamine precursor for menthol, a process operated by Takasago [200]. The synthesis of menthol is given in Scheme 1.57. The key reaction is the enantioselective isomerization of the allylamine to the asymmetric enamine. The amine by-product blocks the cationic catalyst in this case.

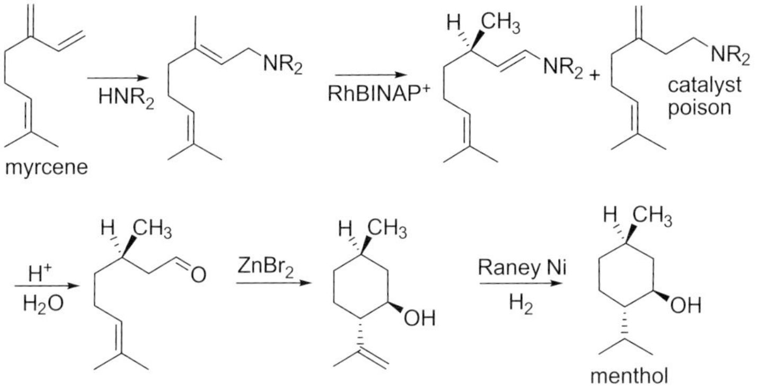

Scheme 1.57  Preparation of menthol according to Takasago.

## 1.9.2
## Dimer Formation

Active, monomeric catalyst species may be involved in the formation of inactive dimers. When this equilibrium is reversible it only leads to a reduction in the amount of catalyst available and does not bring the catalysis to a full stop.

A well-known example is the formation of the so-called orange dimer from HRh(PPh$_3$)$_3$CO, already reported by Wilkinson [201]. This will occur at low pressures of hydrogen and high rhodium concentrations (see Scheme 1.58) [202]. The reaction is reversible.

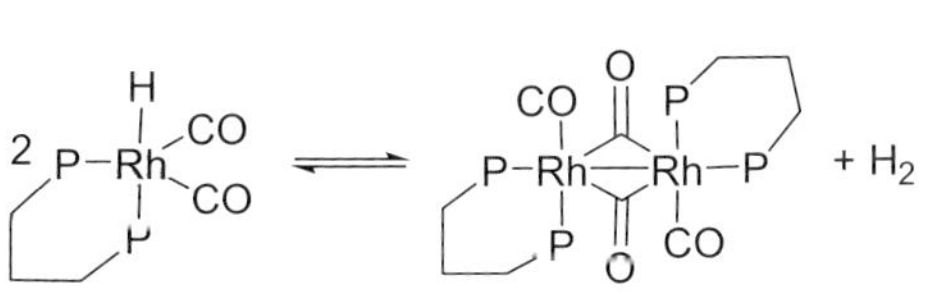

Scheme 1.58  Rhodium dimer formation in rhodium catalysis.

Formation of zerovalent palladium species during carbonylation reactions catalyzed by divalent palladium leads to the formation of formally monovalent palladium [203]. A few types that have been observed are shown in Scheme 1.59.

When dimer formation becomes dominant, one might attempt to destabilize the dimer relative to the monomer. For instance, by making the ligand very bulky one might prevent dimer formation. Another approach is so-called "site isolation" by immobilization of the catalyst, as was described by Grubbs [204], or by encapsulation

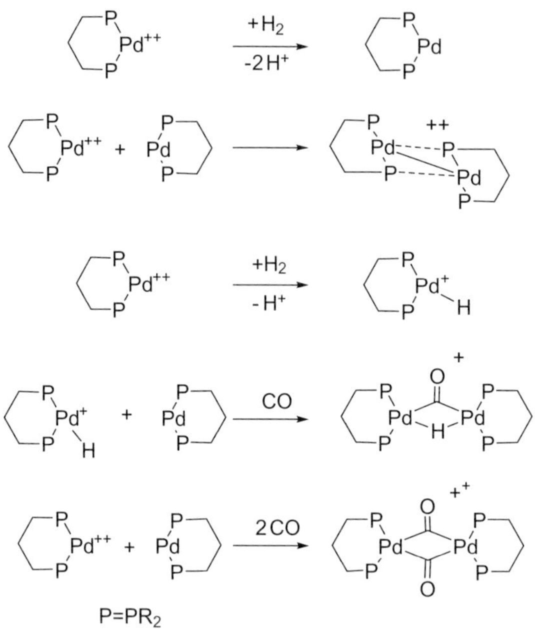

**Scheme 1.59** Formation of palladium(I) dimers in carbonylation catalysts.

in a dendrimer [205]. Grubbs's well-known example concerns a titanocene catalyst that is used as a hydrogenation catalyst. The intermediate titanium hydride is converted almost completely to a dimer rendering the catalyst to be of low activity. Immobilization of the catalyst on a resin support prevents dimerization and an active catalyst is obtained. In oxidation catalysis involving metal porphyrins or metal phthalocyanines dimer formation is a common phenomenon, for instance for Mn(III) phthalocyanine as shown in Figure 1.5 [206].

### 1.9.3
### Ligand Metallation

In Scheme 1.55 we have shown an example of metallation of the ligand that led to an inactive zirconium catalyst. In late transition metal chemistry the same reactions

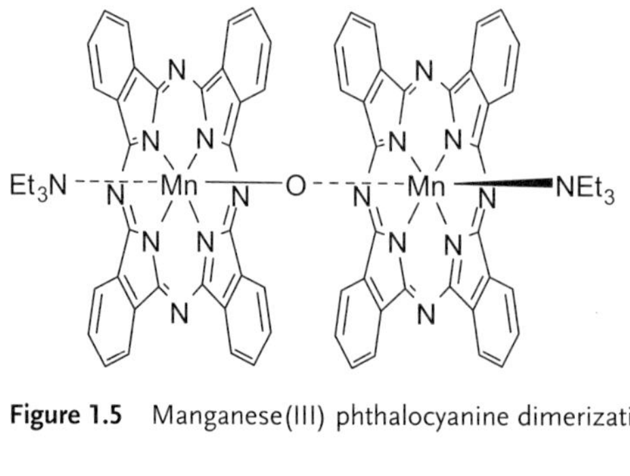

**Figure 1.5** Manganese(III) phthalocyanine dimerization via an oxygen bridge.

occur, but now the complexes formed represent a dormant site and catalyst activity can often be restored. Work-up of rhodium-phosphite catalyst solutions after hydroformylation often shows the partial formation of metallated species, especially when bulky phosphites are used [207]. Dihydrogen elimination or alkane elimination may lead to the metallated complex. The reaction is reversible for rhodium and thus the metallated species could function as a stabilized form of rhodium during a catalyst recycle. Many metallated phosphite complexes have been reported, but we mention only two, one for triphenyl phosphite and rhodium [208] (see Scheme 1.60) and one for a bulky diimine and palladium (Scheme 1.61) [209].

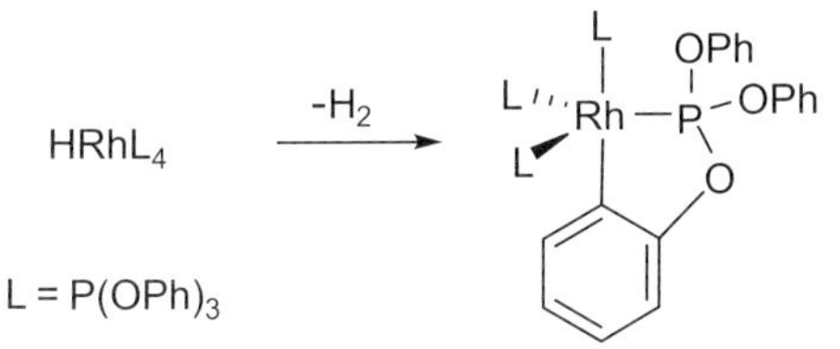

**Scheme 1.60** A metallated rhodium phosphite complex.

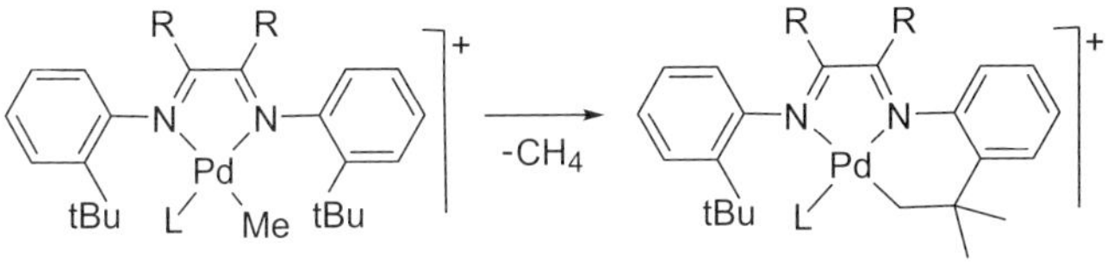

**Scheme 1.61** Intramolecular C–H activation in a cationic palladium α-diimine complex.

## References

1 (a) Diéguez, M., Ruiz, A., and Claver, C. (2002) *J. Org. Chem.*, **67**, 3796–3801; (b) Buisman, G.J.H., Martin, M.E., Vos, E.J., Klootwijk, A., Kamer, P.C.J., and van Leeuwen, P.W.N.M. (1995) *Tetrahedron: Asymmetry*, **6**, 719–738.

2 van Leeuwen, P.W.N.M. (ed.) (2008) *Supramolecular Catalysis*, Wiley VCH Verlag GmbH, Weinheim.

3 Goudriaan, P.E., Jang, X.-B., Kuil, M., Lemmens, R., van Leeuwen, P.W.N.M., and Reek, J.N.H. (2008) *Eur. J. Org. Chem.*, 6079–6092.

4 Breit, B. (2008) *Pure Appl. Chem.*, **80**, 855.

5 Heller, D., de Vries, A.H.M., and de Vries, J.G. (2007) Catalyst inhibition and deactivation in homogeneous hydrogenation, in *Handbook of Homogeneous Hydrogenation*, Part 3 (eds J.G. de Vries and C.J. Elsevier), Wiley-VCH Verlag GmbH, Weinheim, pp. 1483–1516.

6 (a) Moulijn, J.A., van Diepen, A.E., and Kapteijn, F. (2001) *Appl. Catal. A: General*, **212**, 3–16; (b) Moulijn, J.A. (ed.) (2001) *Appl. Catal. A: General*, **212** (1–2), 1–271 (Special issue; Catalyst Deactivation).

7 van Leeuwen, P.W.N.M. (2001) *Appl. Catal. A: General*, **212**, 61–81.

8 (a) Freitas, E.R. and Gum, C.R. (1979) *Chem. Eng. Proc.*, **73**, 70, Ref. 4; (b) Keim, W. (1990) *Angew. Chem., Int. Ed.*, **29**, 235.

9 van Leeuwen, P.W.N.M. and Freixa, Z. (2006) in *Carbon Monoxide as a Chemical Feedstock: Carbonylation Catalysis* (ed. W.B. Tolman), Wiley-VCH Verlag GmbH, Weinheim, Germany, pp. 319–356.

10 Lewis F L.N. (1990) *J. Am. Chem. Soc.*, **112**, 5998.

11 Smidt, J. (1962) *Angew. Chem. Int. Ed.*, **1**, 80.

12 Henry, P.M. (1980) *Palladium-Catalyzed Oxidation of Hydrocarbons*, Reidel, Dordrecht.

13 Beletskaya, I.P. (1982) *J. Organomet. Chem.*, **250**, 551.

14 van Strijdonck, G.P.F., Boele, M.D.K., Kamer, P.C.J., de Vries, J.G., and van Leeuwen, P.W.N.M. (1999) *Eur. J. Inorg. Chem.*, 1073.

15 Albisson, D.A., Bedford, R.B., Lawrence, S.E., and Scully, P.N. (1998) *Chem. Commun.*, 2095.

16 (a) de Meijere, A. and Diederich, F. (eds) (2004) *Metal-Catalyzed Cross-Coupling Reactions*, 2nd edn, Wiley-VCH Verlag GmbH, Weinheim; (b) Hartwig, J.F. (2002) in *Handbook of Organopalladium Chemistry for Organic Synthesis*, vol. 1 (ed. E.I. Negishi,), Wiley-Interscience, New York, p. 1051 and 1097; (c) Jiang, L. and Buchwald, S.L. (2004) in *Metal-Catalyzed Cross-Coupling Reactions*, 2nd edn (eds A. De Meijere and F. Diederich), Wiley-VCH Verlag GmbH, Weinheim, Germany, p. 699.

17 Tromp F M., Sietsma, J.R.A., van Bokhoven, J.A., van Strijdonck, G.P.F., van Haaren, R.J., van der Eerden, A.M.J., van Leeuwen, P.W.N.M., and Koningsberger, D.C. (2003) *Chem. Commun.*, 128–129.

18 Astruc, D. (ed.) (2008) *Nanoparticles and Catalysis*, Wiley-VCH Verlag GmbH, Weinheim.

19 Schmid, G. (2004) *Nanoparticles, from Theory to Application*, Wiley-VCH Verlag GmbH, Weinheim.

20 Starkey Ott, L. and Finke, R.G. (2007) *Coord. Chem. Rev.*, **251**, 1075–1100.

21 Guari, Y., Thieuleux, C., Mehdi, A., Reyé, C.,F C.;., Corriu, R.J.P., Gomez-Gallardo, S., Philippot, K., Chaudret, B., and Dutartre, R. (2001) *Chem. Commun.*, 1374.

22 Dupont, J., de Souza, R.F., and Suarez, P.A.Z. (2002) *Chem. Rev.*, **102**, 3667.

23 Calandra, P., Giordano, C., Longo, A., and Turco Liveri, V. (2006) *Mater. Chem. Phys.*, **98**, 494.

24 (a) Zhao, M., Sun, L., and Crooks, R.M. (1998) *J. Am. Chem. Soc.*, **120**, 4877; (b) Boisselier, E., Diallo, A.K., Salmon, L., Ruiz, J., and Astruc, D. (2008) *Chem. Commun.*, 4819–4821; (c) Balogh, L. and Tomalia, D.A. (1998) *J. Am. Chem. Soc.*, **120**, 7355.

25 Astruc, D., Fu, J., and Aranzaes, J.R. (2005) *Angew. Chem. Int. Ed.*, **44**, 7852–7872.

26 Sablong, R., Schlotterbeck, U., Vogt, D., and Mecking, S. (2003) *Adv. Synth. Catal.*, **345**, 333.

27 Vargaftik, M.N., Zagorodnikov, V.P., Stolarov, I.P., Moiseev, I.I., Kochubey, D.I., Likholobov, V.A., Chuvilin, A.L., and Zamaraev, K.I. (1989) *J. Mol. Catal.*, **53**, 315.

28 (a) de Vries, J.G. (2006) *Dalton Trans.*, 421; (b) Phan, N.T.S., van der Sluys, M., and Jones, C.W. (2006) *Adv. Synth. Catal.*, **348**, 609–679; (c) Durand, J., Teuma, E., and Gomez, M. (2008) *Eur. J. Inorg. Chem.*, 3577–3586; (d) Duran Pachon, L. and Rothenberg, G. (2008) *Applied Organomet. Chem.*, **22**, 288–299; (e) Trzeciak, A.M. and Ziólkowski, J.J. (2007) *Coord. Chem. Rev.*, **251**, 1281–1293; (f) Djakovitch, L., Koehler, K., and de Vries, J.G. (2008) in *Nanoparticles and Catalysis* (ed. D. Astruc), Wiley-VCH Verlag GmbH, Weinheim, pp. 303–348.

29 (a) Studer, M., Blaser, H.-U., and Exner, C. (2003) *Adv. Synth. Catal.*, **345**, 45; (b) Bönnemann, H. and Braun, G.A. (1996) *Angew. Chem., Int. Ed. Engl.*, **35**, 1992; (c) Bönnemann, H. and Braun, G.A. (1997) *Chem. Eur. J.*, **3**, 1200; (d) Zuo, X., Liu, H., Guo, D., and Yang, X. (1999) *Tetrahedron*, **55**, 7787; (e) Köhler, J.U. and Bradley, J.S. (1997) *Catal. Lett.*, **45**, 203; (f) Köhler, J.U. and Bradley, J.S. (1998) *Langmuir*, **14**, 2730.

30 Tamura, M. and Fujihara, H. (2003) *J. Am. Chem. Soc.*, **125**, 15742.

31 (a) Jansat, S., Gómez, M., Phillipot, K., Muller, G., Guiu, E., Claver, C., Castillón, S., and Chaudret, B. (2004) *J. Am. Chem.*

*Soc.*, **126**, 1592; (b) Favier, I., Gómez, M., Muller, G., Axet, M.A., Castillón, S., Claver, C., Jansat, S., Chaudret, B., and Philippot, K. (2007) *Adv. Synth. Catal.*, **349**, 2459.

32 Klabunovskii, E., Smith, G.V., and Zsigmond, A. (2006) in *Heterogeneous Enantioselective Hydrogenation, Theory and Practice, Catalysis by Metal Complexes*, vol. 31 (eds B.R. James and P.W.N.M. van Leeuwen), Springer, Dordrecht, The Netherlands.

33 Dieguez F M., Pamies, O., Mata, Y., Teuma, E., Gomez, M., Ribaudo, F., and van Leeuwen, P.W.N.M. (2008) *Adv. Synth. Catal.*, **350**, 2583–2598.

34 Wada, K., Yano, K., Kondo, T., and Mitsudo, T. (2006) *Catal. Lett.*, **112**, 63–67.

35 Iwasawa, T., Tokunaga, M., Obora, Y., and Tsuji, Y. (2004) *J. Am. Chem. Soc.*, **126**, 6554–6555.

36 Landau, R. and Saffer, A. (1968) *Chem. Eng. Prog.*, **64**, 20.

37 ten Brink, G.-J., Arends, I.W.C.E., and Sheldon, R.A. (2002) *Adv. Synth. Catal.*, **344**, 355.

38 Challa, G., Chen, W., and Reedijk, J. (1992) *Makromol. Chem. Makromol. Symp.*, **59**, 59.

39 (a) Hage, R., Iburg, J.E., Kerschner, J., Koek, J.H., Lempers, E.L.M., Martens, R.J., Racherla, U.S., Russell, S.W., Swarthoff, T., van Vliet, R.P., Warnaar, J.B., van der Wolf, L., and Krijnen, B. (1994) *Nature*, **369**, 637; (b) Que JJr., L. (2007) *Acc. Chem. Res.*, **40**, 493–500; (c) Klopstra, M., Hage, R., Kellogg, R.M., and Feringa, B.L. (2003) *Tetrahedron Lett.*, **44**, 4581–4584.

40 Collins, T.J. (1994) *Acc. Chem. Res.*, **27**, 279.

41 Landau, R., Sullivan, G.A., and Brown, D. (1979) *ChemTech*, 602.

42 Hill, J.G., Sharpless, K.B., Exon, C.M., and Regeneye, R. (1985) *Org. Synth.*, **63**, 66.

43 (a) Colladon, M., Scarso, A., Sgarbossa, P., Michelin, R.A., and Strukul, G. (2007) *J. Am. Chem. Soc.*, **129**, 7680–7689; (b) Strukul, G. and Michelin, R.A. (1985) *J. Am. Chem. Soc.*, **107**, 563–569.

44 Blann, K., Bollmann, A., Dixon, J.T., Hess, F.M., Killian, E., Maumela, H., Morgan, D.H., Neveling, A., Otto, S., and Overett, M.J. (2005) *Chem. Commun.*, 620–621.

45 Tucker, C.E. and de Vries, J.G. (2002) *Top. Catal.*, **19**, 111–118.

46 (a) Garrou, P.E. (1985) *Chem. Rev.*, **85**, 171; (b) Garrou, P.E., Dubois, R.A., and Jung, C.W. (1985) *ChemTech*, 123.

47 Parkins, A.W. (2006) *Coord. Chem. Rev.*, **250**, 449–467.

48 Macgregor, S.A. (2007) *Chem. Soc. Rev.*, **36**, 67–76.

49 Cervilla, A., Perez-Pla, F., Llopis, E., and Piles, M. (2006) *Inorg Chem.*, **45**, 7357–7366.

50 Aresta, M., Dibenedetto, A., and Tommasi, I. (2001) *Eur. J. Inorg. Chem.*, 1801–1806.

51 Larpent, C., Dabard, R., and Patin, H. (1987) *Inorg. Chem.*, **26**, 2922–2924.

52 Nicholas, K.M. (1980) *J. Organomet. Chem.*, **188**, C10–C12.

53 Ropartz, L., Meeuwenoord, N.J., van der Marel, G.A., van Leeuwen, P.W.N.M., Slawin, A.M.Z., and Kamer, P.C.J. (2007) *Chem. Commun.*, 1556–1558.

54 (a) Buckler, S.A., Doll, L., Lind, F.K., and Epstein, M. (1962) *J. Org. Chem.*, **27**, 794–798; (b) Zhao, Y.-L., Flora, J.W., Thweatt, W.D., Garrison, S.L., Gonzalez, C., Houk, K.N., and Marquez, M. (2009) *Inorg. Chem.*, **48**, 1223–1231; (c) Dobbie, R.C. (1971) *J. Chem. Soc. (A)*, 2894–2897; (d) Lim, M.D., Lorkovic, I.M., and Ford, P.C. (2002) *Inorg. Chem.*, **41**, 1026–1028; (e) Odom, J.D. and Zozulin, A.J. (1981) *Phosphorus Sulfur Silicon Relat. Elem.*, **9**, 299–305.

55 Wehman, P., van Donge, H.M.A., Hagos, A., Kamer, P.C.J., and van Leeuwen, P.W.N.M. (1997) *J. Organomet. Chem.*, **535**, 183–193.

56 Otsuka, S., Aotani, Y., Tatsuno, Y., and Yoshida, T. (1976) *Inorg Chem.*, **15**, 656–660.

57 Segarra-Maset, M.D., Freixa, Z., and van Leeuwen, P.W.N.M.,(2010) *Eur. J. Inorg. Chem.*, 2075–2078.

58 Burns, J.A., Butler, J.C., Moran, J., and Whitesides, G.M. (1991) *J. Org. Chem.*, **56**, 2648–2650, and references therein.

59 Levison, M.E., Josephson, A.S., and Kirschenbaum, D.M. (1969) *Experientia*, **25**, 126–127.

60 Storjohann, L., Holst, B., and Schwartz, T.W. (2008) *Biochemistry*, **47**, 9198–9207.

61 Faucher, A.-M. and Grand-Maitre, C. (2003) *Synth. Commun.*, **33**, 3503–3511.

62 (a) Hewertson, W. and Watson, H.R. (1962) *J. Chem. Soc.*, 1490; (b) van Doorn, J.A., Frijns, J.H.G., and Meijboom, N. (1991) *Rec. Trav. Chim. Pays-Bas*, **110**, 441–449.

63 Phadnis, P.P., Dey, S., Jain, V.K., Nethaji, M., and Butcher, R.J. (2006) *Polyhedron*, **25**, 87–94.

64 (a) Chini, P., Martinengo, S., and Garlaschelli, G. (1972) *J. Chem. Soc., Chem. Commun.*, 709; (b) Dubois, R.A. and Garrou, P.E. (1986) *Organometallics*, **5**, 466; (c) Harley, A.D., Guskey, G.J., and Geoffroy, G.L. (1983) *Organometallics*, **2**, 53.

65 Billig, E., Jamerson, J.D., and Pruett, R.L. (1980) *J. Organomet. Chem.*, **192**, C49.

66 Mason, R., Søtofte, I., Robinson, S.D., and Uttley, M.F. (1972) *J. Organomet. Chem.*, **46**, C61.

67 Goel, A.B. (1984) *Inorg. Chim. Acta*, **84**, L25.

68 Abatjoglou, A.G., Billig, E., and Bryant, D.R. (1984) *Organometallics*, **3**, 923–926.

69 Sakakura, T. (1984) *J. Organometal. Chem.*, **267**, 171.

70 Kong, K.–C. and Cheng, C–H. (1991) *J. Am. Chem. Soc.*, **113**, 6313.

71 (a) Sisak, A., Ungváry, F., and Kiss, G. (1983) *J. Mol. Catal.*, **18**, 223–235; (b) van Leeuwen, P.W.N.M., Roobeek, C.F., Frijns, J.H.G., and Orpen, A.G. (1990) *Organometallics*, **9**, 1211.

72 Heyn, R.H. and Görbitz, C.H. (2002) *Organometallics*, **21**, 2781–2784.

73 (a) Bennett, M.A., Berry, D.E., and Beveridge, K.A. (1990) *Inorg. Chem.*, **29**, 4148–4152; (b) Dekker, G.P.C.M., Elsevier, C.J., Poelsma, S.N., Vrieze, K., van Leeuwen, P.W.N.M., Smeets, W.J.J., and Spek, A.L. (1992) *Inorg. Chim. Acta*, **195**, 203–210.

74 (a) Albinati, A., Lianza, F., Pasquali, M., Sommovigo, M., Leoni, P., Pregosin, P.S., and Regger, H.S. (1991) *Inorg. Chem.*, **30**, 4690; (b) Budzelaar, P.H.M.,

van Leeuwen, P.W.N.M., Roobeek, C.F., and Orpen, A.G. (1992) *Organometallics*, **11**, 23; (c) Murahashi, T., Otani, T., Okuno, T., and Kurosawa, H. (2000) *Angew. Chem. Int. Ed.*, **39**, 537.

75 Tschan, M.J.-L., Cherioux, F., Karmazin-Brelot, L., and Suess-Fink, G. (2005) *Organometallics*, **24**, 1974–1981.

76 Bradford, C.W. and Nyholm, R.S. (1973) *J. Chem. Soc., Dalton Trans.*, 529.

77 Bruce, M.I., Humphrey, P.A., Okucu, S., Schmutzler, R., Skelton, B.W., and White, A.H. (2004) *Inorg. Chim. Acta*, **357**, 1805–1812.

78 Franken, A., McGrath, T.D., and Stone, F.G.A. (2006) *J. Am. Chem. Soc.*, **128**, 16169–16177.

79 (a) Geoffroy, G.L., Rosenberg, S., Shulman, P.M., and Whittle, R.R. (1984) *J. Am.Chem. Soc.*, **106**, 1519; (b) Shulman, P.M., Burkhardt, E.D., Lundquist, E.G., Pilato, R.S., Geoffroy, G.L., and Rheingold, A.L. (1987) *Organometallics*, **6**, 101.

80 Sicree, S.A. and Tamborski, C. (1992) *J. Fluor. Chem.*, **59**, 269–273.

81 (a) van der Veen, L.A., Keeven, P.H., Schoemaker, G.C., Reek, J.N.H., Kamer, P.C.J., van Leeuwen, P.W.N.M., Lutz, M., and Spek, A.L. (2000) *Organometallics*, **19**, 872–883; (b) Levy, J.B., Walton, R.C., Olsen, R.E., and SymmesJr., C. (1996) *Phosphorus, Sulfur*, **109–110**, 545–548.

82 Oae, S. and Uchida, Y. (1991) *Acc. Chem. Res.*, **24**, 202–208.

83 Budzelaar, P.H.M. (1998) *J. Org. Chem.*, **63**, 1131–1137.

84 Miyamoto, T.K., Matsuura, Y., Okude, K., Lchida, H., and Sasaki, Y. (1989) *J. Organomet. Chem.*, **373**, C8–C12.

85 Desponds, O. and Schlosser, M. (1996) *J. Organomet. Chem.*, **507**, 257–261.

86 (a) Kikukawa, K., Takagi, M., and Matsuda, T. (1979) *Bull. Chem. Soc. Jpn*, **52**, 1493; (b) Bouaoud, S-E., Braunstein, P., Grandjean, D., and Matt, D. (1986) *Inorg. Chem.*, **25**, 3765; (c) Alcock, N.W., Bergamini, P., Kemp, T.J., and Pringle, P.G. (1987) *J. Chem. Soc., Chem. Commun.*, 235; (d) van Leeuwen, P.W.N.M., Roobeek, C.F., and Orpen, A.G. (1990) *Organometallics*, **9**, 2179.

87 Yasuda, H., Maki, N., Choi, J.–C., Abla, M., and Sakakura, T. (2006) *J. Organometal. Chem.*, **691**, 1307–1310.

88 Alcock, N.W., Bergami, P., Kemp, T.J., and Pringle, P.G. (1987) *J.Chem. Soc., Chem. Commun.*, 235.

89 Ruiz, J., Garcia-Granda, S., Diaz, M.R., and Quesada, R. (2006) *Dalton Trans.*, 4371–4376.

90 Kohl, S.W., Heinemann, F.W., Hummert, M., Bauer, W., and Grohmann, A. (2006) *Chem. Eur. J.*, **12**, 4313–4320.

91 Hartwig, J.F., Bergman, R.G., and Anderson, R.A. (1990) *J. Organomet. Chem.*, **394**, 417.

92 Kaneda, K., Sano, K., and Teranishi, S. (1979) *Chem. Lett.*, 821–822.

93 Chan, A.S.C., Caroll, W.E., and Willis, D.E. (1983) *J. Mol. Catal.*, **19**, 377.

94 Sakamoto, M., Shimizu, I., and Yamamoto, A. (1995) *Chem. Lett.*, 1101.

95 Grushin, V.V. (2000) *Organometallics*, **19**, 1888–1900.

96 Cooper, M.K., Downes, J.M., and Duckworth, P.A. (1989) *Inorg. Synth.*, **25**, 129–133.

97 (a) Marcoux, D. and Charette, A.B. (2008) *J. Org. Chem.*, **73**, 590–593; (b) de la Torre, G., Gouloumis, A., Vazquez, P., and Torres, T. (2001) *Angew. Chem. Int. Ed.*, **40**, 2895–2898; (c) Migita, T., Nagai, T., Kiuchi, K., and Kosugi, M. (1983) *Bull. Chem. Soc. Jpn.*, **56**, 2869–70.

98 (a) Kwong, F.Y. and Chan, K.S. (2000) *Chem. Commun.*, 1069–1070; (b) Kwong, F.Y., Lai, C.W., and Chan, K.S. (2002) *Tetrahedron Lett.*, **43**, 3537–3539; (c) Kwong, F.Y. and Chan, K.S. (2001) *Organometallics*, **20**, 2570–2596; (d) Wang, Y., Lai, C.W., Kwong, F.Y., Jia, W., and Chan, K.S. (2004) *Tetrahedron*, **60**, 9433–9439.

99 (a) Morita, D.K., Stille, J.K., and Norton, J.R. (1995) *J. Am. Chem. Soc.*, **117**, 8576–8581; (b) Segelstein, B.E., Butler, T.W., and Chenard, B.L. (1995) *J. Org. Chem.*, **60**, 12; (c) Goodson, F.E., Wallow, T.I., and Novak, B.M. (1997) *J. Am. Chem. Soc.*, **119**, 12441.

100 (a) O'Keefe, D.F., Dannock, M.C., and Marcuccio, S.M. (1992) *Tetrahedron Lett.*, **33**, 6679; (b) Herrmann, W.A., Brossmer, C., Öfele, K., Beller, M., and Fischer, H. (1995) *J. Organomet. Chem.*, **491**, C1; (c) Batsanov, A.S., Knowles, J.P., and Whiting, A. (2007) *J. Org. Chem.*, **72**, 2525–2532; (d) Leriche, P., Aillerie, D., Roquet, S., Allain, M., Cravino, A., Frere, P., and Roncali, J. (2008) *Org. Biomol. Chem.*, **6**, 3202–3207.

101 Goodson, F.E., Wallow, T.I., and Novak, B.M. (1998) *Macromolecules*, **31**, 2047.

102 Green, M.L.H., Smith, M.J., Felkin, H., and Swierczewski, G. (1971) *J. Chem. Soc. D*, 158.

103 Mohtachemi, R., Kannert, G., Schumann, H., Chocron, S., and Michman, M. (1986) *J. Organomet. Chem.*, **310**, 107.

104 Vierling, P. and Riess, J.G. (1986) *Organometallics*, **5**, 2543.

105 Macgregor, S.A. and Wondimagegn, T. (2007) *Organometallics*, **26**, 1143–1149.

106 Marshall, W.J. and Grushin, V.V. (2003) *Organometallics*, **22**, 555–562.

107 (a) Geldbach, T.J. and Pregosin, P.S. (2002) *Eur. J. Inorg. Chem.*, 1907; (b) Geldbach, T.J., Pregosin, P.S., and Albinati, A. (2003) *Organometallics*, **22**, 1443; (c) den Reijer, C.J., Dotta, P., Pregosin, P.S., and Albinati, A. (2001) *Can. J. Chem.*, **79**, 693; (d) Geldbach, T.J. and Pregosin, P.S. (2001) *Organometallics*, **20**, 2990–2997.

108 Goodman, J. and Macgregor, S.A. (2010) *Coord. Chem. Rev.*, **254**, 1295–1306.

109 Renard, P.-Y., Vayron, P., Leclerc, E., Valleix, A., and Mioskowski, C. (2003) *Angew. Chem. Int. Ed.*, **42**, 2389–2392.

110 Tolman, C.A., McKinney, R.J., Seidel, W.C., Druliner, J.D., and Stevens, W.R. (1985) *Adv. Catal.*, **33**, 1.

111 Pruett, R.L. and Smith, J.A. (1969) *J. Org. Chem.*, **34**, 327.

112 (a) van Leeuwen, P.W.N.M. and Roobeek, C.F. (1983) *J. Organometal. Chem.*, **258**, 343; (b) van Leeuwen, P.W.N.M. and Roobeek, C.F., Brit. Pat. 2,068,377. US Pat. 4,467,116 (to Shell Oil);(1984) *Chem. Abstr.*, **101**, 191142.

113 Billig, E., Abatjoglou, A.G., and Bryant, D.R. (1987) (to Union Carbide Corporation) U.S. Pat. 4,769,498; U.S. Pat. 4,668,651; U.S. Pat. 4,748,261; (1987) *Chem. Abstr.*, **107**, 7392.

**114** (a) Brill, T.B. and Landon, S.J. (1984) *Chem. Rev.*, **84**, 577; (b) Werener, H. and Feser, R. (1979) *Z. Anorg. Allg. Chem.*, **458**, 301.

**115** (a) Billig, E., Abatjoglou, A.G., Bryant, D.R., Murray, R.E., and Maher, J.M. (1988) (to Union Carbide Corporation) U.S. Pat. 4,717,775;(1989) *Chem. Abstr.*, **109**, 233177.

**116** Ramirez, F., Bhatia, S.B., and Smith, C.P. (1967) *Tetrahedron*, **23**, 2067.

**117** Simpson, R.D. (1997) *Organometallics*, **16**, 1797–1799.

**118** Nagaraja, C.M., Nethaji, M., and Jagirdar, B.R. (2004) *Inorg. Chem. Commun.*, **7**, 654–656.

**119** (a) de Meijere, A. and Diederich, F. (eds) (2004) *Metal-Catalyzed Cross-Coupling Reactions*, 2nd edn, Wiley-VCH Verlag GmbH, Weinheim; (b) Hartwig, J.F. (2002) in *Handbook of Organopalladium Chemistry for Organic Synthesis*, vol. 1 (ed. E.I. Negishi), Wiley-Interscience, New York, p. 1051 and 1097; (c) Jiang, L. and Buchwald, S.L. (2004) in *Metal-Catalyzed Cross-Coupling, Reactions*, 2nd edn (eds A. De Meijere and F. Diederich), Wiley-VCH Verlag GmbH, Weinheim, p. 699.

**120** Tietze, L.F., Spiegl, D.A., Stecker, F., Major, J., Raith, C., and Große, C. (2008) *Chem. Eur. J.*, **14**, 8956–8963.

**121** Najera, C., Gil-Molto, J., and Karlstrçma, S. (2004) *Adv. Synth. Catal.*, **346**, 179–1811.

**122** Saito, B. and Fu, G.C. (2008) *J. Am. Chem. Soc.*, **130**, 6694–6695.

**123** Drent, E. and Budzelaar, P.H.M. (1996) *Chem. Rev.*, **96**, 663.

**124** (a) Small, B.L. and Brookhart, M. (1998) *J. Am. Chem. Soc.*, **120**, 4049 and 7143; (b) Britovsek, G.J.P., Gibson, V.C., Kimberly, B.S., Maddox, P.J., McTavish, S.J., Solan, G.A., White, A.J.P., and Williams, D.J. (1998) *Chem. Commun.*, 849.

**125** Rijnberg, E., Boersma, J., Jastrzebski, J.T.B.H., Lakin, M.T., Spek, A.L., and van Koten, G. (1997) *Organometallics*, **16**, 3158.

**126** James, B.R. (1997) *Catal. Today*, **37**, 209.

**127** (a) Glorius, F. (ed.) (2007) *N-Heterocyclic Carbenes in Transition Metal Catalysis: Top. Organomet. Chem*, vol. 28, Springer-Verlag, Berlin/Heidelberg;

(b) Herrmann, W.A. (2002) *Angew. Chem. Int. Ed.*, **41**, 1290.

**128** (a) Trnka, T.M. and Grubbs, R.H. (2001) *Acc. Chem. Res.*, **34**, 18–29; (b) Connon, S.J. and Blechert, S. (2003) *Angew. Chem. Int. Ed.*, **42**, 1900–1923; (c) Schrock, R.R. and Hoveyda, A.H. (2003) *Angew. Chem. Int. Ed.*, **42**, 4592–4633; (d) Grubbs, R.H. (2004) *Tetrahedron*, **60**, 7117–7140; (e) Donohoe, T.J., Orr, A.J., and Bingham, M. (2006) *Angew. Chem. Int. Ed.*, **45**, 2664–2670; (f) Vougioukalakis, G.C. and Grubbs, R.H. (2008) *Chem. Eur. J.*, **14**, 7545–7556.

**129** (a) Würtz, S. and Glorius, F. (2008) *Acc. Chem. Res.*, **41**, 1523–1533;(b)Diez-Gonzalez, S. and Nolan, S.P., pages 47–82 in reference 122a.

**130** Clement, N.D., Routaboul, L., Grotevendt, A., Jackstell, R., and Beller, M. (2008) *Chem. Eur. J.*, **14**, 7408–7420, and references therein.

**131** McGuinness, D.S., Gibson, V.C., Wass, D.F., and Steed, J.W. (2003) *J. Am. Chem. Soc.*, **125**, 12716–12717.

**132** (a) Jimenez, M.V., Perez-Torrente, J.J., Bartolome, M.I., Gierz, V., Lahoz, F.J., and Oro, L.A. (2008) *Organometallics*, **27**, 224–234; (b) Wolf, J., Labande, A., Daran, J.–C., and Poli, R. (2007) *Eur. J. Inorg. Chem.*, **32**, 5069–5079.

**133** Veige, A.S. (2008) *Polyhedron*, **27**, 3177–3189.

**134** Marion, N. and Nolan, S.P. (2008) *Chem. Soc. Rev.*, **37**, 1776–1782.

**135** (a) Castarlenas, R., Esteruelas, M.A., and Onate, E. (2008) *Organometallics*, **27**, 3240–3247; (b) Hauwert, P., Maestri, G., Sprengers, J.W., Catellani, M., and Elsevier, C.J. (2008) *Angew. Chem. Int. Ed.*, **47**, 3223–3226.

**136** Zhang, T. and Shi, M. (2008) *Chem. Eur. J.*, **14**, 3759–3764.

**137** Biffis, A., Tubaro, C., Buscemi, G., and Basato, M. (2008) *Adv. Synth. Catal.*, **350**, 189–196.

**138** Strassner, T. (2007) *Topics in Organometallic Chemistry, Organometallic Oxidation Catalysis*, vol. 22, Springer GmbH, Berlin, pp. 125–148.

**139** Diez-Gonzalez, S. and Nolan, S.P. (2008) *Angew. Chem. Int. Ed.*, **47**, 8881–8884.

**140** Tekavec, T.N. and Louie, J. (2008) *J. Org. Chem.*, **73**, 2641–2648.

**141** (a) McGuinness, D.S., Suttil, J.A., Gardiner, M.G., and Davies, N.W. (2008) *Organometallics*, **27**, 4238–4247; (b) McGuinness, D.S., Mueller, W., Wasserscheid, P., Cavell, K.J., Skelton, B.W., White, A.H., and Englert, U. (2002) *Organometallics*, **21**, 175–181.

**142** Crudden, C.M. and Allen, D.P. (2004) *Coord. Chem. Rev.*, **248**, 2247–2273.

**143** (a) Lappert, M.F., Cardin, D.J., Cetinkaya, B., Manojlovic-Muir, L., and Muir, K.W. (1971) *J. Chem. Soc. D: Chem. Commun.*, 400–401; (b) Lappert, M.F. (2005) *J. Organomet. Chem.*, **690**, 5467–5473.

**144** Bourissou, D., Guerret, O., Gabbaï, F.P., and Bertrand, G. (2000) *Chem. Rev.*, **100**, 39–91.

**145** Arduengo, A.J., Harlow, R.L., and Kline, M. (1991) *J. Am. Chem. Soc.*, **113**, 361–363.

**146** Magill, A.M., Cavell, K.J., and Yates, B.F. (2004) *J. Am. Chem. Soc.*, **126**, 8717–8724.

**147** Diez-Gonzalez, S. and Nolan, S.P. (2007) *Coord. Chem. Rev.*, **251**, 874–883.

**148** (a) Hahn, F.E. and Jahnke, M.C. (2008) *Angew. Chem., Int. Ed.*, **47**, 3122; (b) Kaufhold, O. and Hahn, F.E. (2008) *Angew. Chem., Int. Ed.*, **47**, 4057; (c) Chen, J.C.C. and Lin, I.J.B. (2000) *J. Chem. Soc., Dalton Trans.*, 839; (d) Peris, E., Loch, J.A., Mata, J., and Crabtree, R.H. (2001) *Chem. Commun.*, 201.

**149** Cardin, D.J., Doyle, M.J., and Lappert, M.F. (1972) *J. Chem. Soc., Chem. Commun.*, 927.

**150** Hahn, F.E., Langenhahn, V., Meier, N., Lügger, T., and Fehlhammer, W.P. (2003) *Chem. Eur. J.*, **9**, 704–712.

**151** (a) McGuinness, D.S., Green, M.J., Cavell, K.J., Skelton, B.W., and White, A.H. (1998) *J. Organomet. Chem.*, **565**, 165; (b) Magill, A.M., McGuinness, D.S., Cavell, K.J., Britovsek, G.J.P., Gibson, V.C., White, A.J.P., Williams, D.J., White, A.H., and Skelton, B.W. (2001) *J. Organomet. Chem*, **617–618**, 546.

**152** (a) McGuinness, D.S., Cavell, K.J., Skelton, B.W., and White, A.H. (1999) *Organometallics*, **18**, 1596; (b) McGuinness, D.S. and Cavell, K.J. (2000) *Organometallics*, **19**, 741; (c) McGuinness, D.S. and Cavell, K.J. (2000) *Organometallics*, **19**, 4918.

**153** Cavell, K.J. and McGuinness, D.S. (2004) *Coord. Chem. Rev.*, **248**, 671–681.

**154** (a) Caddick, S., Cloke, F.G.N., Hitchcock, P.B., Leonard, J., Lewis, A.K., McKerrecher, D., and Titcomb, L. R., (2002) *Organometallics*, **21**, 4318–4319; (b) Marshall, W.J., Grushin, V.V., (2003) *Organometallics*, **22**, 1591–1593.

**155** McGuinness, D.S., Saendig, N., Yates, B.F., and Cavell, K.J. (2001) *J. Am. Chem. Soc.*, **123**, 4029–4040.

**156** Hartwig, J.F. (2007) *Inorg. Chem.*, **46**, 1936.

**157** Graham, D.C., Cavell, K.J., and Yates, B.F. (2006) *Dalton Trans.*, **14**, 1768–1775.

**158** (a) Barañano, D. and Hartwig, J.F. (1995) *J. Am. Chem. Soc.*, **117**, 2937; (b) Widenhoefer, R.A., Zhong, H.A., and Buchwald, S.L. (1997) *J. Am. Chem. Soc.*, **119**, 6787; (c) Driver, M.S. and Hartwig, J.F. (1997) *J. Am. Chem. Soc.*, **119**, 8232.

**159** Weskamp, T., Boehm, V.P.W., and Herrmann, W.A. (1999) *J. Organomet. Chem.*, **585**, 348.

**160** Carmichael, A.J., Earle, M.J., Holbrey, J.D., McCormac, P.B., and Seddon, K.R. (1999) *Org. Lett.*, **1**, 997.

**161** Clement, N.D., Cavell, K.J., Jones, C., and Elsevier, C.J. (2004) *Angew. Chem. Int. Ed.*, **43**, 1277.

**162** (a) Clement, N.D. and Cavell, K.J. (2004) *Angew. Chem., Int. Ed.*, **43**, 3845; (b) Cavell, K.J. (2008) *Dalton Trans.*, **47**, 6676–6685.

**163** (a) Waltman, A.W., Ritter, T., and Grubbs, R.H. (2006) *Organometallics*, **25**, 4238–4239; (b) Pelegri, A.S., Elsegood, M.R.J., McKee, V., and Weaver, G.W. (2006) *Org. Lett.*, **8**, 3049–3051.

**164** Pugh, D., Boyle, A., and Danopoulos, A.A. (2008) *Dalton Trans.*, 1087–1094.

**165** Vehlow, K., Gessler, S., and Blechert, S. (2007) *Angew. Chem. Int. Ed.*, **46**, 8082–8085.

**166** Hong, S.H., Chlenov, A., Day, M.W., and Grubbs, R.H. (2007) *Angew. Chem. Int. Ed.*, **46**, 5148–5151.

**167** Chung, C.K. and Grubbs, R.H. (2008) *Org. Lett.*, **10**, 2693–2696.

**168** Ulman, M. and Grubbs, R.H. (1999) *J. Org. Chem.*, **64**, 7202–7207.

169 Hansen, S.M., Rominger, F., Metz, M., and Hofmann, P. (1999) *Chem. Eur. J.*, **5**, 557–566.

170 (a) Hong, S.H., Wenzel, A.G., Salguero, T.T., Day, M.W., and Grubbs, R.H. (2007) *J. Am. Chem. Soc.*, **129**, 7961–7968; (b) Hong, S.H., Day, M.W., and Grubbs, R.H. (2004) *J. Am. Chem. Soc.*, **126**, 7414–7415.

171 Ivin, K.J. and Mol, J.C. (1997) *Olefin Metathesis and Metathesis Polymerization*, Academic Press, San Diego, CA.

172 Schrock, R.R., Murdzek, J.S., Bazan, G.C., Robbins, J., DiMare, M., and O'Regan, M. (1990) *J. Am. Chem. Soc.*, **112**, 3875.

173 Wengrovius, J.H., Schrock, R.R., Churchill, M.R., Missert, J.R., and Youngs, W.J. (1980) *J. Am. Chem. Soc.*, **102**, 4515.

174 Salameh, A., Coperet, C., Basset, J.-M., Boehm, V.P.W., and Roeper, M. (2007) *Adv. Synth. Catal.*, **349**, 238–242.

175 Natta, G., Dall'Asta, G., and Porri, L. (1965) *Makromol. Chem.*, **81**, 253.

176 (a) Nubel, P.O. and Hunt, C.L. (1999) *J. Mol. Catal. A: Chem.*, **145**, 323; (b) Dinger, M.B. and Mol, J.C. (2002) *Adv. Synth. Catal.*, **344**, 671.

177 (a) Baker, R. and Crimmin, M.J. (1977) *Tetrahedron Lett.*, 441; (b) Verkuijlen, E. and Boelhouwer, C. (1974) *J. Chem. Soc., Chem. Commun.*, 793–794.

178 (a) Dinger, M.B. and Mol, J.C. (2003) *Organometallics*, **22**, 1089; (b) Fürstner, A., Ackermann, L., Gabor, B., Goddard, R., Lehmann, C.W., Mynott, R., Stelzer, F., and Thiel, O.R. (2001) *Chem. Eur. J.*, **7**, 3236; (c) Fogg, D.E., Amoroso, D., Drouin, S.D., Snelgrove, J., Conrad, J., and Zamanian, F. (2002) *J. Mol. Catal. A: Chem.*, **190**, 177–184; (d) Drouin, S.D., Zamanian, F., and Fogg, D.E. (2001) *Organometallics*, **20**, 5495–5497; (e) Louie, J., Bielawski, C.W., and Grubbs, R.H. (2001) *J. Am. Chem. Soc.*, **123**, 11312–11313; (f) Werner, H., Grünwald, C., Stüer, W., and Wolf, J. (2003) *Organometallics*, **22**, 1558–1560.

179 (a) Dinger, M.B. and Mol, J.C. (2003) *Eur. J. Inorg. Chem.*, 2827; (b) Banti, D. and Mol, J.C. (2004) *J. Organomet. Chem.*, **689**, 3113–3116.

180 Kim, M., Eum, M.-S., Jin, M.Y., Jun, K.-W., Lee, C.W., Kuen, K.A., Kim, C.H., and Chin, C.S. (2004) *J. Organomet. Chem.*, **689**, 3535–3540.

181 (a) Tsang, W.C.P., Schrock, R.R., and Hoveyda, A.H. (2001) *Organometallics*, **20**, 5658–5669; (b) van der Eide, E.F., Romero, P.E., and Piers, W.E. (2008) *J. Am. Chem. Soc.*, **130**, 4485–4491.

182 Leduc, A.-M., Salameh, A., Soulivong, D., Chabanas, M., Basset, J.-M., Copéret, C., Solans-Monfort, X., Clot, E., Eisenstein, O., Böhm, V.P.W., and Röper, M. (2008) *J. Am. Chem. Soc.*, **130**, 6288–6297.

183 van Rensburg, W.J., Steynberg, P.J., Meyer, W.H., Kirk, M.M., and Forman, G.S. (2004) *J. Am. Chem. Soc.*, **126**, 14332–14333.

184 Blanc, F., Berthoud, R., Cope ret, C., Lesage, A., Emsley, L., Singh, R., Kreickmann, T., and Schrock, R.R. (2008) *Proc. Natl. Acad. Sci*, **105**, 12123–12127.

185 Wengrovius, J.H., Sancho, J., and Schrock, R.R. (1981) *J. Am. Chem. Soc.*, **103**, 3932.

186 (a) Freudenberger, J.H. and Schrock, R.R. (1986) *Organometallics*, **5**, 1411; (b) Mortreux, A., Petit, F., Petit, M., and Szymanska-Buzar, T. (1995) *J. Mol. Catal. A: Chem.*, **96**, 95–105.

187 Schrock, R.R. (1986) *Acc. Chem. Res.*, **19**, 342–348.

188 Perrin, D.D. and Armarego, W.L.F. (1988) *Purification of Laboratory Chemicals*, Pergamon Press, Oxford.

189 Horváth, I.T. and Joó, F. (eds) (1995) *Aqueous Organometallic Chemistry and Catalysis, NATO ASI Series*, vol. **5**, Kluwer, Dordrecht, p. 3.

190 Drent, E. and Budzelaar, P.H.M. (1996) *Chem. Rev.*, **96**, 663.

191 Zuideveld, M.A., Kamer, P.C.J., van Leeuwen, P.W.N.M., Klusener, P.A.A., Stil, H.A., and Roobeek, C.F. (1998) *J. Am. Chem. Soc.*, **120**, 7977.

192 Reuben, B. and Wittcoff, H. (1988) *J. Chem. Educ.*, **65**, 605–7.

193 Tolman, C.A., McKinney, R.J., Seidel, W.C., Druliner, J.D., and Stevens, W.R. (1985) *Adv. Catal.*, **33**, 1.

194 Goertz, W., Keim, W., Vogt, D., Englert, U., Boele, M.K.D., van der Veen, L.A., Kamer, P.C.J., and van Leeuwen,

P.W.N.M. (1998) *J. Chem. Soc., Dalton Trans.*, 2981.

**195** Eshuis, J.J.W., Tan, Y.Y., Meetsma, A., and Teuben, J.H. (1992) *Organometallics*, **11**, 362.

**196** Kaminsky, W. (1995) *Macromol. Symp.*, **97**, 79.

**197** Yang, X., Stern, C.L., and Marks, T.J. (1994) *J. Am. Chem. Soc.*, **116**, 10015.

**198** Albizzati, E., Galimberti, M., Giannini, U., and Morini, G. (1991) *Macromol. Symp.*, **48/49**, 223–238.

**199** Barbè, P.C., Cecchin, G., and Noristi, L. (1987) *Adv. Polym. Sci.*, **81**, 1–81.

**200** Akutagawa, S. (1992) Chapter 16, in *Chirality in Industry* (eds A.N. Collins, G.N. Sheldrake, and J. Crosby), John Wiley & Sons.

**201** Evans, D., Yagupsky, G., and Wilkinson, G. (1968) *J. Chem. Soc. (A)*, 2660.

**202** Castellanos-Páez, A., Castillón, S., Claver, C., van Leeuwen, P.W.N.M., and de Lange, W.G.J. (1998) *Organometallics*, **17**, 2543.

**203** (a) Budzelaar, P.H.M., van Leeuwen, P.W.N.M., Roobeek, C.F., and Orpen, A.G. (1992) *Organometallics*, **11**, 23; (b) Portnoy, M., Frolow, F., and Milstein, D. (1991) *Organometallics*, **10**, 3960.

**204** Bonds, W.D., Brubaker, C.H., Chandrasekaran, E.S., Gibsons, C., Grubbs, R.H., and Kroll, L.C. (1975) *J. Am. Chem. Soc.*, **97**, 2128.

**205** Mueller, C., Ackerman, L.J., Reek, J.N.H., Kamer, P.C.J., and van Leeuwen, P.W.N.M. (2004) *J. Am. Chem. Soc.*, **126**, 14960–14963.

**206** Lever, A.B.P., Wilshire, J.P., and Quan, S.K. (1979) *J. Am. Chem. Soc.*, **101** (13), 3668–3669.

**207** Trzeciak, A.M. and Ziolkowski, J.J. (1987) *J. Mol. Catal.*, **43**, 15–20.

**208** (a) Parshall, G.W., Knoth, W.H., and Schunn, R.A. (1969) *J. Am. Chem. Soc.*, **91**, 4990; (b) Coolen, H.K.A.C., van Leeuwen, P.W.N.M., and Nolte, R.J.M. (1995) *J. Organomet. Chem.*, **496**, 159.

**209** Tempel, D.J., Johnson, L.K., Huff, R.L., White, P.S., and Brookhart, M. (2000) *J. Am. Chem. Soc.*, **122**, 6686–6700.

# 2
# Early Transition Metal Catalysts for Olefin Polymerization

## 2.1
## Ziegler–Natta Catalysts

### 2.1.1
### Introduction

In order to put the state of the art regarding the activation and deactivation of homogeneous catalysts for olefin polymerization into perspective, it is instructive to first consider Ziegler–Natta catalysts, which for more than 50 years have dominated the polyolefin industry. These are heterogeneous systems, but they are strongly linked to the more recently developed homogeneous systems with regard to the organometallic chemistry involved and the polymerization procedures applied. It should also be taken into account that widespread implementation of a homogeneous catalyst in slurry and gas-phase processes for polyolefins requires heterogenization of the catalyst on a solid support.

Ziegler–Natta catalysts used in slurry and gas-phase processes for the production of polyethylene (PE) and polypropylene (PP) typically comprise an inorganic support (magnesium chloride or silica) and a transition metal precatalyst component such as a titanium chloride. Polypropylene catalysts also contain a Lewis base, termed an internal donor, the function of which is to control the amount and distribution of $TiCl_4$ on an $MgCl_2$ support, thereby influencing the catalyst selectivity [1]. Such catalysts are used in bulk (liquid monomer) and gas-phase processes for PP and have replaced the $TiCl_3$ catalysts used in early slurry processes.

The activity profile of a Ziegler–Natta catalyst during the course of polymerization is dependent on both chemical and physical factors. Chemical activation is generally rapid, but in certain cases there can be a significant induction period before a maximum rate is obtained. Ziegler–Natta catalysts typically have particle sizes in the range 10–100 μm. Each particle contains millions or even billions of primary crystallites with sizes up to approximately 15 nm. At the start of polymerization, cocatalyst and monomer diffuse through the catalyst particle and polymerization takes place on the surface of each primary crystallite within the particle. As solid,

*Homogeneous Catalysts: Activity – Stability – Deactivation*, First Edition. Piet W.N.M. van Leeuwen and John C. Chadwick.
© 2011 Wiley-VCH Verlag GmbH & Co. KGaA. Published 2011 by Wiley-VCH Verlag GmbH & Co. KGaA.

crystalline polymer is formed, the crystallites are pushed apart and the particle starts to grow. The particle shape is retained (replicated), and the diameter of a spherical particle may increase by as much as 50-fold, depending on the catalyst productivity. However, the ease of particle fragmentation and growth is dependent on the nature of the support, magnesium chloride typically giving easier fragmentation than silica [2, 3]. The rate of polymerization may also be affected by the relatively slow diffusion of the monomer through the crystalline polymer formed on the particle surface, a mass transfer limitation phenomenon which is frequently encountered in ethene homopolymerization [4, 5]. The diffusion limitation can be alleviated by introducing a comonomer, which leads to easier monomer diffusion through the less crystalline copolymer formed, or by carrying out a prepolymerization with propene or another α-olefin prior to an ethene homopolymerization [6–8]. Productivity can also be increased by incorporating into the catalyst system a nickel diimine, giving branched polyethylene; formation of the latter reduces the monomer diffusion limitation, thereby increasing the productivity of the main (e.g., Ti) catalyst component [9].

## 2.1.2
### Effect of Catalyst Poisons

In olefin polymerization, particularly with early transition metal catalysts, it is necessary to avoid the presence of polar impurities in the monomer or solvent (more accurately: diluent, in the case of a slurry process). Rigorous drying of monomer and solvents to remove traces of water is an obvious requirement, but even more potent poisons for Ziegler–Natta catalysts are CO, $CO_2$, $CS_2$, acetylene and allene [10]. Such impurities are also, in general, less amenable to removal by reaction (scavenging) with the alkylaluminum cocatalyst.

The addition of carbon monoxide during the course of an olefin polymerization immediately kills the catalyst activity, as a result of rapid coordination to the active transition metal center, followed by (slower) insertion into the transition metal–carbon bond. Its effectiveness in catalyst deactivation is such that it has been widely used in active center determination. Keii estimated the number of active centers in an $MgCl_2$-supported $TiCl_4$ catalyst by measuring the (virtually instantaneous) drop in propene polymerization rate when a defined amount of CO was added [11]. A more widely investigated technique for active center determination involves [14]CO radiotagging, developed by Yermakov and Zakharov in the 1970s [12]. Polymerization is immediately suppressed on contact with [14]CO, but successful radiotagging requires its incorporation into the polymer chain. This requires a certain contact time and it has been proposed that only a minor part of the active centers can be determined by this method [13]. On the other hand, too long a contact time can result in multiple insertions of [14]CO via olefin/CO copolymerization [14]. An alternative to [14]CO radiotagging is active center determination via quenching of the polymerization with tritiated alcohol [15], which reacts with propagating centers as follows:

$$Pol-Ti + RO^3H \rightarrow Pol-{}^3H + RO-Ti$$

The active center content can be obtained by radiochemical analysis of the polymer. However, an important complication is the presence of radiolabeled polymer resulting from chain transfer to the aluminum alkyl cocatalyst:

$$\text{Pol}-\text{AlR}_2 + \text{RO}^3\text{H} \rightarrow \text{Pol}-^3\text{H} + \text{RO}-\text{AlR}_2$$

The method therefore requires the determination of the metal (i.e., Ti or Al)–polymer bond (MPB) content at various polymerization times. If the Ti–polymer bond content remains constant while the Al–polymer bond content increases linearly with time, the Ti–polymer bond content can be derived via extrapolation of a plot of [MPB] versus time ($t$) to $t = 0$. A bonus of this approach is that the rate of chain transfer with aluminum alkyl can be determined.

## 2.1.3
### TiCl$_3$ Catalysts

Despite the fact that TiCl$_3$ catalysts have been replaced by high-activity MgCl$_2$/TiCl$_4$-based catalysts, it is interesting to briefly review the main factors influencing their activity and stability. It has been noted by Kissin [10] that the active centers of TiCl$_3$-based Ziegler–Natta catalysts are, potentially, extremely stable, even at 80 °C. However, in many cases deactivation is observed during polymerization. The use of AlEt$_3$ as cocatalyst can lead to catalyst deactivation via (over)reduction of titanium to divalent species, which generally have little or no activity in propene polymerization. AlEt$_2$Cl gives lower activity than AlEt$_3$, but significantly higher stereoselectivity and, for this reason, was selected as cocatalyst in polypropylene production using TiCl$_3$ catalysts. The reducing power of AlEt$_2$Cl is less than that of AlEt$_3$, but catalyst deactivation during polymerization can still take place. First-generation TiCl$_3$ catalysts typically had the composition TiCl$_3$·0.33AlCl$_3$, the aluminum trichloride resulting from the reduction of TiCl$_4$ with Al or AlR$_3$. The presence of AlCl$_3$ in the solid catalyst can lead to the formation of AlEtCl$_2$ via Et/Cl exchange between AlCl$_3$ and AlEt$_2$Cl:

$$\text{AlCl}_3 + \text{AlEt}_2\text{Cl} \rightarrow 2\text{AlEtCl}_2$$

AlEtCl$_2$ is a potent catalyst poison [16]. Its rate lowering effect is explained by a dynamic adsorption/desorption equilibrium between surface TiCl$_3$ sites and AlEtCl$_2$ [17].

$$\text{Ti(active)} + \text{AlEtCl}_2 \leftrightarrow \text{Ti} \cdot \text{AlEtCl}_2 \text{(inactive)}$$

Catalyst activity can be increased by the addition of a small amount of a Lewis base (e.g., dibutyl ether) able to form a complex with AlEtCl$_2$, thereby shifting the above equilibrium to the left. Alternatively, a hindered phenol may be used. Reaction of 4-methyl-2,6-di-*tert*-butyl-phenol with AlEt$_2$Cl generates the phenoxide species ArOAl(Et)Cl, which is able to interact with AlEtCl$_2$ to form the mixed dimer [EtClAl(OAr)(Cl)AlEtCl] [18]. The effectiveness of such hindered phenols is due to the fact that the bulky *tert*-butyl groups prevent the formation of inactive dimers of type [EtClAl(OAr)$_2$AlEtCl]; the phenoxide is only able to form a dimer with the least sterically demanding Lewis acid in the system, EtAlCl$_2$. The use of a sterically hindered phenoxide derived from the reaction of AlEt$_2$Cl with 4-methyl-2,6-di-*tert*-butyl-phenol has also been shown to be effective in increasing the activity of a β-TiCl$_3$ catalyst in isoprene polymerization [19].

In the 1970s, an improved TiCl$_3$ catalyst for PP was developed by Solvay [20]. The catalyst preparation involved reduction of TiCl$_4$ using AlEt$_2$Cl, followed by treatment with an ether and TiCl$_4$. The ether treatment results in removal of AlCl$_3$ from TiCl$_3$·$n$AlCl$_3$, while treatment with TiCl$_4$ effects a phase transformation from β to δ-TiCl$_3$ at a relatively mild temperature (<100 °C) [17]. Using catalysts of this type, it was possible to obtain PP yields in the range 5–20 kg (g cat.)$^{-1}$ in 1–4 h polymerization in liquid monomer, during which time the catalyst activity remained almost constant [20]. Commercial implementation of second-generation catalysts was, however, overshadowed by the advent of third- and later-generation magnesium chloride-supported catalysts, described below.

### 2.1.4
### MgCl$_2$-supported Catalysts

The basis for the development of the high-activity supported catalysts lay in the discovery, in the late 1960s, of "activated" MgCl$_2$ able to support TiCl$_4$ and give high catalyst activity, and the subsequent discovery, in the mid 1970s, of electron donors (Lewis bases) capable of increasing the stereospecificity of the catalyst so that (highly) isotactic polypropylene could be obtained [21–24].

#### 2.1.4.1 MgCl$_2$/TiCl$_4$/Ethyl Benzoate Catalysts
In the early stages of MgCl$_2$-supported catalyst development, activated magnesium chloride was prepared by ball milling in the presence of ethyl benzoate, leading to the formation of very small (≤3 nm thick) primary crystallites within each particle [17]. Nowadays, however, the activated support is prepared by chemical means such as via complex formation of MgCl$_2$ and an alcohol or by reaction of a magnesium alkyl or alkoxide with a chlorinating agent or TiCl$_4$. Many of these approaches are also effective for the preparation of catalysts having controlled particle size and morphology. The function of the internal donor (in this case ethyl benzoate, EB) in MgCl$_2$-supported catalysts is twofold. One function is to stabilize small primary crystallites of magnesium chloride; the other is to control the amount and distribution of TiCl$_4$ in the final catalyst. Activated magnesium chloride has a disordered structure comprising very small lamellae. Giannini has indicated that, on preferential lateral cleavage surfaces, the magnesium atoms are coordinated with 4 or 5 chlorine atoms, as opposed to 6 chlorine atoms in the bulk of the crystal [22]. These lateral cuts correspond to what have generally been denoted as the (110) and (100) faces of MgCl$_2$, as shown in Figure 2.1, and the coordination of TiCl$_4$ and the donor takes place on these surfaces. Recent results by Busico have indicated that the five-coordinate surface is the dominant lateral termination in both large crystals and in ball-milled MgCl$_2$ and that this surface should be indexed as (104) rather than (100) [25]. The

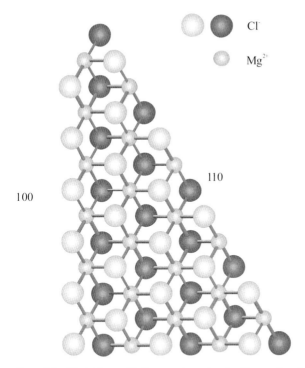

Cl⁻

Mg²⁺

110

100

**Figure 2.1** Model for an MgCl₂ monolayer showing the most probable cleavage cuts.

presence of a donor in support preparation can, however, lead to preferential formation of the (110) face [26].

Ziegler–Natta catalysts comprising $MgCl_2$, $TiCl_4$ and ethyl benzoate are referred to as third-generation PP catalysts. They are used in combination with an aluminum trialkyl cocatalyst such as $AlEt_3$ and an "external" electron donor such as methyl p-toluate or ethyl p-ethoxybenzoate, added in the polymerization. PP yields obtained with this type of catalyst are typically in the range 15–30 kg (g cat.)$^{-1}$. The initial catalyst activity in propene polymerization is very high, but this system also exhibits rapid deactivation during polymerization, such that the activity after 1 h may be as little as 10% of the initial activity.

The rapid decay in activity of third-generation catalysts can, at least partially, be ascribed to the use of an ester as an external as well as an internal donor. Esters are rapidly reduced by metal hydrides [27]. Hydrogen is used as a chain transfer agent in olefin polymerization with Ziegler–Natta catalysts and the Ti–H species formed in the chain transfer reaction can react with esters to form Ti–O bonds inactive for chain propagation [28].

$$Ti - H + RCOOR' \rightarrow Ti - OCH(R)OR'$$

Esters also react with aluminum alkyls. Lewis acid–Lewis base complexation between $AlEt_3$ and ethyl benzoate leads to a shift in the C=O stretching frequency from 1720–1730 cm$^{-1}$ in the free ester to 1655–1670 cm$^{-1}$ in the complex [24]. In the presence of excess $AlEt_3$, alkyl addition to the carbonyl group takes place, resulting in the formation of an aluminum alkoxide [29–31].

$$PhCOOEt \cdot AlEt_3 + AlEt_3 \rightarrow PhEt_2 COAlEt_2 + AlEt_2OEt$$

It has been suggested that the aluminum alkoxides formed in such reactions may play a significant role in controlling the stereoselectivity of the catalyst [30, 32]. However, this can be discounted, as the most effective ester external donors are those most resistant to alkylation, and separate addition of alkoxide decomposition products has no effect on catalyst activity or selectivity [33].

The high activity of $MgCl_2$-supported catalysts is partly due to the fact that the propagation rate constant in propene polymerization is around an order of magnitude greater than that with $TiCl_3$-based catalysts [34]. The decay in the rate of propene polymerization with the catalyst system $MgCl_2/TiCl_4/EB–AlEt_3$ can be described by a second-order deactivation of active centers, which has been suggested to result in the reduction of $Ti^{3+}$ to $Ti^{2+}$, brought about by the presence of $AlEt_3$ [11, 34]. Deactivation in ethene and propene polymerization with the catalyst system $MgCl_2/TiCl_4–AlEt_3$ was also explained on the basis of reduction of $Ti^{3+}$ to lower oxidation states [35]. However, Chien reported that contact of a $MgCl_2/TiCl_4/EB$ type catalyst with $AlEt_3$ in the presence of methyl *p*-toluate resulted in only partial reduction and proposed the formation of $Ti–C–C–Ti$ species from adjacent active species on the support surface (with elimination of a polymer chain but without change in the Ti valency) as an explanation for catalyst deactivation [36]. A transformation from first- to second-order decay kinetics as the $AlEt_3$ cocatalyst concentration was increased in ethene polymerization with $MgCl_2/TiCl_4/EB$ was reported by Hsu, who suggested that this was due to progressive activation of adjacent Ti species as the Al/Ti ratio was increased [37].

The most important factor affecting the activity/decay/selectivity balance of third-generation catalysts for propene polymerization, however, is the concentration of the ester used as external donor, in relation to the cocatalyst concentration [24, 38]. Increasing the external donor concentration (i.e., lowering the $AlEt_3$/ester ratio) leads to increased stereoselectivity but a decrease in activity. The deactivation of active centers via interaction with the base is largely irreversible, although a small increase in activity can be observed if further $AlEt_3$ is added at an intermittent stage during the polymerization; this can be explained by simple Lewis acid–Lewis base complexation, shifting the following equilibrium to the right:

$$Ti \cdot ester(inactive) + AlEt_3 \leftrightarrow Ti(active) + AlEt_3 \cdot ester$$

### 2.1.4.2 $MgCl_2/TiCl_4$/Diester Catalysts

Significant improvements in catalyst stability resulted from the development of fourth-generation Ziegler–Natta catalysts containing a diester such as diisobutyl phthalate as internal donor. These catalysts are used in combination with an alkoxysilane external donor of type $RR'Si(OMe)_2$ or $RSi(OMe)_3$ [39]. The combination $MgCl_2/TiCl_4$/phthalate ester–$AlR_3$–alkoxysilane is currently the most widely used catalyst system in PP manufacture. The most effective alkoxysilane donors for high isospecificity are methoxysilanes containing relatively bulky alkyl groups branched at the position alpha to the silicon atom [40–42]. Comparison with the catalysts containing ethyl benzoate as internal donor reveals that the phthalate/silane-based systems exhibit lower initial activity in propene polymerization [43], but their much lower decay rate leads to much greater productivity, giving yields of up to 80 kg PP (g cat.)$^{-1}$ as opposed to around 30 kg g$^{-1}$ for the third-generation catalysts.

In contrast to catalysts containing ethyl benzoate as internal donor, deactivation of phthalate-containing catalysts in propene polymerization can be described by first-order kinetics; Soga *et al.* suggested that the active Ti species in ester-free and monoester systems interact with neighboring Ti species, whereas those in the diester system do not [44]. Another difference between mono- and diester-based systems is that the presence of the external donor (an alkoxysilane) in the latter systems leads not only to an increase in stereoselectivity but also to increased activity [45]. This is related to replacement of the internal by the external donor when the components are mixed in the presence of the cocatalyst [45, 46] and the active involvement of the external donor in the generation of highly active, isospecific sites [47]. A too high concentration of the external donor can, however, lead to reduced activity due to the poisoning of both weakly and highly isospecific centers [45, 48].

The optimum polymerization temperature for $MgCl_2$-supported catalysts is generally around 70 °C. At temperatures greater than 80 °C catalyst productivity can be limited by relatively rapid decay in activity. For example, large decreases in activity were noted by Kojoh *et al.* when the polymerization temperature was increased from 70 to 100 °C [49]. The decrease in activity was larger with AlEt$_3$ than with Al*i*Bu$_3$ and a significant drop in polypropylene melting point was observed using AlEt$_3$ at 100 °C, which was ascribed to incorporation into the polymer chain of ethene, formed from the decomposition of AlEt$_3$. Decomposition of Al*i*Bu$_3$ (to Al*i*Bu$_2$H and isobutene), although relatively rapid at such temperatures, has little effect on the polymer due to the negligible (co)polymerization ability of isobutene.

### 2.1.4.3 MgCl$_2$/TiCl$_4$/Diether Catalysts

The most active $MgCl_2$-supported catalysts for polypropylene contain a diether, typically a 2,2-disubstituted-1,3-dimethoxypropane, as internal donor [50, 51]. In contrast to the catalysts containing ethyl benzoate or a phthalate ester as internal donor, the diether is not removed from the support when the catalyst is brought into contact with the AlR$_3$ cocatalyst. These catalysts therefore have high stereoselectivity, even in the absence of an external donor, giving polypropylene yields exceeding 100 kg (g cat.)$^{-1}$. Catalyst systems of type $MgCl_2$/TiCl$_4$/diether–AlR$_3$ also give relatively stable polymerization kinetics, which can, at least partly, be ascribed to the absence of an ester as either internal or external donor.

### 2.1.5
### Ethene Polymerization

The preceding sections have dealt largely with factors affecting catalyst deactivation in propene polymerization using Ziegler–Natta systems. Polymerization rate/time profiles provide reliable information on the rates of catalyst activation and decay, but this is not the case in ethene polymerization. Propagation rate constants for ethene polymerization are higher than those for propene polymerization, but the overall rate of polymerization can be severely retarded by mass transfer limitations. In ethene homopolymerization using a heterogeneous catalyst, the rate of polymerization can be impeded by relatively slow diffusion of the monomer through the crystalline polyethylene formed on the particle surface [4, 5], and by difficult polymer particle growth in the case of catalyst supports having low friability [52]. The polymerization kinetics are often characterized by a gradual increase in activity as progressive fragmentation of the catalyst particle resulting from the hydraulic forces generated by the growing polymer exposes new catalyst centers within the particle. Regular particle growth, resulting from easier monomer diffusion, can be obtained with catalysts having relatively large pores [53]. The presence of hydrogen generally leads to decreased activity in ethene polymerization [54, 55]. The reason for this is as yet unclear. Kissin has proposed that deactivation by hydrogen can be explained by the formation of "dormant" Ti-CH$_2$CH$_3$ species formed by insertion of ethene into Ti–H after chain transfer with hydrogen [56–60]. The low propagation activity of the Ti–Et species was ascribed to $\beta$-agostic stabilization. However, this proposal was not supported by Garoff, who found that increasing the amount of hydrogen present in polymerization tended to slow down the activation of catalysts in which the titanium was present as Ti(IV) [61]. Unusual stepwise increases in activity during ethene polymerizations carried out in the presence of hydrogen have been reported by Pennini and coworkers, who showed that this effect was caused by particle disintegration to expose new active centers [53].

Significant increases in ethene polymerization activity using heterogeneous catalysts are frequently observed in the presence of an $\alpha$-olefin comonomer. The increased activity in copolymerization is generally referred to as the comonomer activation effect, which can be ascribed to easier monomer diffusion through the less crystalline copolymer. Comonomer activation is particularly prevalent in systems in which low friability of the support impedes effective particle fragmentation and growth [52]. In such cases, ethene homopolymerization leads to the pores of an unfragmented catalyst becoming filled with polymer, resulting in limited monomer diffusion into the particle.

## 2.2
## Metallocenes

### 2.2.1
### Introduction

Metallocenes have been known since the 1950s and early work by Natta and Breslow showed that ethene could be polymerized, albeit with low activity, using Cp$_2$TiCl$_2$ and an aluminum alkyl [62, 63]. Reichert and Meyer then found that minute quantities of water activated ethene polymerization with Cp$_2$TiEtCl/AlEtCl$_2$ [64]. Breslow and Long ascribed the effect of water to the formation of aluminoxanes resulting from partial hydrolysis of the aluminum alkyl [65]. The breakthrough in metallocene-catalyzed polymerization came in the late 1970s, when Kaminsky and Sinn discovered that extremely high ethene polymerization activity could be obtained using $Cp_2ZrMe_2$ or $Cp_2ZrCl_2$ in combination with methylaluminoxane (MAO) [66, 67]. Leading up to this breakthrough was the observation that, whereas aluminum trialkyls are ineffective cocatalysts for metallocenes, $AlMe_3$ became highly effective when traces of water were present. The product of this reaction, methylaluminoxane, was represented in early publications as an oligomer $-[Al(Me)O]_n-$ of undefined composition. The structure of MAO is still not properly resolved; cage and tube structures have been proposed [68], and the structure is further complicated by the presence of associated $AlMe_3$ [69].

Metallocene activation using MAO involves a combination of alkylation and anion ($Cl^-$ or $Me^-$) abstraction, to give a cationic active species such as $[Cp_2ZrMe]^+$, illustrated in Scheme 2.1. The realization that a weakly coordinating anion is crucial for high catalytic activity led to the development of other (non-MAO) activators capable of generating cationic active species. The role of the cocatalyst in the generation of active species has been reviewed by Chen and Marks, with special reference to the borane and borate activators shown in Scheme 2.2 [70].

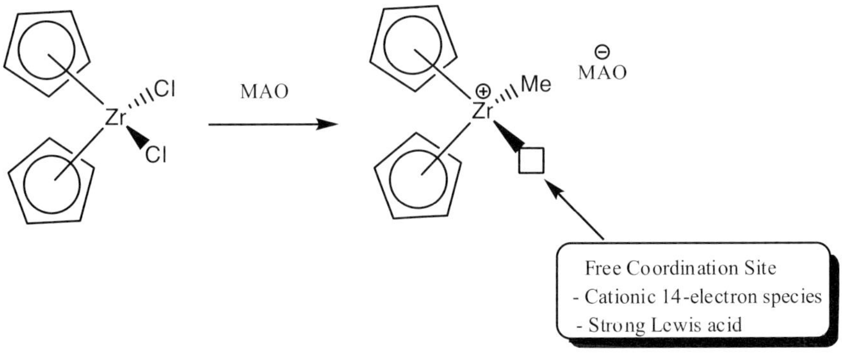

Free Coordination Site
- Cationic 14-electron species
- Strong Lewis acid

**Scheme 2.1** Metallocene activation using methylalumoxane (MAO).

Chain propagation in metallocene-catalyzed polymerization proceeds via chain migratory insertion into the metal–carbon bond, illustrated in Scheme 2.3. As is the case for heterogeneous Ziegler–Natta catalysts, polymer molecular weight can be controlled by the addition of hydrogen as chain transfer agent. In fact, metallocenes are considerably more responsive to chain transfer with hydrogen than most Ziegler–Natta catalysts. In the absence of hydrogen, the usual chain transfer mechanism involves β-H elimination of the growing polymer chain, resulting in a polyethylene chain with a vinyl end-group. Interestingly, the reaction proceeds via β-H transfer from the polymer chain to the monomer rather than to the metal. Chain transfer to aluminum can also take place, particularly with MAO containing significant amounts of $AlMe_3$. The various chain transfer processes in ethene polymerization are illustrated in Scheme 2.4.

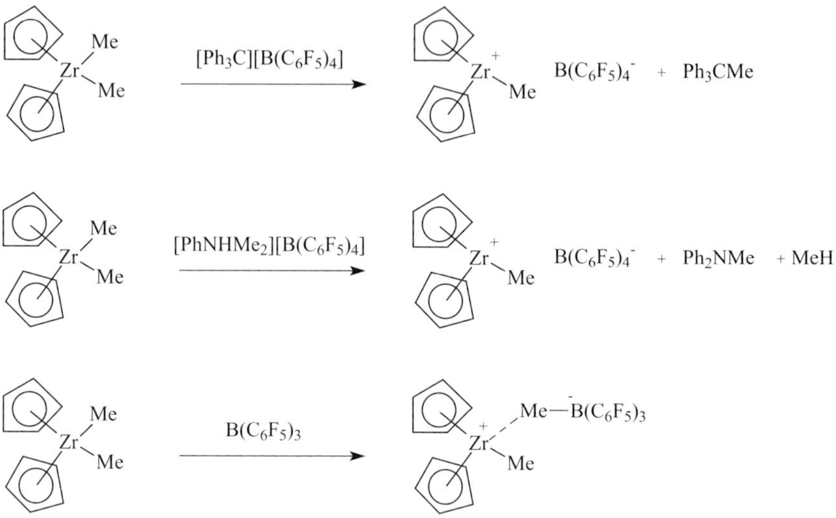

$[Ph_3C][B(C_6F_5)_4]$      $B(C_6F_5)_4^-$   +   $Ph_3CMe$

$[PhNHMe_2][B(C_6F_5)_4]$      $B(C_6F_5)_4^-$   +   $Ph_2NMe$   + MeH

$B(C_6F_5)_3$      $Me-B(C_6F_5)_3$

**Scheme 2.2** Metallocene activation using boranes and borates.

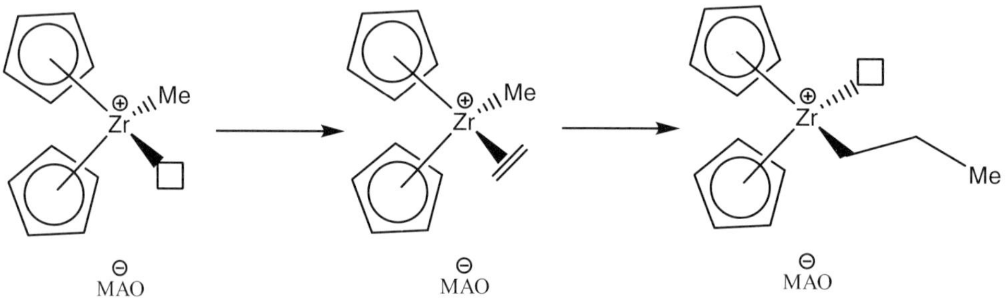

**Scheme 2.3** Chain propagation step in metallocene-catalyzed ethene polymerization.

$Zr\text{-}CH_2\text{-}CH_2\text{-}R \ + \ H_2$   →   $Zr\text{-}H \ + \ CH_3\text{-}CH_2\text{-}R$

$Zr\text{-}CH_2\text{-}CH_2\text{-}R \ + \ CH_2{=}CH_2$   →   $Zr\text{-}CH_2\text{-}CH_3 \ + \ CH_2{=}CH\text{-}R$

$Zr\text{-}CH_2\text{-}CH_2\text{-}R \ + \ AlMe_3$   →   $Zr\text{-}Me \ + \ AlMe_2\text{-}CH_2\text{-}CH_2\text{-}R$

**Scheme 2.4** Chain termination in ethene polymerization.

The discovery of the zirconocene/MAO systems led to an enormous increase in research, in both academia and industry, on the development of homogeneous catalysts for olefin polymerization. The well-defined, single-center nature of such catalysts allows much greater control over polymer micro- and macro-structure than could hitherto be achieved using multi-center Ziegler–Natta catalysts. A further stimulus for research on metallocene catalysts has been the discovery that stereo-selectivity in propene polymerization can be driven to an unprecedented extent,

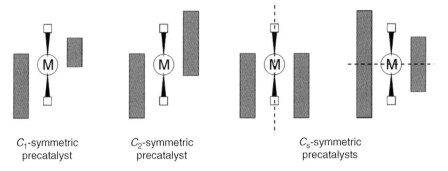

$C_1$-symmetric
precatalyst

$C_2$-symmetric
precatalyst

$C_s$-symmetric
precatalysts

**Figure 2.2** Symmetries in Group 4 transition metal-catalyzed olefin polymerization.

following the development of metallocenes in which the cyclopentadienyl rings are linked by a dimethylsilyl or other bridge, preventing ring rotation, and in which the introduction of substituent groups creates chirality around the metal center [71]. A schematic representation of common symmetries in Ti, Zr and Hf catalysts for propene polymerization is shown in Figure 2.2. At a $C_2$-symmetric active center, steric hindrance imparted by a ligand (indicated by a gray rectangle in Figure 2.2) can force the polymer chain towards the open sector of the chiral coordination sphere. The incoming propene monomer then adopts the enantiofacial orientation which places the methyl group trans with respect to the polymer chain. A simplified representation of this is shown in Figure 2.3 for both a metallocene and a Ziegler–Natta catalyst. These are very different systems, but the underlying mechanism of stereocontrol (enantiomorphic site control) in propene polymerization to give isotactic polypropylene is the same [72]. Syndiotactic polypropylene is obtained with $C_s$-symmetric metallocenes having chiral propagating sites (Figure 2.2, furthest right), while $C_1$-symmetric metallocenes can give either hemi-isotactic or isotactic polymers, dependent on the steric bulk of the cyclopentadienyl ring substituents, as illustrated in Figure 2.4 [73]. The evolution of metallocene catalysts for propene polymerization is described in an extensive review by Resconi et al. [74].

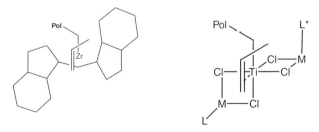

**Figure 2.3** Stereocontrol in isospecific propene polymerization.

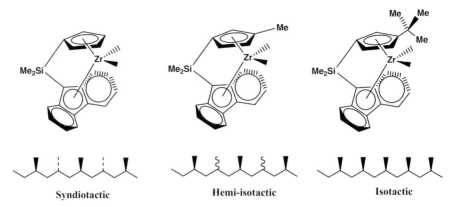

| Syndiotactic | Hemi-isotactic | Isotactic |

**Figure 2.4** Effect of metallocene symmetry on polypropylene microstructure.

### 2.2.2
### Metallocene/MAO Systems

In Group 4 metallocene-catalyzed olefin polymerization, the order of activity is generally $Zr > Hf > Ti$. Titanocenes deactivate rapidly above around 50 °C, most probably due to the reduction of Ti(IV) to Ti(III) [75, 76]. However, as described in Chapter 4, stable polymerization activity can be achieved by immobilization of the titanocene on a support such as $MgCl_2$. In addition to the effect of a support in preventing bimolecular decomposition, a stabilizing effect of $MgCl_2$ with regard to aluminum alkyl-induced reduction of titanium has been demonstrated for Ziegler–Natta systems containing $TiCl_4$ and $MeOTiCl_3$, and has been attributed to an electronic interaction between the support and the titanium chloride [77].

Most early studies on metallocene-catalyzed olefin polymerization were carried out with zirconocene dichloride, $Cp_2ZrCl_2$. On contacting $Cp_2ZrCl_2$ with methylaluminoxane, evolution of $\geq 1$ mol methane per mol Zr is observed [78]. Kaminsky proposed that this arises from the formation of $Zr-CH_2-Al$ structures as follows:

$$Zr-CH_3 + CH_3-Al(CH_3)-O- \rightarrow Zr-CH_2-Al(CH_3)-O- + CH_4$$

This condensation reaction is much faster with MAO than with $AlMe_3$. The reaction product is inactive in polymerization, although addition of excess MAO can reactivate the system via the formation of $Al-CH_2-Al$ and $Zr-CH_3$ structures [79].

Decay kinetics in propene polymerization under homogeneous conditions (toluene solution) with $Cp_2ZrCl_2$/MAO have been interpreted in terms of reversible and irreversible deactivation [80, 81]. At 40 °C, a rapid decay followed by almost constant activity was observed, whereas at 60 °C rapid initial decay was followed by a slow (irreversible) catalyst deactivation. The kinetic profile of $Cp_2ZrCl_2$/MAO-catalyzed polymerization can also be dependent on the composition of the methylaluminox-ane. MAO generally contains an appreciable quantity of associated trimethylalumi-num, which can suppress the polymerization rate by the formation of dormant alkyl-bridged species of the type $[Cp_2Zr(\mu\text{-}R)(\mu\text{-}Me)AlMe_2]^+$ **1**. A positive effect of AlMe$_3$, however, has been reported by Sivaram, who noted a change from a decay type to a build-up kinetic profile on adding AlMe$_3$ to MAO in ethene polymerization using Cp$_2$ZrCl$_2$ [82]. Chien [83] also observed no loss of activity when MAO was partially replaced by AlMe$_3$ in ethene polymerization with Cp$_2$ZrCl$_2$, but the partial replace-ment of MAO by AlMe$_3$ in propene polymerization with *rac*-Et(Ind)$_2$ZrCl$_2$**2** led to large decreases in catalyst activity [84]. A further complicating factor in olefin polymerization with homogeneous conditions is that, as pointed out by Rieger and Janiak [85], the system is only truly homogeneous at the very start of polymerization. Even at a temperature of 110 °C in ethene polymerization, a highly viscous solution with partial polymer precipitation forms rapidly. Occlusion of the catalyst in a solid mass of polymer can then lead to a significant monomer diffusion barrier.

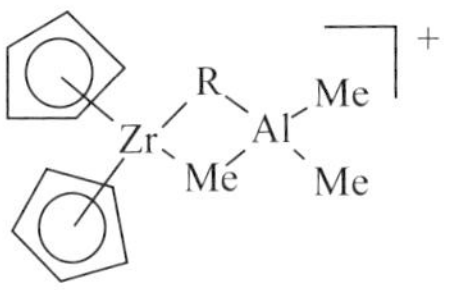

**1** 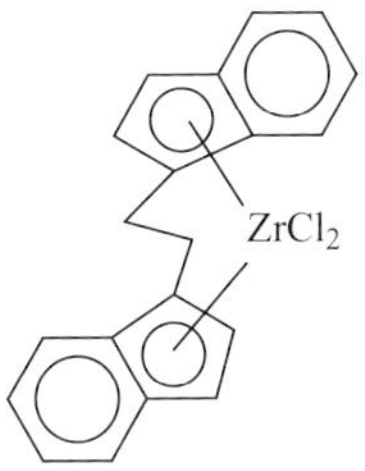                    **2**

In homogeneous polymerization, high MAO concentrations are required to obtain maximum catalyst activity. Jüngling and Mülhaupt concluded that an [Al] concen-tration of 5 mmol L$^{-1}$ is likely to convert most metallocenes into the active cationic species, but at a higher concentration the MAO could compete with the olefin monomer for the vacant coordination site of the metallocenium complex [86]. The MAO concentration appears to be a more important parameter than the MAO/Zr ratio [87]. Fink and coworkers found that the optimum MAO concentration was dependent on both the metallocene and the monomer. Maximum polymerization activity was obtained at a higher MAO concentration in ethene as opposed to propene polymerization, and it was also observed that *rac*-Me$_2$Si(Ind)$_2$ZrCl$_2$**3** required a higher MAO concentration than Me$_2$C(Cp)(Flu)ZrCl$_2$**4** [88]. These differences were attributed to greater shielding of the Zr atom by the bulkier bis(indenyl) ligand, limiting the formation of a tight ion pair comprising the Zr cation and the MAO-derived anion. It was also observed that partial replacement of MAO by AlR$_3$ led to decreased activity with R = Me or Et, but increased activity with R = *i*-Bu or *n*-Bu, indicating that the formation of $[Cp_2Zr(\mu\text{-}R)(\mu\text{-}Me)AlMe_2]^+$ via interaction of the active zirconocenium center with the AlMe$_3$ present in MAO could be reduced by the formation of mixed Al$_2$R$_6$ dimers from AlMe$_3$ and Al*i*Bu$_3$ or Al*n*Bu$_3$.

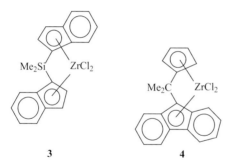

3                                    4

Not surprisingly, metallocene-catalyzed polymerization is strongly inhibited by the presence of oxygen or $CO_2$. However, Kallio *et al.* have reported that irradiation with visible light led to complete recovery of activity in ethene polymerization with $(n\text{-BuCp})_2ZrCl_2$/MAO after exposure to $O_2$ or $CO_2$ [89]. An equilibrium between the cationic metal center and coordinated $O_2$ or $CO_2$ was proposed, the activating effect of light being ascribed to dissociation of the $O_2$ or $CO_2$ from the metal center. However, chemical interactions between $O_2$ or $CO_2$ and the zirconocene and/or MAO would seem more likely than simple (reversible) coordination.

A particularly convenient method for the investigation of zirconocene alkylation and activation by aluminoxanes is UV–visible spectroscopy [90–95]. The ligand to metal charge transfer bands in the UV–vis spectrum of a zirconocene are sensitive to changes in the electron density on zirconium. Replacement of a chlorine by a methyl group leads to a hypsochromic shift (to lower wavelength) due to a decrease in net positive charge on the metal atom, whereas a bathochromic shift (to higher wavelength) is indicative of the formation of cationic active species. An increase in wavelength on going from MAO to ethyl(isobutyl)aluminoxane (EBAO) in the activation of *rac*-Et(Ind)$_2$ZrCl$_2$ was attributed to a looser ion pair in the case of EBAO [93]. It was also concluded that a more rapid rate of deactivation in ethene polymerization with *rac*-Et(Ind)$_2$ZrCl$_2$/EBAO was due to the relatively loose, unprotected ion pair.

UV–vis spectroscopy has also been used to gain insight into the relatively rapid deactivation observed with titanocene catalysts [76]. Compared to zirconocenes, alkylation and activation of a titanocene required only very modest quantities of MAO. However, whereas zirconocene-derived cationic activation products are stable for about a week at room temperature, the cationic complexes formed from *rac*-Et (Ind)$_2$TiCl$_2$ and MAO were unstable. An absorption band at 606 nm associated with the contact ion pair [*rac*-Et(Ind)$_2$TiMe$^+$...Me-MAO$^-$] began to decay within minutes at room temperature and was only stable at temperatures below $-20\,^\circ$C [76]. On the other hand, the AlMe$_3$-containing outer-sphere ion pair [*rac*-Et(Ind)$_2$Ti($\mu$-Me)$_2$Al-Me$_2$]$^+$[Me-MAO]$^-$ was stable for several hours at room temperature. EPR studies revealed that rapid deactivation in titanocene-catalyzed polymerization arises from reduction of Ti(IV) to Ti(III) species. Addition of Al*i*Bu$_3$ to *rac*-Et(Ind)$_2$TiCl$_2$ in toluene led to immediate and complete reduction to Ti(III). It was proposed that hydride species such as Al*i*Bu$_2$H (present in Al*i*Bu$_3$) are responsible for the rapid titanocene reduction and deactivation. Catalyst deactivation was also found to be accelerated in the presence of propene.

A possible role of reduction from the tetravalent to the trivalent oxidation state has also been investigated as a possible deactivation pathway for zirconocene catalysts [96]. EPR studies showed that $(2\text{-PhInd})_2\text{ZrCl}_2$ was more prone to reduction than was $rac\text{-Me}_2\text{Si(Ind)}_2\text{ZrCl}_2$, and that MAO modified with $\text{Al}i\text{Bu}_3$ (MMAO) was a stronger reducing agent than MAO. Addition of monomer also facilitated the reduction of Zr(IV) to Zr(III). Despite the fact that zirconocene reduction appears to be more rapid with MMAO, it has been found that partial replacement of MAO by $\text{Al}i\text{Bu}_3$ can lead to significant improvements in catalyst stability in zirconocene-catalyzed olefin polymerization [97]. The stabilizing effect of $\text{Al}i\text{Bu}_3$ may be due to the negligible tendency of an isobutyl group to bridge between Zr and Al, avoiding deactivation via the formation of alkylidene species of type $[\text{L}_2\text{Zr}(\mu\text{-CH}_2)(\mu\text{-Me})\text{AlMe}_2]^+$ that can arise in zirconocene/MAO systems [98]. Steric interactions facilitating the expulsion of $\text{AlMe}_n i\text{Bu}_{3-n}$ from a cation of type $[\text{L}_2\text{Zr}(\mu\text{-Me})_2\text{AlR}_2]^+$ in exchange for an olefin substrate were suggested to contribute to the exceptionally high polymerization activities that can be obtained with sterically crowded catalysts such as $rac\text{-Me}_2\text{Si(2-Me-4-Ph-Ind)}_2\text{ZrCl}_2\mathbf{5}$. A particularly high (initial) activity in propene polymerization with $rac\text{-Me}_2\text{Si(2-Me-4-}t\text{Bu-Cp)}_2\text{ZrCl}_2$/MAO has similarly been attributed to the fact that the sterically shielded cation $[rac\text{-Me}_2\text{Si(2-Me-4-}t\text{Bu-Cp)}_2\text{ZrMe}]^+$ does not form the $\text{AlMe}_3$ adduct $rac\text{-Me}_2\text{Si(2-Me-4-}t\text{Bu-Cp)}_2\text{Zr}(\mu\text{-Me})_2\text{AlMe}_2]^+$ [99]. However, this system also exhibits rapid decay, attributed to the lack of protection against catalyst deactivation which is provided by $\text{AlMe}_3$ adduct formation.

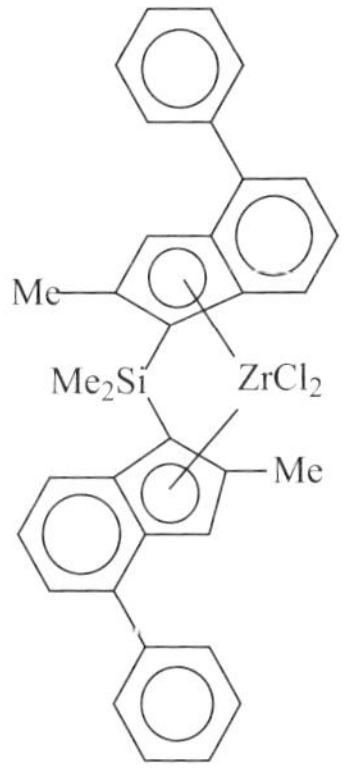

Me—

$\text{Me}_2\text{Si}$      $\text{ZrCl}_2$

—Me

**5**

It will be apparent from the above studies that the presence or absence of $\text{AlMe}_3$ is a major factor influencing the activity and stability in olefin polymerization with metallocene/MAO catalysts. On the basis of NMR studies, it has been proposed that a complex of type $[\text{L}_2\text{Zr}(\mu\text{-Me})_2\text{AlMe}_2]^+[\text{Me-MAO}]^-$ is the main precursor of the active centers in MAO-based systems [100]. Polymerization activity will depend on the amount of $\text{AlMe}_3$ present and the ease with which associated $\text{AlMe}_3$ can be displaced by the monomer. An effective method for the removal of "free" $AlMe_3$ from MAO involves the addition of a sterically hindered phenol such as 2,6-di-*tert*-butylphenol or its 4-Me-substituted homolog [101]. The effect of this on polymerization activity and stability is dependent on both the catalyst and the polymerization conditions. Rytter and coworkers [102] have observed that the removal of $AlMe_3$ from MAO resulted in increased activity in propene polymerization with $Ph_2C(Cp)(Flu)$ $ZrCl_2$ at 40 °C, but not at 80 °C, which was due to rapid decay in activity in the absence of $AlMe_3$ at the higher polymerization temperature. A more complex behavior was observed with *rac*-$Me_2Si(2$-Me-4-*t*-Bu$)_2ZrCl_2$, where the presence of $AlMe_3$ led to higher activities but without any influence on stability. It was suggested that, in this system, due to steric hindrance, MAO could be much less effective than $AlMe_3$ in the first step in the activation process, which involves methylation of the metallocene to form $L_2ZrMeCl$.

## 2.2.3
### Metallocene/Borate Systems

As illustrated in Scheme 2.2, metallocene alkyls can be activated by perfluoroaryl boranes and borates such as $B(C_6F_5)_3$, $[Ph_3C][B(C_6F_5)_4]$ and $[PhNHMe_2][B(C_6F_5)_4]$. Of these, the borates give higher activities than the borane, because the alkyl group in the $[RB(C_6F_5)_3]^-$ anion formed from $L_2ZrR_2$ and $B(C_6F_5)_3$ forms a relatively strong bridge between Zr and B, in contrast to the weakly coordinating $[B(C_6F_5)_4]^-$ anion. Bochmann and Lancaster reported that zirconium dibenzyl complexes reacted with $[Ph_3C][B(C_6F_5)_4]$ to give cationic complexes of type $[Cp'_2Zr(CH_2Ph)]^+[B(C_6F_5)_4]^-$ which were more stable than the related methyl complexes [103]. The increased thermal stability was attributed to $\eta^2$-coordination of the benzyl ligand, satisfying to some extent the electron requirement of the electron-deficient metal center. It was also proposed that the same stabilizing influence was responsible for relatively low catalytic activity at low temperatures, on the assumption that an $\eta^2 \rightarrow \eta^1$ rearrangement of the benzyl ligand is required before olefin coordination can take place. Reaction of $Cp_2ZrMe_2$ with $[Ph_3C][B(C_6F_5)_4]$ at low temperature ($-60$ °C) led to the formation of the dinuclear complex $[\{Cp_2ZrMe\}_2(\mu$-Me$)]^+$, indicating that adduct formation with $Cp_2ZrMe_2$ stabilized $[Cp_2ZrMe]^+$ more effectively than a solvent molecule or anion coordination [104]. The presence of $AlMe_3$ resulted in the formation of the dormant species $[Cp_2Zr(\mu$-Me$)_2AlMe_2]^+[B(C_6F_5)_4]^-$ and it was confirmed that the addition of excess $AlMe_3$ led to a significant decrease in polymerization activity. However, it has been noted by Talsi *et al.* that the presence of $AlMe_3$ led to increased stability of the cationic species derived from $Cp_2ZrMe_2$ and $[Ph_3C][B(C_6F_5)_4]$ [105]. Mutual transformations of the cationic complexes **6**, **7** and **8** (Figure 2.5) were investigated using $^1H$ NMR spectroscopy. The binuclear cation **6** was formed when $Cp_2ZrMe_2$ and $[Ph_3C][B(C_6F_5)_4]$ were mixed in the proportion 2:1, whereas a 1:1 Zr/B ratio gave the cationic complex **7**. Addition of a 20-fold excess of $AlMe_3$ to **7** gave rise to the formation of the heterodinuclear cation **8**. The results indicated that complexes **6**, **7** and **8** had similar energy levels and could be readily interconverted by changing the ratio between $Cp_2ZrMe_2$,

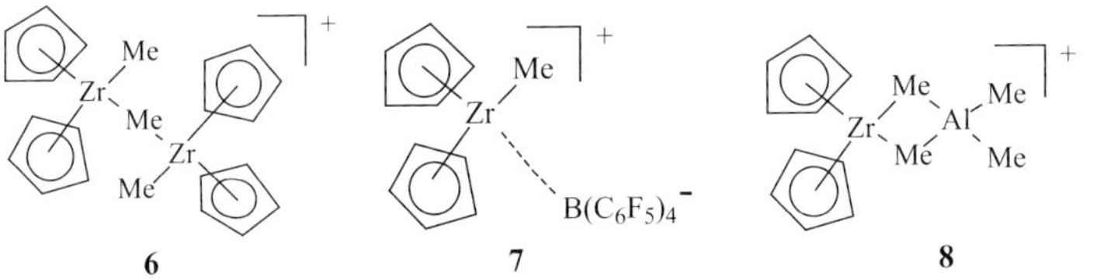

**Figure 2.5** Zirconocenium species formed from Cp$_2$ZrMe$_2$, [Ph$_3$C][B(C$_6$F$_5$)$_4$] and AlMe$_3$ [105].

[Ph$_3$C][B(C$_6$F$_5$)$_4$] and AlMe$_3$. Their activities in 1-hexene polymerization were reported to be of the same order of magnitude, suggesting that the olefin was able to replace Cp$_2$ZrMe$_2$, [B(C$_6$F$_5$)$_4$] and AlMe$_3$, respectively, to form the reactive intermediate [Cp$_2$ZrMe(olefin)]$^+$.

Studies by Brintzinger and coworkers have indicated that optimum polymerization activity is obtained with equimolar proportions of the metallocene and the borate [106]. The results indicated that the mononuclear zirconocene cation formed from equimolar amounts of *rac*-Me$_2$Si(Ind)$_2$ZrMe$_2$ and [*n*-Bu$_3$NH][B(C$_6$F$_5$)$_4$] was the sole catalytic species, even at higher Zr/B ratios, although formation of the binuclear cation [{*rac*-Me$_2$Si(Ind)$_2$ZrMe}$_2$(μ-Me)]$^+$ in the presence of excess zirconocene could give increased stability against deactivation. At a Zr/B ratio of 1:1, large decreases in the rate of propene polymerization were observed on increasing the temperature in the range 15–45 °C, indicating a strong effect of temperature on catalyst deactivation.

The possibility of catalyst deactivation via the formation of μ-CH$_2$ complexes, discussed in the previous section for metallocene/MAO systems, has been demonstrated by reaction of a fulvalene complex with [Ph$_3$C][B(C$_6$F$_5$)$_4$], which resulted in the facile elimination of methane to give the dinuclear μ-CH$_2$ complex shown in Scheme 2.5 [107].

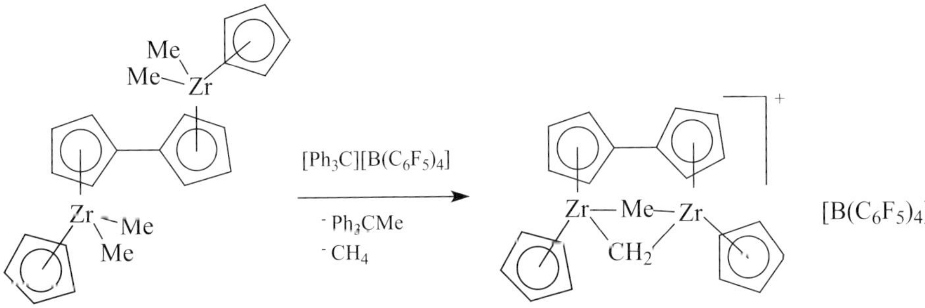

**Scheme 2.5** Catalyst deactivation via μ-CH$_2$ complex formation [107].

A binary catalyst system comprising a metallocene alkyl and a borate activator has the disadvantage that, in the absence of an aluminum alkyl, the active species may be deactivated by impurities present in the polymerization. Metallocene dialkyls are also more expensive and less readily obtainable than the parent dichlorides. There has, therefore, been a significant effort directed at the development of ternary systems comprising a metallocene dichloride, an aluminum alkyl and a borate, where the presence of the $AlR_3$ leads to *in situ* alkylation of the metallocene [108–111]. Alkyl exchange between $AlR_3$ and zirconocene dichlorides has been studied by Beck and Brintzinger, who found only monomethylation using $AlMe_3$, giving $Cp_2ZrMeCl$, whereas uptake of two alkyl groups, followed by alkane evolution, was observed with $AlEt_3$ and $AliBu_3$ [112]. In ternary catalyst systems, the aluminum alkyl may react not only with the metallocene but also with the borate. Bochmann and Sarsfield have observed that $AlMe_3$ reacts with $[Ph_3C][B(C_6F_5)_4]$ to give $Ph_3CMe$ and a transient species $[AlMe_2]^+[B(C_6F_5)_4]^-$, which immediately decomposes to $AlMe_2(C_6F_5)$ and $B(C_6F_5)_3$ [113]. Further ligand exchange can eventually lead to the formation of $Al(C_6F_5)_3$ and $BMe_3$. Reaction of $[Ph_3C][B(C_6F_5)_4]$ with $AliBu_3$ is much faster than that with $AlMe_3$ and involves hydride abstraction and elimination of isobutene, generating $Ph_3CH$ and the transient species $[AliBu_2]^+[B(C_6F_5)_4]^-$, which decomposes to $AliBu_n(C_6F_5)_{3-n}$ and $iBu_{3-n}B(C_6F_5)_n$. It was considered that the formation of $Al\text{-}C_6F_5$ species could represent a deactivation pathway in ternary catalyst systems. $Cp_2ZrMe_2$ was found to react with $AlMe_2(C_6F_5)$ to give $Cp_2ZrMe(C_6F_5)$, but it was noted that if conditions giving rise to the formation of "$[AlR_2]^+$" (such as premixing $[Ph_3C][B(C_6F_5)_4]$ with $AlR_3$) are avoided, degradation of the $[B(C_6F_5)_4]^-$ anion can be prevented [113]. Nevertheless, metallocene/borate/$AlR_3$ systems appear to be less stable than metallocene/MAO. In ethene polymerizations carried out in toluene solution at high temperature (100–140 °C) and pressure, Dornik *et al.* reported that the half-lives of the active species resulting from MAO activation were around twice those which were obtained using $[PhNHMe_2][B(C_6F_5)_4]/AliBu_3$ [114]. The deactivation kinetics were first order and it was also found that deactivation was less rapid with bridged metallocenes, $Me_2Si(Cp)_2ZrCl_2$, giving active species with twice the half-life of those from $Cp_2ZrCl_2$.

The thermal stability of various cationic zirconocene complexes has been investigated by Marks and coworkers [115, 116]. The stability of $[L_2ZrMe]^+[MeB(C_6F_5)_3]^-$ was found to be sensitive to the type of ancillary ligands surrounding the metal center. A solution of $[Cp_2ZrMe]^+[MeB(C_6F_5)_3]^-$ in toluene was stable for days at room temperature, whereas $[\{1,3\text{-}(SiMe_3)_2Cp\}_2ZrMe]^+[MeB(C_6F_5)_3]^-$ had a half-life of only $\approx$10 h, decomposing to $\{1,3\text{-}(SiMe_3)_2Cp\}_2ZrMe(C_6F_5)$ and $MeB(C_6F_5)_2$. The 1,3-di-*t*-butyl-substituted homolog rapidly decomposed, with evolution of 1 equiv of $CH_4$, to give the metallacyclic product **9**, apparently via intramolecular C–H activation. In contrast to complexes containing as anion $[MeB(C_6F_5)_3]^-$ or $[B(C_6F_4Si\text{-}tBuMe_2)_4]^-$, zirconium methyl complexes with a $[B(C_6F_5)_4]^-$ anion were not stable at room temperature. It was also noted that prolonged standing of a solution of $[Cp_2''ZrMe]^+[MeB(C_6F_5)_3]^-$ ($Cp'' = 1,2\text{-}Me_2C_5H_3$) in benzene for several weeks at 25 °C resulted in the formation of the fluoride-bridged complex $[Cp_2''MeZr(\mu\text{-}F)ZrMeCp_2'']^+[MeB(C_6F_5)_3]^-$ **10**, which is an adduct of $[Cp_2''ZrMe]^+[MeB(C_6F_5)_3]^-$ and $Cp_2''ZrMeF$ [115]. However, later studies by Arndt, Rosenthal and coworkers showed that the formation of a zirconium fluoride complex is not necessarily a deactivation mechanism, as a fluoride of type $L_2ZrF_2$ can be converted to a catalytically active zirconocene hydride in the presence of $R_2AlH$ [117].

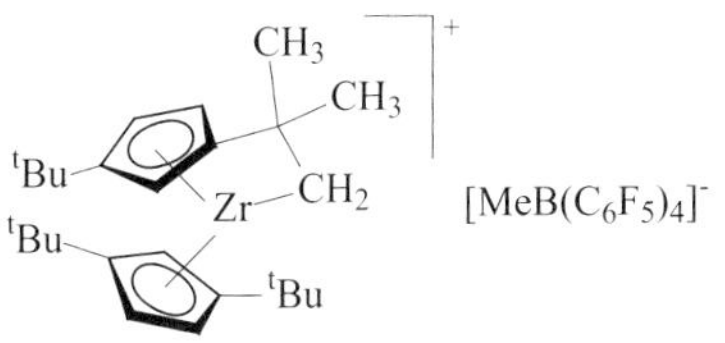

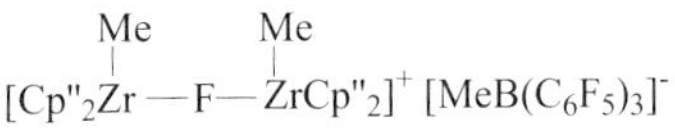

**9**                                                    **10**

Borate activation has resulted in very high activity, not only in zirconocene- but also in hafnocene-catalyzed olefin polymerization. The use of a hafnocene in combination with MAO generally leads to much lower activity than is obtained with ziconocenes, but Rieger and coworkers found that the replacement of MAO by $[Ph_3C][B(C_6F_5)_4]$ as activator in hafnocene-catalyzed propene polymerization gave activities which even exceeded those obtained with the corresponding zirconocene [118]. The reason for low activity when using MAO as activator in hafnocene-catalyzed polymerization is related to the presence of associated $AlMe_3$ in MAO, which generates species of type $[Cp_2Hf(\mu\text{-}Me)_2AlMe_2]^+[MeMAO]^-$. Recent studies have shown that such species, as well as species such as $[Cp_2Hf(\mu\text{-}Me)_2AlMe_2]^+[B(C_6F_5)_4]^-$ formed from contact of a hafnocene with $AlMe_3$ and $[Ph_3C][B(C_6F_5)_4]$, have significantly higher stability and lower olefin polymerization activity than their zirconocene counterparts, due to stronger binding of $AlMe_3$ [119, 120]. Hafnocene activation using $AliBu_3$ and $[Ph_3C]$ $[B(C_6F_5)_4]$, on the other hand, generates much more active hydrido-bridged species of type $[Cp_2Hf(\mu\text{-}H)_2AliBu_2]^+[B(C_6F_5)_4]^-$ or $[Cp_2Hf(\mu\text{-}H)_2Al(H)iBu]^+[B(C_6F_5)_4]^-$.

## 2.3
## Other Single-Center Catalysts

### 2.3.1
### Constrained Geometry and Half-Sandwich Complexes

Since the discovery of highly active metallocene-based systems, ever-increasing numbers of alternative single-center catalysts for olefin polymerization have been developed [121, 122]. An example of a metallocene analog that has received considerable commercial attention is the "Constrained Geometry" class of complexes developed concurrently by Dow and Exxon, which contain an ansa-amide-cyclopentadienyl ligand, first described by Bercaw and coworkers [123]. The generalized structure of such catalysts is shown below.

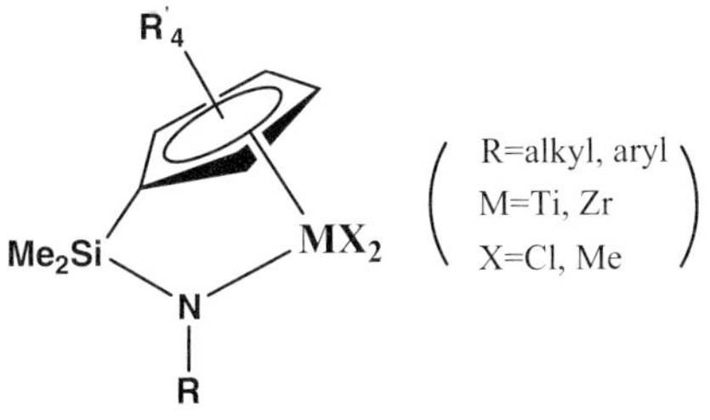

A key feature of such catalysts is the open nature of the active site, which allows the incorporation of other olefins, including vinyl-terminated polymer chains, into the main chain, resulting in the formation of long-chain branching. Constrained geometry catalysts are also much more stable than bis(cyclopentadienyl) metallocenes, allowing their use in polymerization processes carried out at relatively high temperatures ($>100\,°C$) [124].

Reactions of constrained geometry dibenzyl complexes $(CGC)M(CH_2Ph)_2$, where CGC represents the ligand $t$-BuNSiMe$_2$(C$_5$Me$_4$), with $B(C_6F_5)_3$ and $[Ph_3C][B(C_6F_5)_4]$ have been investigated by Chen and Marks [125]. Low-temperature activation of $(CGC)Zr(CH_2Ph)_2$ with the borate generated $[(CGC)ZrCH_2Ph]^+[B(C_6F_5)_4]^-$, whereas the corresponding Ti complex gave the intramolecular metallation products **11** and **12**, respectively, on reaction with $B(C_6F_5)_3$ and $[Ph_3C]$ $[B(C_6F_5)_4]$. In contrast, reaction of $(CGC)TiMe_2$ with $B(C_6F_5)_3$ cleanly generated $[(CGC)TiMe]^+[MeB(C_6F_5)_3]^-$.

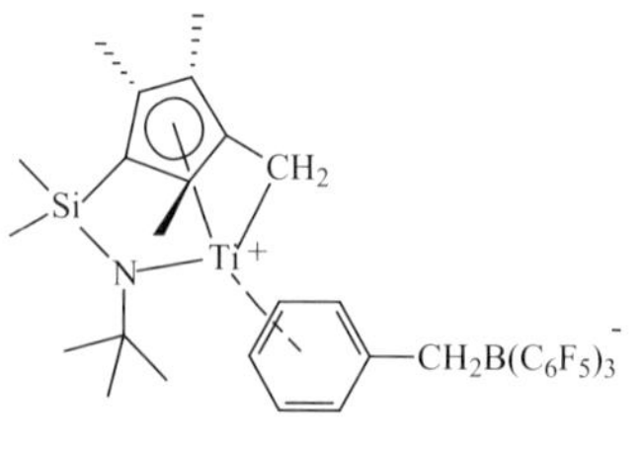

**11** 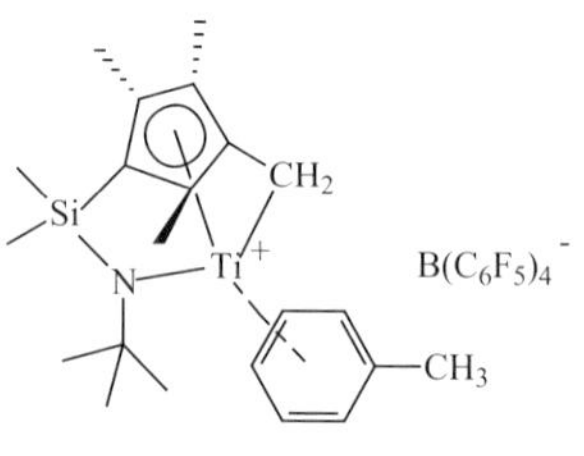                    **12**

A potentially tridentate variation of the CGC ligand was introduced by Okuda and coworkers, who synthesized complexes such as **13**, in which the additional two-electron donor function X (representing NMe$_2$ or OMe) was designed to act as a semilabile ligand that would temporarily stabilize highly electrophilic metal centers [126]. The hydrazido complex **14** was synthesized by Park *et al.* [127]. The strong donor nature of the hydrazido moiety was expected to stabilize the catalytically active cationic species in polymerization, but it was found that in the presence of an aluminum alkyl the $-NMe_2$ nitrogen atom coordinated to Al rather than to Ti, leading to thermal instability and poor polymerization activity.

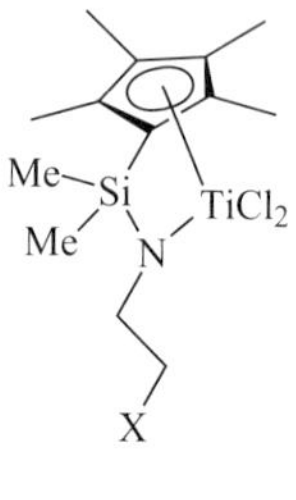

**13** 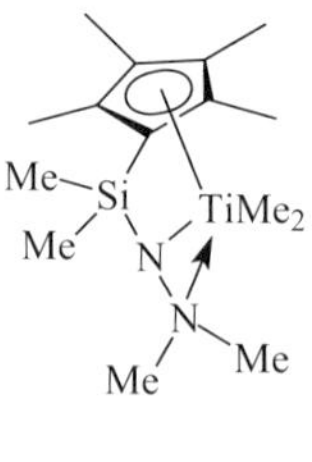                    **14**

Cocatalyst effects in propene polymerization with [*t*-BuNSiMe$_2$(C$_5$Me$_4$)]TiMe$_2$ have been investigated by Shiono and coworkers, who observed living polymerization using as cocatalyst a sample of MAO from which AlMe$_3$ had been removed by vacuum drying [128]. The activity was an order of magnitude higher than that obtained with untreated MAO, with which rapid deactivation was apparent. The highest activities were obtained with [Ph$_3$C][B(C$_6$F$_5$)$_4$]. Typically, a stoichiometric 1:1 borate to catalyst ratio is used in metallocene activation with [Ph$_3$C][B(C$_6$F$_5$)$_4$]. However, Bochmann and coworkers have reported that the activity of a number of zirconocenes in olefin polymerization increases in the presence of excess borate [129]. With the ternary system [*t*-BuNSiMe$_2$(C$_5$Me$_4$)] TiCl$_2$/Al*i*Bu$_3$/[Ph$_3$C][B(C$_6$F$_5$)$_4$], a 10-fold increase in the borate concentration led to an order of magnitude increase in catalyst productivity. The origin of this effect was not clear.

Theoretical investigations of possible deactivation routes in olefin polymerization have been carried out by Ziegler and coworkers, using density functional theory and molecular mechanics [130, 131]. Activation energies for aryl group (C$_6$F$_5$) transfer from a borate anion to the metal center were calculated, indicating that Ti complexes were more stable than their Zr analogs in this respect, apparently due to the larger size of the Zr atom. The results also indicated that aryl group transfer was less likely with catalysts having sterically demanding substituents around the metal center, and that deactivation pathways could be avoided by the presence of electron-donating substituents on the ancillary ligands.

Non-bridged amido cyclopentadienyl complexes of titanium have been investigated by Okuda and coworkers [132]. Complex **15**, in combination with MAO, gave poor activity in ethene polymerization, whereas efficient polymerization of styrene was achieved with a related non-bridged system. These results indicated the critical role of a covalent link between the cyclopentadienyl ligand and the amido group to stabilize the Ti(IV) alkyl cation for ethene polymerization. The absence of the bridge in complexes such as **15** leads to reduction to Ti(III) species, which are active in syndiospecific styrene polymerization [133]. Nomura considered that the deactivation of complex **15** in ethene polymerization, resulting from dissociation of the amide ligand from the metal center by AlMe$_3$ present in MAO, could be suppressed somewhat by the introduction of a bulky cyclohexyl group, and found that complex **16** indeed gave higher ethene polymerization activity [134]. Relatively high activities at temperatures of 60–70 °C were achieved with (cyclopentadienyl)(aryloxy)Ti(IV) complexes such as **17** [135]. The combination of both the methyl substitution in the Cp* ligand and the 2,6-diisopropyl substitution in the phenoxy group was found to be indispensable for high activity, underlining the role of the steric effect in the stabilization of catalytically active species. Reactions of Cp(OAr)TiMe$_2$ with B(C$_6$F$_5$)$_3$ have been investigated by Phomphrai *et al.*, who found that with OAr = OC$_6$H$_3$*i*Pr$_2$-2,6 the main product was Cp(OAr)TiMe(C$_6$F$_5$), formed by C$_6$F$_5$ transfer from B to Ti [136].

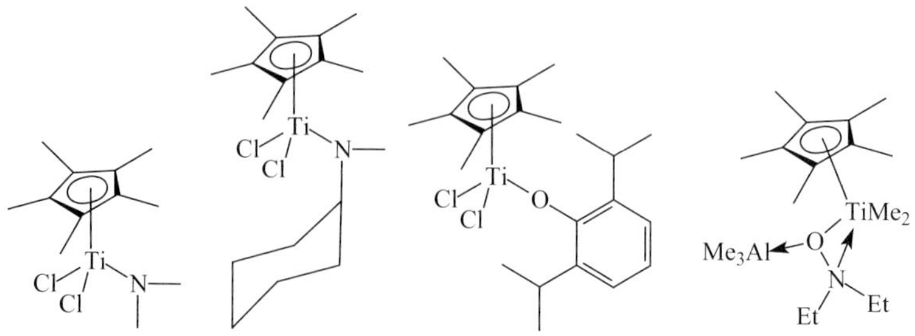

**15**   **16**   **17**   **18**

Recently, hydroxylaminato complexes such as Cp*TiMe$_2$(ONEt$_2$) have been investigated by Waymouth and coworkers, who found that addition of AlMe$_3$ gave the stable adduct **18** [137]. However, addition of Al$i$Bu$_3$ or AlMe$_3$ to the cation [Cp*TiMe (ONR$_2$)]$^+$[B(C$_6$F$_5$)$_4$]$^-$ led to decomposition to unidentified products, accompanied by liberation of isobutene and methane, respectively.

Single-center titanium catalysts containing a phosphinimide ligand have been developed by Stephan and coworkers [138, 139]. A typical example is ($t$Bu$_3$P = N) CpTiCl$_2$. The steric bulk of the phosphinimide ligand, similar to that of the cyclopentadienyl ligand, was considered to preclude deactivation pathways such as cation dimerization and interaction with Lewis acids leading to C—H activation [140, 141]. In contrast to their titanium analogs, zirconium phosphinimide complexes were found to have low catalytic activity, attributed to the larger ionic radius of Zr facilitating deactivation pathways over chain propagation [142]. Aryl group transfer from B(C$_6$F$_5$)$_3$ to the metal was observed with Zr but not with Ti, while phosphinimide ligand abstraction and the formation of clustered Zr species took place in the presence of AlMe$_3$.

Piers and coworkers have investigated the activation and deactivation of ketimide complexes of type Cp[$t$Bu(R)C=N]TiMe$_2$, using B(C$_6$F$_5$)$_3$ as activator [143]. It was found that deactivation takes place via the elimination of methane from a contact ion pair, as illustrated in Scheme 2.6. A similar phenomenon is observed when a dimethyl derivative of a constrained geometry catalyst is treated with B(C$_6$F$_5$)$_3$. Elimination of methane was also observed on contacting Cp[$t$Bu$_2$C=N]TiMe$_2$ with half an equivalent of [Ph$_3$C][B(C$_6$F$_5$)$_4$], as a result of a σ-bond metathesis reaction involving a C—H bond of the bridging methyl group in the dimeric species {[CpLTiMe]$_2$-(µ-Me)}$^+$[B(C$_6$F$_5$)$_4$]$^-$ [144].

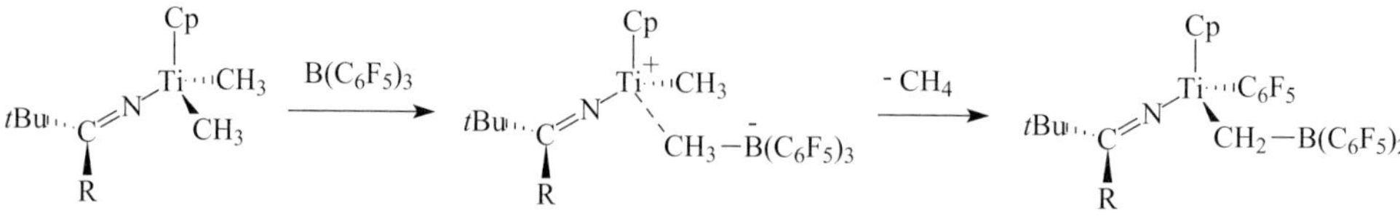

**Scheme 2.6** Reaction of a Cp-ketimide complex with B(C$_6$F$_5$)$_3$ [143].

2.3.2
## Octahedral Complexes

A fundamental difference between $MgCl_2/TiCl_4$-based Ziegler–Natta catalysts and the metallocenes and half-metallocenes described above concerns the coordination sphere around the transition metal atom: octahedral in the case of the Ziegler–Natta systems as opposed to tetrahedral for the metallocenes and related complexes. Recently, however, there has been growing interest in the development of octahedral single-center early transition metal catalysts.

One important new family of octahedral complexes for olefin polymerization comprises the FI catalysts developed by Mitsui Chemicals [145–147]. The name FI derives from the Japanese term for a phenoxy-imine ligand. The general structure of such catalysts is as follows:

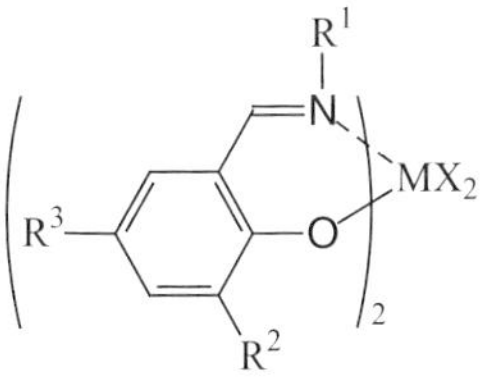

The ethene polymerization activities of Ti-, Zr- and Hf-FI catalysts were varied by changing the ligand structures. A bulky substituent in the ortho-position with respect to the phenoxy-oxygen is required to achieve high activity. The large substituent ($R^2$) provides steric bulkiness against electrophilic attack by Lewis acids in the polymerization medium, and increases the activity by creating effective ion separation between the cationic species and an anionic cocatalyst. FI catalysts are distorted octahedral complexes having two oxygen atoms in trans-positions and two nitrogen atoms and two chlorine atoms in cis-positions. With MAO as activator, the ethene polymerization activity of complex **19** was found to decrease at temperatures above $40\,^\circ C$ [146]. The activity decrease at higher temperatures was ascribed to decomposition of the active species through loss of the ligand(s). Ligand transfer from the transition metal center to Al has been confirmed by $^1H$ NMR studies, which indicated that deactivation of FI catalysts involves ligand transfer to $AlMe_3$ present in MAO, forming $LAlMe_2$ [148, 149]. Increased stability of the active species can, therefore, be obtained by removal of $AlMe_3$ from MAO, as well as by the introduction of bulky ligand substituents close to the metal center. In order to strengthen the metal–ligand bonds, electron-donating groups were introduced at the para-position $R^3$ and on the imine nitrogen. The combination of $R^1 = n$-hexyl or cyclohexyl and $R^3 = OMe$ led to greater stability and enhanced activity at $75\,^\circ C$ [150]. Extremely high polymerization activities of around $800\,000\,kg\,mol^{-1}\,h^{-1}\,bar^{-1}$ were obtained with, in addition, an adamantyl or cumyl substituent at position $R^2$.

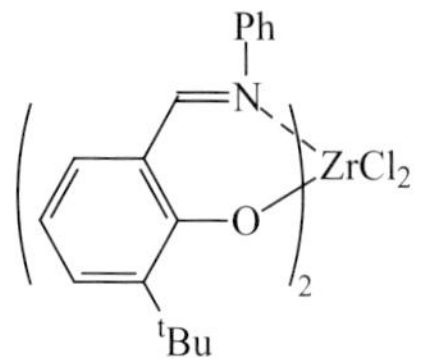

19 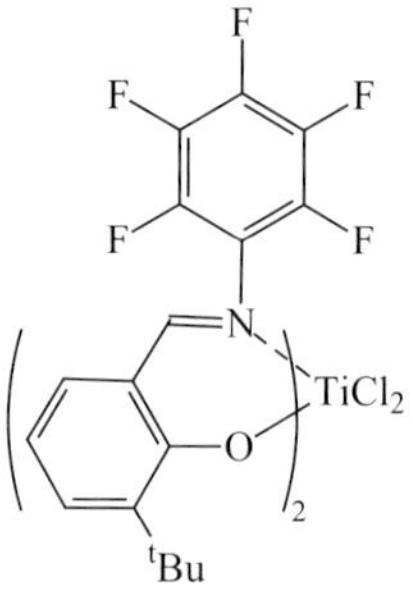                                    20 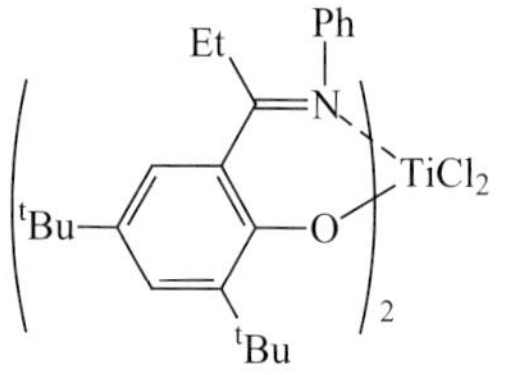                                    21

Titanium FI catalysts generally have lower activity than their zirconium analogs, but have somewhat higher stability. Their activity can be increased by the introduction of strongly electron-withdrawing substituents on an $R^1$-phenyl group, thereby increasing the electrophilicity of the active Ti species. For example, replacement of Ph by $3,5\text{-}(CF_3)_2C_6H_3$ gave a fourfold increase in ethene polymerization activity [151]. Ti FI catalysts can also exhibit low rates of chain termination, such that living polymerization is possible. This was demonstrated with complex **20**, with which a linear increase in polyethylene molecular weight with time was observed in MAO-activated polymerization at 50 °C [152]. It was found that living polymerization requires the presence of at least one fluorine in the 2,6-positions of the phenyl group [153]. The fluorine adjacent to the imine nitrogen was proposed to suppress β-hydrogen transfer from the growing chain, as a result of electrostatic interaction between the negatively charged fluorine and the positively charged β-H [153, 154]. However, Coates and coworkers found that living polymerization could also be achieved with non-fluorinated complexes such as **21**, suggesting that other factors besides a β-H/σ-F interaction could play a role [155].

The relationship between catalyst structure and living polymerization was addressed by Busico *et al.*, with special reference to propene polymerization [156]. Computational studies revealed that the main effect of an ortho-F atom in giving living polymerization is steric, destabilizing the transition state for β-H transfer to the monomer, which is the predominant chain transfer pathway with bis(phenoxy-imine) catalysts.

Living polymerization of 1-hexene has been demonstrated by Kol and coworkers, who developed a new family of dianionic tetradentate [ONNO]-type chelating ligands [157]. Dependent on the substituents present in these Salan ligands, $C_2$-symmetric complexes able to polymerize α-olefins to isotactic polymers can be synthesized. Busico *et al.* carried out propene polymerization with these complexes and found that the mechanism of stereocontrol was dependent on the steric hindrance imparted mainly by the alkyl substituents $R_1$ in the following complex, where Bn represents benzyl [158]. Very high isotacticity can be obtained with $R_1$ = adamantyl and $R_2$ = methyl and the preparation of block copolymers of propene and ethene via living polymerization has been achieved [159, 160]. The steric bulk of the substituent at the $R_1$ position is important, not only for high enantioselectivity (giving highly isotactic polypropylene) but also it destabilizes the transition state for chain transfer to propene, thereby giving living polymerization in the absence of other chain transfer pathways [156].

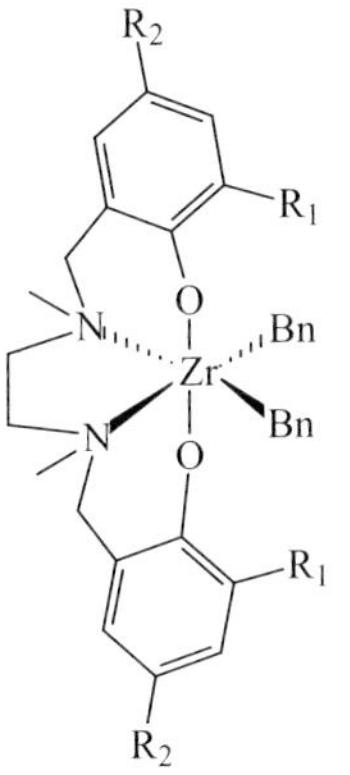

## 2.3.3
## Diamide and Other Complexes

High activities in $\alpha$-olefin polymerization have been obtained (for short periods of time) with diamide complexes of type [ArN(CH$_2$)$_3$NAr]TiX$_2$, where $X =$ Cl or Me [161]. Catalysts of type **22**, bearing 2,6-diisopropyl substituents in the aryl rings, are more active than their 2,6-dimethyl substituted analogs, illustrating again the importance of steric factors. The presence of small amounts of toluene causes a significant decrease in polymerization activity, indicating that toluene can effectively compete with an $\alpha$-olefin for coordination to the cationic active species. The dimethyl derivatives of these complexes (i.e., $X =$ Me) can be activated by B(C$_6$F$_5$)$_3$ and living polymerization of 1-hexene has been achieved at ambient temperature [162]. Notable increases in activity were observed when the polymerization was carried out in the presence of dichloromethane rather than pentane. This effect was ascribed to greater charge separation between the cationic metal center and the borate counterion due to the polarity of CH$_2$Cl$_2$.

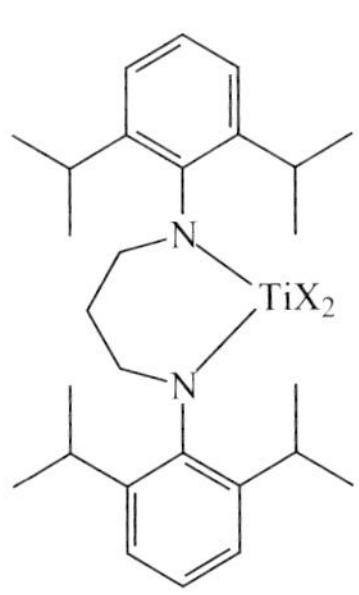

22                          23

The shielding effect of a bulky ligand in improving catalyst stability, and, therefore, productivity, apparent in many of the preceding examples, is also apparent in complexes of type $ROZrCl_3$, where the highest ethene polymerization activities are obtained with complexes such as **23**, derived from fenchyl alcohol [163].

## 2.4
## Vanadium-Based Catalysts

Vanadium-based catalysts have been developed for the production of ethene copolymers, notably ethene-propene-diene rubbers (EPDM). Homogeneous vanadium catalysts give polymers with relatively narrow molecular weight distributions and typically give easier and more random comonomer incorporation in ethene copolymerization than can be achieved with titanium-based Ziegler–Natta catalysts. However, unsupported vanadium catalysts generally give low productivity in olefin polymerization, as a result of rapid decay. Typical precatalysts are $VOCl_3$, $VCl_4$ or V $(acac)_3$, used in combination with an aluminum alkyl cocatalyst such as $AlEt_2Cl$ or $Al_2Et_3Cl_3$ and a halocarbon promoter. It is widely accepted that deactivation results from reduction of active V(III) species to poorly active or inactive V(II) and that the function of the promoter (e.g., a chlorinated ester) is to reoxidize the divalent species to the active, trivalent species [164–167].

The rate of catalyst deactivation in vanadium-catalyzed polymerization is, as may be expected, dependent on temperature. At very high temperature (160 °C), more than 90% of the initial catalytic activity obtained with the catalyst system $VCl_3 \cdot 0.33AlCl_3$–$AlEt_3$ was lost within the first minute of polymerization in the absence of a halocarbon promoter [168]. A more than 10-fold increase in productivity was noted in the presence of $CH_3CCl_3$ [169]. In contrast, living polymerization is possible at very low temperatures. Doi *et al.* demonstrated the living polymerization of propene, giving syndiotactic polypropylene, at −65 °C using $V(acac)_3$/$AlEt_2Cl$, and subsequently found that, on changing the ligand system from acetylacetonate to 2-methyl-1,3-butanedionate, it was possible to obtain higher activity and living polymerization at −40 °C [170, 171]. Syndiospecific propagation in these systems results from a secondary (2,1-) propene insertion mechanism with chain-end control [172]. The synthesis of block copolymers is possible using $V(acac)_3$ and related complexes, but is not a commercial proposition in view of the low catalyst activities and the need for very low temperatures. However, block copolymers comprising an ethene and an ethene/propene block have been produced in a tubular plug flow reactor with side stream injection, using $VCl_4$/$Al_2Et_3Cl_3$ [173]. The reactor temperature was around 20 °C and the polymerization time of several seconds for each block was matched to the chain growth time and catalyst decay rate.

The poor thermal stability of β-diketonate vanadium catalysts has been shown to be due to leaching of the ligand by the aluminum alkyl cocatalyst, generating vanadium (II)/aluminum bimetallic species having negligible polymerization activity, irrespective of the substituents in the acac ligand [174].

Attempts to improve the thermal stability of vanadium catalysts have involved the synthesis of complexes containing ligands such as amide, amidinate, imide and phenolate. Moderate ethene polymerization activities have been obtained with V(IV) amides of type $(R_2N)_2VCl_2$ (R = *i*-Pr or cyclohexyl) [175]. In the presence of $Al_2Et_3Cl_3$ or other Al alkyl cocatalyst, deactivation was suggested to be due to reduction of tetravalent to trivalent vanadium. Activity could be restored by addition of benzyl chloride, 1,2-dichloroethane or $CHCl_3$. Similar activities have been obtained with the diamide complex **24**, which was also active in propene polymerization [176].

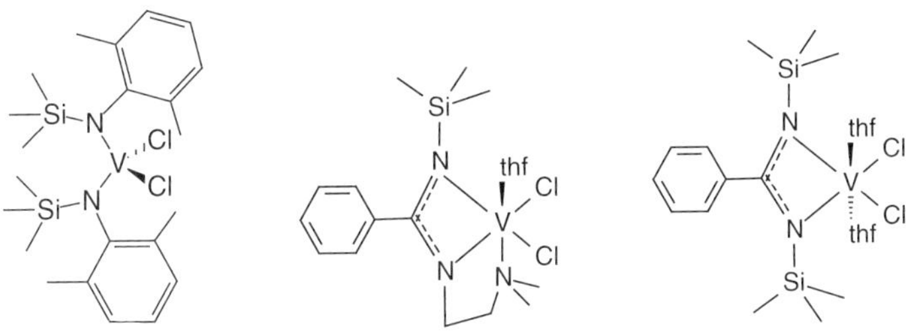

**24**                    **25**                    **26**

Vanadium(III) amidinate complexes **25** and **26** have been synthesized by Hessen and coworkers [177]. In combination with $AlEt_2Cl$, these catalysts were active in ethene polymerization in the temperature range 30–80 °C. Complex **25**, containing a pendant amine functionality, was most active at 30 °C. At 80 °C rapid deactivation was observed and the polyethylene molecular weight distribution became bimodal, signifying a deviation from single-center catalysis. Complex **26** gave higher polyethylene molecular weight and was somewhat more stable at 50–80 °C, indicating that the introduction of the pendant amine functionality in **25** did not improve the thermal stability of the active species.

Ethene polymerization using vanadium(IV) complexes with tetradentate [ONNO] salen-type ligands has also been investigated [178]. Under homogeneous conditions (in toluene, with $AlEt_2Cl$ or $AlEtCl_2$ as cocatalyst), a large decrease in polymerization activity was observed on increasing the temperature in the range 20–50 °C. Poor catalyst activity at elevated temperatures was ascribed to reduction of vanadium to lower and inactive oxidation states.

Various (arylimido)(aryloxo)vanadium(V) complexes of type $VCl_2(NAr)(OAr')$ have been synthesized by Nomura and coworkers [179]. Using as cocatalyst MAO from which $AlMe_3$ had been removed, the highest ethene polymerization activities with complex **27** were obtained at 25 °C. Much higher catalytic activities were obtained when $AlEt_2Cl$ was used as cocatalyst, but again higher polymerization temperatures led to greatly decreased productivity, indicating deactivation of active species. The more sterically hindered complex **28**, however, gave stable polymerization activity at 60 °C using methylisobutylaluminoxane as cocatalyst. Dibenzyl complexes of type V $(CH_2Ph)_2(NAr)(OAr')$ initiated the ring-opening polymerization of norbornene in the absence of any cocatalyst. Activities in ethene/norbornene copolymerizations at 0 °C with the system **27**/$AlEt_2Cl$ were higher than those obtained in ethene homopolymerization, but increasing the polymerization temperature to 25 °C gave greatly decreased productivity [180]. The addition of a halocarbon (CCl$_3$CO$_2$Et) in ethene polymerization catalyzed by VCl$_2$(NAr)(OAr$'$) gave decreased rather than increased activity, indicating that the catalytic species in this system differ from those derived from V(III) and V(IV) complexes [181].

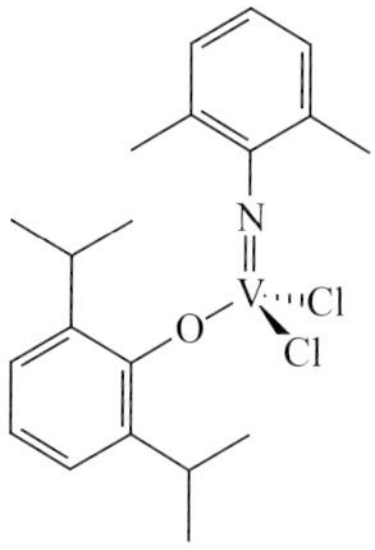

**27** 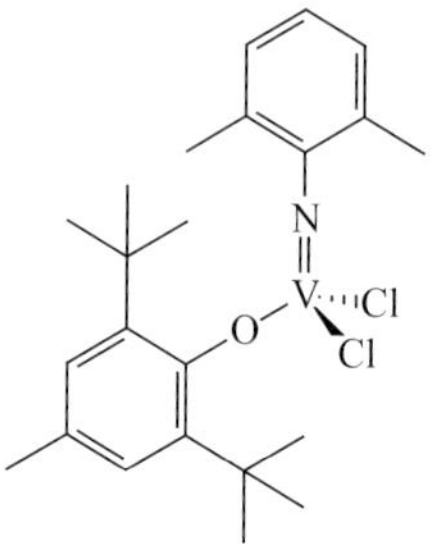                **28** 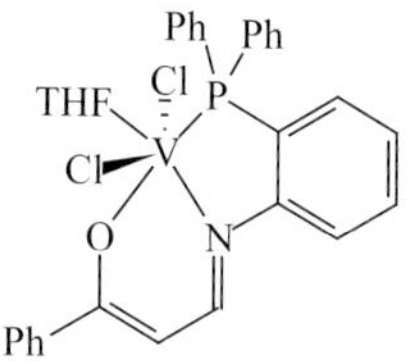                **29**

Complex **29**, containing a tridentate β-enaminoketonato ligand, on activation with AlEt$_2$Cl and ethyl trichloroacetate, has been found to give reasonable stability in ethene polymerization [182]. A 30% decrease in activity was observed over the course of 30 min of polymerization at 50 °C. Similar deactivation profiles were observed with other complexes containing soft (P or S) donor-containing ligands, whereas complexes having pendant O- or N-donor ligands lost around 70% of their activity within 30 min polymerization.

High activities in vanadium-catalyzed ethene polymerization have been obtained with complexes of type [V(O)(μ$_2$-(OPr)L)]$_2$, where L represents the deprotonated form of a di- or tri-phenol [183]. Using AlMe$_2$Cl as cocatalyst, it was found that the presence of the re-activator CCl$_3$CO$_2$Et led to a sevenfold increase in activity. The combination of AlMe$_2$Cl and CCl$_3$CO$_2$Et was also effective in ethene polymerization with vanadium calixarenes [184]. An increase in polymerization temperature from 25 to 80 °C resulted in higher activity, indicating relatively high thermal stability. Vanadium complexes **30** and **31**, containing bis(benzimidazole)amine ligands, have also been shown to give unusually robust, single-center catalysts for ethene homo- and copolymerization [185]. A constant polymerization activity of around 36 000 kg mol$^{-1}$ h$^{-1}$ bar$^{-1}$ was noted over a period of 1 h at 60 °C with the catalyst system **31**/AlMe$_2$Cl/CCl$_3$CO$_2$Et. The presence of hydrogen was found to decrease the polymer molecular weight without affecting the overall catalyst performance.

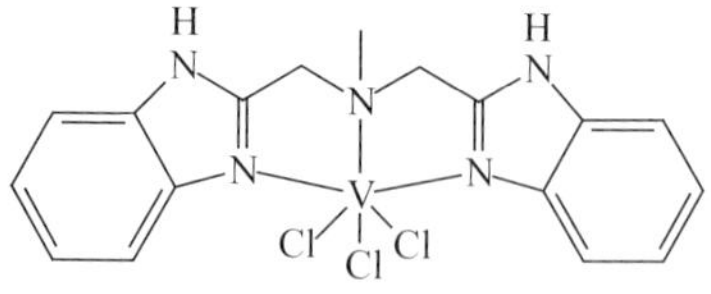

**30** 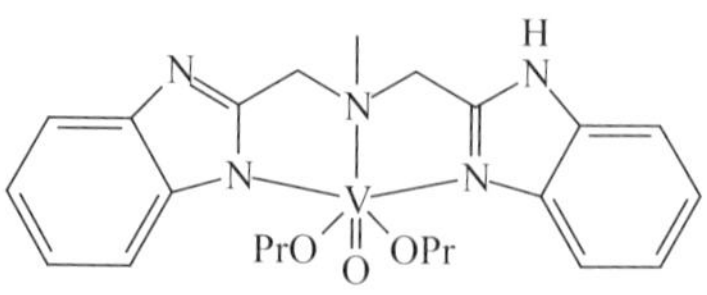                                    **31**

An interesting example of a single-component catalyst giving high molecular weight polyethylene has recently been reported by Gambarotta, Duchateau and coworkers [186]. The divalent vanadocene-type complex **32**, synthesized by reaction of $VCl_3(THF)_3$ with 2,5-dimethylpyrrole in the presence of $AlMe_3$, proved to be thermally robust and did not decompose upon heating in toluene solution at 100 °C. This complex was able to initiate ethene polymerization at 75 °C in the absence of any cocatalyst. It was proposed that the presence of ethene triggers a reorganization leading to the transfer of an alkyl group from Al to V, as illustrated in Scheme 2.7.

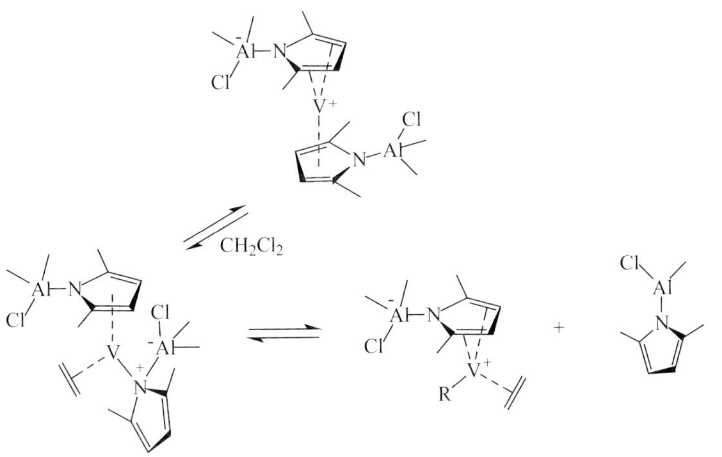

**Scheme 2.7** Proposed formation of active species in a single-component vanadium catalyst for ethene polymerization [186].

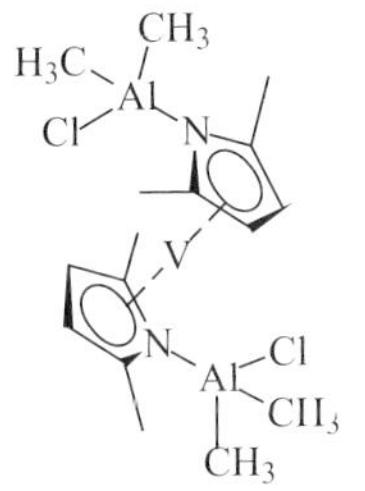

**32**

Earlier studies by Gambarotta involved the active species generated from the bis (imino)pyridyl vanadium complex **33** (Ar = 2,6-diisopropylphenyl) [187]. Analogous iron complexes are highly active in ethene polymerization, as described in Chapter 3. In toluene solution, using MAO as cocatalyst, the activity of complex **33** in ethene polymerizations decreased noticeably in the temperature range 50–85 °C and the complex became completely inactive within 1 min at 140 °C. A color change from red to green was observed when **33** was contacted with MAO or MeLi and it was shown that in each case the product was complex **34**, resulting from alkylation of the pyridine ring. This reaction, leading to a decrease in the metal coordination number, was suggested to be a key factor contributing to the polymerization activity obtained using complex **33** in combination with MAO, and it was found that complex **34** indeed gave similar polymerization activities and polymer characteristics.

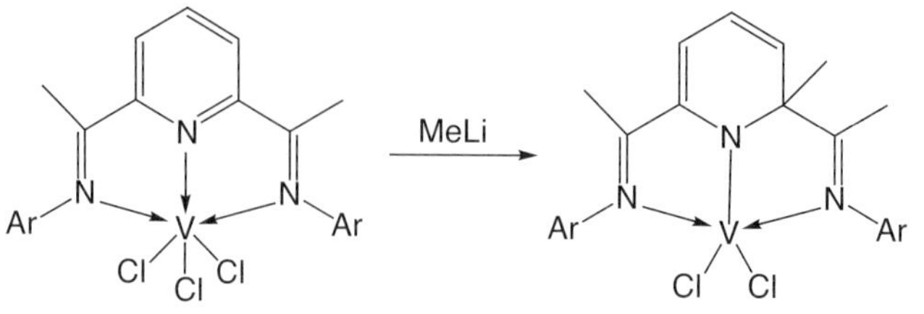

The stability of many vanadium polymerization catalysts can be greatly improved by immobilization on magnesium chloride supports. For example, stable ethene polymerization activity at 70 °C has been observed after immobilization of complex **33** on $MgCl_2$ [188]. This, along with other examples of the positive effects of immobilization in preventing catalyst deactivation, is discussed in more detail in Chapter 4.

## 2.5
## Chromium-Based Catalysts

Phillips catalysts, comprising chromium oxide supported on silica, are important industrial catalysts and account for about one-third of current global HDPE production. An unusual property of these heterogeneous catalysts is the ability to polymerize ethene in the absence of any activator. However, as is the case for Ziegler–Natta catalysts, the nature of the possible active species is still a matter of debate [189–191]. These catalysts give polyethylene with very broad molecular weight distribution and the catalyst activity increases with temperature up to 130 °C [192].

Various attempts have been made to synthesize homogeneous chromium catalysts for ethene polymerization, some of which were regarded as models for Phillips-type ($CrO_x/SiO_2$) or Union Carbide-type ($Cp_2Cr/SiO_2$) heterogeneous systems [193]. Most interest has centered on Cr(III) complexes. MAO activation of triazacyclohexane complexes (**35**) has given ethene polymerization activities of up to around 700 kg $mol^{-1} h^{-1} bar^{-1}$ at 40 °C [194]. Activation with $[PhNHMe_2][B(C_6F_5)_4]/AliBu_3$ has also been investigated [195]. Decomposition of the active complex in toluene was found to occur via transfer of the triazacyclohexane to aluminum, leading to [(triazacyclohexane)$AliBu_2$][$B(C_6F_5)_4$] and the Cr(I) complex [(arene)$_2$Cr][$B(C_6F_5)_4$].

Jolly and coworkers have obtained an ethene polymerization activity of approximately 3000 kg $mol^{-1} h^{-1} bar^{-1}$ with the amino-substituted cyclopentadienyl chromium(III) complex **36** at around 30 °C [196, 197]. Related complexes, of type **37**, were reported by Huang and coworkers, who obtained higher activities at 25 than at 50 °C, indicative of deactivation at elevated temperature [198]. An unusual observation in this work was that ethene/1-hexene copolymerization gave polymers with higher molecular weights than were obtained in ethene homopolymerization. Enders and coworkers introduced a rigid spacer group between the cyclopentadienyl ligand and the nitrogen donor atom, as in complex **38**, in order to increase the stability of these Cr (III) half-sandwich compounds [199, 200]. No decomposition was detected when such complexes were contacted with MAO in toluene and the resulting solutions were active in ethene polymerization at temperatures up to 110 °C. In propene polymerization, these complexes generate amorphous polymers with microstructures similar to those of ethene/propene copolymers [201]. The unusual microstructure results from a "chain-walking" mechanism whereby a 2,1-inserted propene unit is converted to a 3,1-unit in the polymer chain, as illustrated in Scheme 2.8.

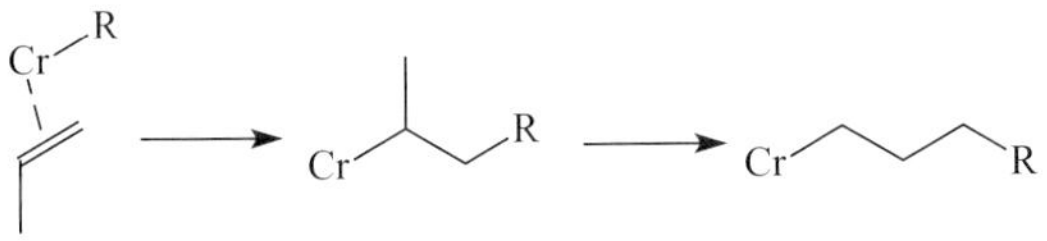

**Scheme 2.8** Chain-walking after a 2,1-insertion of propene [201].

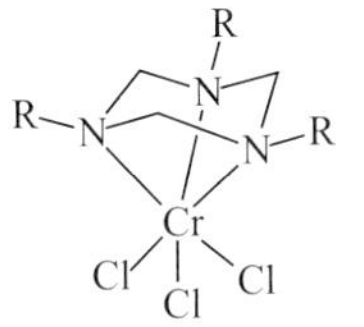

35

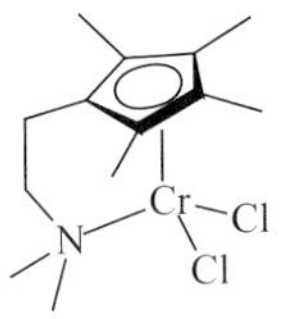

36

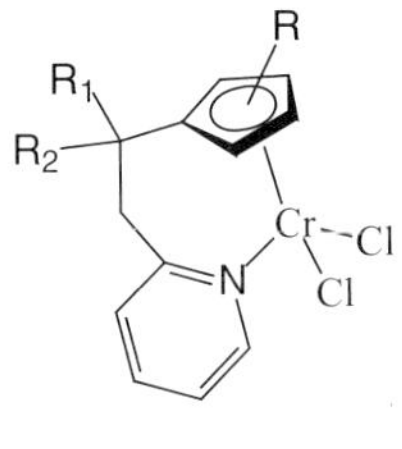

37

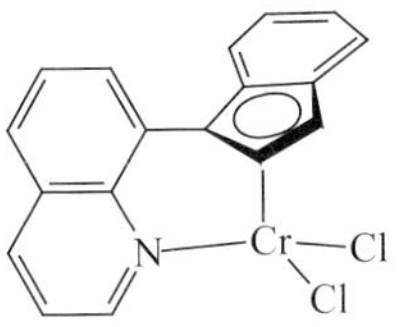

38

Bis(imino)pyridyl chromium(III) complexes such as **39** and **40** have been synthesized by Esteruelas *et al.* [202]. Ethene polymerization under homogeneous conditions, using MAO as activator, resulted in the formation of waxes and low molecular weight polymers. An increase in temperature in the range 60–90 °C led to decreased activity, but it was noted that the stability of these chromium complexes at 70 °C was greater than that of analogous iron and cobalt complexes. Stable polymerization activity at 80 °C was obtained after immobilization of the complex on MAO-impregnated silica. Small *et al.* reported the formation of mixtures of oligomers and polymers using a range of bis(imino)pyridyl complexes and provided evidence that the active species derived from Cr(II) and Cr(III) complexes were identical [203].

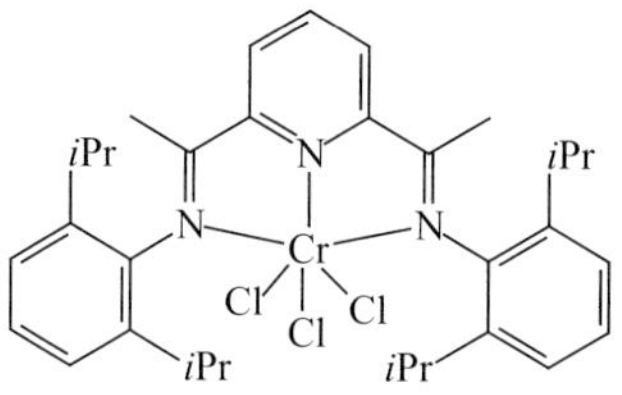

**39**

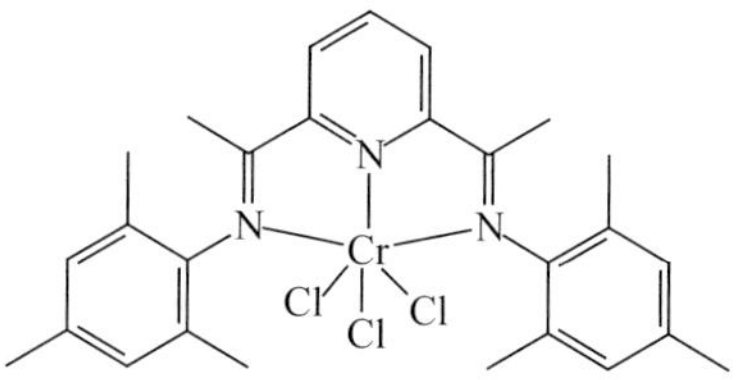

**40**

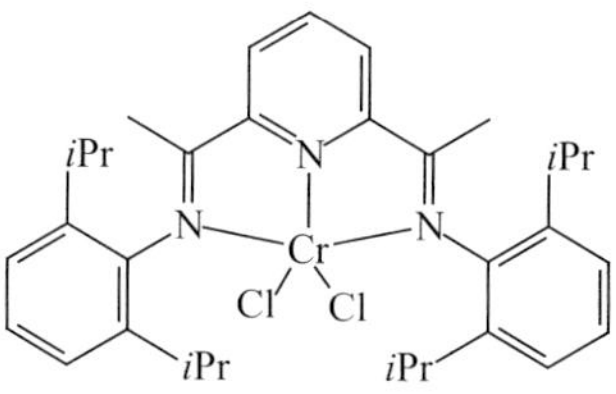

**41**

The issue of oxidation state in the chromium-catalyzed oligomerization and polymerization of ethene has been addressed by Gambarotta and Duchateau [204, 205]. Treatment of a trivalent bis(imino)pyridyl chromium complex with an aluminum alkyl leads to reduction to the divalent state [206]. In order to investigate the oxidation state of the catalytically active species, the divalent $LCrCl_2$ complex **41** was synthesized and its reactions with various reducing and alkylating agents investigated [205]. Reaction with NaH yielded LCrCl, which, in combination with MAO, was a potent polymerization catalyst. On reaction with excess $AlMe_3$, LCrCl gave a mixture of $LCr(\mu\text{-}Cl)_2AlMe_2$ and LCrMe. These are formally monovalent Cr complexes, but ligand backbone deformations suggested reduction of the ligand by one electron, so that the complexes could be better described as Cr(II) centers bound to ligand radical anions. It was concluded that overall reduction of the system, with electron density mainly located on the ligand, enhances the catalytic activity. Addition of isobutylaluminoxane to LCrCl led to partial transfer of the ligand from chromium to aluminum. It was proposed that this transmetallation reaction represents the deactivation pathway for bis(imino)pyridyl chromium complexes.

## 2.6
## Conclusions

The activity and stability of homogeneous catalysts in olefin polymerization is dependent on a number of different factors, the relative importance of which is dependent on the transition metal. In zirconocene-catalyzed polymerization, steric protection of the metal center plays an important role. Bulky ligands (L) can hinder the coordination of $AlMe_3$ to the active zirconocenium cation $[L_2ZrR]^+$, which leads to easier displacement of associated $AlMe_3$ by the monomer. This gives higher catalyst activity but can also lead to lower stability, due to the lack of protection against deactivation which is provided by $AlMe_3$ adduct formation. The overall effect of $AlMe_3$ present in MAO is therefore dependent on the catalyst and the polymerization conditions. Hafnocenes give low activity in the presence of $AlMe_3$, due to the formation of stable, low-activity species of type $[L_2Hf(\mu\text{-Me})_2AlMe_2]^+$, but give high activity with $AliBu_3/[Ph_3C][B(C_6F_5)_4]$. Deactivation via alkane elimination, giving $Zr-CH_2-Zr$ or $Zr-CH_2-Al$ species, is also important in zirconocene-catalyzed polymerization. The formation of such species may be avoided or reduced by the (partial) replacement of MAO or $AlMe_3$ by $AliBu_3$, due to the much lower tendency of an isobutyl as opposed to a methyl group to bridge between Zr and Al. In the case of borate-activated polymerization, for example, with systems such as $L_2ZrCl_2/[Ph_3C][B(C_6F_5)_4]/AliBu_3$, it is important to avoid precontact of the borate with the aluminum alkyl, which can lead to the formation of $Al-C_6F_5$ species and deactivation via transfer of the pentafluorophenyl group to Zr.

Reduction to lower, inactive or poorly active oxidation states is a dominant deactivation pathway for homogeneous titanium and vanadium catalysts. Titanocenes deactivate via reduction from the active tetravalent to the trivalent state. However, the relatively small ionic radius of Ti, compared to Zr, can lead to greater stability and therefore higher activity for half-sandwich complexes such as phosphinimides. Vanadium catalysts for ethene polymerization are generally trivalent complexes, which are rapidly reduced by reduction to divalent species. Polymerization is therefore carried out in the presence of a halocarbon promoter in addition to a cocatalyst such as $AlEt_2Cl$. The function of the promoter is to reoxidize divalent vanadium species to the active, trivalent state.

Decomposition of the active species through loss of the ligand has been noted for various catalysts, including phenoxy-imine complexes, vanadium diketonates and chromium triazacyclohexane and bis(imino)pyridyl complexes. In such cases, deactivation takes place via transmetallation reactions involving transfer of the ligand from the transition metal to aluminum.

An important consideration regarding catalyst stability in olefin polymerization is that the decay in activity frequently observed in homogeneous polymerization can, in many cases, be prevented by immobilization of the catalyst on a support. Furthermore, most polyolefin production processes require the use of heterogeneous, supported catalysts. Examples of the effects of catalyst immobilization on activity and stability are given in Chapter 4.

## References

1 Albizzati, E., Cecchin, G., Chadwick, J.C., Collina, G., Giannini, U., Morini, G., and Noristi, L. (2005) *Polypropylene Handbook*, 2nd edn (ed. N. Pasquini), Hanser Publishers, Munich, pp. 15–106.

2 McKenna, T.F. and Soares, J.B.P. (2001) *Chem. Eng. Sci.*, **56**, 3931–3949.

3 Abboud, M., Denifl, P., and Reichert, K.-H. (2005) *J. Appl. Polym. Sci.*, **98**, 2191–2200.

4 Floyd, S., Mann, G.E., and Ray, W.H. (1986) (eds T. Keii and K. Soga), *Catalytic Polymerization of Olefins*, Elsevier, Amsterdam, pp. 339–367.

5 Soga, K., Yanagihara, H., and Lee, D. (1989) *Makromol. Chem.*, **190**, 995–1006.

6 Tait, P.J.T. and Berry, I.G. (1994) (eds K. Soga and M. Terano), *Catalyst Design for Tailor-Made Polyolefins*, Elsevier, Amsterdam, pp. 55–72.

7 Kou, B., McCauley, K.B., Hsu, J.C.C., and Bacon, D.W. (2005) *Macromol. Mater. Eng.*, **290**, 537–557.

8 Zakharov, V.A., Bukatov, G.D., and Barabanov, A.A. (2004) *Macromol. Symp.*, **213**, 19–28.

9 Huang, R., Koning, C.E., and Chadwick, J.C. (2007) *Macromolecules*, **40**, 3021–3029.

10 Kissin, Y.V. (1985) *Isospecific Polymerization of Olefins*, Springer-Verlag, New York.

11 Keii, T., Suzuki, E., Tamura, M., Murata, M., and Doi, Y. (1982) *Makromol. Chem.*, **183**, 2285–2304.

12 Yermakov, Yu.I. and Zakharov, V.A. (1975) *Coordination Polymerization* (ed. J.C.W. Chien), Academic Press, New York, pp. 91–133.

13 Mejzlik, J., Lesna, M., and Majer, J. (1983) *Makromol. Chem.*, **184**, 1975–1985.

14 Busico, V., Guardasole, M., Margonelli, A., and Segre, A.L. (2000) *J. Am. Chem. Soc.*, **122**, 5226–5227.

15 Yaluma, A.K., Tait, P.J.T., and Chadwick, J.C. (2006) *J. Polym. Sci., Part A: Polym. Chem.*, **44**, 1635–1647.

16 Caunt, A.D. (1963) *J. Polym. Sci., Part C*, **4**, 49–69.

17 Goodall, B.L. (1990) *Polypropylene and other Polyolefins* (ed. S. van der Ven), Elsevier, Amsterdam, pp. 1–133.

18 Goodall, B.L. (1988) *Transition Metals and Organometallics as Catalysts for Olefin Polymerization* (eds W. Kaminsky and H. Sinn), Springer-Verlag, Berlin, pp. 361–370.

19 Chadwick, J.C. and Goodall, B.L. (1984) Eur. Patent 107871.

20 Bernard, A. and Fiasse, P. (1990) *Catalytic Olefin Polymerization* (eds T. Keii and K. Soga), Kodansha, Tokyo, pp. 405–423.

21 Kashiwa, N. (2004) *J. Polym. Sci., Part A: Polym. Chem.*, **42**, 1–8.

22 Giannini, U. (1981) *Makromol. Chem., Suppl.*, **5**, 216–229.

23 Galli, P., Luciani, L., and Cecchin, G. (1981) *Angew. Makromol. Chem.*, **94**, 63–89.

24 Barbè, P.C., Cecchin, G., and Noristi, L. (1987) *Adv. Polym. Sci.*, **81**, 1–81.

25 Busico, V., Causà, M., Cipullo, R., Credendino, R., Cutillo, F., Friederichs, N., Lamanna, R., Segre, A., and Van Axel Castelli, V. (2008) *J. Phys. Chem. C*, **112**, 1081–1089.

26 Andoni, A., Chadwick, J.C., Niemantsverdriet, J.W., and Thüne, P.C. (2008) *J. Catal.*, **257**, 81–86.

27 Zakharin, L.I. and Khorlina, I.M. (1962) *Tetrahedron Lett.*, **14**, 619–620.

28 Albizzati, E., Galimberti, M., Giannini, U., and Morini, G. (1991) *Macromol. Symp.*, **48/49**, 223–238.

29 Pasynkiewicz, S., Kozerski, L., and Grabowski, B. (1967) *J. Organometal. Chem.*, **8**, 233–238.

30 Goodall, B.L. (1983) *Transition Metal Catalyzed Polymerizations: Alkenes and Dienes* (ed. R.P. Quirk), Harwood Academic Publishers, New York, pp. 355–378.

31 Spitz, R., Lacombe, J.-L., and Primet, M. (1984) *J. Polym. Sci.: Polym. Chem. Ed.*, **22**, 2611–2624.

32 Kissin, Y.V. and Sivak, A.J. (1984) *J. Polym. Sci.: Polym. Chem. Ed.*, **22**, 3747–3758.

33 Tashiro, K., Yokoyama, M., Sugano, T., and Kato, K. (1984) *Contemp. Top. Polym. Sci.*, **4**, 647–662.

34 Doi, Y., Murata, M., Yano, K., and Keii, T. (1982) *Ind. Eng. Chem., Prod. Res. Dev.*, **21**, 580–585.

35 Busico, V., Corradini, P., Ferraro, A., and Proto, A. (1986) *Makromol. Chem.*, **187**, 1125–1130.

36 Chien, J.C.W., Weber, S., and Hu, Y. (1989) *J. Polym. Sci., Part A: Polym. Chem.*, **27**, 1499–1514.

37 Dusseault, J.J.A. and Hsu, C.C. (1993) *J. Appl. Polym. Sci.*, **50**, 431–447.

38 Spitz, R., Lacombe, J.L., and Guyot, A. (1984) *J. Polym. Sci.: Polym. Chem. Ed.*, **22**, 2625–2640, 2641–2650.

39 Parodi, S., Nocci, R., Giannini, U., Barbè, P.C., and Scatà, U. (1981) Eur. Patent 45977.

40 Seppälä, J.V., Härkönen, M., and Luciani, L. (1989) *Makromol. Chem.*, **190**, 2535–2550.

41 Härkönen, M., Seppälä, J.V., and Väänänen, T. (1990) *Catalytic Olefin Polymerization* (eds T. Keii and K. Soga), Elsevier, Amsterdam, pp. 87–105.

42 Proto, A., Oliva, L., Pellecchia, C., Sivak, A.J., and Cullo, L.A. (1990) *Macromolecules*, **23**, 2904–2907.

43 Spitz, R., Bobichon, C., and Guyot, A. (1989) *Makromol. Chem.*, **190**, 707–716.

44 Soga, K., Shiono, T., and Doi, Y. (1988) *Makromol. Chem.*, **189**, 1531–1541.

45 Sacchi, M.C., Forlini, F., Tritto, I., Mendichi, R., Zannoni, G., and Noristi, L. (1992) *Macromolecules*, **25**, 5914–5918.

46 Noristi, L., Barbè, P.C., and Baruzzi, G. (1991) *Makromol. Chem.*, **192**, 1115–1127.

47 Chadwick, J.C. (1995) *Ziegler Catalysts. Recent Scientific Innovations and Technological Improvements* (eds G. Fink, R. Mülhaupt, and H.H. Brintzinger), Springer-Verlag, Berlin, pp. 427–440.

48 Bukatov, G.D., Goncharov, V.S., Zakharov, V.A., Dudchenko, V.K., and Sergeev, S.A. (1994) *Kinet. Catal.*, **35**, 358–362.

49 Kojoh, S., Kioka, M., and Kashiwa, N. (1999) *Eur. Polym. J.*, **35**, 751–755.

50 Albizzati, E., Giannini, U., Morini, G., Smith, C.A., and Zeigler, R.C. (1995) *Ziegler Catalysts. Recent Scientific Innovations and Technological Improvements* (eds G. Fink, R. Mülhaupt, and H.H. Brintzinger), Springer-Verlag, Berlin, pp. 413–425.

51 Albizzati, E., Giannini, U., Morini, G., Galimberti, M., Barino, L., and Scordamaglia, R. (1995) *Macromol. Symp.*, **89**, 73–89.

52 Hammawa, H. and Wanke, S.E. (2007) *J. Appl. Polym. Sci.*, **104**, 514–527.

53 Hassan Nejad, M., Ferrari, P., Pennini, G., and Cecchin, G. (2008) *J. Appl. Polym. Sci.*, **108**, 3388–3402.

54 Guastalla, G. and Giannini, U. (1983) *Makromol. Chem., Rapid Commun.*, **4**, 519–527.

55 Pasquet, V. and Spitz, R. (1993) *Makromol. Chem.*, **194**, 451–461.

56 Kissin, Y.V., Mink, R.I., Nowlin, T.E., and Brandolini, A.J. (1999) *Top. Catal.*, **7**, 69–88.

57 Kissin, Y.V., Mink, R.I., Nowlin, T.E., and Brandolini, A.J. (1999) *J. Polym. Sci., Part A: Polym. Chem.*, **37**, 4255–4272.

58 Kissin, Y.V. and Brandolini, A.J. (1999) *J. Polym. Sci., Part A: Polym. Chem.*, **37**, 4273–4280.

59 Kissin, Y.V., Mink, R.I., Nowlin, T.E., and Brandolini, A.J. (1999) *J. Polym. Sci., Part A: Polym. Chem.*, **37**, 4281–4294.

60 Kissin, Y.V. (2002) *Macromol. Theory Simul.*, **11**, 67–76.

61 Garoff, T., Johansson, S., Pesonen, K., Waldvogel, P., and Lindgren, D. (2002) *Eur. Polym. J.*, **38**, 121–132.

62 Natta, G., Pino, P., Mazzanti, G., and Giannini, U. (1957) *J. Am. Chem. Soc.*, **79**, 2975–2976.

63 Breslow, D.S. and Newburg, N.R. (1957) *J. Am. Chem. Soc.*, **79**, 5073–5074.

64 Reichert, K.H. and Meyer, K.R. (1973) *Makromol. Chem.*, **169**, 163–176.

65 Long, W.P. and Breslow, D.S. (1975) *Liebigs Ann. Chem.*, **3**, 463–469.

66 Kaminsky, W. and Arndt, M. (1997) *Adv. Polym. Sci.*, **127**, 143–187.

67 Kaminsky, W. (2004) *J. Polym. Sci., Part A: Polym. Chem.*, **42**, 3911–3921.

68 Linnolahti, M., Severn, J.R. and Pakkanen, T.A. (2006) *Angew. Chem. Int. Ed.*, **45**, 3331–3334; (2008) *Angew. Chem. Int. Ed.*, **47**, 9279–9283.

69 Tritto, I., Méalares, C., Sacchi, M.C., and Locatelli, P. (1997) *Macromol. Chem. Phys.*, **198**, 3963–3977.

70 Chen, E.Y.-X. and Marks, T.J. (2000) *Chem. Rev.*, **100**, 1391–1434.

71 Brintzinger, H.H., Fischer, D., Mülhaupt, R., Rieger, B., and Waymouth, R.M. (1995) *Angew. Chem. Int. Ed.*, **34**, 1143–1170.

72 Corradini, P., Guerra, G., and Cavallo, L. (2004) *Acc. Chem. Res.*, **37**, 231–241.

73 Miller, S.A. and Bercaw, J.E. (2002) *Organometallics*, **21**, 934–945.

74 Resconi, L., Cavallo, L., Fait, A., and Piemontesi, F. (2000) *Chem. Rev.*, **100**, 1253–1346.

75 Chien, J.C.W. (1959) *J. Am. Chem. Soc.*, **81**, 86–92.

76 Bryliakov, K.P., Babushkin, D.E., Talsi,
E.P., Voskoboynikov, A.Z., Gritzo, H.,
Schröder, L., Damrau, H.-R.H., Wieser,
U., Schaper, F., and Brintzinger, H.H.
(2005) *Organometallics*, **24**, 894–904.

77 Ivanchev, S.S., Baulin, A.A., and
Rodionov, A.G. (1980) *J. Polym. Sci.:
Polym. Chem. Ed.*, **18**, 2045–2050.

78 Kaminsky, W. and Steiger, R. (1988)
*Polyhedron*, **7**, 2375–2381.

79 Kaminsky, W., Bark, A., and Steiger, R.
(1992) *J. Mol. Catal.*, **74**, 109–119.

80 Fischer, D. and Mülhaupt, R. (1991)
*J. Organometal. Catal.*, **417**, C7–C11.

81 Fischer, D., Jüngling, S., and Mülhaupt,
R. (1993) *Makromol. Chem., Macromol.
Symp.*, **66**, 191–202.

82 Srinivasa Reddy, S., Shashidhar, G., and
Sivaram, S. (1993) *Macromolecules*, **26**,
1180–1182.

83 Chien, J.C.W. and Wang, B.-P. (1988)
*J. Polym. Sci., Part A: Polym. Chem.*, **26**,
3089–3102.

84 Chien, J.C.W. and Sugimoto, R. (1991)
*J. Polym. Sci., Part A: Polym. Chem.*, **29**,
459–470.

85 Rieger, B. and Janiak, C. (1994) *Angew.
Macromol. Chem.*, **215**, 35–46.

86 Jüngling, S. and Mülhaupt, R. (1995)
*J. Organometal. Chem.*, **497**, 27–32.

87 Koltzenburg, S. (1997) *J. Mol. Catal. A:
Chem.*, **116**, 355–363.

88 Kleinschmidt, R., van der Leek, Y., Reffke,
M., and Fink, G. (1999) *J. Mol. Catal. A:
Chem.*, **148**, 29–41.

89 Kallio, K., Wartmann, A., and Reichert,
K.-H. (2002) *Macromol. Rapid Commun.*,
**23**, 187–190.

90 Pieters, P.J.J., van Beek, J.A.M., and van
Tol, M.F.H. (1995) *Macromol. Rapid
Commun.*, **16**, 463–467.

91 Coevoet, D., Cramail, H., and Deffieux, A.
(1998) *Macromol. Chem. Phys.*, **199**,
1451–1457 1459–1464.

92 Pédeutour, J.N., Coevoet, D., Cramail, H.,
and Deffieux, A. (1999) *Macromol. Chem.
Phys.*, **200**, 1215–1221.

93 Wang, Q., Song, L., Zhao, Y., and Feng, L.
(2001) *Macromol. Rapid Commun.*, **22**,
1030–1034.

94 Mäkelä-Vaarne, N.I., Linnolahti, M.,
Pakkanen, T.A., and Leskelä, M.A. (2003)
*Macromolecules*, **36**, 3854–3860.

95 Alonso-Moreno, C., Antiñolo, A.,
Carillo-Hermosilla, F., Carrión, P.,
Rodríguez, A.M., Otero, A., and Sancho,
J. (2007) *J. Mol. Catal. A: Chem.*, **261**,
53–63.

96 Lyakin, O.Y., Bryliakov, K.P., Panchenko,
V.N., Semikolenova, N.V., Zakharov, V.A.,
and Talsi, E.P. (2007) *Macromol. Chem.
Phys.*, **208**, 1168–1175.

97 Seraidaris, T., Löfgren, B., Mäkelä-
Vaarne, N., Lehmus, P., and Stehling, U.
(2004) *Macromol. Chem. Phys.*, **205**,
1064–1069.

98 Babushkin, D.E. and Brintzinger, H.H.
(2007) *Chem. Eur. J.*, **13**, 5294–5299.

99 Schröder, L., Brintzinger, H.H.,
Babushkin, D.E., Fischer, D., and
Mülhaupt, R. (2005) *Organometallics*, **24**,
867–871.

100 Bryliakov, K.P., Semikolenova, N.V.,
Yudaev, D.V., Zakharov, V.A., Brintzinger,
H.H., Ystenes, M., Rytter, E., and Talsi,
E.P. (2003) *J. Organometal. Chem.*, **683**,
92–102.

101 Busico, V., Cipullo, R., Cutillo, F.,
Friederichs, N., Ronca, S., and Wang, B.
(2003) *J. Am. Chem. Soc.*, **125**,
12402–12403.

102 Tynys, A., Eilertsen, J.L., and Rytter, E.
(2006) *Macromol. Chem. Phys.*, **207**,
295–303.

103 Bochmann, M. and Lancaster, S.J. (1993)
*Organometallics*, **12**, 633–640.

104 Bochmann, M. and Lancaster, S.J. (1994)
*Angew. Chem. Int. Ed.*, **33**, 1634–1637.

105 Talsi, E.P., Eilertsen, J.L., Ystenes, M., and
Rytter, E. (2003) *J. Organometal. Chem.*,
**677**, 10–14.

106 Beck, S., Prosenc, M.-H., Brintzinger,
H.-H., Goretzki, R., Herfert, N., and Fink,
G. (1996) *J. Mol. Catal. A: Chem.*, **111**,
67–79.

107 Bochmann, M., Cuenca, T., and Hardy,
D.T. (1994) *J. Organometal. Chem.*, **484**,
C10–C12.

108 Chien, J.C.W. and Xu, B. (1993) *Makromol.
Chem., Rapid Commun.*, **14**, 109–114.

109 Chien, J.C.W. and Tsai, W.-M. (1993)
*Makromol. Chem., Macromol. Symp.*, **66**,
141–156.

110 Chien, J.C.W., Song, W., and Rausch,
M.D. (1994) *J. Polym. Sci., Part A: Polym.
Chem.*, **32**, 2387–2393.

111 Götz, C., Rau, A., and Luft, G. (2002) *J. Mol. Catal. A: Chem.*, **184**, 95–110.

112 Beck, S. and Brintzinger, H.H. (1998) *Inorg. Chim. Acta*, **270**, 376–381.

113 Bochmann, M. and Sarsfield, M.J. (1998) *Organometallics*, **17**, 5908–5912.

114 Dornik, H.P., Luft, G., Rau, A., and Wieczorek, T. (2004) *Macromol. Mater. Eng.*, **289**, 475–479.

115 Yang, X., Stern, C.L., and Marks, T.J. (1994) *J. Am. Chem. Soc.*, **116**, 10015–10031.

116 Jia, L., Yang, X., Stern, C.L., and Marks, T.J. (1997) *Organometallics*, **16**, 842–857.

117 Arndt, P., Jäger-Fiedler, U., Klahn, M., Baumann, W., Spannenberg, A., Burlakov, V.V., and Rosenthal, U. (2006) *Angew. Chem. Int. Ed.*, **45**, 4195–4198.

118 Rieger, B., Troll, C., and Preuschen, J. (2002) *Macromolecules*, **35**, 5742–5743.

119 Bryliakov, K.P., Talsi, E.P., Voskoboynikov, A.Z., Lancaster, S.J., and Bochmann, M. (2008) *Organometallics*, **27**, 6333–6342.

120 Busico, V., Cipullo, R., Pellecchia, R., Talarico, G., and Razavi, A. (2009) *Macromolecules*, **42**, 1789–1791.

121 Britovsek, G.J.P., Gibson, V.C., and Wass, D.F. (1999) *Angew. Chem. Int. Ed.*, **38**, 429–447.

122 Gibson, V.C. and Spitzmesser, S.K. (2003) *Chem. Rev.*, **103**, 283–315.

123 Shapiro, P.J., Bunel, E.E., Schaefer, W.P., and Bercaw, J.E. (1990) *Organometallics*, **9**, 867–869.

124 McKnight, A.L. and Waymouth, R. (1998) *Chem. Rev.*, **98**, 2587–2598.

125 Chen, Y.-X. and Marks, T.J. (1997) *Organometallics*, **16**, 3649–3647.

126 Amor, F., Butt, A., du Plooy, K.E., Spaniol, T.P., and Okuda, J. (1998) *Organometallics*, **17**, 5836–5849.

127 Park, J.T., Yoon, S.C., Bae, B.-J., Seo, W.S., Suh, I.-H., Han, T.K., and Park, J.R. (2000) *Organometallics*, **19**, 1269–1276.

128 Ioku, A., Hasan, T., Shiono, T., and Ikeda, T. (2002) *Macromol. Chem. Phys.*, **203**, 748–755.

129 Song, F., Cannon, R.D., Lancaster, S.J., and Bochmann, M. (2004) *J. Mol. Catal. A: Chem.*, **218**, 21–28.

130 Wondimagegn, T., Xu, Z., Vanka, K., and Ziegler, T. (2004) *Organometallics*, **23**, 3847–3852.

131 Wondimagegn, T., Xu, Z., Vanka, K., and Ziegler, T. (2005) *Organometallics*, **24**, 2076–2085.

132 Sinnema, P.-J., Spaniol, T.P., and Okuda, J. (2000) *J. Organometal. Chem.*, **598**, 179–181.

133 Grassi, A., Zambelli, A., and Laschi, F. (1996) *Organometallics*, **15**, 480–482.

134 Nomura, K. and Fujii, K. (2003) *Macromolecules*, **36**, 2633–2641.

135 Nomura, K., Naga, N., Miki, M., and Yanagi, K. (1998) *Macromolecules*, **31**, 7588–7597.

136 Phomphrai, K., Fenwick, A.E., Sharma, S., Fenwick, P.E., Caruthers, J.M., Delgass, W.N., Abu-Omar, M.M., and Rothwell, I.P. (2006) *Organometallics*, **25**, 214–220.

137 Dove, A.P., Kiesewetter, E.T., Ottenwaelder, X., and Waymouth, R.M. (2009) *Organometallics*, **28**, 405–412.

138 Stephan, D.W., Stewart, J.C., Guérin, F., Courtenay, S., Kickham, J., Hollink, E., Beddle, C., Hoskin, A., Graham, T., Wei, P., Spence, R.E.v.H., Xu, W., Koch, L., Gao, X., and Harrison, D.G. (2003) *Organometallics*, **22**, 1937–1947.

139 Beddle, C., Hollink, E., Wei, P., Gauld, J., and Stephan, D.W. (2004) *Organometallics*, **23**, 5240–5251.

140 Yue, N.L.S. and Stephan, D.W. (2001) *Organometallics*, **20**, 2303–2308.

141 Kickham, J.F., Guérin, F., Stewart, J.C., Urbanska, E., and Stephan, D.W. (2001) *Organometallics*, **20**, 1175–1182.

142 Yue, N., Hollink, E., Guérin, F., and Stephan, D.W. (2001) *Organometallics*, **20**, 4424–4433.

143 Zhang, S., Piers, W.E., Gao, X., and Parvez, M. (2000) *J. Am. Chem. Soc.*, **122**, 5499–5509.

144 Zhang, S. and Piers, W.E. (2001) *Organometallics*, **20**, 2088–2092.

145 Matsui, S., Mitani, M., Saito, J., Tohi, Y., Makio, H., Matsukawa, N., Takagi, Y., Tsuru, K., Nitabaru, M., Nakano, T., Tanaka, H., Kashiwa, N., and Fujita, T. (2001) *J. Am. Chem. Soc.*, **123**, 6847–6856.

146 Makio, H., Kashiwa, N., and Fujita, T. (2002) *Adv. Synth. Catal.*, **344**, 477–493.

147 Mitani, M., Saito, J., Ishii, S.-I., Nakayama, Y., Makio, H., Matsukawa, N., Matsui, S., Mohri, J.-I., Furuyama, R., Terao, H., Bando, H., Tanaka, H., and Fujita, T. (2004) *The Chemical Record*, **4**, 137–158.

148 Makio, H. and Fujita, T. (2004) *Macromol. Symp.*, **213**, 221–223.

149 Kravtsov, E.A., Bryliakov, K.P., Semikolenova, N.V., Zakharov, V.A., and Talsi, E.P. (2007) *Organometallics*, **26**, 4810–4815.

150 Matsukawa, N., Matsui, S., Mitani, M., Saito, J., Tsuru, K., Kashiwa, N., and Fujita, T. (2001) *J. Mol. Catal.*, **169**, 99–104.

151 Ishii, S.-I., Saito, J., Mitani, M., Mohri, J.-I., Matsukawa, N., Tohi, Y., Matsui, S., Kashiwa, N., and Fujita, T. (2002) *J. Mol. Catal.*, **179**, 11–16.

152 Saito, J., Mitani, M., Mohri, J.I.-, Yoshida, Y., Matsui, S., Ishii, S.-I., Kojoh, S.-I., Kashiwa, N., and Fujita, T. (2001) *Angew. Chem. Int. Ed.*, **40**, 2918–2920.

153 Mitani, M., Mohri, J.-I., Yoshida, Y., Saito, J., Ishii, S., Tsuru, K., Matsui, S., Furuyama, R., Nakano, T., Tanaka, H., Kojoh, S.-I., Matsugi, T., Kashiwa, N., and Fujita, T. (2002) *J. Am. Chem. Soc.*, **124**, 3327–3336.

154 Furuyama, R., Saito, J., Ishii, S., Makio, H., Mitani, M., Tanaka, H., and Fujita, T. (2005) *J. Organometal. Chem.*, **690**, 4398–4413.

155 Reinartz, S., Mason, A.F., Lobkovsky, E.B., and Coates, G.W. (2003) *Organometallics*, **22**, 2542–2544.

156 Busico, V., Talarico, G., and Cipullo, R. (2005) *Macromol. Symp.*, **226**, 1–16.

157 Tshuva, E.Y., Goldberg, I., and Kol, M. (2000) *J. Am. Chem. Soc.*, **122**, 10706–10707.

158 Busico, V., Cipullo, R., and Ronca, S. (2001) *Macromol. Rapid Commun.*, **22**, 1405–1410.

159 Busico, V., Cipullo, R., Friederichs, N., Ronca, S., and Togrou, M. (2003) *Macromolecules*, **36**, 3806–3808.

160 Busico, V., Cipullo, R., Friederichs, N., Ronca, S., Talarico, G., Togrou, M., and Wang, B. (2004) *Macromolecules*, **37**, 8201–8203.

161 Scollard, J.D., McConville, D.H., Payne, N.C., and Vittal, J.J. (1996) *Macromolecules*, **29**, 5241–5243.

162 Scollard, J.D. and McConville, D.H. (1996) *J. Am. Chem. Soc.*, **118**, 10008–10009.

163 Mitani, M., Oouchi, K., Hayakawa, M., Yamada, T., and Mukaiyama, T. (1995) *Polym. Bull.*, **34**, 199–202.

164 Evens, G.G., Pijpers, E.M.J., and Seevens, R.H.M. (1988) *Transition Metal Catalyzed Polymerizations* (ed. R.P. Quirk), Cambridge University Press, Cambridge, pp. 782–798.

165 Hagen, H., Boersma, J., and van Koten, G. (2002) *Chem. Soc. Rev.*, **31**, 357–364.

166 Gambarotta, S. (2003) *Coord. Chem. Rev.*, **237**, 229–243.

167 D'Agnillo, L., Soares, J.B.P., and van Doremaele, G.H.J. (2005) *Macromol. Mater. Eng.*, **290**, 256–271.

168 Adisson, E., Deffieux, A., and Fontanille, M. (1993) *J. Polym. Sci., Part A: Polym. Chem.*, **31**, 831–839.

169 Adisson, E., Deffieux, A., Fontanille, M., and Bujadoux, K. (1994) *J. Polym. Sci., Part A: Polym. Chem.*, **32**, 1033–1041.

170 Doi, Y., Ueki, S., and Keii, T. (1979) *Macromolecules*, **12**, 814–819.

171 Doi, Y., Suzuki, S., and Soga, K. (1986) *Macromolecules*, **19**, 2896–2900.

172 Zambelli, A., Sessa, I., Grisi, F., Fusco, R., and Accomazzi, P. (2001) *Macromol. Rapid Commun.*, **22**, 297–310.

173 Ver Strate, G., Cozewith, C., West, R.K., Davis, W.M., and Capone, G.A. (1999) *Macromolecules*, **32**, 3837–3850.

174 Ma, Y., Reardon, D., Gambarotta, S., Yap, G., Zahalka, H., and Lemay, C. (1999) *Organometallics*, **18**, 2773–2781.

175 Desmangles, N., Gambarotta, S., Bensimon, C., Davis, S., and Zahalka, H. (1998) *J. Organometal. Chem.*, **562**, 53–60.

176 Cuomo, C., Milione, S., and Grassi, A. (2006) *J. Polym. Sci.: Part A, Polym. Chem.*, **44**, 3279–3289.

177 Brandsma, M.J.R., Brussee, E.A.C., Meetsma, A., Hessen, B., and Teuben, J.H. (1998) *Eur. J. Inorg. Chem.*, 1867–1870.

178  Białek, M. and Czaja, K. (2008) *J. Polym. Sci., Part A: Polym. Chem.*, **46**, 6940–6949.

179  Nomura, K., Sagara, A., and Imanishi, Y. (2002) *Macromolecules*, **35**, 1583–1590.

180  Wang, W. and Nomura, K. (2005) *Macromolecules*, **38**, 5905–5913.

181  Wang, W. and Nomura, K. (2006) *Adv. Synth. Catal.*, **348**, 743–750.

182  Wu, J.-Q., Pan, L., Li, Y.-G., Liu, S.-R., and Li, Y.-S. (2009) *Organometallics*, **28**, 1817–1825.

183  Redshaw, C., Warford, L., Dale, S.H., and Elsegood, M.R.J. (2004) *Chem. Commun.*, 1954–1955.

184  Redshaw, C., Rowan, M.A., Warford, L., Homden, D.M., Arbaoui, A., Elsegood, M.R.J., Dale, S.H., Yamato, T., Casas, C.P., Matsui, S., and Matsuura, S. (2007) *Chem. Eur. J.*, **13**, 1090–1107.

185  Tomov, A.K., Gibson, V.C., Zaher, D., Elsegood, M.R.J., and Dale, S.H. (2004) *Chem. Commun*, 1956–1957.

186  Jabri, A., Korobkov, I., Gambarotta, S., and Duchateau, R. (2007) *Angew. Chem. Int. Ed.*, **46**, 6119–6122.

187  Reardon, D., Conan, F., Gambarotta, S., Yap, G., and Wang, Q. (1999) *J. Am. Chem. Soc.*, **121**, 9318–9325.

188  Huang, R., Kukalyekar, N., Koning, C.E., and Chadwick, J.C. (2006) *J. Mol. Catal. A: Chem.*, **260**, 135–143.

189  McDaniel, M.P. (1985) *Adv. Catal.*, **33**, 47–98.

190  Groppo, E., Lamberti, C., Bordiga, S., Spoto, G., and Zecchina, A. (2005) *Chem. Rev.*, **105**, 115–183.

191  Fang, Y., Liu, B., Hasebe, K., and Terano, M. (2005) *J. Polym. Sci., Part A: Polym. Chem.*, **43**, 4632–4641.

192  van Kimmenade, E.M.E., Loos, J., Niemantsverdriet, J.W., and Thüne, P.C. (2006) *J. Catal.*, **210**, 39–16.

193  Theopold, K.H. (1998) *Eur. J. Inorg. Chem.*, 15–24.

194  Köhn, R.D., Haufe, M., Mihan, S., and Lilge, D. (2000) *Chem. Commun.*, 1927–1928.

195  Köhn, R.D., Smith, D., Mahon, M.F., Prinz, M., Mihan, S., and Kociok-Köhn, G. (2003) *J. Organometal. Chem.*, **683**, 200–208.

196  Döhring, A., Göhre, J., Jolly, P.W., Kryger, B., Rust, J., and Verhovnik, G.P.J. (2000) *Organometallics*, **19**, 388–402.

197  Int. Patent WO 98/04570 (1998) Studiengesellschaft Kohle m.b.H., invs.; Jolly, P.W., Jonas, K., Verhovnik, G.P.J., Döring, A., Göhre, J., and Weber, J.C. (1998) *Chem. Abstr.*, **128**, 167817v.

198  Zhang, H., Ma, J., Qian, Y., and Huang, J. (2004) *Organometallics*, **23**, 5681–5688.

199  Enders, M., Fernández, P., Ludwig, G., and Pritzkow, H. (2001) *Organometallics*, **20**, 5005–5007.

200  Enders, M., Kohl, G., and Pritzkow, H. (2004) *Organometallics*, **23**, 3832–3839.

201  Derlin, S. and Kaminsky, W. (2008) *Macromolecules*, **41**, 6280–6288.

202  Esteruelas, M.A., López, A.M., Méndez, L., Oliván, M., and Oñate, E. (2003) *Organometallics*, **22**, 395–406.

203  Small, B.L., Carney, M.J., Holman, D.M., O'Rourke, C.E., and Halfen, J.A. (2004) *Macromolecules*, **37**, 4375–4386.

204  Crewdson, P., Gambarotta, S., Djoman, M.-C., Korobkov, I., and Duchateau, R. (2005) *Organometallics*, **24**, 5214–5216.

205  Vidyaratne, I., Scott, J., Gambarotta, S., and Duchateau, R. (2007) *Organometallics*, **26**, 3201–3211.

206  Sugiyama, H., Aharonian, G., Gambarotta, S., Yap, G.P.A., and Budzelaar, P.H.M. (2002) *J. Am. Chem. Soc.*, **124**, 12268–12274.

# 3
# Late Transition Metal Catalysts for Olefin Polymerization

## 3.1
## Nickel- and Palladium-based Catalysts

### 3.1.1
### Diimine Complexes

Interest in ethene polymerization with late transition metal catalysts received a large boost following the discovery in 1995 by Brookhart and coworkers of cationic Ni(II) and Pd(II) α-diimine catalysts [1]. A typical example of a nickel diimine catalyst precursor is shown in Figure 3.1. The ortho-substituents in the aryl rings, which lie roughly perpendicular to the square plane, block the axial approach of olefins, thereby retarding the rate of associative displacement and chain transfer illustrated in Scheme 3.1. High molecular weight polymers are therefore accessible with these systems, as opposed to the dimers/oligomers typically formed using nickel catalysts.

Special features of nickel and palladium diimine catalysts are lower oxophilicity than early transition metal catalysts, allowing the copolymerization of ethene with polar monomers such as acrylates [2], and the formation of polyethylenes with substantial chain branching. The formation of methyl and longer branches takes place via a process of chain walking (Scheme 3.2), analogous to that first described by Fink [3]. The extent of branching increases with increasing temperature and decreasing ethene pressure and is also dependent on the catalyst structure. Reducing the steric bulk of the diimine ligand by replacing the ortho-isopropyl groups by ortho methyl groups results in a less branched, more linear polymer with decreased molecular weight [1, 4].

An increase in temperature from 35 to 60 °C in nickel-catalyzed ethene polymerization was found to lead to decreased activity as a result of catalyst decay. At 85 °C, the activity decreased sharply [4]. C−H activation of an ortho-alkyl substituent was considered as a possible deactivation pathway, although not supported by the fact that similar deactivation was observed with an ortho-$C_6F_5$ substituent in each aryl ring.

The above polymerizations were carried out using MAO as cocatalyst, but it has also been shown that $AlEt_2Cl$ is an efficient activator for nickel catalysts [5, 6].

*Homogeneous Catalysts: Activity – Stability – Deactivation*, First Edition. Piet W.N.M. van Leeuwen and John C. Chadwick.
© 2011 Wiley-VCH Verlag GmbH & Co. KGaA. Published 2011 by Wiley-VCH Verlag GmbH & Co. KGaA.

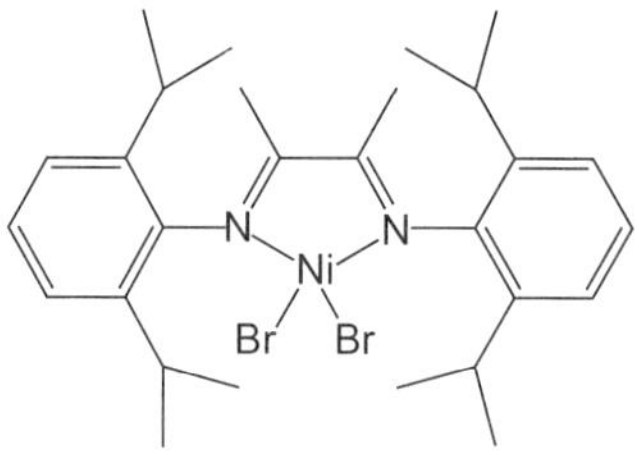

**Figure 3.1** 2,3-Bis(2,6-diisopropylphenylimino)butane nickel(II) dibromide.

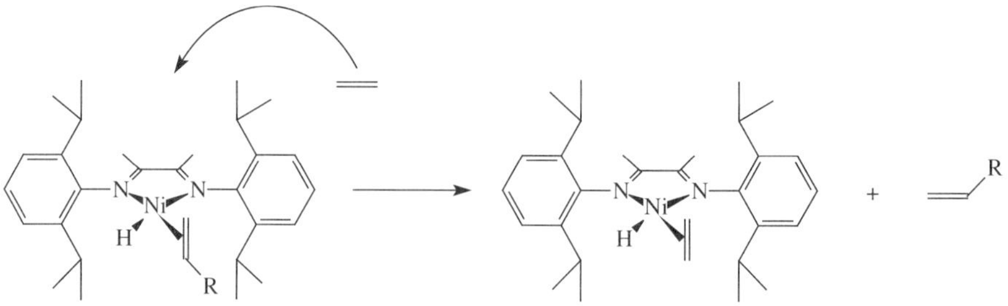

**Scheme 3.1** Chain transfer by associative exchange, retarded by steric blocking of the axial positions relative to the metal center.

Decay-type kinetics were observed at 25 °C with both AlEt$_2$Cl and with MAO and the rate of decay increased with decreasing Al/Ni ratio [6]. A borate activator, NaB[3,5-(CF$_3$)$_2$C$_6$H$_3$]$_4$ was used in a study of the effect of electron-donating and -withdrawing substituents in Pd(II) α-diimine catalysts [7]. The more electron-deficient catalysts gave less chain branching and were observed to be less thermally stable and less

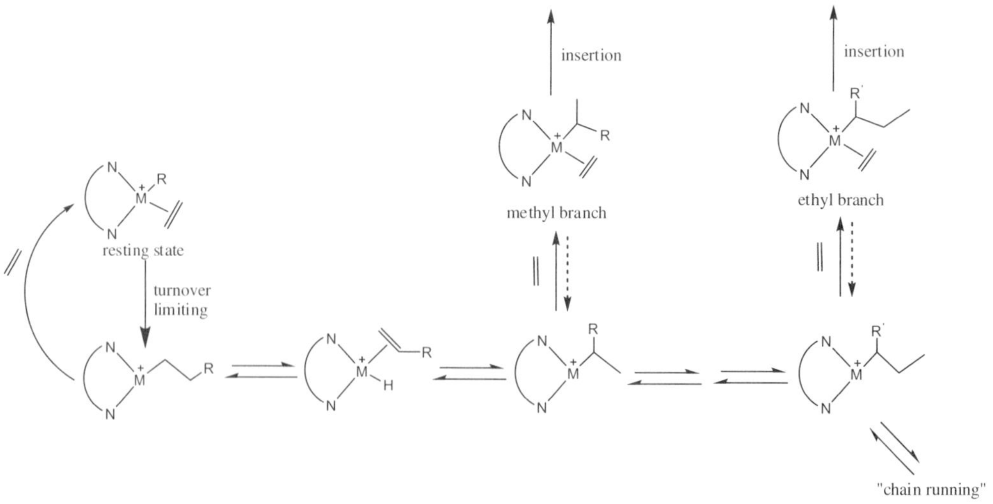

**Scheme 3.2** Polyethylene branch formation with Ni and Pd α-diimine complexes.

tolerant to the presence of a polar comonomer. The lesser tendency of more electron-rich complexes to bind to electronegative groups such as ester functionalities facilitates the copolymerization of ethene with a monomer such as methyl acrylate. The cationic palladium systems are less electrophilic and thus less sensitive to protic solvents and comonomer functional groups than their nickel analogues.

At low temperature (around $-10$ to $+5\,°C$), Ni(II) and Pd(II) α-diimine catalysts can initiate the living polymerization of α-olefins [8–12]. However, the living character was lost at temperatures higher than ambient, as a result of chain transfer and deactivation reactions. Guan and coworkers found, however, that a cyclophane-based nickel α-diimine catalyst was more stable and discovered that complex 1, activated by MMAO, gave living polymerization of propene at a temperature as high as $50\,°C$ [13, 14]. The improved performance at high temperature was attributed to the cyclophane framework completely blocking the axial faces of the metal, leaving only two cis-coordination sites for monomer entry and polymer chain growth. The lack of rotational flexibility prevents deactivation via C–H activation, as was proposed to occur with the 2,6-di-isopropyl substituted ligand of the acyclic catalyst shown in Figure 3.1. Brookhart and coworkers have reported decomposition of an ether adduct of a cationic Pd(II) α-diimine complex via intramolecular C–H activation of an ortho-substituent on the aryl ring, as illustrated in Scheme 3.3, but commented that it was not clear whether such decomposition pathways are responsible for catalyst decay [15]. Recently, Guan and coworkers have reported particularly high stability for Ni(II) catalysts bearing a fluorinated cyclophane ligand [16]. Little loss in activity was observed during 70 min ethylene polymerization at $105\,°C$ and the high thermal stability was attributed to stabilization of the reactive $14\ e^-$ intermediate 2 by donation of a fluorine lone pair to the metal center. The presence of fluorine in the ligand also had an inhibiting effect on chain walking, most probably by reducing the rate of β-hydride elimination, leading to relatively low branching density in the resulting polyethylene.

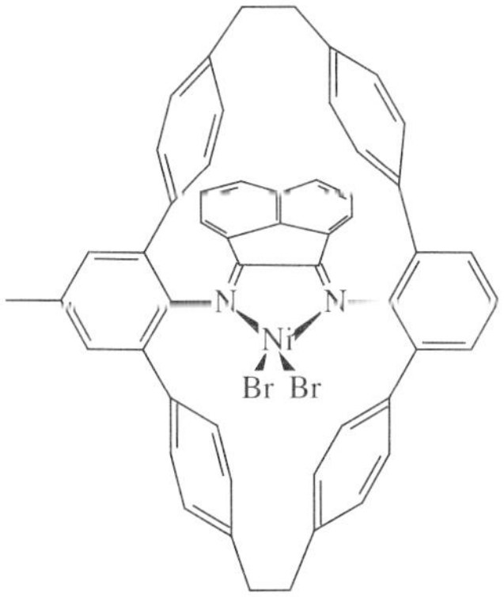

1 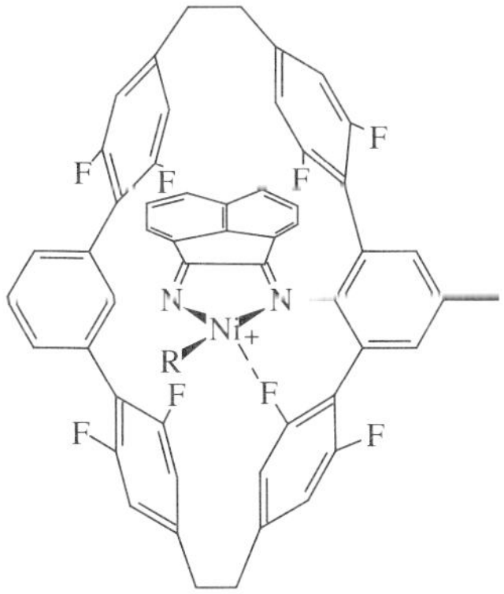                              2

Cramail and coworkers have used UV–visible spectroscopy to study activation and deactivation in nickel diimine/MAO systems [17]. Contact of the Ni(II) diimine 3

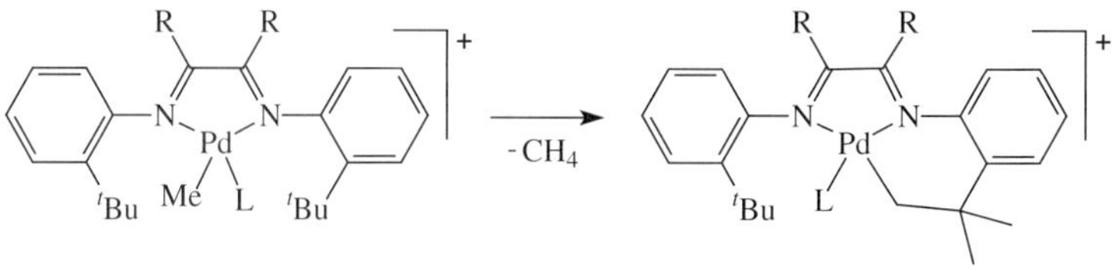

**Scheme 3.3** Intramolecular C−H activation in a cationic palladium α-diimine complex [15].

with MAO in toluene gave absorption bands at 480–510 and 710 nm. On standing, the intensity of the 710 nm band increased at the expense of the absorption in the range 480–510 nm. These changes were accompanied by a rapid deactivation of the catalytic system and precipitation of a black solid, and were attributed to structural modification of the soluble complex as well as to partial reduction to Ni(0). Deactivation was less rapid in the presence of 1-hexene, indicating a stabilizing effect of α-olefin coordination to the active species. In agreement with the previous reports of living polymerization at low temperature, the UV–vis spectrum remained unchanged over several days at −10 °C, in contrast to the changes observed at 20 °C. Evidence for deactivation via reduction of active Ni(II) species to a lower oxidation state is apparent from the patent literature, where it has been reported that significant increases in the productivity of both homogeneous and supported nickel(II) diimine complexes can be obtained by polymerization in the presence of an oxidizing agent such as iodine or an active halocarbon [18]. This approach resembles the use of halocarbon oxidizing agents for reactivation of reduced vanadium species in olefin polymerization, described in Chapter 2.

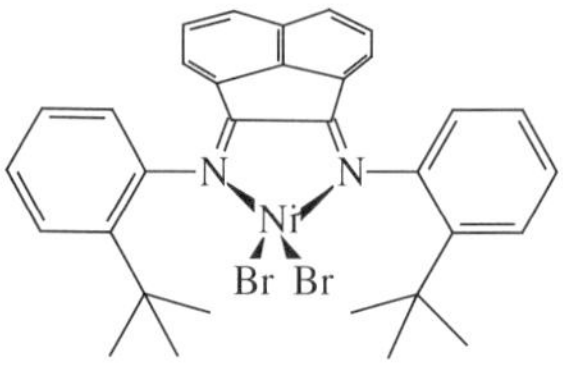

**3**

## 3.1.2
## Neutral Nickel(II) Complexes

In view of the fact that cationic nickel systems are more electrophilic and, therefore, more sensitive to polar impurities and functional groups than the palladium systems, there has been significant interest in the development of neutral nickel complexes. Neutral salicylaldiminato Ni(II) complexes of type 4 have been synthesized by Grubbs and coworkers [19]. It was proposed that the introduction of bulky substituents on the ketimine nitrogen and the phenolic ring might block the axial faces of the metal center, retarding associative displacement, and decrease the rate of catalyst deactivation. Ethene polymerization was accomplished in the presence of either $Ni(COD)_2$ or $B(C_6F_5)_3$, which act as phosphine scavengers. Subsequent studies revealed that

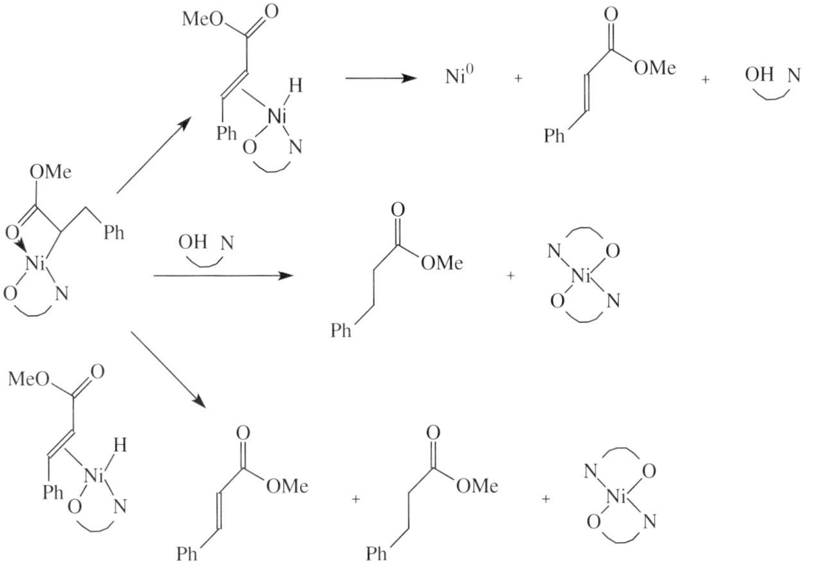

**Scheme 3.4** Deactivation of a neutral nickel complex in the presence of a functionalized olefin [21].

high-activity single-component catalysts could be obtained by replacement of PPh$_3$ by the more labile ligand acetonitrile [20]. These neutral systems were also found to have high tolerance to functional groups, high activity being maintained in the presence of ethers, ketones and esters.

The deactivation of a neutral nickel(II) ethene polymerization catalyst **5** in the presence of methyl acrylate has been proposed to proceed as illustrated in Scheme 3.4, starting from the nickel enolate product of a 2,1-insertion of methyl acrylate into the Ni–Ph bond [21].

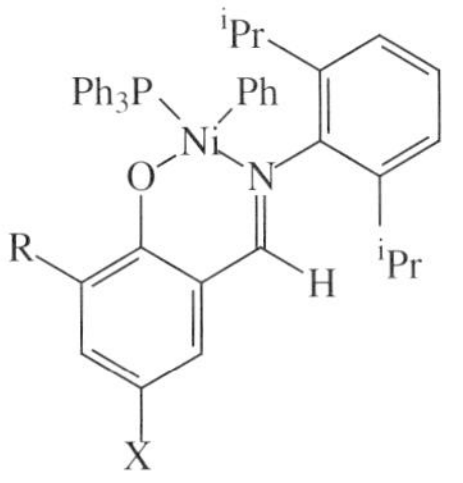

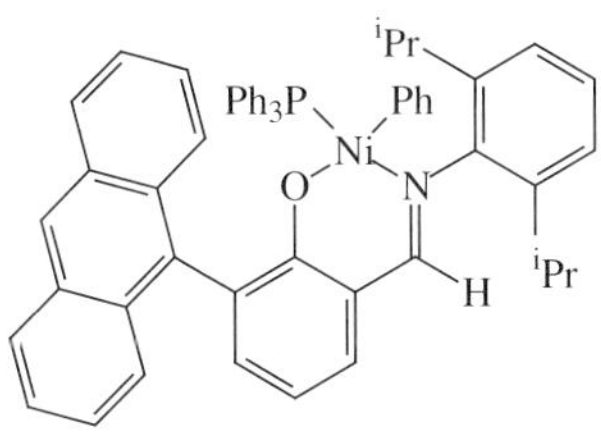

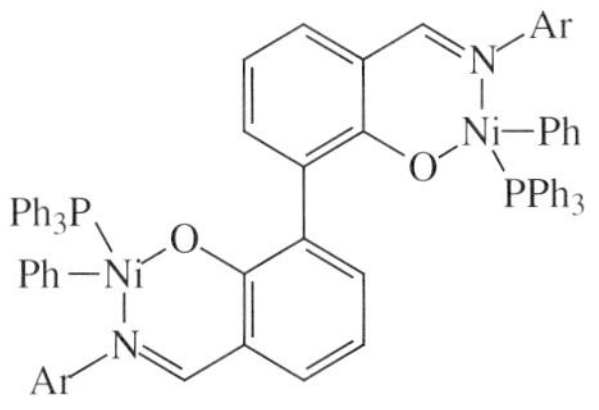

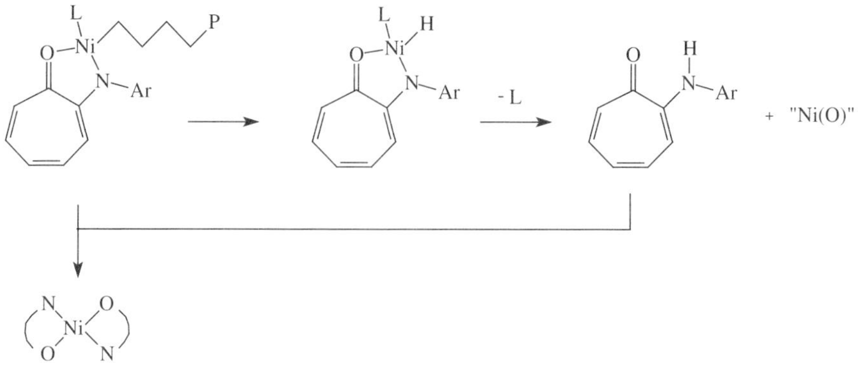

**Scheme 3.5** Decomposition via nickel hydride formation and reductive elimination [26].

Li and coworkers have found that the binuclear complex **6** (Ar = 2,6-$i$Pr$_2$C$_6$H$_3$) can be used as a single-component catalyst for ethene polymerization, giving polyethylene containing mainly methyl branches [22]. At 43 °C, polymerization activity remained stable over a period of 2 h. The activities were higher than those obtained with mononuclear complexes of type **4** and **5**, and it was noted that each salicylaldimine unit in **6** plays the role of a bulky group at the C-3 position of the other unit.

Neutral nickel complexes of type **7** and **8**, containing a five- rather than a six-membered chelate ring, have been synthesized by Brookhart and coworkers [23–26]. The anilinotropone-based catalyst (**7**; Ar = 2,6-$i$Pr$_2$C$_6$H$_3$) exhibited a high initial activity but also rapid decay in ethene polymerization at 80 °C. This catalyst did not require the presence of a phosphine scavenger; addition of further PPh$_3$ lowered the activity but increased the lifetime of the catalyst. The fate of the catalyst upon deactivation was determined by quenching the polymerization with methanol and isolating the decomposition product, which was found to be the bis(anilinotropone) Ni(II) complex **9**. Its formation was explained by reductive elimination from a nickel hydride intermediate to form free ligand, followed by attack of the free ligand on a Ni(II) species present in the catalytic cycle, as illustrated in Scheme 3.5 [25, 26]. More stable kinetics was observed in ethene polymerization with the anilinoperinaphthenone complex **8**. At 80 °C, the catalyst half-life was approximately 20–30 min [24]. Analysis of the deactivation products revealed only the free ligand. The weaker acidity of the anilinoperinaphthenone N–H proton was proposed to explain the slow rate of bis-ligand Ni complex formation in this system.

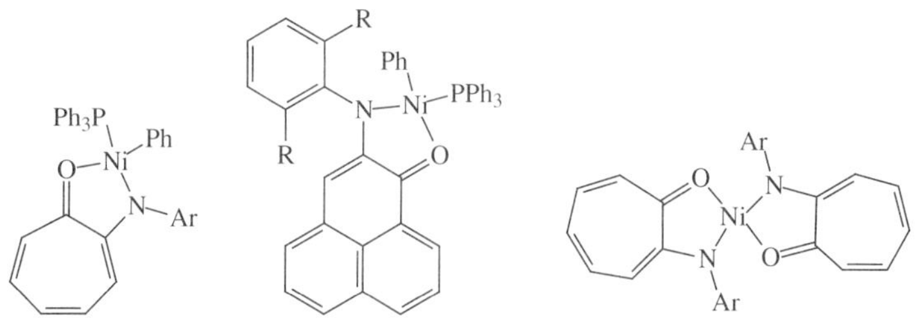

7          8          9

Further development of neutral Ni(II) catalysts by the Brookhart group led to the synthesis of complexes of type **10**, derived from aniline-substituted enone ligands bearing trifluoromethyl and trifluoroacetyl substituents [27]. To initiate polymerization, ethene must replace $PPh_3$ to yield a Ni(Ph)ethene complex, which then undergoes migratory insertion to generate the growing polymer chain. With this complex, the replacement of $PPh_3$ by ethene was greatly enhanced by the addition of $Ni(COD)_2$ or $B(C_6F_5)_3$. Deactivation during polymerization was apparent at 60 °C, but at 35 °C the catalyst was remarkably stable.

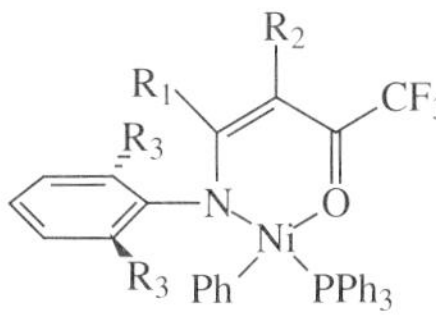

**10** 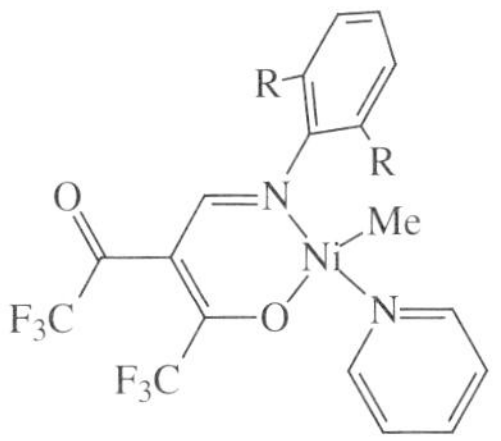                                          **11**

Neutral nickel(II) catalysts with high stability in olefin polymerization have been reported by Mecking and coworkers [28–30]. Complex **11** showed almost constant activity over a period of 4 h in ethene polymerization at 60 °C in toluene. Despite the electron-deficient nature of the metal center, resulting from the presence of the strongly electron-withdrawing $CF_3$ and $CF_3CO$ groups, the catalyst stability was also sufficient to allow polymerization in an aqueous emulsion, albeit that activity was reduced 5–10-fold. Most recently, deactivation pathways of neutral nickel(II) polymerization catalysts have been investigated using the dimethyl sulfoxide (DMSO)-coordinated complex **12** [31]. It was shown that, after dissociation of DMSO, [N,O]Ni(II)−$CH_3$ species can undergo bimolecular deactivation, generating ethane, and that higher alkyl species [N,O]Ni(II)−R and the hydride [N,O]Ni(II)−H (both of which are intermediates in the catalytic cycle) react to give the alkane RH. This bimolecular alkane elimination represented the major deactivation pathway for the system investigated. It was observed that bimolecular deactivation should be less prevalent in systems with sterically demanding substituents in the (N,O) ligand framework. Also, no decomposition of the complex was observed when DMSO was replaced by the more strongly coordinating pyridine.

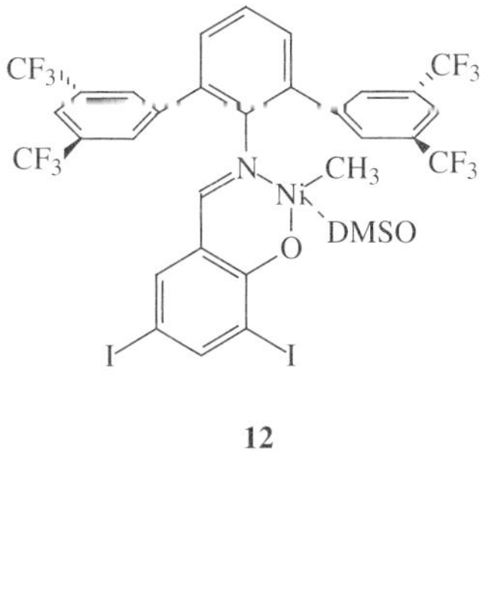

**12**

### 3.1.3
### Other Nickel(II) and Palladium(II) Complexes

Ni(II) and Pd(II) catalysts [13; $Ar_f = 3,5\text{-}(CF_3)_2C_6H_3$] based on a bidentate P,O chelating ligand have also been investigated [32]. These complexes are cationic analogues of the well-known SHOP-type systems for ethene oligomerization, which can be modified to give high molecular weight polyethylene as well as being effective for the copolymerization of ethene with polar monomers and carbon monoxide [33–35]. A complex in which $Ar = Ar' = 2,4,6\text{-}Me_3C_6H_2$ gave polyethylene, whereas replacement of one or both of these groups by Ph gave predominantly 1-butene, illustrating the importance of steric bulk in retarding chain transfer. However, these catalysts had poor thermal stability and exhibited rapid decay.

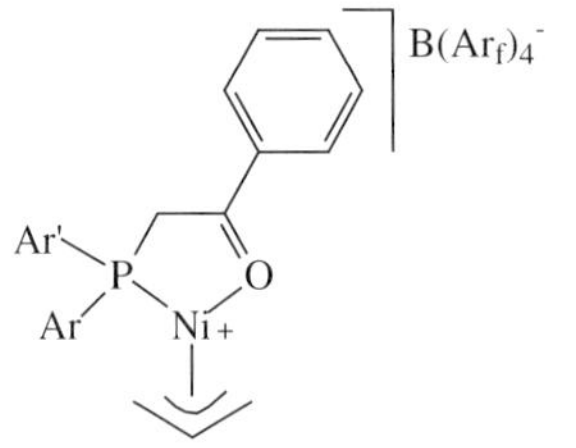

**13**

Nickel catalysts containing a PNP-type ligand and showing relatively high resistance to traditional poisons for olefin polymerization catalysts have been described by Wass and coworkers [36]. The activity of these catalysts, of type $[Ar_2PN(Me)PAr_2]$ $NiBr_2$, in which Ar represents a 2-substituted phenyl group, was found to be dependent on the $AlMe_3$ content of the MAO used as activator. The presence of free $AlMe_3$ deactivated the catalyst. A similar deactivation effect of $AlMe_3$ has been reported by Carlini *et al.*, using a bis(salicylaldiminate) nickel(II)-based catalyst [37]. The deactivating effect of $AlMe_3$ was attributed to reduction of the transition metal. Increased activity was obtained by reducing the content of $AlMe_3$ in MAO by reaction with 2,6-di-*tert*-butylphenol.

### 3.2
### Iron- and Cobalt-based Catalysts

### 3.2.1
### Bis(imino)Pyridyl Complexes

Following the advent of the nickel α-diimine complexes, a second major advance in the area of late transition metal catalysts for ethene polymerization took place in 1998, when highly active bis(imino)pyridyl iron catalysts were discovered independently by the Brookhart and Gibson groups [38, 39]. One of the most active precatalysts of this family is complex **14**. In contrast to the nickel and palladium complexes, these catalysts produce highly linear, high-density polyethylene, the molecular weight of which is dependent on the steric bulk of the substituents present in the imino-aryl rings. For example, the 2,6-diisopropyl substituted complex **15** gives higher molecular weight polyethylene than **14**, whereas the presence of only a single ortho-substituent in each aryl ring, as in complex **16**, leads to oligomerization and the formation of linear α-olefins [40]. The effect on polyethylene molecular weight of steric bulk in the *ortho*-aryl position is ascribed to restricted rotation around the nitrogen–aryl bond [41]. Bis(imino)pyridyl cobalt complexes are generally less active and give lower molecular weights than their iron analogues [42], but cobalt complexes having an *ortho*-CF$_3$ substituent in each aryl have given oligomerization activities matching those of iron systems [43]. Longer catalyst lifetimes were observed, suggesting that the trifluoromethyl group not only increases the electrophilicity of the metal center but also improves catalyst stability.

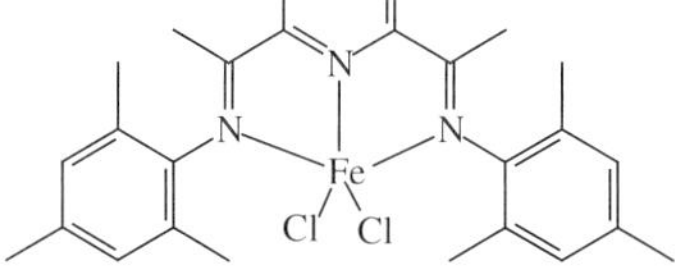

**14** 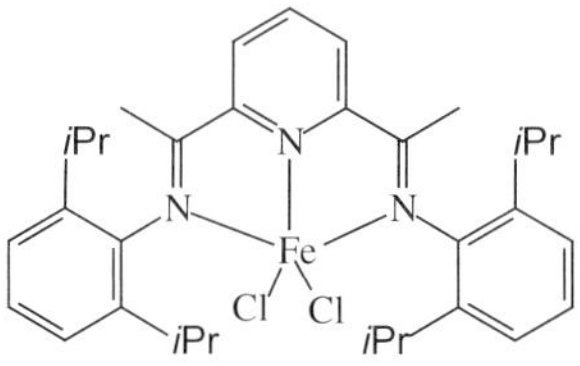                    **15**

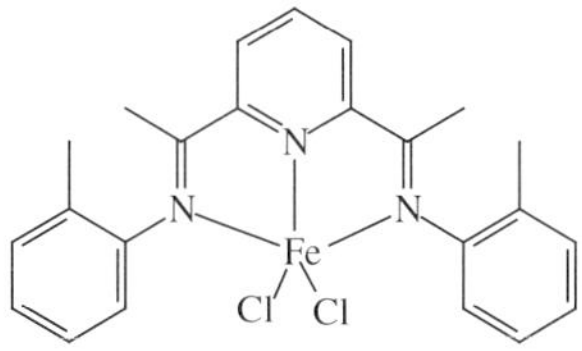

**16**

In contrast to the great majority of homogeneous catalysts for olefin polymerization, the above complexes typically give polyethylene with relatively broad molecular weight distribution. In many cases a bimodal distribution is obtained and evidence has been presented that, for systems activated with MAO, this is caused by the formation of a low molecular weight fraction resulting from chain transfer to aluminum, particularly in the early stages of polymerization [42]. However, chain transfer to aluminum is not the only reason for the broad polydispersities obtained with these systems. Strong evidence for the presence of different active species has been provided by Barabanov *et al.* [44], who used [14]CO radiotagging to determine the numbers of active centers and propagation rate constants in homogeneous polymerization. The results obtained indicated the presence of highly reactive but unstable centers producing a low molecular weight polymer fraction, as well as less active but more stable species producing higher molecular weight polymer. Iron-based precatalysts can be activated not only by MAO but also by common aluminum alkyls such as AlEt$_3$ and AliBu$_3$ [6, 45]. It has been reported that polyethylene with narrow molecular weight distribution can be obtained using Ali Bu$_3$ [6, 46] or $i$Bu$_2$AlOAli Bu$_2$ [47] as cocatalyst.

Unless immobilized on a suitable support (see Chapter 4), bis(imino)pyridyl iron catalysts deactivate relatively rapidly during the course of polymerization. At 35 °C, catalyst activity has been observed to decrease between 3- and 5-fold within 10 min [44], and increasing the polymerization temperature from 35 to 70 °C results in significant decreases in productivity [42]. In ethene oligomerization at 120 °C, the lifetime of the active species derived from complex **16** and MMAO is less than 3 min, but lifetimes of 10–20 min can be achieved with complexes having *meta*-aryl substituents in the imino aryl rings [48]. The reasoning behind this modification was that the thermal stability of these complexes would be improved by increasing remotely the steric protection around the metal.

The nature of the active species in iron-catalyzed polymerization is not well understood, as a result of the ability of the ligand to undergo a number of different transformations, including alkylation at any position of the pyridine ring [49]. Reactions between the free ligand and various lithium, magnesium and zinc alkyls have been shown to give *N*-alkylated products arising from attack of the metal alkyl at the pyridine nitrogen atom [50]. Recent work by Budzelaar and coworkers has revealed that reactions of the free ligand with aluminum alkyls are unexpectedly complex, alkyl additions to the imine carbon and at the 2- and 4-positions of the pyridine ring being observed [51]. It has also been shown that the iron atom in a bis (imino)pyridyl iron complex can be displaced by reaction with an aluminum alkyl, but the resulting Al complex is catalytically inactive, indicating that this reaction represents a deactivation pathway [52]. There is even uncertainty as to the oxidation state of iron in the active species derived from activation of LFeCl$_2$ (L = bis(imino)pyridine) with MAO, with conflicting claims for the $+3$ and $+2$ oxidation states [53, 54].

Active intermediates in LFeCl$_2$/AlR$_3$ and LFeCl$_2$/MAO systems have been investigated by Bryliakov *et al.*, who concluded that neutral species of type LFe(II)Cl ($\mu$-R)$_2$AlR$_2$ or LFe(II)R($\mu$-R)$_2$AlR$_2$ dominate in LFeCl$_2$/AlR$_3$ systems, whereas with MAO the ion pairs [LFe(II)($\mu$-Me)($\mu$-Cl)AlMe$_2$]$^+$[Me-MAO]$^-$ and [LFe(II)($\mu$-Me)$_2$Al-Me$_2$]$^+$[Me-MAO]$^-$ predominate [55]. Similar polymerization activities were obtained with these systems, but it was observed that the activity of LFeCl$_2$/AlR$_3$ decayed more rapidly than that of the LFeCl$_2$/MAO system.

It will be clear from this chapter that the behavior of nickel and iron catalysts in ethene oligomerization and polymerization is highly dependent on ligand steric bulk. The importance of bulky aryl ligands in late transition metal complexes has been underlined by Gibson and coworkers, who synthesized a series of hybrid ligands containing a bulky arylimino substituent on one side and a relatively unhindered heterocyclic donor on the other [56]. Complex **17**, derived from a tridentate imino-bipyridine ligand and used in combination with MAO, had much lower activity than the parent bis(imino)pyridyl complex **15** and produced mainly 1-butene and 1-hexene rather than high molecular weight polyethylene. Apparently, the absence of steric protection on one side has an even more dramatic effect than having a small aryl substituent on each side, such as in the oligomerization catalyst **16**. The low activity of **17** was attributed to relatively rapid deactivation, as a result of the lack of steric protection, outweighing the effect of easier access of the monomer to the active center. It was proposed that the catalyst resting state is an iron alkyl species and that the rate determining step in propagation is ethene coordination to the iron center.

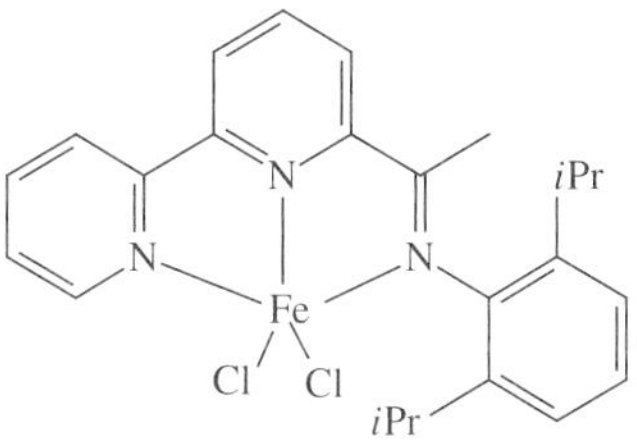

**17**

## 3.3
## Conclusions

The properties of late transition metal catalysts for ethene polymerization are strongly influenced by ligand steric bulk. The chain microstructure and the molecular weight of polyethylene synthesized with aryl-substituted α-diimine nickel(II) complexes are dependent on the presence of ortho-substituents on the aryl rings, which block the axial approach of olefins and retard the rate of chain transfer by associative displacement. Isopropyl substituents give branched polymers with relatively high molecular weight. Effective activators include both MAO and AlEt$_2$Cl, but the activity of cationic nickel diimine complexes decays relatively rapidly, particularly at elevated polymerization temperatures. Evidence for deactivation via intramolecular C$-$H activation of an *ortho*-alkyl substituent has been presented, as well as deactivation as a result of reduction of active Ni(II) species to a lower oxidation state. Improved performance at high temperature can, however, be obtained with a cyclophane-based complex in which the axial faces of the nickel are completely blocked by the cyclophane framework. The presence of bulky ligand substituents is also an important factor affecting the stability of neutral nickel(II) polymerization catalysts.

Bis(imino)pyridyl iron complexes, activated by MAO or AlR$_3$, are very active catalysts for ethene polymerization, but under homogeneous polymerization conditions their activity decays rapidly. Deactivation is particularly rapid for complexes lacking steric protection of the active center. In contrast to the nickel or palladium systems, polyethylene with negligible chain branching and relatively broad molecular weight distribution is obtained. The broad molecular weight distribution results from the presence of more than one type of active center, but the nature of the active centers in iron-catalyzed polymerization is not well resolved. The bis(imino)pyridyl ligand can undergo a range of alkylation reactions with metal alkyls and transfer of the ligand from iron to aluminum, leading to loss of activity, can also take place. However, stable polymerization kinetics in iron-catalyzed polymerization can be achieved by immobilization of the iron complex on a support, which is described in Chapter 4.

## References

1 Johnson, L.K., Killian, C.M., and Brookhart, M. (1995) *J. Am. Chem. Soc.*, **117**, 6414–6415.

2 Johnson, L.K., Mecking, S., and Brookhart, M. (1996) *J. Am. Chem. Soc.*, **118**, 267–268.

3 Möring, V.M. and Fink, G. (1985) *Angew. Chem. Int. Ed.*, **24**, 1001–1003.

4 Gates, D.P., Svejda, S.A., Oñate, E., Killian, C.M., Johnson, L.K., White, P.S., and Brookhart, M. (2000) *Macromolecules*, **33**, 2320–2334.

5 Pappalardo, D., Mazzeo, M., and Pellecchia, C. (1997) *Macromol. Rapid Commun.*, **18**, 1017–1023.

6 Kumar, K.R. and Sivaram, S. (2000) *Macromol. Chem. Phys.*, **201**, 1513–1520.

7 Popeney, C. and Guan, Z. (2005) *Organometallics*, **24**, 1145–1155.

8 Killian, C.M., Tempel, D.J., Johnson, L.K., and Brookhart, M. (1996) *J. Am. Chem. Soc.*, **118**, 11664–11665.

9 Gottfried, A.C. and Brookhart, M. (2001) *Macromolecules*, **34**, 1140–1142.

10 Gottfried, A.C. and Brookhart, M. (2003) *Macromolecules*, **36**, 3085–3100.

11 Yuan, J.-C., Silva, L.C., Gomes, P.T., Valerga, P., Campos, J.M., Ribeiro, M.R., Chien, J.C.W., and Marques, M.M. (2005) *Polymer*, **46**, 2122–2132.

12 Rose, J.M., Cherian, A.E., and Coates, G.W. (2006) *J. Am. Chem. Soc.*, **128**, 4186–4187.

13 Camacho, D.H., Salo, E.V., Ziller, J.W., and Guan, Z. (2004) *Angew. Chem. Int. Ed.*, **43**, 1821–1825.

14 Camacho, D.H. and Guan, Z. (2005) *Macromolecules*, **38**, 2544–2546.

15 Tempel, D.J., Johnson, L.K., Huff, R.L., White, P.S., and Brookhart, M. (2000) *J. Am. Chem. Soc.*, **122**, 6686–6700.

16 Popeney, C.S., Rheingold, A.L., and Guan, Z. (2009) *Organometallics*, **28**, 4452–4463.

17 Peruch, F., Cramail, H., and Deffieux, A. (1999) *Macromolecules*, **32**, 7977–7983.

18 Arthur, S., Teasley, M.F., Kerbow, D.L., Fusco, O., Dall'Occo, T., and Morini, G. (2001) Int. Patent WO 01/68725;(2001) *Chem. Abstr*, **135**, 257599.

19 Wang, C., Friedrich, S., Younkin, T.R., Li, R.T., Grubbs, R.H., Bansleben, D.A., and Day, M.W. (1998) *Organometallics*, **17**, 3149–3151.

20 Younkin, T.R., Connor, E.F., Henderson, J.I., Friedrich, S.K., Grubbs, R.H., and Bansleben, D.A. (2000) *Science*, **287**, 460–462.

21 Waltman, A.W., Younkin, T.R., and Grubbs, R.H. (2004) *Organometallics*, **23**, 5121–5123.

22 Hu, T., Tang, L.-M., Li, X.-F., Li, Y.-S., and Hu, N.-H. (2005) *Organometallics*, **24**, 2628–2632.

23 Hicks, F.A. and Brookhart, M. (2001) *Organometallics*, **20**, 3217–3219.

24 Jenkins, J.C. and Brookhart, M. (2003) *Organometallics*, **22**, 250–256.

25 Hicks, F.A., Jenkins, J.C., and Brookhart, M. (2003) *Organometallics*, **22**, 3533–3545.

26 Jenkins, J.C. and Brookhart, M. (2004) *J. Am. Chem. Soc.*, **23**, 5827–5842.

27 Zhang, L., Brookhart, M., and White, P.S. (2006) *Organometallics*, **25**, 1868–1874.

28 Zuideveld, M.A., Wehrmann, P., Rühr, C., and Mecking, S. (2004) *Angew. Chem. Int. Ed.*, **43**, 869–873.

29 Wehrmann, P. and Mecking, S. (2006) *Macromolecules*, **39**, 5963–5964.

30 Yu, S.-M., Berkeveld, A., Göttker-Schnetmann, I., Müller, G., and Mecking, S. (2007) *Macromolecules*, **40**, 421–428.

31 Berkefeld, A. and Mecking, S. (2009) *J. Am. Chem. Soc.*, **131**, 1565–1574.

32 Malinoski, J.M. and Brookhart, M. (2003) *Organometallics*, **22**, 5324–5335.

33 Keim, W., Kowalt, F.H., Goddard, R., and Krüger, C. (1978) *Angew. Chem. Int. Ed.*, **17**, 466–467.

34 Klabunde, U. and Ittel, S.D. (1987) *J. Mol. Catal.*, **41**, 123–134.

**35** Ostoja Starzewski, K.A. and Witte, J. (1987) *Angew. Chem. Int. Ed.*, **26**, 63–64.

**36** Cooley, N.A., Green, S.M., Wass, D.F., Heslop, K., Orpen, A.G., and Pringle, P.G. (2001) *Organometallics*, **20**, 4769–4771.

**37** Carlini, C., De Luise, V., Fernandes, E.G., Martinelli, M., Raspolli Galletti, A.M., and Sbrana, G. (2005) *Macromol. Rapid Commun.*, **26**, 808–812.

**38** Small, B.L., Brookhart, M., and Bennett, A.M.A. (1998) *J. Am. Chem. Soc.*, **120**, 4049–4050.

**39** Britovsek, G.J.P., Gibson, V.C., Kimberley, B.S., Maddox, P.J., McTavish, S.J., Solan, G.A., White, A.J.P., and Williams, D.J. (1998) *Chem. Commun.*, 849–850.

**40** Small, B.L. and Brookhart, M. (1998) *J. Am. Chem. Soc.*, **120**, 7143–7144.

**41** Britovsek, G.J.P., Mastroianni, S., Solan, G.A., Baugh, S.D., Redshaw, C., Gibson, V.C., White, A.J.P., Williams, D.J., and Elsegood, M.R.J. (2000) *Chem. Eur. J.*, **6**, 2221–2231.

**42** Britovsek, G.J.P., Bruce, M., Gibson, V.C., Kimberley, B.S., Maddox, P.J., Mastroianni, S., McTavish, S.J., Redshaw, C., Solan, G.A., Strömberg, S., White, A.J.P., and Williams, D.J. (1999) *J. Am. Chem. Soc.*, **121**, 8728–8740.

**43** Tellmann, K.P., Gibson, V.C., White, A.J.P., and Williams, D.J. (2005) *Organometallics*, **24**, 280–286.

**44** Barabanov, A.A., Bukatov, G.D., Zakharov, V.A., Semikolenova, N.V., Echevskaja, L.G., and Matsko, M.A. (2005) *Macromol. Chem. Phys.*, **206**, 2292–2298.

**45** Wang, Q., Yang, H., and Fan, Z. (2002) *Macromol. Rapid Commun.*, **23**, 639–642.

**46** Radhakrishnan, K., Cramail, H., Deffieux, A., François, P., and Momtaz, A. (2003) *Macromol. Rapid Commun.*, **24**, 251–254.

**47** Wang, Q., Li, L., and Fan, Z. (2005) *J. Polym. Sci., Part A: Polym. Chem.*, **43**, 1599–1606.

**48** Ionkin, A.S., Marshall, W.J., Adelman, D.J., Bonik Fones, B., Fish, B.M., and Schiffhauwer, M.F. (2006) *Organometallics*, **25**, 2978–2992.

**49** Scott, J., Gambarotta, S., Korobkov, I., and Budzelaar, P.H.M. (2005) *J. Am. Chem. Soc.*, **127**, 13019–13029.

**50** Blackmore, I.J., Gibson, V.C., Hitchcock, P.B., Rees, C.W., Williams, D.W., and White, A.J.P. (2005) *J. Am. Chem. Soc.*, **127**, 6012–6020.

**51** Knijnenburg, Q., Smits, J.M.M., and Budzelaar, P.H.M. (2006) *Organometallics*, **25**, 1036–1046.

**52** Scott, J., Gambarotta, S., Korobkov, I., Knijnenburg, Q., de Bruin, B., and Budzelaar, P.H.M. (2005) *J. Am. Chem. Soc.*, **127**, 17204–17206.

**53** Britovsek, G.J.P., Clentsmith, G.K.B., Gibson, V.C., Goodgame, D.M.L., McTavish, S.J., and Pankhurst, Q.A. (2002) *Catal. Commun.*, **3**, 207–211.

**54** Bryliakov, K.P., Semikolenova, N.V., Zudin, V.N., Zakharov, V.A., and Talsi, E.P. (2004) *Catal. Commun.*, **5**, 45–48.

**55** Bryliakov, K.P., Semikolenova, N.V., Zakharov, V.A., and Talsi, E.P. (2004) *Organometallics*, **23**, 5375–5378.

**56** Britovsek, G.J.P., Baugh, S.P.D., Hoarau, O., Gibson, V.C., Wass, D.F., White, A.J.P., and Williams, D.J. (2003) *Inorg. Chim. Acta*, **345**, 279–291.

# 4
# Effects of Immobilization of Catalysts for Olefin Polymerization

## 4.1
## Introduction

It will be apparent from Chapters 2 and 3 that intensive research aimed at the discovery and development of single-center catalyst systems for olefin polymerization has resulted in an ever increasing number of novel, homogeneous catalysts. However, widespread application of these catalysts in commercial gas- and slurry-phase processes for polyolefin production requires their immobilization on a suitable support material in order to prevent reactor fouling. The challenge here is to achieve immobilization without altering the single-center nature of the active species, and without a significant decrease in catalyst activity. This is not an easy task. Many different supports and immobilization methods have been investigated, but it is frequently observed that after immobilization the catalyst activity is much lower than the activity that was obtained under homogeneous polymerization conditions [1, 2]. Nevertheless, there are notable exceptions to this trend, most importantly when catalyst immobilization results in stabilization of the active species, preventing the deactivation which is often apparent in homogeneous polymerization.

To perform in particle-forming processes, soluble catalytic species must be deposited on a suitable carrier and, more importantly, remain strongly associated with the carrier throughout the polymerization [3]. There are two main methods for catalyst immobilization: physical impregnation and chemical tethering. Simple impregnation or deposition of an organometallic complex on a support can often give rise to serious fouling problems, especially in slurry systems where the solvent can dissolve the active catalyst. An obvious solution to the leaching problem is to chemically tether the catalyst to the support, but this frequently involves a complicated synthetic procedure. An alternative is to immobilize the activator on the support, an example being the use of MAO-impregnated silica. Silica is by far the most widely used support material for single-center catalyst immobilization, but supports such as $MgCl_2$, $Al_2O_3$ and microspheroidal polymers have also received considerable attention [1–9]. Techniques for the self-immobilization of metallocenes containing an olefin function that can act as a comonomer in the polymerization

*Homogeneous Catalysts: Activity – Stability – Deactivation*, First Edition. Piet W.N.M. van Leeuwen and John C. Chadwick.
© 2011 Wiley-VCH Verlag GmbH & Co. KGaA. Published 2011 by Wiley-VCH Verlag GmbH & Co. KGaA.

process have also been developed [10]. Another approach involves an emulsion process in which droplets of a zirconocene/MAO/toluene solution in a fluorocarbon nonsolvent are converted to spherical catalyst particles by raising the temperature to effect transfer of toluene to the fluorocarbon phase [11].

The effect of immobilization on catalyst activity and productivity is influenced by a complex mixture of chemical and physical factors. For example, with $SiO_2$ supports the number and nature of surface $Si-OH$ groups need to be taken into account. Direct reaction of a surface hydroxy group with a transition metal precatalyst should generally be avoided. On the other hand, surface hydroxy groups can be utilized to anchor MAO or another Al alkyl. With MAO, a beneficial effect of such reactions can be preferential reaction of $Si-OH$ with the $AlMe_3$ present in MAO [12]. The presence on the support of Lewis or Brønsted acidic surface species having sufficiently high acidity to generate the active, cationic transition metal species required for olefin polymerization is another important chemical factor. Physical factors which have an important effect on catalyst productivity are the ability of a support to undergo progressive fragmentation during the course of polymerization. A low support friability can lead to the pores of an unfragmented catalyst becoming filled with polymer, resulting in limited monomer diffusion into the particle [13]. High support porosity and, in particular, the presence of relatively large pores aids monomer diffusion in ethene polymerization, leading to regular particle growth [14].

The purpose of this chapter is not to present a comprehensive review of catalyst immobilization techniques, a subject that is already well documented [1–9], but rather to pinpoint how immobilization affects the activity and stability of the various types of early and late transition metals described in Chapters 2 and 3.

## 4.2
## Metallocenes and Related Complexes

### 4.2.1
### Immobilized MAO/Metallocene Systems

The most common techniques used to support a zirconocene on silica comprise either immobilization of MAO on the support, followed by addition of the zirconocene, or precontacting the zirconocene with MAO before immobilization on the support. Direct contact of the metallocene with silica before the introduction of MAO generally leads to low activity. Kaminsky found that the latter procedure gave much lower activities than were obtained in homogeneous polymerization, but observed very significant increases in polypropylene molecular weight and isotacticity [15]. The immobilized catalysts required much lower MAO/Zr ratios than were needed in homogeneous polymerization.

As described in Chapter 2, deactivation of metallocene catalysts can take place by the following type of $\alpha$-H transfer mechanism, generating inactive $Zr-CH_2-Al$ or $Zr-CH_2-Zr$ species and methane.

$$Zr-CH_3 + CH_3-Al(CH_3)-O- \rightarrow Zr-CH_2-Al(CH_3)-O- + CH_4$$

It was found that immobilization of a zirconocene on $SiO_2$/MAO effectively suppressed this reaction [16]. No significant difference in methane generation was observed between the system $SiO_2$/MAO/$Cp_2ZrCl_2$ and a control experiment with $SiO_2$/MAO. Stable activity during ethene polymerization was noted. Subsequent studies also demonstrated that immobilization of a zirconocene on $SiO_2$/MAO leads to increased stability during polymerization [17, 18]. In these experiments the solid $SiO_2$/MAO/zirconocene catalyst was used in combination with Al$i$Bu$_3$, which is an effective scavenger of any impurities present in polymerization but is not an effective activator for zirconocenes. Separate addition of MAO in polymerization was avoided, as this could cause leaching of the zirconocene from the support and the generation of active species in solution. Polypropylenes prepared with the immobilized zirconocene $rac$-Me$_2$Si[2-Me-4-(1-naphthyl)Ind]$_2$ZrCl$_2$1 were found to have somewhat lower melting points than those prepared under homogeneous conditions. This was attributed to a lower local monomer concentration at the catalytic centers within the growing particle. In metallocene-catalyzed propene polymerization, a decrease in monomer concentration leads to an increase in the content of chain irregularities formed by a process of chain epimerization [19].

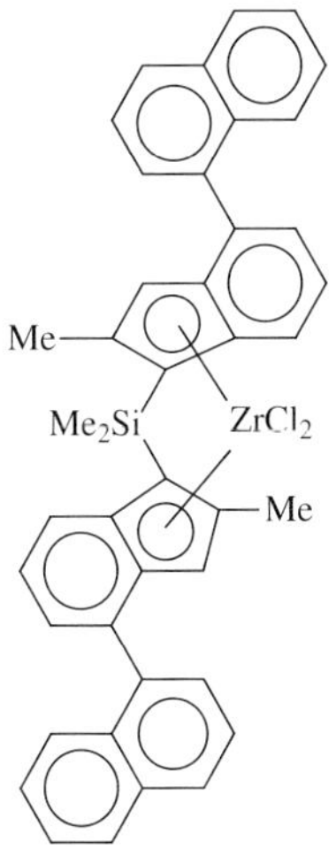

**1** 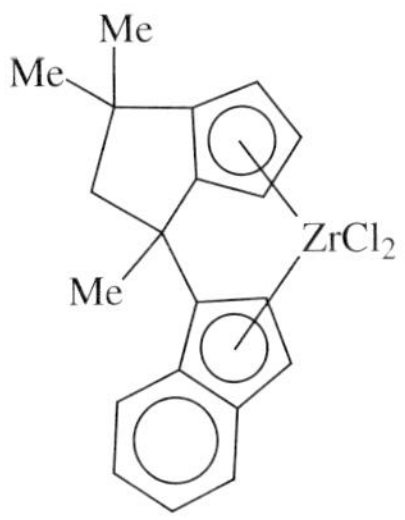                    **2**

High stability in ethene polymerization, under both homogeneous and heterogeneous conditions, has been reported for pentalenyl-derived bridged zirconocenes such as complex **2** [20]. Gas-phase polymerization after immobilization on $SiO_2$/MAO gave much lower activity but (at 90 °C) higher molecular weight than was obtained in solution in toluene.

The activity/decay profile of an immobilized metallocene in olefin polymerization can also be influenced by the presence of different aluminum alkyls. In gas-phase ethene homo- and copolymerization with ($n$-BuCp)$_2$ZrCl$_2$/MAO immobilized on porous polymer supports, it has been found that the presence of $AlEt_3$ or $AliBu_3$ decreases the initial polymerization rate but leads to more stable polymerization kinetics, resulting in increased productivity [21]. $AlEt_3$ had a stronger effect than $AliBu_3$ on suppressing the initial rate, which could be attributed to the easier formation of dormant species of type $[(n\text{-}BuCp)_2Zr(\mu\text{-}R)_2AlR_2]^+$.

Physical aspects pertaining to olefin polymerization with silica-supported zirconocenes have been investigated by Fink and coworkers. A slow build up in the rate of propene polymerization with $rac\text{-}Me_2Si(Ind)_2ZrCl_2$ immobilized on $SiO_2$/MAO was attributed to gradual fragmentation of the support in the early stages of polymerization [22]. The particle growth model adopted involved a shell-by-shell fragmentation, progressing from the outside to the inside of the particle. Particularly long induction times were apparent when polymerization was carried out at relatively low temperature (30 °C) [23]. It was also found that the melting point of polypropylene obtained at intermediate stages of polymerization decreased until the end of the induction period, then increased at longer polymerization times. This provided a clear indication of a monomer diffusion limitation in this system. Incomplete fragmentation in the early stages of polymerization results in the formation of a polymer layer on the particle surface which restricts monomer diffusion. The low monomer concentration at active centers within the particle then results in a less stereoregular polymer with lower melting point. A 1-octene prepolymerized catalyst gave much higher polypropylene yields, as a result of easier monomer diffusion through amorphous poly(1-octene) as opposed to crystalline polypropylene [23, 24].

The characteristics of particle growth in olefin polymerization with immobilized catalysts can be determined by scanning electron microscopy investigation on cross-sections of particles obtained at various stages of polymerization [25, 26]. An interesting illustration of monomer diffusion effects on particle growth was the finding by Fink and coworkers that, in ethene/1-hexene copolymerization with a silica-supported zirconocene, the formation of a copolymer shell around the silica particles in the early stages of polymerization resulted in what was termed a filter effect [27]. Ethene was able to diffuse through this outer shell relatively easily whereas the larger 1-hexene diffused more slowly. Thus, a copolymer shell is formed around a polymer mass consisting mainly of ethene homopolymer, resulting in a broader copolymer chemical composition distribution than is obtained in homogeneous polymerization. Subsequent observations of a broadening in chemical composition with increasing polymerization time, as a result of the gradual formation of a relatively high molecular weight, ethene-rich polymer fraction, provided support for Fink's filter model [28]. Diffusion limitations in supports with low porosity can also lead to a broadening in polyethylene molecular weight distribution [29].

Immobilization of MAO and metallocene on a silica support is often carried out using the "incipient wetness" method, in which a metallocene/MAO solution in toluene is added to the support in an amount which is equivalent to the pore volume of the support, after which the solvent is removed under vacuum. A variation of this method has been reported by Rytter and coworkers, who impregnated the pores of a silica support with a cold ($-40$ °C) solution of $(n\text{-}BuCp)_2ZrCl_2$ and MAO in 1-hexene, after which the mixture was heated to room temperature to effect polymerization of 1-hexene within the pores of the support [30, 31]. This method gave a uniform distribution of the zirconocene throughout the support and resulted in a catalyst which was robust towards exposure to air.

As indicated above, direct immobilization of a metallocene on silica, in the absence of MAO, generally leads to low activity. An example of this is $SiO_2/rac$-Et(Ind)$_2$ZrCl$_2$, which gave much lower activity than $SiO_2/(n$-BuCp)$_2$ZrCl$_2$ in ethene/1-hexene copolymerizations with MAO as cocatalyst [32, 33]. The low activity of the *ansa*-metallocene was ascribed to the steric hindrance of the silica support surface.

### 4.2.2
### Immobilized Borane and Borate Activators

An alternative to the use of MAO as cocatalyst is activation using an immobilized borane or borate activator. For example, immobilization of $B(C_6F_5)_3$ on silica involves the following reaction with surface hydroxy groups [34]:

$$Si - OH + B(C_6F_5)_3 \rightarrow [Si - OB(C_6F_5)_3]^- H^+$$

In contrast to ethene polymerization with the homogeneous catalyst system $Cp_2ZrCl_2/AliBu_3/B(C_6F_5)_3$, in which a peak in activity is followed by rapid decay, the use of $SiO_2/B(C_6F_5)_3$ in combination with $Cp_2ZrCl_2/AliBu_3$ resulted in lower but more stable activity [35].

Treatment of $SiO_2/B(C_6F_5)_3$ with a tertiary amine such as $Et_2NPh$ forms an anilinium borate species which is able to interact with dialkylzirconocenes to generate active zirconocenium species [36, 37]:

$$[Si - OB(C_6F_5)_3]^- H^+ + NR_3 \rightarrow [Si - OB(C_6F_5)_3]^- [HNR_3]^+$$

$$[Si - OB(C_6F_5)_3]^- [HNR_3]^+ + Cp_2 ZrMe_2 \rightarrow$$
$$[Si - OB(C_6F_5)_3]^- [Cp_2 ZrMe]^+ + CH_4 + NR_3$$

Alternatively, the silica support may be treated with BuLi, followed by addition of $B(C_6F_5)_3$ and trityl chloride to form an immobilized trityl borate activator [38]:

$$Si - OH + BuLi \rightarrow Si - OLi + BuH$$

$$Si - OLi + B(C_6F_5)_3 + Ph_3 CCl \rightarrow [Si - OB(C_6F_5)_3]^- [Ph_3C]^+ + LiCl$$

$$[Si - OB(C_6F_5)_3]^- [Ph_3C]^+ + Cp_2 ZrMe_2 \rightarrow$$
$$[Si - OB(C_6F_5)_3]^- [Cp_2 ZrMe]^+ + Ph_3 CMe$$

Direct reaction of trityl chloride with $SiO_2/B(C_6F_5)_3$ has also been reported to form the above active species [39, 40]. Higher and relatively stable polymerization activities were obtained when the silica was pretreated with $AliBu_3$ to effect partial conversion of surface Si–OH species to Si–O $AliBu_2$.

4.2.3
**Superacidic Supports**

Activation of a metallocene on an inorganic support can be achieved even without the use of MAO or a borate activator, if the support contains sites having high Lewis or Brønsted acidity. Sulfated alumina is an example of a support containing strong Brønsted acid surface sites [41, 42]. Marks and coworkers proposed that activation of a $Cp^*_2ZrMe_2$ on sulfated alumina proceeds as shown in Scheme 4.1 and found that nearly all the zirconium in such systems was catalytically active [41].

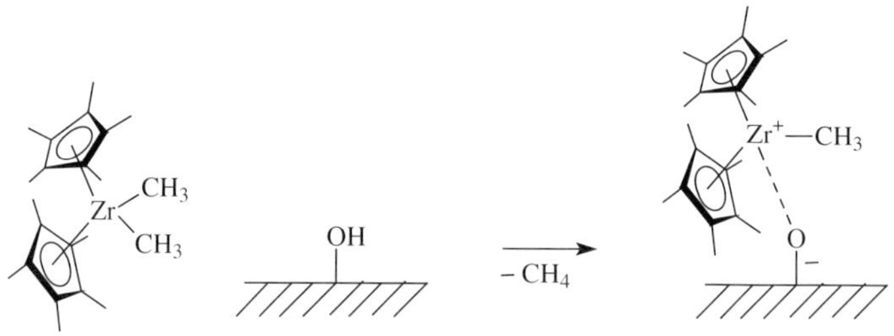

**Scheme 4.1**  Zirconocene activation via protonolysis on Brønsted acid support sites [41].

Jones and coworkers have reported the preparation of sulfonic acid functionalized silica containing Brønsted acidic sites [43]. Functionalization was effected via reaction of a silica support with a fluorinated sulfone precursor, as illustrated in Scheme 4.2. In combination with the resulting support and $AlMe_3$, $Cp^*_2ZrMe_2$ gave ethene polymerization activities approaching those obtained using $SiO_2/MAO$. It was suggested that the active species was formed via interaction of the metallocene with a combined Brønsted acid – $AlMe_3$ adduct, as illustrated in Figure 4.1.

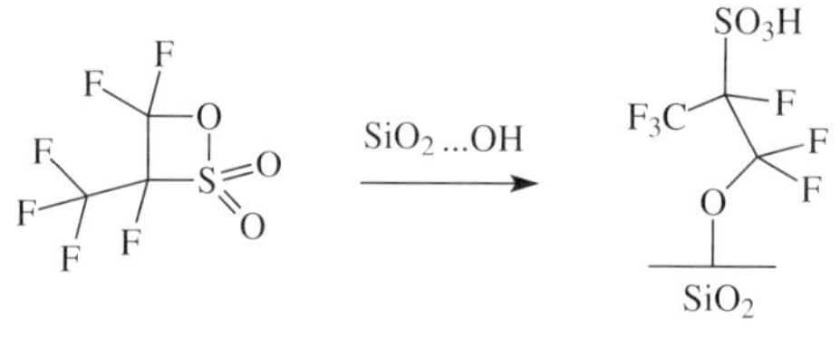

**Scheme 4.2**  Synthesis of tethered Brønsted acid [43].

4.2.4
**MgCl₂-Supported Systems**

The presence of surface acidic sites is also an important factor determining the performance of magnesium chloride as support material for the immobilization and activation of metallocenes and other single-center catalysts. The presence of Lewis acidic surface sites in magnesium chloride can enable catalyst activation without the use of MAO or borate. This has been reported by Marks, who demonstrated that

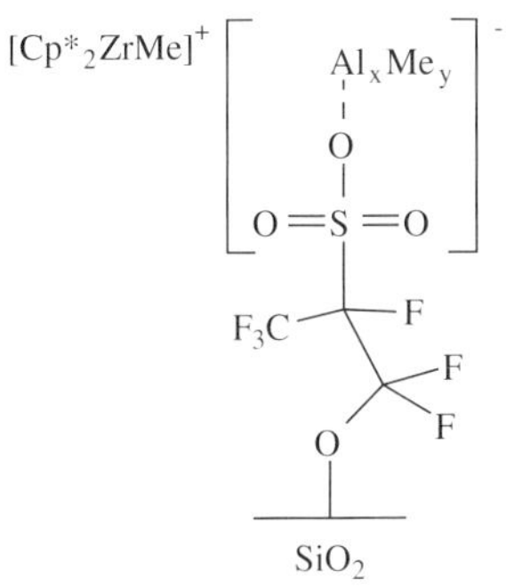

$[Cp^*_2ZrMe]^+$

$$\left[ \begin{array}{c} Al_xMe_y \\ | \\ O \\ | \\ O{=}S{=}O \end{array} \right]^-$$

$F_3C$ — F
— F
— F
O
|
$SiO_2$

**Figure 4.1** Proposed active species on perfluoroalkanesulfonic acid-functionalized silica [43].

$MgCl_2$ was able to activate $Cp^*_2ThMe_2$ by abstraction of a methide anion, generating a catalytically active center $[Cp^*_2ThMe]^+$, as illustrated in Figure 4.2 [44]. The presence of surface acidic sites has been demonstrated for $MgCl_2$ prepared by reaction of magnesium with excess *n*-BuCl in refluxing heptane [45]. A surface acidic site concentration of approximately $170\,\mu mol\,g^{-1}$ was reported, which corresponded to the amounts of titanium and iron catalysts that could effectively be immobilized [46].

Magnesium chloride supports have been used together with MAO for zirconocene immobilization and activation [47–50], but an important incentive for the use of magnesium chloride is that activation of many single-center catalysts may be achieved using simple cocatalysts such as $AlEt_3$ or $Al i Bu_3$. However, most attempts to activate $MgCl_2$-supported zirconocenes using these cocatalysts have been relatively unsuccessful. Immobilization of $Cp_2ZrCl_2$ on a support of type $MgCl_2/AlR_n(OEt)_{3-n}$ gave only a third of the activity in ethene polymerization that was obtained with $Cp_2TiCl_2$ [51]. Kissin *et al.* have used mixtures of $MgR_2$ and $AlEt_2Cl$ to generate $MgCl_2$ and $AlR_3$ *in situ*, obtaining activities 5–10 times lower than those obtained with MAO [52]. Kaminaka and Soga reported that ball-milled $MgCl_2/rac$-$Et(IndH4)_2ZrCl_2$ was active in propene polymerization with $AlMe_3$ or $AlEt_3$ as cocatalyst, although the activity was about an order of magnitude less than that obtained in homogeneous polymerization using MAO [53]. However, a linear increase in polymer yield with

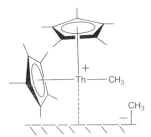

Support Acidic Surface

**Figure 4.2** Generation of active species via methide transfer to a Lewis acidic surface site on $MgCl_2$ [44].

increasing polymerization time indicated high catalyst stability [54]. A key role of the MgCl$_2$ support was inferred from the fact that no activity was obtained with the systems *rac*-Et(IndH4)$_2$ZrCl$_2$ – AlMe$_3$ or SiO$_2$/*rac*-Et(IndH4)$_2$ZrCl$_2$ – AlMe$_3$ [53]. Echevskaya *et al.* have immobilized *rac*-Me$_2$Si(Ind)$_2$ZrCl$_2$ on highly dispersed MgCl$_2$, prepared by the reaction of MgBu$_2$·*n*AlEt$_3$ with CCl$_4$. Using Al*i*Bu$_3$ as cocatalyst, they obtained an ethene polymerization activity similar to that obtained using a SiO$_2$/ MAO support, albeit that the activity was much lower than that obtained in homogeneous polymerization with MAO as cocatalyst [55].

Recently, it has been shown that, for MgCl$_2$-immobilized zirconocenes of type (RCp)$_2$ZrCl$_2$, activated using AlEt$_3$ or Al*i*Bu$_3$, the substituent R on the cyclopenta-dienyl ring has a remarkable effect on catalyst activity [56]. Low activity was obtained with R = H or Et, but longer alkyl substituents, notably *n*-Pr or *n*-Bu, gave more than an order of magnitude increase in activity in ethene polymerization. Particularly high activity was obtained at a low loading of zirconocene on the support, indicating the presence of a limited number of active species. Further studies revealed that high polymerization activities were only obtained with supports of type MgCl$_2$/ AlR$_n$(OEt)$_{3-n}$ prepared from the reaction of AlEt$_3$ or Al*i*Bu$_3$ with MgCl$_2$·1.1EtOH [57]. These supports, in contrast to other chemically activated supports, including those prepared from MgCl$_2$/EtOH adducts with higher ethanol contents, are characterized by additional peaks in the X-ray diffraction pattern, indicating the presence of a crystalline structure which is absent in the other supports. Effective zirconocene activation implies the presence of highly Lewis acidic sites on the support which are able to generate the active metallocenium species. It was also demonstrated that, whereas the use of AlEt$_3$ as cocatalyst resulted in a significant decay in activity during polymerization at 70 °C, relatively stable activity was obtained with Al*i*Bu$_3$ [56]. An increase in Al*i*Bu$_3$ concentration led to a further stabilization of the polymerization activity. The stabilizing effect of Al*i*Bu$_3$ may be due to the negligible tendency of an isobutyl group to bridge between Zr and Al, avoiding deactivation via the formation of alkylidene species analogous to those of type [L$_2$Zr($\mu$-CH$_2$)($\mu$-Me)AlMe$_2$]$^+$ that can arise in zirconocene/MAO systems [58].

The stabilizing effect of magnesium chloride in metallocene-catalyzed polymeri-zation is particularly evident with titanocenes. As indicated in Chapter 2, Cp$_2$TiCl$_2$ deactivates rapidly during polymerization, most probably due to reduction to inactive Ti(III) species. However, Satyanarayana and Sivaram found that stable ethene poly-merization activity could be obtained after immobilization of Cp$_2$TiCl$_2$ on an MgCl$_2$ support, using Al*i*Bu$_3$ as cocatalyst [59]. Very stable polymerization activity has been obtained with CpTiCl$_3$ and related complexes immobilized on MgCl$_2$/AlEt$_n$(OEt)$_{3-n}$ supports and activated with Al*i*Bu$_3$ [60]. The activities obtained were orders of magnitude higher than those reported [61] for MAO-activated polymerization under homogeneous conditions. GPC analysis of polyethylenes obtained with MgCl$_2$-supported CpTiCl$_3$ and its analogues indicated narrow molecular weight distribution, but subsequent rheological characterization of these polymers revealed a deviation from the expected Schulz–Flory distribution, indicating the presence of more than one type of active species [62]. This was in contrast to MgCl$_2$-supported Zr, V and Cr complexes, which retained their single-center characteristics after immobilization.

## 4.3
## Other Titanium and Zirconium Complexes

### 4.3.1
### Constrained Geometry Complexes

Immobilization and activation of constrained geometry complexes on silica has been achieved by treatment of the silica support with MAO or $AlR_3$, followed by reaction of the pretreated support with $[HNEt_3][B(C_6F_5)_3(C_6H_4\text{-}4\text{-}OH]$, resulting in tethering of the borate to the support via reaction of a Si–O–Al–R moiety with the active hydrogen of the borate, as illustrated in Scheme 4.3 [63]. The resulting immobilized borate is an effective activator for constrained geometry catalysts such as $t\text{-}BuNSiMe_2(C_5Me_4)$ $TiMe_2$. The use of analogous trialkylammonium borates having long hydrocarbon chains on the ammonium cation, and consequently having improved solubility in toluene, has also been described [64, 65]. A similar approach has been used for immobilization of $[HNEt_3][B(C_6F_5)_3(C_6H_4\text{-}4\text{-}OH]$ on a support of type $MgCl_2/$ $AlEt_n(OEt)_{3-n}$, resulting in significantly higher activities in metallocene-catalyzed polymerization than were obtained using the same borate immobilized on a silica support [66]. The improved performance with $MgCl_2$ was attributed to easier fragmentation of the support during the course of polymerization.

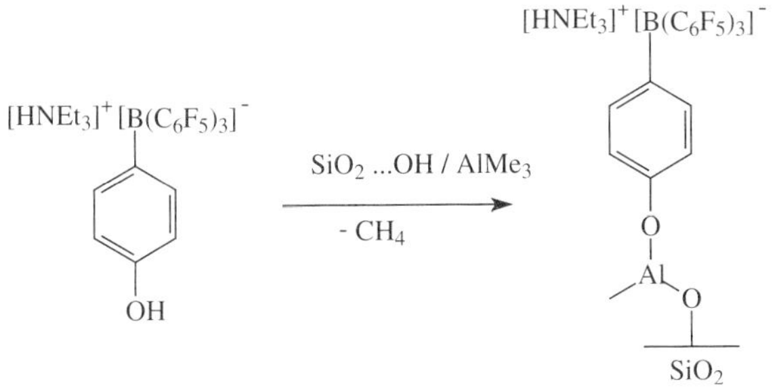

**Scheme 4.3** Immobilization of a borate activator on silica [63].

Tethering of a constrained geometry catalyst to a silica support has been reported by Eisen and coworkers, who contacted complex **3** with the support to effect immobilization via reaction of the $-Si(OMe)_3$ moiety with surface Si–OH groups [67]. Using MAO as cocatalyst, it was found that deactivation in ethene polymerization at $20\,^{\circ}C$ followed first-order kinetics, no bimolecular deactivation pathways being possible for the heterogeneous catalyst.

In order to avoid steric crowding in the tethering of a constrained geometry catalyst to a silica support, Jones and coworkers first contacted the support with a bulky patterning molecule, as illustrated in Scheme 4.4 [68, 69]. The large trityl groups on the patterning agent prevent incorporation of the silane on adjacent surface sites. Unreacted silanols are then pacified using hexamethyldisilazane, followed by hydrolysis of the imine bond of the patterning molecule to give the amine-functionalized silica, which is then used as a scaffold for the synthesis of the constrained geometry catalyst, as shown in Scheme 4.5. The resulting catalyst,

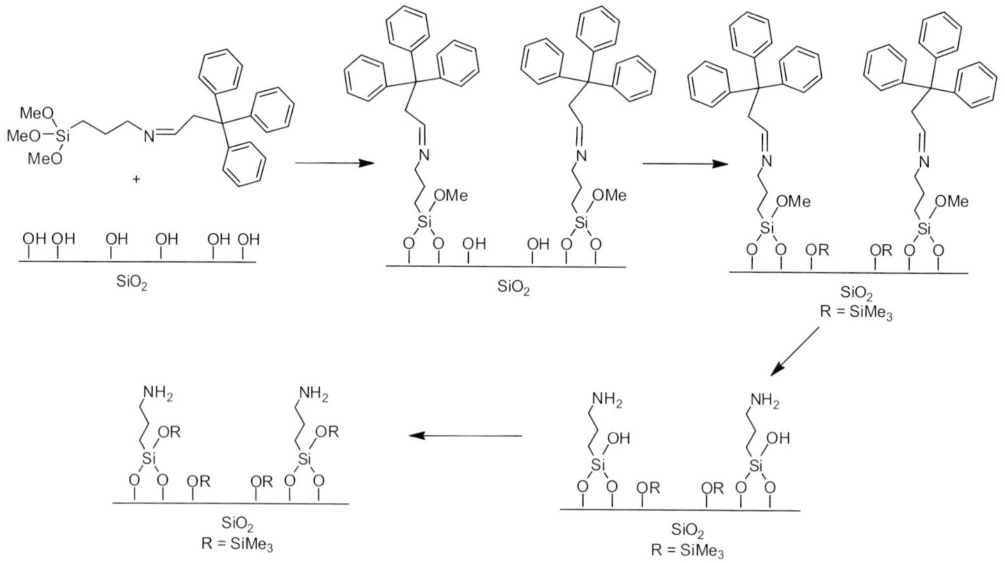

**Scheme 4.4** Silica surface functionalization using a patterning technique [70].

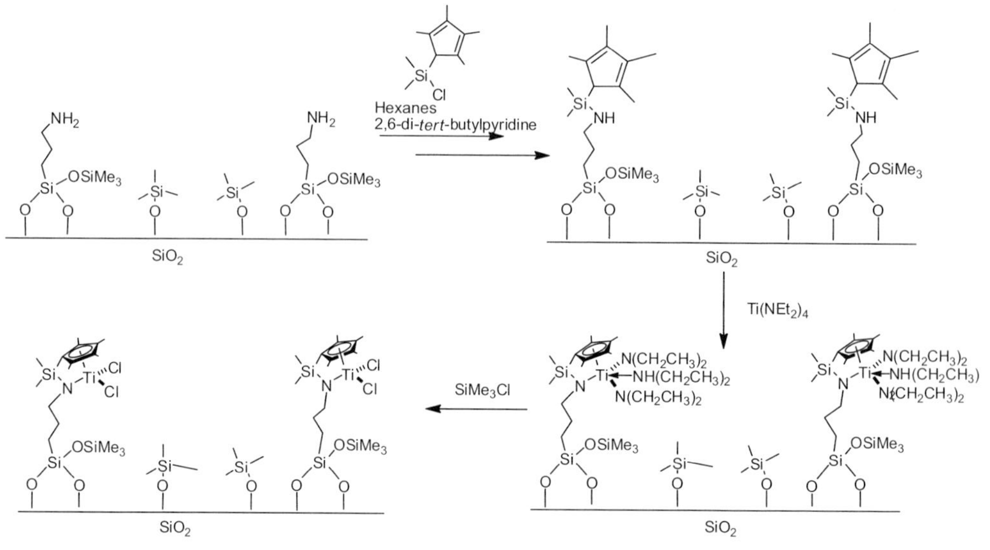

**Scheme 4.5** Constrained geometry catalyst synthesis on a patterned silica support [70].

activated using AliBu$_3$/B(C$_6$F$_5$)$_3$, gave higher activity than was obtained in homogeneous polymerization. The high activity was attributed to the uniform and isolated nature of the active species, as well as avoidance of unwanted interactions of the complex with the oxide surface.

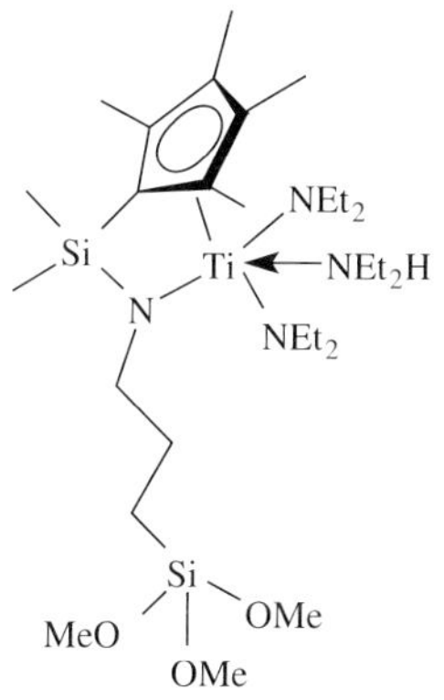

**3**

## 4.3.2
## Octahedral Complexes

Homogeneous and heterogeneous ethene polymerizations using Zr(salen) Cl$_2$(THF), where salen represents the *N,N'*-ethenebis(salicylideneiminato) ligand, have been carried out by Repo *et al.* [71]. Using MAO as activator, homogeneous polymerization in toluene at 80 °C gave only moderate activity. Higher activities were obtained when the salen complex **4** was first immobilized on a SiO$_2$ support. This behavior differs from that with zirconocenes, which exhibit lower activity after direct immobilization on silica. The activity increased with decreasing catalyst loading on the support and it was reported that deactivation during polymerization was reduced by immobilization. The inverse effect of catalyst loading on activity was attributed to greater separation of the active species, decreasing the possibility for bimolecular deactivation. Białek *et al.* have reported stable ethene polymerization activity at 50 and 60 °C with the related titanium complex **5** immobilized on a support prepared by reaction of MgCl$_2$·3.4EtOH with AlEt$_2$Cl [72]. The highest activities were obtained with AlMe$_3$ as cocatalyst. Much lower polymer yields were obtained under homogeneous polymerization conditions, or on a silica support, using MAO or AlEt$_2$Cl as activator.

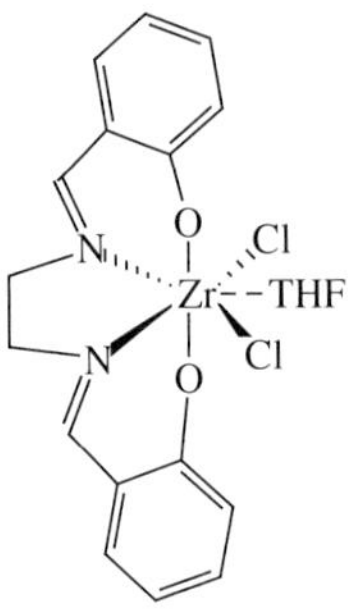

4 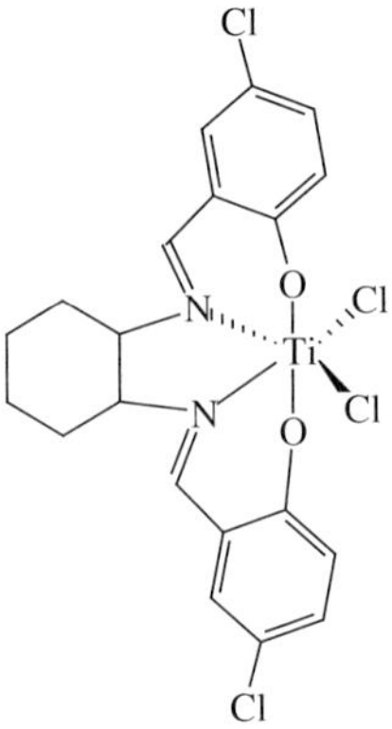                                    5

Fujita and coworkers have found that bis(phenoxy-imine) complexes (FI catalysts) can be effectively immobilized and activated using $MgCl_2/iBu_nAl(OR)_{3-n}$ supports prepared *in situ* via reaction of $AliBu_3$ with a solution of a 1:3 adduct of $MgCl_2$ and 2-ethyl-1-hexanol in decane [73]. Ethene polymerization at 50 °C with the titanium complex **6** gave an activity of approximately $4000\,kg\,mol^{-1}\,bar^{-1}\,h^{-1}$, which was around 80% of the activity obtained under homogeneous conditions using MAO. A steady uptake of ethene was noted over the 30 min duration of the polymerization. It was suggested that the effectiveness of the $MgCl_2$-based activator was related to the presence in the bis(phenoxy-imine) complex of O and N heteroatoms capable of electronic interaction with the support. Living polymerization in propene polymerization at 25 °C, giving highly syndiotactic polypropylene, was demonstrated with the fluorinated bis(phenoxy-imine) complex **7** immobilized on $MgCl_2/iBu_nAl$ $(OR)_{3-n}$ [74]. Particularly high ethene polymerization activities were obtained with zirconium FI catalysts. Complex **8**, when used with $MgCl_2/iBu_nAl(OR)_{3-n}$ in ethene polymerization at 50 °C, gave an activity of $202\,000\,kg\,mol^{-1}\,bar^{-1}\,h^{-1}$, which was higher than the activity obtained using MAO [75]. Well-defined polymer particle morphology could also be obtained using these systems [76, 77].

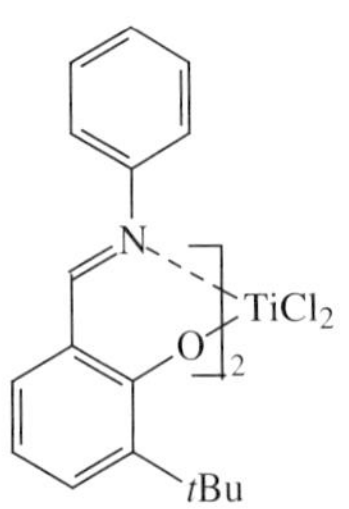

6 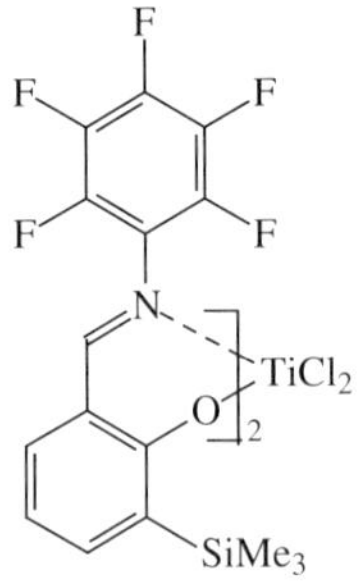                                    7 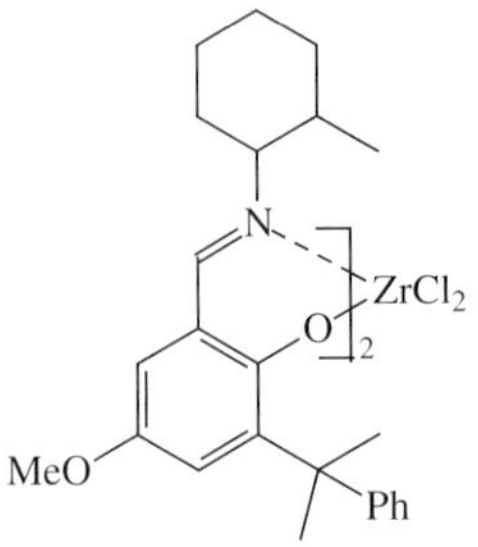                                    8

## 4.4
## Vanadium Complexes

As described in Chapter 2, homogeneous vanadium catalysts generally give low productivity in olefin polymerization, as a result of rapid decay due to reduction of active V(III) to poorly active or inactive V(II) species, which can be reoxidized and activated using a halocarbon promoter. Reduction is particularly rapid at high temperatures, but in ethene polymerization at 160 °C with $VCl_3$ supported on $AlCl_3$, using $AlEt_3$ as cocatalyst, it has been found that the presence of $CH_3CCl_3$ can give an up to 20-fold increase in productivity [78]. However, the activating effect of $CH_3CCl_3$ was much smaller when $VCl_3$ supported on $CrCl_3$ was used, due to competition between vanadium and chromium in the reduction–reoxidation cycle. In the absence of a halocarbon, productivities obtained with $VCl_3/CrCl_3$ were around 5 times higher than those obtained with unsupported $VCl_3$ [79]. The higher productivities in these supported systems resulted from higher initial activities and not from slower decay rates.

Various investigations of the use of $MgCl_2$-based supports for the immobilization of vanadium complexes have been carried out. Stable activity in ethene polymerization at 22 °C has been reported for a catalyst prepared by ball milling $MgCl_2(THF)_2$ with $VOCl_3$ and used in combination with $AlEt_2Cl$ [80]. The activity was higher than that obtained with $TiCl_4$ and the polyethylene molecular weight was also significantly higher. Subsequent studies showed that the high stability in polymerization was due to much greater resistance to reduction to V(II) than is the case with unsupported $VOCl_3$ or $VCl_4$ [81].

In ethene polymerization, the activity of $MgCl_2$-supported vanadium catalysts has been found to be highly sensitive to the presence of hydrogen, which leads to greater decreases in polymerization activity than are observed with titanium systems [82, 83]. Spitz et al. found that, in the presence of a promoter ($CF_2Cl–CCl_3$), deactivation by hydrogen was reversible and that activity could be restored by the removal of hydrogen [84]. It was postulated that hydrogen might contribute to the reduction of vanadium to inactive species and that this is reversed by the promoter. Czaja and Białek have observed that the effect of hydrogen on lowering the polymerization activity increases in the series $TiCl_4 \ll VOCl_3 < VCl_4$ [85]. It was found that hydrogen had no effect on kinetic stability; catalysts supported on $MgCl_2(THF)_2$ showed stable polymerization activity, irrespective of the presence of hydrogen. The lower activity in the presence of hydrogen was attributed to a lower concentration of active centers. Mikenas et al. reported 2- to 6-fold decreases in the activity of $VCl_4/MgCl_2$ and $VOCl_3/MgCl_2$ in ethene polymerization in the presence of hydrogen, and observed that activity also deceased with increasing cocatalyst ($AliBu_3$) concentration [86]. It was proposed that the vanadium hydride species formed by chain transfer with hydrogen undergo an exchange reaction with $AliBu_3$, generating $AliBu_2H$:

$$V-Pol + H_2 \rightarrow V-H + Pol-H$$

$$V-H + AlR_3 \rightarrow V-R + AlR_2H$$

The alkylaluminum hydride species may then block active sites via strong coordination to give species of type $V(\mu\text{-Pol})(\mu\text{-H})AlR_2$. Removal of hydrogen was suggested to result in reactivation via reaction of $AlR_2H$ with ethene to give $AlR_3$. Increased activity in the presence of hydrogen was observed when $MgCl_2$ or $AlCl_3$ was added as an additional component in the system with the intention of binding alkylaluminum hydrides [87]:

$$MCl_n + AlR_2 H \rightarrow (MCl_n \cdot AlR_2 H)$$

Determinations of the number of active centers in ethene polymerization with $MgCl_2$-supported $VCl_4$ revealed that the number of vanadium–polymer bonds corresponded to 6% of the vanadium present, irrespective of the presence of hydrogen [88]. However, the propagation rate constant was much lower when hydrogen was present. These results provided further support for the formation of dormant $V(\mu\text{-Pol})(\mu\text{-H})AlR_2$ species in the presence of hydrogen.

Ethene polymerization with $MgCl_2/VCl_4$ in the presence of hydrogen has been found to give polyethylene with a broad, bimodal molecular weight distribution [89]. It was proposed that a broadening in the distribution with increasing hydrogen concentration was due to the presence of two types of active center, only one of which was susceptible to chain transfer with hydrogen.

In contrast to the above Ziegler–Natta type systems, based mainly on the use of $VCl_4$ or $VOCl_3$, more attention is now being paid to the development and immobilization of single-center vanadium catalysts for ethene polymerization. Recently, it has been reported that the bis(phenoxy-imine) vanadium complex **9**, immobilized and activated using $MgCl_2/iBu_nAl(OR)_{3-n}$, gives high and stable activity in ethene polymerization at both 50 and 75 °C, whereas with $VOCl_3$ an increase in polymerization temperature led to rapid decay and a decrease in productivity [75–77, 90]. At 75 °C, the immobilized FI catalyst **9** gave a polymerization activity of 65 100 kg mol$^{-1}$ bar$^{-1}$ h$^{-1}$.

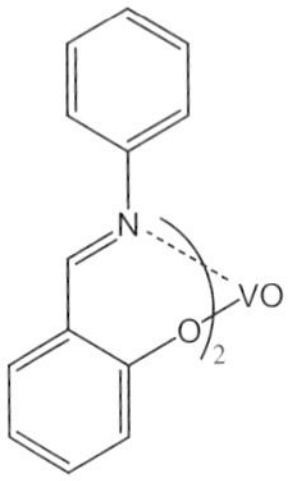

**9** 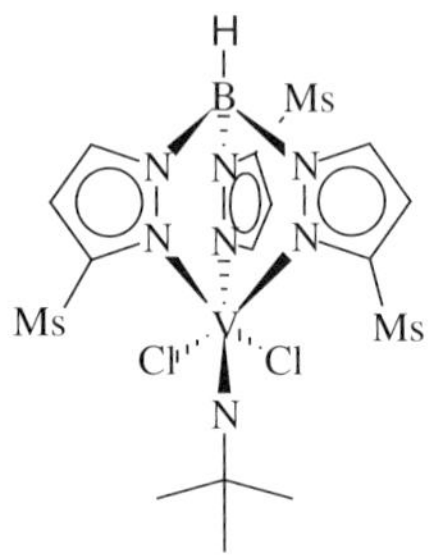                    **10**

Tris(pyrazolyl)borate vanadium catalysts such as **10** (Ms = 2,4,6-trimethylphenyl) have been used together with various inorganic supports [91]. Moderate activities, exceeding those obtained in homogeneous polymerization, were obtained when $SiO_2/MAO$ and the catalyst were added separately to the polymerization reactor in the presence of cocatalyst (AlR$_3$ or MAO) [92]. An increase in polymerization temperature from 30 to 60 °C led to a decrease in activity, indicating limited stability at the higher temperature.

Polystyrene supports have also been used for vanadium polymerization catalysts [93]. Treatment of amino-functionalized polystyrene with CpV(N$t$Bu)Cl$_2$ resulted in the formation of an immobilized imidovanadium complex (Scheme 4.6). Ethene polymerization at 50 °C, using AlEt$_2$Cl as cocatalyst, resulted in low but stable activity during the course of a 1 h polymerization whereas, in homogeneous polymerization, catalysts of type CpV(NR)Cl$_2$ deactivate within minutes due to a reductive bimolecular decomposition process [94]. However, polymerization with the polystyrene-immobilized catalyst at 75 °C led to significantly decreased productivity, which was ascribed to catalyst deactivation at elevated temperature [93].

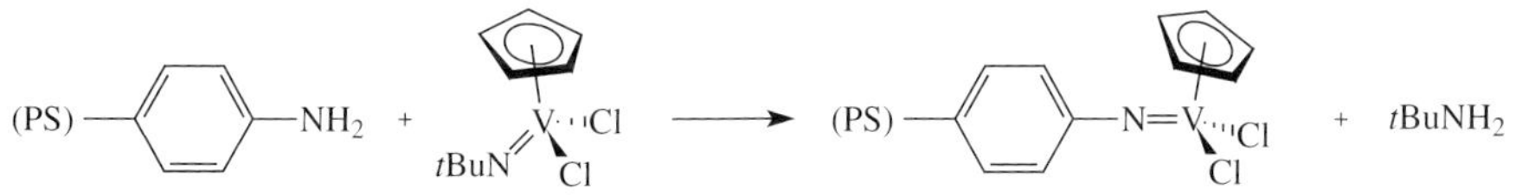

**Scheme 4.6** Immobilization of an imidovanadium complex on amino-functionalized polystyrene [93].

Very stable activity in ethene polymerization at elevated polymerization temperatures has been obtained with a wide range of vanadium complexes immobilized on supports of type MgCl$_2$/AlR$_n$(OEt)$_{3-n}$, prepared by reaction of AlEt$_3$ or Al$i$Bu$_3$ with solid MgCl$_2$/EtOH adducts. Ethene polymerization at 50 and 70 °C with the bis (imino)pyridyl complex **11** (Ar = 2,6-diisopropylphenyl) immobilized on such a support, using AlEt$_3$ as cocatalyst, resulted in activities of 560 and 2140 kg mol$^{-1}$ bar$^{-1}$ h$^{-1}$, respectively [95]. No deactivation was apparent during the course of 1 h polymerization. Furthermore, high molecular weight polyethylene with narrow (Schulz–Flory) molecular weight distribution was obtained, confirming the single-center characteristic of this catalyst system, in contrast to the relatively broad molecular weight distributions obtained with analogous bis(imino)pyridyl iron complexes, both supported and unsupported. Similar results were obtained with complex **12**, indicating that the precursor for the active species in this system is likely to originate via alkylation of the pyridine ring, leading to a decrease in the metal coordination number [96].

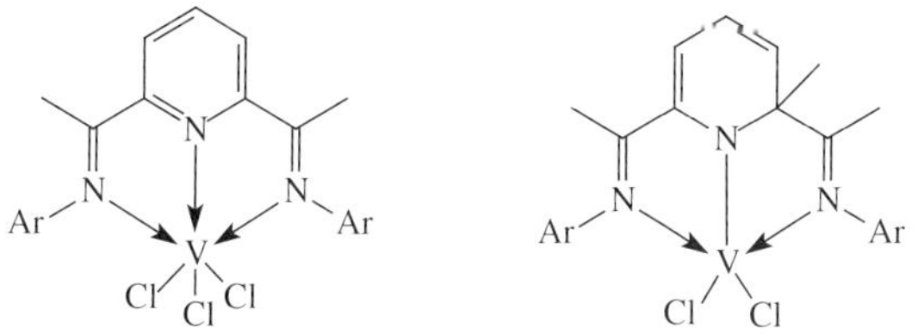

11                              12

The use of a support of composition $MgCl_2 \cdot 0.24AlEt_{2.3}(OEt)_{0.7}$ for the immobilization and activation of the vanadium(III) amidinate complexes **13** and **14** resulted in significantly higher activities (1500–3100 kg mol$^{-1}$ bar$^{-1}$ h$^{-1}$ at 50 °C) than had previously been obtained in homogeneous polymerization, again illustrating the stabilizing effect of a $MgCl_2$ support [97]. Recently, it has been demonstrated that the productivities of vanadium phebox and pincer catalysts increase by orders of magnitude when they are immobilized on a $MgCl_2/AlR_n(OEt)_{3-n}$ support [98]. Under homogeneous polymerization conditions, with MAO as cocatalyst, complex **15**, containing the 2,6-bis(2′-oxazolinyl)phenyl ligand (phebox), deactivated rapidly and gave an activity of around 100 kg mol$^{-1}$ bar$^{-1}$ h$^{-1}$. After immobilization, activities in the range 500–2000 kg mol$^{-1}$ bar$^{-1}$ h$^{-1}$ were obtained. The NCN pincer complex **16** also gave poor activity (130 kg mol$^{-1}$ bar$^{-1}$ h$^{-1}$ at 70 °C) in homogeneous polymerization, but very high activity (up to 30 000 kg mol$^{-1}$ bar$^{-1}$ h$^{-1}$) when immobilized on $MgCl_2$ supports [57, 98]. Furthermore, the $MgCl_2$-immobilized vanadium complexes **13–16** all gave narrow molecular weight distribution polyethylene, in contrast to their titanium counterparts which gave more polydisperse polymers. This fundamental difference between the single-center behavior of the immobilized vanadium catalysts, as opposed to the deviation from single-center catalysis in the case of titanium, was confirmed by rheological characterization of the resulting polymers [62]. Vanadium catalysts also give greater comonomer incorporation in ethene/α-olefin copolymerization than their titanium counterparts, which together with their retention of single-center properties after immobilization (leading to random comonomer incorporation) makes them interesting candidates for the synthesis of copolymers with narrow chemical composition distribution.

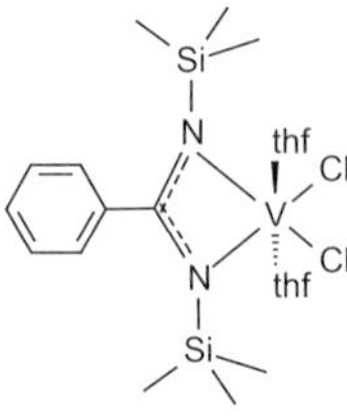

**13** 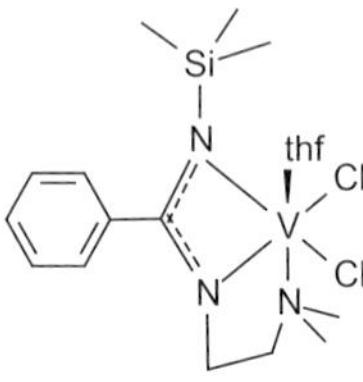                    **14**

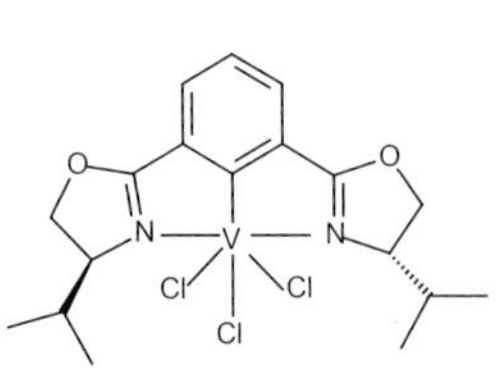

**15** 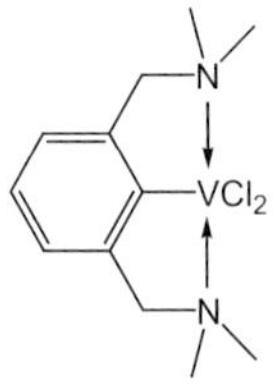                    **16**

## 4.5
## Chromium Complexes

Recent developments in the synthesis of well-defined, single-center chromium catalysts for ethene polymerization have been described in Chapter 2. Catalyst deactivation is a common phenomenon in homogeneous polymerization, but stable activity can be obtained by immobilization of a chromium complex on a suitable support. Uozumi and coworkers reported moderate but stable activity in ethene polymerization with the catalyst system $MgCl_2/Cr(acac)_3$–$AlEt_2Cl$, giving high molecular weight polyethylene with narrow molecular weight distribution [99]. Esteruelas *et al.* obtained stable polymerization activity at $80\,°C$ after immobilization of a bis(imino)pyridyl chromium(III) complex on MAO-impregnated silica [100].

Monoi, Yasuda and coworkers have observed that immobilization of chromium complexes on a silica support can lead to significant increases in activity in aluminoxane-activated polymerization, although in most cases polymers with high polydispersities were obtained [101]. Narrow molecular weight distribution polyethylene has been synthesized using the half-sandwich complex **17** immobilized on $MgCl_2/AlR_n(OEt)_{3-n}$ supports obtained by the reaction of various aluminum trialkyls with an adduct $MgCl_2 \cdot 1.1EtOH$ [102]. At $50\,°C$, using $AliBu_3$ as cocatalyst/scavenger, polymerization activities in the range $1900$–$2700\,kg\,mol^{-1}\,bar^{-1}\,h^{-1}$ were obtained, comparable to activities obtained under homogeneous conditions [103, 104] with MAO as cocatalyst. Confirmation of single-center catalysis, giving high molecular weight polyethylene with narrow molecular weight distribution, was obtained by characterization of the melt rheological properties of the polymers [102]. The use of controlled-morphology $MgCl_2$-based supports resulted in polymers with spherical particle morphology, as shown in Figure 4.3, with no evidence of reactor fouling.

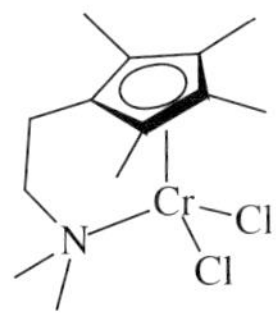

17 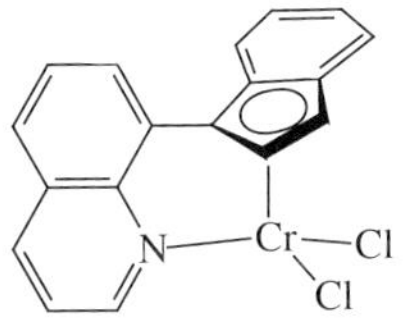                                    18

High and stable activities have also been obtained with 1-(8-quinolyl)indenyl chromium(III) dichloride **18** immobilized on $MgCl_2/AlR_n(OEt)_{3-n}$ supports [105]. This complex has also been used in combination with a bis(imino)pyridyl iron catalyst [106–108]. Co-immobilization of the two catalysts on a single support generates an intimate blend of high- (Cr-catalyzed) and relatively low-molecular weight (Fe-catalyzed) polyethylene. The crystalline structure and properties of such bimodal polymers are strongly influenced by shear-induced orientation during processing in the melt.

**Figure 4.3** Scanning electron micrograph of polyethylene prepared using complex **17** immobilized on a $MgCl_2/AlR_n(OEt)_{3-n}$ support [102].

## 4.6
## Nickel Complexes

Various approaches for the immobilization of nickel diimine polymerization catalysts have been followed. $SiO_2$/MAO-based systems, analogous to those widely used in zirconocene immobilization and activation, have involved either the use of MAO-impregnated silica or contact of silica with a Ni diimine/MAO solution [109–111]. However, the ethene polymerization activities of such silica-supported systems have been found to be much lower than activities obtained under homogeneous conditions and the supported catalyst also gave less chain walking than was observed in homogeneous polymerization, resulting in polymers with less short-chain branching [112]. It has, however, been noted that an increase in polymerization temperature leads to a stronger increase in chain branching than is seen in homogeneous polymerization; this effect was ascribed to the solubility of branched polyethylene in the polymerization medium at temperatures exceeding 55 °C [113].

Preishuber-Pflugl and Brookhart considered that the low activities obtained with silica-supported nickel diimines resulted from the use of MAO as cocatalyst [114]. They synthesized covalently-supported catalysts via the reaction of a Ni(II) diimine complex containing hydroxy or amino functionality with $AlMe_3$-pretreated silica, as illustrated in Scheme 4.7. Ethene polymerization with these supported catalysts was carried out using $Al_2Et_3Cl_3$ or $AlMeCl_2$ as cocatalyst. The activities, up to 2330 kg $mol^{-1} bar^{-1} h^{-1}$ at 60 °C, were more than an order of magnitude higher than those obtained with $SiO_2$/MAO. $Al_2Et_3Cl_3$ gave higher activity than $AlMeCl_2$ at 60 °C but at 80 °C this trend was reversed, which was attributed to faster deactivation with $Al_2Et_3Cl_3$ at high temperature. Further studies showed that productivities of up to

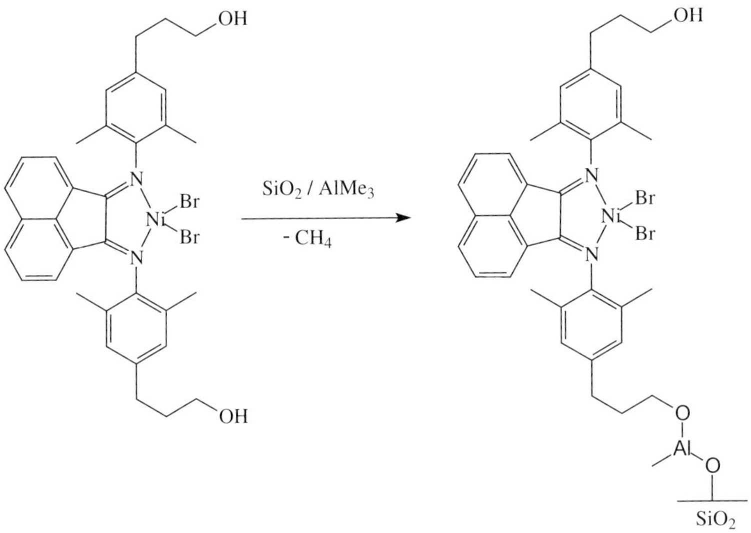

**Scheme 4.7** Immobilization of a nickel diimine complex on silica [114].

6 kg polymer per g (catalyst + support) could be obtained at 80 °C and high (48 bar) ethene pressure, with a 3 wt% loading of Ni on the support [115]. Related studies have been reported by Wang and coworkers [116].

Neutral nickel(II) complexes containing a hydroxy functionality have also been supported on silica pretreated with AlMe₃, as illustrated in Scheme 4.8. Li and coworkers reported that this approach resulted in heterogeneous catalysts having about half the activity of their homogeneous counterparts and gave polyethylenes with higher molecular weight and a lower degree of branching [117].

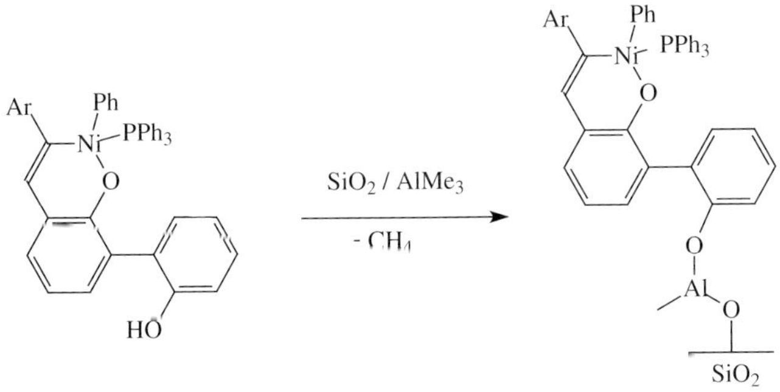

**Scheme 4.8** Immobilization of a salicylaldimine-based nickel complex on silica [117].

Magnesium chloride supports have also proved to be very effective for the immobilization and activation of nickel complexes. Immobilization of various nickel(II) diimine complexes on a support of composition $MgCl_2 \cdot 0.24AlEt_{2.3}(OEt)_{0.7}$, prepared by the reaction of $AlEt_3$ with $MgCl_2 \cdot 2.1EtOH$, followed by ethene polymerization at $50\,^\circ C$ in the presence of Ali$Bu_3$, gave activities of up to about $7000\,kg\,mol^{-1}\,bar^{-1}\,h^{-1}$ [118]. These activities were significantly higher than those previously reported for homogeneous polymerization, or for nickel diimine complexes immobilized on silica, but decay in activity during polymerization was observed. The productivity in the second 30 min of polymerization was around a quarter of that in the first 30 min. Stable polymerization kinetics have, however, been reported for a nickel diimine immobilized on a magnesium chloride support prepared by thermal dealcoholation of an adduct $MgCl_2 \cdot 2.6EtOH$ and containing 2 wt% residual ethanol [119].

Significant increases in the productivity of various iron-, chromium- and titanium-based heterogeneous catalysts for ethene polymerization have been achieved by incorporation into the solid catalyst of a limited amount of a nickel diimine [105, 120]. The formation of nickel-catalyzed branched polyethylene during the early stages of polymerization reduces the monomer diffusion limitation inherent in ethene homopolymerization, thereby increasing the productivity of the main (linear polyethylene-producing) catalyst component. The final products are essentially linear, high-density polyethylenes containing very small amounts of branched polymer.

## 4.7
## Iron Complexes

Substantial improvements in the productivity of bis(imino)pyridyl iron complexes ($LFeCl_2$) in ethene polymerization can be achieved by immobilization on a support. In contrast to the rapid deactivation observed in homogeneous polymerization, particularly at elevated temperatures, stable polymerization activity can be obtained when the iron complex is immobilized on supports such as $SiO_2$ and $MgCl_2$. Semikolenova *et al.* found that $SiO_2/LFeCl_2$–Ali$Bu_3$ gave stable polymerization activity at $70\,^\circ C$, whereas rapid decay was apparent even at $35\,^\circ C$ in the homogeneous system $LFeCl_2$–Ali$Bu_3$ [121]. Similar improvements in catalyst stability have been noted for bis(imino)pyridyl iron complexes chemically tethered to a silica support surface [122–124].

Diffuse-reflectance IR spectroscopy (DRIFTS) has been used to study the interaction of the iron complex **19** with silica and alumina [125]. Only a small proportion of the surface hydroxy groups in silica was able to interact with $LFeCl_2$, giving supported catalysts with low iron loadings. Higher loadings were obtained in the system $Al_2O_3/LFeCl_2$ as a result of interaction of the iron complex with Lewis acidic sites on alumina and hydrogen bond formation between the ligand L and Al-OH groups. Ray and Sivaram concluded, on the basis of XPS studies, that there was no strong interaction between $LFeCl_2$ and the surface of silica [126]. They also observed that, while the silica-supported catalyst gave much more stable polymerization kinetics than was obtained under homogeneous conditions, complete elimination of hydroxy groups from silica via pretreatment with MAO had an adverse effect on catalyst activity. A lower degree of stabilization of active centers by the support was suggested to explain this effect.

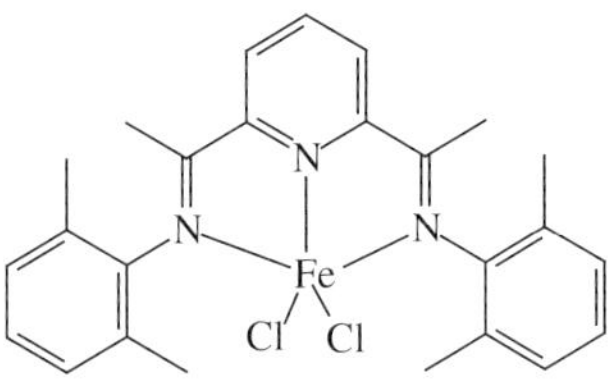

**19** 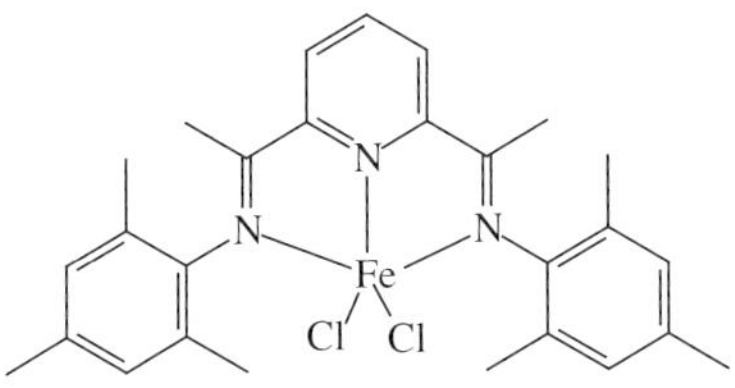                    **20**

Determinations of the numbers of active centers in supported bis(imino)pyridyl iron catalysts, activated with Al$i$Bu$_3$ as cocatalyst, have been carried out by Barabanov *et al.* using $^{14}$CO inhibition [127]. Whereas previous determinations carried out on homogeneous systems [128] had revealed that up to 41% of the total iron present was catalytically active, only 2–4% of the iron immobilized on SiO$_2$, Al$_2$O$_3$ or MgCl$_2$ was active. The support composition affected neither the molecular weight nor the molecular weight distribution of the polyethylene produced, leading to the conclusion that the nature of the support had little effect on the structure of the active centers.

Magnesium chloride is a particularly effective support material for iron catalysts. MgCl$_2$ prepared by reaction of magnesium with excess *n*-BuCl was found to give higher activity than SiO$_2$ or Al$_2$O$_3$ supports [127, 129, 130]. The maximum amount of LFeCl$_2$ which could be immobilized was around 150 µmol g$^{-1}$, close to the concentration of Lewis acidic support sites and similar to the maximum amount of TiCl$_4$ closely bound to the same support in a MgCl$_2$/TiCl$_4$ catalyst. Huang *et al.* have developed the use of controlled-morphology supports of composition MgCl$_2$/AlEt$_n$(OEt)$_{3-n}$, prepared by reaction of AlEt$_3$ with MgCl$_2$/EtOH adducts having spherical particle morphology [95, 131, 132]. The ethene polymerization activity of complex **20** immobilized on such supports, at 70 °C and with AlEt$_3$ as cocatalyst, was an order of magnitude higher than in homogeneous polymerization [131]. Polyethylene with relatively broad molecular weight distribution was obtained, but there was no indication of the bimodality (ascribed to chain transfer to aluminum) often observed in homogeneous, MAO-activated polymerization [95, 132]. The polymer molecular weights were higher than those obtained in homogeneous polymerization, although the effect of catalyst immobilization on increasing the polymer molecular weight was less spectacular than is frequently observed with early-transition metal catalysts.

## 4.8
## Conclusions

Immobilization on a particulate support material is a vital requirement for widespread implementation of homogeneous catalysts in polyolefin production processes. The effects of immobilization on catalyst activity are dependent on both chemical and physical factors. Decreased catalyst productivity, particularly in ethene polymerization, can result if the support does not undergo easy fragmentation to facilitate particle replication and growth during polymerization. Limited monomer diffusion into the growing particle can occur with supports having low friability and narrow pore size. Chemical interaction of the catalyst with hydroxy or other functional groups on the support surface can also lead to decreased activity. On the other hand, significant increases in catalyst activity can be obtained in cases where immobilization of a catalyst on a support greatly improves the stability of the active species in polymerization.

The immobilization of a metallocene on a silica support, typically via impregnation of the support with a solution of MAO and zirconocene in toluene followed by removal of the solvent, generally leads to lower activity than is obtained in homogeneous polymerization with zirconocene and MAO. However, it has been shown that immobilization leads to increased stability during polymerization and that $\alpha$-H transfer to give inactive $Zr-CH_2-Al$ or $Zr-CH_2-Zr$ species is suppressed. Furthermore, immobilized zirconocenes require much less MAO than their homogeneous counterparts, for which extremely high MAO/Zr ratios are required for optimum activity. Al$i$Bu$_3$ is often used to scavenge impurities in polymerizations involving immobilized zirconocenes and other single-center catalysts and its presence can also lead to a further improvement in catalyst stability.

Dramatic improvements in the stability of a wide range of titanium, vanadium and iron catalysts can be achieved by immobilization on a support. Magnesium chloride is particularly effective in this respect, and in the case of titanium and vanadium complexes it is likely that immobilization on $MgCl_2$ stabilizes the active species, preventing reduction to lower, less active oxidation states. For example, high and stable activities in ethene polymerization at 70–75 °C have been obtained with vanadium complexes containing bis(phenoxy-imine) and NCN-pincer ligands, in contrast to the rapid decay and poor activity generally obtained with vanadium catalysts in homogeneous polymerization. The single-center characteristics of zirconium, vanadium and chromium polymerization catalysts are retained after immobilization, whereas this is not the case for the corresponding titanium complexes immobilized on $MgCl_2$.

## References

1 Hlatky, G.G. (2000) *Chem. Rev.*, **100**, 1347–1376.
2 Severn, J.R., Chadwick, J.C., Duchateau, R., and Friederichs, N. (2005) *Chem. Rev.*, **105**, 4073–4147.
3 Severn, J.R. and Chadwick, J.C. (eds) (2008) *Tailor-made Polymers via Immobilization of Alpha-Olefin Polymerization Catalysts*, Wiley-VCH Verlag GmbH, Weinheim.
4 Ribeiro, M.R., Deffieux, A., and Portela, M.F. (1997) *Ind. Eng. Chem. Res.*, **36**, 1224–1237.

5 Chien, J.C.W. (1999) *Top. Catal.*, **7**, 23–36.
6 Kaminsky, W. and Winkelbach, H. (1999) *Top. Catal.*, **7**, 61–67.
7 Kristen, M.O. (1999) *Top. Catal.*, **7**, 89–95.
8 Carnahan, E.M. and Jacobsen, G.B. (2000) *CatTech*, **4**, 74–88.
9 Wang, W. and Wang, L. (2003) *J. Polym. Mater.*, **20**, 1–8.
10 Alt, H.G. (1999) *J. Chem. Soc., Dalton Trans.*, 1703–1709.
11 Bartke, M., Oksman, M., Mustonen, M., and Denifl, P. (2005) *Macromol. Mater. Eng.*, **290**, 250–255.

12 Panchenko, V.N., Semikolenova, N.V., Danilova, I.G., Paukshtis, E. A., and Zakharov, V.A. (1999) *J. Mol. Catal.*, **142**, 27–37.

13 Hammawa, H. and Wanke, S.E. (2007) *J. Appl. Polym. Sci.*, **104**, 514–527.

14 Hassan Nejad, M., Ferrari, P., Pennini, G., and Cecchin, G. (2008) *J. Appl. Polym. Sci.*, **108**, 3388–3402.

15 Kaminsky, W. and Renner, F. (1993) *Makromol. Chem., Rapid Commun.*, **14**, 239–243.

16 Kaminsky, W. and Strübel, C. (1998) *J. Mol. Catal. A: Chem.*, **128**, 191–200.

17 Arrowsmith, D., Kaminsky, W., Laban, A., and Weingarten, U. (2001) *Macromol. Chem. Phys.*, **202**, 2161–2167.

18 Frediani, M. and Kaminsky, W. (2003) *Macromol. Chem. Phys.*, **204**, 1941–1947.

19 Busico, V. and Cipullo, R. (1994) *J. Am. Chem. Soc.*, **116**, 9329–9330.

20 Kaminsky, W., Müller, F., and Sperber, O. (2005) *Macromol. Mater. Eng.*, **290**, 347–352.

21 Hammawa, H., Mannan, T.M., Lynch, D.T., and Wanke, S.E. (2004) *J. Appl. Polym. Sci.*, **92**, 3549–3560.

22 Bonini, F., Fraaije, V., and Fink, G. (1995) *J. Polym. Sci., Part A: Polym. Chem.*, **33**, 2393–2402.

23 Zechlin, J., Hauschild, K., and Fink, G. (2000) *Macromol. Chem. Phys.*, **201**, 597–603.

24 Zechlin, J., Steinmetz, B., Tesche, B., and Fink, G. (2000) *Macromol. Chem. Phys.*, **201**, 515–524.

25 Knoke, S., Korber, F., Fink, G., and Tesche, B. (2003) *Macromol. Chem. Phys.*, **204**, 607–617.

26 Zheng, X., Smit, M., Chadwick, J.C., and Loos, J. (2005) *Macromolecules*, **38**, 4673–4678.

27 Przybyla, C., Tesche, B., and Fink, G. (1999) *Macromol. Rapid Commun.*, **20**, 328–332.

28 Smit, M., Zheng, X., Brüll, R., Loos, J., Chadwick, J.C., and Koning, C.E. (2006) *J. Polym. Sci., Part A: Polym. Chem.*, **44**, 2883–2890.

29 Hammawa, H. and Wanke, S.E. (2006) *Polym. Int.*, **55**, 426–434.

30 Kamfjord, T., Wester, T.S., and Rytter, E. (1998) *Macromol. Rapid Commun.*, **19**, 505–509.

31 Rytter, E. and Ott, M. (2001) *Macromol. Rapid Commun.*, **22**, 1427–1431.

32 Galland, G.B., Seferin, M., Guimarães, R., Rohrmann, J.A., Stedile, F.C., and dos Santos, J.H.Z. (2002) *J. Mol. Catal. A: Chem.*, **189**, 233–240.

33 Guimarães, R., Stedile, F.C., and dos Santos, J.H.Z. (2003) *J. Mol. Catal. A: Chem.*, **206**, 353–362.

34 Tian, J., Wang, S., Feng, Y., Li, J., and Collins, S. (1999) *J. Mol. Catal. A: Chem.*, **144**, 137–150.

35 Charoenchaidet, S., Chavadej, S., and Gulari, E. (2002) *J. Mol. Catal. A: Chem.*, **185**, 167–177.

36 (a) Walzer, J.F. (1999) US Patent 5,972,823; (1999) *Chem. Abstr.*, **131**, 310941. (b) Walzer, J.F. US Patent (1997) 5,643,847; (1997) *Chem. Abstr.*, **127**, 122104.

37 Bochmann, M., Jiménez Pindado, G., and Lancaster, S.J. (1999) *J. Mol. Catal. A: Chem.*, **146**, 179–190.

38 Ward, D.G. and Carnahan, E.M. (1999) US Patent 5,939,347; (1999) *Chem. Abstr.*, **131**, 158094.

39 Charoenchaidet, S., Chavadej, S., and Gulari, E. (2002) *Macromol. Rapid Commun.*, **23**, 426–431.

40 Charoenchaidet, S., Chavadej, S., and Gulari, E. (2002) *J. Polym. Sci., Part A: Polym. Chem.*, **40**, 3240–3248.

41 Nicholas, C.P., Ahn, H., and Marks, T.J. (2003) *J. Am. Chem. Soc.*, **125**, 4325–4331.

42 McDaniel, M.P., Jensen, M.D., Jayaratne, K., Collins, K.S., Benham, E.A., McDaniel, N.D., Das, P.K., Martin, J.L., Yang, Q., Thorn, M.G., and Masino, A.P. (2008) in *Tailor-made Polymers via Immobilization of Alpha-Olefin Polymerization Catalysts* (eds J.R. Severn and J.C. Chadwick), Wiley-VCH Verlag GmbH, Weinheim, pp. 171–210.

43 Hicks, J.C., Mullis, B.A., and Jones, C.W. (2007) *J. Am. Chem. Soc.*, **129**, 8426–8427.

44 Marks, T.J. (1992) *Acc. Chem. Res.*, **25**, 57–65.

**45** Zakharov, V.A., Paukshtis, E.A., Mikenas, T.B., Volodin, A.M., Vitus, E.N., and Potapov, A.G. (1995) *Macromol. Symp.*, **89**, 55–61.

**46** Mikenas, T.B., Zakharov, V.A., Echevskaya, L.G., and Matsko, M.A. (2005) *J. Polym. Sci., Part A: Polym. Chem.*, **43**, 2128–2133.

**47** Guan, Z., Zheng, Y., and Jiao, S. (2002) *J. Mol. Catal. A: Chem.*, **188**, 123–131.

**48** Cho, H.S. and Lee, W.Y. (2003) *J. Mol. Catal. A: Chem.*, **191**, 155–165.

**49** Ochędzan-Siodłak, W. and Nowakowska, M. (2005) *Eur. Polym. J.*, **41**, 941–947.

**50** Aragón Sáez, P.J., Carrillo-Hermosilla, F., Villaseñor, E., Otero, A., Antiñolo, A., and Rodríguez, A.M. (2008) *Eur. J. Inorg. Chem.*, 330–337.

**51** Severn, J.R. and Chadwick, J.C. (2004) *Macromol. Rapid Commun.*, **25**, 1024–1028.

**52** Kissin, Y.V., Nowlin, T.E., Mink, R.I., and Brandolini, A.J. (2000) *Macromolecules*, **33**, 4599–4601.

**53** Kaminaka, M. and Soga, K. (1991) *Makromol. Chem., Rapid Commun.*, **12**, 367–372.

**54** Soga, K. and Kaminaka, M. (1993) *Makromol. Chem.*, **194**, 1745–1755.

**55** Echevskaya, L.G., Zakharov, V.A., Semikolenova, N.V., Mikenas, T.B., and Sobolev, A.P. (2001) *Polym. Sci., Ser. A*, **43**, 220–227.

**56** Huang, R., Duchateau, R., Koning, C.E., and Chadwick, J.C. (2008) *Macromolecules*, **41**, 579–590.

**57** Huang, R., Malizia, F., Pennini, G., Koning, C.E., and Chadwick, J.C. (2008) *Macromol. Rapid Commun.*, **29**, 1732–1738.

**58** Babushkin, D.E. and Brintzinger, H.H. (2007) *Chem. Eur. J.*, **13**, 5294–5299.

**59** Satyanarayana, G. and Sivaram, S. (1993) *Macromolecules*, **26**, 4712–4714.

**60** Severn, J.R. and Chadwick, J.C. (2004) *Macromol. Chem. Phys.*, **205**, 1987–1994.

**61** Kang, K.K., Oh, J.K., Jeong, Y.-T., Shiono, T., and Ikeda, T. (1999) *Macromol. Rapid Commun.*, **20**, 308–311.

**62** Kukalyekar, N., Huang, R., Rastogi, S., and Chadwick, J.C. (2007) *Macromolecules*, **40**, 9443–9450.

**63** Jacobsen, G.B., Wijkens, P., Jastrzebski, J.T.B.H., and van Koten, G. (1998) US Patent 5,834,393;(1998) *Chem. Abstr.*, **130**, 14330.

**64** Jacobsen, G.B., Loix, P.H.H., and Stevens, T.J.P. (2001) US Patent 6,271,165;(2001) *Chem. Abstr.*, **135**, 153238.

**65** Mealares, C.M.-C. and Taylor, M.J. (2002) Int. Patent WO 02/06357;(2002) *Chem. Abstr.*, **136**, 118861.

**66** Smit, M., Severn, J.R., Zheng, X., Loos, J., and Chadwick, J.C. (2006) *J. Appl. Polym. Sci.*, **99**, 986–993.

**67** Galan-Fereres, M., Koch, T., Hey-Hawkins, E., and Eisen, M.S. (1999) *J. Organometal. Chem.*, **580**, 145–155.

**68** McKittrick, M.W. and Jones, C.W. (2004) *J. Am. Chem. Soc.*, **126**, 3052–3053.

**69** Jones, C.W., McKittrick, M.W., Nguyen, J.V., and Yu, K. (2005) *Top. Catal.*, **34**, 67–76.

**70** Hicks, J.C. and Jones, C.W. (2008) in *Tailor-made Polymers via Immobilization of Alpha-Olefin Polymerization Catalysts* (eds J.R. Severn and J.C. Chadwick), Wiley-VCH Verlag GmbH, Weinheim, pp. 239–260.

**71** Repo, T., Klinga, M., Pietikäinen, P., Leskelä, M., Uusitalo, A.-M., Pakkanen, T., Hakka, K., Aaltonen, P., and Löfgren, B. (1997) *Macromolecules*, **30**, 171–175.

**72** Białek, M., Garłovska, A., and Liboska, O. (2009) *J. Polym. Sci., Part A: Polym. Chem.*, **47**, 4811–4821.

**73** Nakayama, Y., Bando, H., Sonobe, Y., Kaneko, H., Kashiwa, N., and Fujita, T. (2003) *J. Catal.*, **215**, 171–175.

**74** Nakayama, Y., Saito, J., Bando, H., and Fujita, T. (2005) *Macromol. Chem. Phys.*, **206**, 1847–1852.

**75** Nakayama, Y., Bando, H., Sonobe, Y., and Fujita, T. (2004) *J. Mol. Catal. A: Chem.*, **213**, 141–150.

**76** Nakayama, Y., Bando, H., Sonobe, Y., and Fujita, T. (2004) *Bull. Chem. Soc. Jpn.*, **77**, 617–625.

**77** Nakayama, Y., Saito, J., Bando, H., and Fujita, T. (2006) *Chem. Eur. J.*, **12**, 7546–7556.

**78** Ribeiro, M.R., Deffieux, A., Fontanille, M., and Portela, M.F. (1995) *Macromol. Chem. Phys.*, **196**, 3833–3844.

**79** Ribeiro, M.R., Deffieux, A., Fontanille, M., and Portela, M.F. (1996) *Eur. Polym. J.*, **32**, 811–819.

**80** Czaja, K. and Białek, M. (1996) *Macromol. Rapid Commun.*, **17**, 253–260.

**81** Czaja, K. and Białek, M. (1998) *Macromol. Rapid Commun.*, **19**, 163–166.

**82** Hsieh, H.L., McDaniel, M.P., Martin, J.L., Smith, P.D., and Fahey, D.R. (1987) in *Advances in Polyolefins* (eds R.B. Seymour and T. Cheng), Plenum Press, New York, pp. 153–169.

**83** Zhou, X., Lin, S., and Chien, J.C.W. (1990) *J. Polym. Sci.: Part A: Polym. Chem.*, **28**, 2609–2632.

**84** Spitz, R., Pasquet, V., Patin, M., and Guyot, A. (1995) in *Ziegler Catalysts. Recent Scientific Innovations and Technological Improvements* (eds G. Fink, R. Mülhaupt, and H.H. Brintzinger), Springer-Verlag, Berlin, pp. 401–411.

**85** Czaja, K. and Białek, M. (2001) *J. Appl. Polym. Sci.*, **79**, 361–365.

**86** Mikenas, T.B., Zakharov, V.A., Echevskaya, L.G., and Matsko, M.A. (2000) *Polimery*, **45**, 349–352.

**87** Mikenas, T.B., Zakharov, V.A., Echevskaya, L.G., and Matsko, M.A. (2001) *Macromol. Chem. Phys.*, **202**, 475–481.

**88** Matsko, M.A., Bukatov, G.D., Mikenas, T.B., and Zakharov, V.A. (2001) *Macromol. Chem. Phys.*, **202**, 1435–1439.

**89** Echevskaya, L.G., Matsko, M.A., Mikenas, T.B., and Zakharov, V.A. (2006) *Polym. Int.*, **55**, 165–170.

**90** Nakayama, Y., Bando, H., Sonobe, Y., Suzuki, Y., and Fujita, T. (2003) *Chem. Lett. (Japan)*, **32**, 766–767.

**91** Casagrande, A.C.A., Tavares, T.T. da R., Kuhn, M.C.A., Casagrande, O.L., dos Santos, J.H.Z., and Teranishi, T. (2004) *J. Mol. Catal. A: Chem.*, **212**, 267–275.

**92** Casagrande, A.C.A., dos Anjos, P.S., Gamba, D., Casagrande, O.L., dos Santos, J.H.Z., and Teranishi, T. (2006) *J. Mol. Catal. A: Chem.*, **255**, 19–24.

**93** Chan, M.C.W., Chew, K.C., Dalby, C.I., Gibson, V.C., Kohlmann, A., Little, I.R., and Reed, W. (1998) *Chem. Commun.*, 1673–1674.

**94** Chan, M.C.W., Cole, J.M., Gibson, V.C., and Howard, J.A.K. (1997) *Chem. Commun.*, 2345–2346.

**95** Huang, R., Kukalyekar, N., Koning, C.E., and Chadwick, J.C. (2006) *J. Mol. Catal. A: Chem.*, **260**, 135–143.

**96** Reardon, D., Conan, F., Gambarotta, S., Yap, G., and Wang, Q. (1999) *J. Am. Chem. Soc.*, **121**, 9318–9325.

**97** Severn, J.R., Duchateau, R., and Chadwick, J.C. (2005) *Polym. Int.*, **54**, 837–841.

**98** Huang, R. (2008) Immobilization and Activation of Early- and Late-Transition Metal Catalysts for Ethene Polymerization using MgCl$_2$-based Supports, Ph.D. thesis, Eindhoven University of Technology.

**99** Takawaki, K., Uozumi, T., Ahn, C.-H., Tian, G., Sano, T., and Soga, K. (2000) *Macromol. Chem. Phys.*, **201**, 1605–1609.

**100** Esteruelas, M.A., López, A.M., Méndez, L., Oliván, M., and Oñate, E. (2003) *Organometallics*, **22**, 395–406.

**101** Ikeda, H., Monoi, T., Ogata, K., and Yasuda, H. (2001) *Macromol. Chem. Phys.*, **202**, 1806–1811.

**102** Severn, J.R., Kukalyekar, N., Rastogi, S., and Chadwick, J.C. (2005) *Macromol. Rapid Commun.*, **26**, 150–154.

**103** Emrich, R., Heinemann, O., Jolly, P.W., Krüger, C., and Verhovnik, G.P.J. (1997) *Organometallics*, **16**, 1511–1513.

**104** Döring, A., Göhre, J., Jolly, P.W., Kryger, B., Rust, J., and Verhovnik, G.P.J. (2000) *Organometallics*, **19**, 388–402.

**105** Huang, R., Koning, C.E., and Chadwick, J.C. (2007) *Macromolecules*, **40**, 3021–3029.

**106** Kukalyekar, N. (2007) Bimodal Polyethylenes from One-Pot Synthesis; Effect of Flow-Induced Crystallization on Physical Properties, Ph.D. thesis, Eindhoven University of Technology.

**107** Kukalyekar, N., Balzano, L., Chadwick, J.C., and Rastogi, S. (2007) *PMSE Preprints*, **96**, 840–841.

108 Balzano, L., Kukalyekar, N., Rastogi, S., Peters, G.W.M., and Chadwick, J.C. (2008) *Phys. Rev. Lett.*, **100**, 048302.

109 Vaughan, G.A., Canich, J.A.M., Matsunaga, P.T., Grindelberger, D.E., and Squire, K.R. (1997) Int. Patent WO 97/48736;(1997) *Chem. Abstr.*, **128**, 89235.

110 Bennett, A.M.A. and McLain, S.D. (1998) Int. Patent WO 98/56832;(1998) *Chem. Abstr.*, **130**, 66907.

111 MacKenzie, P.B., Moody, L.S., Killian, C.M., and Lavoie, G.G. (1999) Int. Patent WO 99/62968;(1999) *Chem. Abstr.*, **132**, 36184.

112 AlObaidi, F., Ye, Z., and Zhu, S. (2003) *Macromol. Chem. Phys.*, **204**, 1653–1659.

113 Simon, L.C., Patel, H., Soares, J.B.P., and de Souza, R.F. (2001) *Macromol. Chem. Phys.*, **202**, 3237–3247.

114 Preishuber-Pfugl, P. and Brookhart, M. (2002) *Macromolecules*, **35**, 6074–6076.

115 Schrekker, H.S., Kotov, V., Preishuber-Pfugl, P., White, P., and Brookhart, M. (2006) *Macromolecules*, **39**, 6341–6354.

116 Jiang, H., Wu, Q., Zhu, F., and Wang, H. (2007) *J. Appl. Polym. Sci.*, **103**, 1483–1489.

117 Hu, T., Li, Y.-G., Lu, J.-Y., and Li, Y.-S. (2007) *Organometallics*, **26**, 2609–2615.

118 Severn, J.R., Chadwick, J.C., and Van Axel Castelli, V. (2004) *Macromolecules*, **37**, 6258–6259.

119 Xu, R., Liu, D., Wang, S., and Mao, B. (2006) *Macromol. Chem. Phys.*, **207**, 779–786.

120 Huang, R. and Chadwick, J.C. (2007) Int. Patent WO 2007/111499.

121 Semikolenova, N.V., Zakharov, V.A., Talsi, E.P., Babushkin, D.E., Sobolev, A.P., Echevskaya, L.G., and Khysniyarov, M.M. (2002) *J. Mol. Catal. A: Chem.*, **182–183**, 283–294.

122 Kaul, F.A.R., Puchta, G.T., Schneider, H., Bielert, F., Mihalios, D., and Herrmann, W.A. (2002) *Organometallics*, **21**, 74–82.

123 Zheng, Z., Liu, J., and Li, Y. (2005) *J. Catal.*, **234**, 101–110.

124 Han, W., Müller, C., Vogt, D., Niemantsverdriet, J.W., and Thüne, P.C. (2006) *Macromol. Rapid Commun.*, **27**, 279–283.

125 Semikolenova, N.V., Zakharov, V.A., Paukshtis, E.A., and Danilova, I.G. (2005) *Top. Catal.*, **32**, 77–82.

126 Ray, S. and Sivaram, S. (2006) *Polym. Int.*, **55**, 854–861.

127 Barabanov, A.A., Bukatov, G.D., Zakharov, V.A., Semikolenova, N.V., Mikenas, T.B., Echevskaja, L.G., and Matsko, M.A. (2006) *Macromol. Chem. Phys.*, **207**, 1368–1375.

128 Barabanov, A.A., Bukatov, G.D., Zakharov, V.A., Semikolenova, N.V., Mikenas, T.B., Echevskaja, L.G., and Matsko, M.A. (2005) *Macromol. Chem. Phys.*, **206**, 2292–2298.

129 Mikenas, T.B., Zakharov, V.A., Echevskaya, L.G., and Matsko, M.A. (2005) *J. Polym. Sci., Part A: Polym. Chem.*, **43**, 2128–2133.

130 Zakharov, V.A., Semikolenova, N.V., Mikenas, T.B., Barabanov, A.A., Bukatov, G.D., Echevskaya, L.G., and Mats'ko, M.A. (2006) *Kinet. Catal.*, **47**, 303–309.

131 Huang, R., Liu, D., Wang, S., and Mao, B. (2004) *Macromol. Chem. Phys.*, **205**, 966–972.

132 Huang, R., Liu, D., Wang, S., and Mao, B. (2005) *J. Mol. Catal. A: Chem.*, **233**, 91–97.

# 5
# Dormant Species in Transition Metal-Catalyzed Olefin Polymerization

## 5.1
## Introduction

Catalyst productivity in transition metal-catalyzed olefin polymerization is dependent on the number, activity and stability of active species. In previous chapters, factors affecting catalyst deactivation have been discussed. However, even if deactivation is negligible and the catalyst remains stable during the course of polymerization, by no means all the catalytic centers may be active at the same time. In many cases, a dynamic equilibrium between active and dormant species is present and the overall activity is dependent on the proportion of species in the active state at any given moment. In metallocene-catalyzed polymerization, a strongly coordinating anion can block effective coordination of the monomer and therefore lead to low activity. The presence of trimethylaluminum can also depress the catalyst activity, via the formation of dormant alkyl-bridged species of type $[Cp_2Zr(\mu\text{-R})(\mu\text{-Me})AlMe_2]^+$, again blocking monomer coordination to the transition metal. In the Ziegler–Natta or metallocene-catalyzed polymerization of propene and higher $\alpha$-olefins, dormant species can be formed by secondary (2,1)- rather than the usual primary (1,2)- monomer insertion. In the case of metallocenes, dormant species can also result from the formation of $\eta^3$-allylic intermediates. This chapter highlights the formation and effects of dormant site formation in olefin polymerization with both early and late transition metal catalysts. An important point to remember in relation to dormant site formation is that if the interconversion between active and dormant species is relatively rapid within the timeframe of polymerization, dormant site formation will affect only the overall propagation rate constant, not catalyst decay.

*Homogeneous Catalysts: Activity – Stability – Deactivation*, First Edition. Piet W.N.M. van Leeuwen and John C. Chadwick.
© 2011 Wiley-VCH Verlag GmbH & Co. KGaA. Published 2011 by Wiley-VCH Verlag GmbH & Co. KGaA.

## 5.2
## Ziegler–Natta Catalysts

### 5.2.1
### Ethene Polymerization

The formation of dormant species has been proposed as an explanation for the lower activity frequently observed in ethene polymerization with Ziegler–Natta catalysts when hydrogen is present. Kissin has proposed that the rate-lowering effect of hydrogen results from the low activity of $Ti-CH_2CH_3$ species formed by ethene insertion into the $Ti-H$ bond after chain transfer with hydrogen [1–4]. The low activity is attributed to $\beta$-agostic interaction between Ti and a $CH_3$ hydrogen atom. The formation of $C_2H_3D$ in polymerization in the presence of deuterium was suggested to arise from decomposition of $Ti-CH_2CH_2D$ to $Ti-H$ and $CH_2=CHD$. An increased probability of chain initiation via $\alpha$-olefin insertion into a $Ti-H$ bond in ethene/$\alpha$-olefin copolymerization was also presented as evidence for the proposed hypothesis. However, subsequent studies by Garoff *et al.* on the effects of hydrogen in ethene polymerization with various Ziegler–Natta catalysts did not support Kissin's hypothesis [5]. Instead, it was proposed that the presence of hydrogen led to slower catalyst activation, particularly for catalysts where Ti was present as Ti(IV).

As already indicated in Chapter 4, vanadium-based Ziegler–Natta catalysts are particularly sensitive to the presence of hydrogen. Mikenas *et al.* have proposed that the decreased activity of $VCl_4/MgCl_2$ or $VOCl_3/MgCl_2$ in the presence of hydrogen arises from an exchange reaction between vanadium hydride species and the $AlR_3$ cocatalyst, generating $AlR_2H$ [6–8]. The aluminum hydride then blocks the active center via the formation of dormant species of type $V(\mu\text{-Pol})(\mu\text{-H})AlR_2$.

### 5.2.2
### Propene Polymerization

In contrast to ethene polymerization, the presence of hydrogen in propene polymerization with Ziegler–Natta catalysts leads to significant increases in catalyst activity. Guastalla and Giannini observed a 150% increase in propene polymerization rate on adding hydrogen to the catalyst system $MgCl_2/TiCl_4-AlEt_3$, whereas in ethene polymerization the activity decreased [9]. There is now general agreement that, at least for Ziegler–Natta catalysts, hydrogen activation in propene polymerization results from the regeneration of active species via chain transfer at dormant, 2,1-inserted sites. In propene polymerization with Ziegler–Natta catalysts, propagation takes place by primary (1,2) monomer insertion. A secondary (2,1) insertion greatly slows down the rate of chain growth, as a result of the steric hindrance of the $Ti-CH(CH_3)-CH_2-R$ moiety at the metal atom. Busico *et al.* estimated that 10–30% of the catalytic sites were in the dormant, 2,1-inserted state at any given time in propene hydrooligomerization carried out at 50 °C with a $MgCl_2/TiCl_4/dioctyl$ $phthalate-AlEt_3-PhSi(OEt)_3$ catalyst system [10]. Tsutsui *et al.* found that hydrogen activation in metallocene-catalyzed propene polymerization was accompanied by a significant increase in the proportion of chains with an $n$-butyl terminal unit, indicative of chain transfer with hydrogen after 2,1-insertion [11]:

$$M - CH(CH_3)CH_2[CH_2CH(CH_3)]_n Pr + H_2 \rightarrow M - H$$
$$+ nBuCH(CH_3)[CH_2CH(CH_3)]_{n-1} Pr$$

Chain transfer with hydrogen after the usual primary (1,2) insertion gives an isobutyl-terminated chain:

$$M - CH_2CH(CH_3)[CH_2CH(CH_3)]_n Pr + H_2 \rightarrow M - H$$
$$+ iBuCH(CH_3)[CH_2CH(CH_3)]_{n-1} Pr$$

Determination of the relative proportions of $n$Bu- and $i$Bu-terminated chains, using $^{13}$C NMR, has proved to be a particularly powerful technique in determining the incidence and importance of regioirregular (i.e., 2,1) insertion in propene polymerization with state-of-the-art $MgCl_2$-supported catalysts. The proportion of $n$Bu chain ends increases with decreasing hydrogen concentration, due to a higher equilibrium concentration of dormant, 2,1-inserted sites [12]. Particularly high proportions of $n$Bu-terminated chains are found in polypropylenes produced with the catalyst system $MgCl_2/TiCl_4$/diether–$AlEt_3$, which explains the high hydrogen response (i.e., sensitivity to chain transfer) of catalysts containing a diether as internal donor [13]. Chain transfer with hydrogen is more probable after a 2,1-insertion as a result of the slowdown in propagation at the sterically hindered site. In the absence of hydrogen, 2,1-inserted units are incorporated into the polymer chain ("dormancy" implies low but not zero activity). The concentration of chain regioirregularity is generally well below the threshold limit for detection by $^{13}$C NMR, but precise measurements of 2,1-insertion can be made by $^{13}$C NMR analysis of copolymers of propene and 1-$^{13}$C-ethene [14, 15]. Ethene inserts much more rapidly than propene at 2,1-inserted centers, so that regioselectivity in propene polymerization can be measured by determination of $^{13}$C-enriched ethene units flanked by propene units. Using this technique, it has been established that occasional regioirregular insertions (typically less than 1 in 1000 for isospecific active species) have a profound impact on the molecular weight distribution of polypropylene produced with Ziegler–Natta catalysts [16]. The presence of (some) highly regiospecific active species (i.e., ratio of primary/secondary insertions $\gg$ 1000) leads to the formation of a high molecular weight polymer fraction and, overall, a broad molecular weight distribution.

The probability of a 2,1-insertion is much greater for the first insertion of propene into a Ti–H bond. When followed by a series of 1,2-insertions, this results in the formation of a 2,3-dimethylbutyl terminal unit, as opposed to $n$-propyl when the first insertion is primary. Significant proportions of 2,3-dimethylbutyl chain-ends have been found in polypropylenes prepared with both metallocene [17, 18] and Ziegler–Natta catalysts [19]. The relatively low propagation rate of the 2,1-inserted species applies equally well to the Ti–isopropyl reaction product of secondary insertion into Ti–H, making a further transfer reaction with hydrogen (generating propane) more probable than chain growth to form a 2,3-dimethylbutyl end-group [19]. It has been suggested [20] that the hydrogen activation effect is related to the hydrogenolysis of the Ti—isopropyl unit, but this seems unlikely as, in this case, hydrogen would be involved in both the formation and removal of such dormant species.

No evidence has been found for internal vinylidene groups, $-[CH_2CH(CH_3)]_m-CH_2C(=CH_2)-[CH_2CH(CH_3)]_n-$, in polypropylene produced with Ziegler–Natta catalysts [19]. As described in Section 5.3.4, such a structure has been observed in polymers prepared using certain metallocene catalysts and results from the formation (with release of $H_2$) of dormant $\eta^3$-allyl species after a $\beta$-H elimination step.

## 5.3
## Metallocenes and Related Early Transition Metal Catalysts

### 5.3.1
### Cation–Anion Interactions

Activation of metallocene catalysts leads to 14-electron cationic complexes of type $[Cp_2ZrR]^+$. However, as observed by Bochmann, a whole range of different resting states and equilibria is involved in metallocene-catalyzed olefin polymerization [21]. Each resting state can be considered as a dormant species and the most important factor is the degree of coordination of the counterion. $[B(C_6F_5)_4]^-$ is considered to be close to being a "non-coordinating anion", but in many cases metal–F interactions are detectable. Marks and coworkers have measured the relative coordination ability of various borate anions to a constrained geometry Zr cation and noted that the polymerization activities were dependent on the tightness of the cation–anion pairing [22]. Arene solvent coordination to the constrained geometry cation was observed only when weakly coordinating $[B(C_6F_5)_4]^-$ was the counterion. Brintzinger and coworkers investigated displacement by $PR_3$ of the anion $[MeB(C_6F_5)_3]^-$ from the contact ion pair $[Cp_2ZrMe^+ \ldots MeB(C_6F_5)_3^-]$ and with regard to olefin coordination concluded that the olefin would displace the borate anion from only a small equilibrium fraction of the zirconocene alkyl cation [23]. The question was raised as to how many consecutive olefin insertions might occur before re-coordination of the anion to give the temporarily dormant contact ion pair. Further studies of $[MeB(C_6F_5)_3]^-$ anion displacement from zirconocene methyl cations were carried out with relatively weakly coordinating Lewis bases such as dimethylaniline [24]. The results indicated that Lewis base–anion exchange proceeds by an $S_N2$-type, associative mechanism and a five-coordinated reaction intermediate with the Lewis base coordinated to zirconium was suggested. An "intermittent" olefin polymerization model was proposed, in which displacement of the counterion by an olefin is an infrequent event that is followed by a burst of propagations before re-coordination of the anion. The rate constant for anion displacement decreased with increasing steric bulk around the metal center and it was predicted that this would lead to observable induction periods in polymerizations, especially with propene, in cases where unfavorable steric interactions would hinder the formation of five-coordinated intermediates. Stopped-flow experiments carried out with *rac*-Me$_2$Si(2-Me-4-Ph-

Ind)$_2$ZrCl$_2$/MAO have indeed revealed an induction period in propene polymerization [25].

Brintzinger and coworkers have also reported that anion displacement and exchange in zirconocene borate systems is around 5000 times faster with [B(C$_6$F$_5$)$_4$]$^-$ than with [MeB(C$_6$F$_5$)$_3$]$^-$, but they proposed that even the very weakly coordinating [B(C$_6$F$_5$)$_4$]$^-$ might exchange by an associative mechanism [26]. In nonpolar solvents, borate anion exchange was proposed to proceed by way of ion quadruples or higher order ionic aggregates rather than via dissociation to solvent-separated ions. Bochmann and coworkers have carried out stopped-flow polymerization of propene with the catalyst systems (SBI)ZrCl$_2$/MAO and (SBI)ZrMe$_2$/Ali$Bu_3$/[Ph$_3$C][CN{B (C$_6$F$_5$)$_3$}$_2$], where SBI represents *rac*-Me$_2$Si(1-indenyl)$_2$ [27]. The kinetic data indicated similar proportions of dormant species, but the propagation rate for the MAO system was about 40 times lower than that found using the borate. It was concluded that the counterion profoundly influences the energetics of the migratory monomer insertion cycle and is closely associated with the cation throughout the insertion sequence.

Landis and coworkers found that addition of [PhNMe$_3$][MeB(C$_6$F$_5$)$_3$] in 1-hexene polymerization with *rac*-Et(Ind)$_2$ZrMe$_2$/B(C$_6$F$_5$)$_3$ had no influence on the propagation rate and concluded that free ions were not the principal propagating species in this system [28]. NMR experiments at low temperature confirmed a continuous rather than an intermittent polymerization mechanism [29]. Taking this and other results into account, Bochmann has proposed the existence of different kinetic regimes for different polymerization systems [30]. The continuous model would apply if monomer insertion is slower than anion exchange, as in 1-hexene polymerization. On the other hand, the intermittent model, in which anion substitution is followed by a series of monomer insertions before chain growth is interrupted by anion re-association, would apply in ethene and propene polymerizations where monomer insertion is faster than anion exchange. The two reaction schemes are illustrated in Schemes 5.1 and 5.2 [31]. In the low-activity "continuous" propagation mode (Scheme 5.1), each monomer insertion is followed by anion re-association. However, a mechanism in which the anion re-coordinates to form a tight (inner-sphere) ion pair with the metal center after each insertion step is not appropriate for a counterion such as [B(C$_6$F$_5$)$_4$]$^-$. In this case, the anion forms an outer-sphere ion pair within the solvent cage, as indicated in Scheme 5.2, and each insertion step involves a

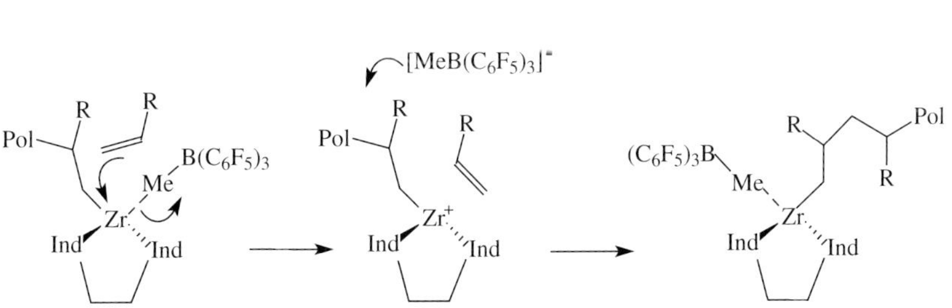

**Scheme 5.1** Inner-sphere ion-pairs in low-activity metallocene-catalyzed polymerization [31].

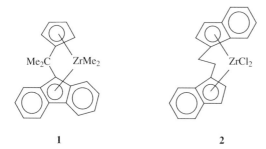

**Scheme 5.2** Outer-sphere ion-pairs in high-activity metallocene-catalyzed polymerization [31].

change in the cation–anion distance. In such a situation, agostic metal–alkyl inter-actions form the catalyst resting states, in preference to anion coordination [32].

Counterion effects in metallocene-catalyzed polymerization are strongly depen-dent on solvent polarity. For example, Marks and coworkers have observed that replacement of octane by 1,3-dichlorobenzene as solvent in propene polymerization with $Me_2C(Cp)(Flu)ZrMe_2$**1** and different activators led to increases in activity and much lower dependence of polypropylene chain stereoregularity on the nature of the activator, as a result of significantly weaker ion pairing effects in a polar solvent [33]. Earlier studies by Deffieux and coworkers showed that replacement of toluene by dichloromethane as solvent in 1-hexene polymerization with *rac*-Et(Ind)$_2$ZrCl$_2$ (**2**) and MAO led to higher activity and allowed effective zirconocene activation at relatively low MAO/Zr ratios [34]. Sacchi and coworkers found that increasing the $CH_2Cl_2$ content in a $CH_2Cl_2$/toluene solvent mixture led to increased 1-hexene incorporation in copolymerization with ethene using *rac*-Et(Ind)$_2$ZrCl$_2$/MAO [35]. This effect was attributed to greater cation/anion separation in the more polar solvent, making the active centers more accessible for the bulkier monomer. Investigation of the solvent effect in propene/1-hexene copolymerization revealed that 1-hexene incorporation was lower in toluene than in *o*-dichlorobenzene [36, 37]. In the more nucleophilic solvent, toluene, the approach to the catalytic site of the bulkier 1-hexene, displacing the coordinated solvent, is more disfavored relative to propene.

$Me_2C$   $ZrMe_2$          $ZrCl_2$

**1**                    **2**

5.3.2
**Effects of AlMe₃**

The effect of $AlMe_3$ in suppressing the rate of metallocene-catalyzed olefin poly-merization, via the formation of dormant, alkyl-bridged species of type [$Cp_2Zr(\mu$-R)

($\mu$-Me)AlMe$_2$]$^+$ **3**, has already been described in Chapter 2. Polymerization activity is dependent on the amount of AlMe$_3$ present in the system, most commonly as a component of MAO, and the ease with which associated AlMe$_3$ can be displaced by the monomer. As already noted, the presence of AlMe$_3$ has a particularly large rate-depressing effect with hafnocenes, due to the high stability and, therefore, low polymerization activity of species such as [Cp$_2$Hf($\mu$-R)($\mu$-Me)AlMe$_2$]$^+$ [38]. DFT calculations have indicated that such species are more stable by 3 kcal mol$^{-1}$ than their zirconocene analogues, leading to a much larger proportion of bridged (dormant) species in hafnocene/MAO-catalyzed polymerization [39]. A simple and effective solution to the problem of "free" AlMe$_3$ being present in MAO is to contact MAO with a sterically hindered phenol such as 2,6-di-*tert*-butylphenol or 2,6-di-*tert*-butyl-4-methylphenol [40]. The phenol is not very reactive with metallocenium ions and its reaction product with trimethylaluminum, MeAl(OAr)$_2$, is itself an effective scavenger in polymerization [41].

Association of an aluminum alkyl with metallocenium species [Cp$_2$M—Pol]$^+$ becomes less favorable with increasing size of the alkyl group. The equilibrium in Equation 5.1 shifts to the left when R = Et rather than Me, and there is no evidence for analogous Zr—Al complexes with Al*i*Bu$_3$, which is, therefore, the standard scavenger used in borate-activated polymerization [30].

$$[Cp_2MR]^+ + AlR_3 \rightleftharpoons [Cp_2M(\mu-R)_2AlR_2]^+ \tag{5.1}$$

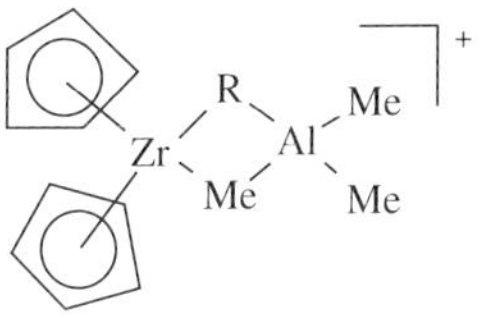

**3**

## 5.3.3
### Effects of 2,1-insertion in Propene Polymerization

In propene polymerization, the monomer can insert either in a 1,2- or a 2,1-fashion, to give a primary (p) or a secondary (s) alkyl product. As is the case for Ziegler–Natta catalysts, the dominant insertion mode for metallocenes is primary (1,2)-insertion. 2,1-Insertions are infrequent but can have a significant influence on the polymerization kinetics. A 2,1-insertion gives a "dormant" site which has low activity for further propene insertion but can readily undergo chain transfer with hydrogen, regenerating the active species [11]. Busico estimated that in propene polymerization at 60 °C with *rac*-Et(Ind)$_2$ZrCl$_2$/MAO, around 90% of the catalytic sites would be in the dormant (2,1-inserted) state at any given time under conditions of negligible chain transfer [42]. As well as slowing down the chain propagation, a 2,1-insertion can lead to both —CH$_2$—CH(Me)—CH(Me)—CH$_2$—CH$_2$—CH(Me)— and CH$_2$—CH

(Me)$-CH_2-CH_2-CH_2-CH_2-CH(Me)-$ regioirregularities in the chain. The latter arise from isomerization of 2,1- to 3,1-units before the next insertion occurs, as follows [43–46]:

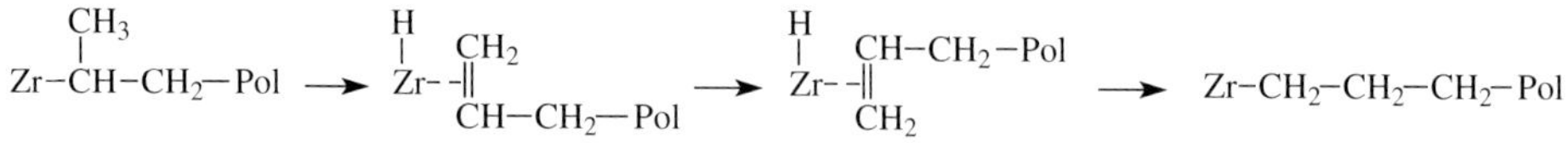

$$
\underset{\text{CH}_3}{\text{Zr}-\text{CH}-\text{CH}_2-\text{Pol}} \longrightarrow \underset{\substack{H \\ \text{Zr}--\| \\ \text{CH}-\text{CH}_2-\text{Pol}}}{\overset{\text{CH}_2}{}} \longrightarrow \underset{\substack{H \\ \text{Zr}--\| \\ \text{CH}_2}}{\overset{\text{CH}-\text{CH}_2-\text{Pol}}{}} \longrightarrow \text{Zr}-\text{CH}_2-\text{CH}_2-\text{CH}_2-\text{Pol}
$$

Large proportions of 3,1-inserted units have been found in polypropylene synthesized with $rac$-Me$_2$Si(3-MeCp)$_2$TiCl$_2$/MAO, whereas the corresponding Zr and Hf complexes gave mainly 2,1-inserted regioirregularities [47]. A large hydrogen activation effect was observed with the titanocene, consistent with a low propagation activity of 2,1-inserted centers and isomerization to a 3,1-unit in the absence of hydrogen. The retarding effect of 2,1-insertion is greater still for monomers bulkier than propene. Large hydrogen activation effects have been observed in 1-butene polymerization, consistent with reactivation of dormant species via chain transfer, and the presence of 4,1- rather than 2,1-inserted units in the polymer chain provides further evidence for particularly high dormancy after a regioirregular insertion, leading to isomerization to a 4,1-unit via a series of β-H eliminations and reinsertions [48].

The extent to which 2,1-insertions slow down propagation not only depends on how frequently they occur (i.e., the ratio $k_{ps}/k_{pp}$ in Scheme 5.3), but also on the rate at which a dormant, 2,1-inserted species can transform to an active species via chain transfer, isomerization or monomer insertion. Even traces of 2,1-regioerrors can effectively inhibit the catalyst activity if the ratio $k_{sp}/k_{ps}$ is low enough [49]. As indicated in Section 5.2.2, ethene can insert into a dormant, 2,1-inserted chain much faster than propene. Propene polymerizations with $rac$-Me$_2$Si(Ind)$_2$ZrCl$_2$**4** and $rac$-Me$_2$Si(2-Me-4-Ph-Ind)$_2$ZrCl$_2$**5** in the presence of small quantities of 1-$^{13}$C-ethene have been carried out by Busico $et\ al.$, in order to determine the extent and effect of regioirregular insertion [49]. The results indicated that dormancy after 2,1-insertion was greater for catalyst **4**, which is in line with the low activity of **4** compared to **5**. A relatively low dormancy for catalyst **5** is also apparent from the fact that regioirregular 2,1-units can be detected in polypropylene prepared in the presence of hydrogen [50]. Negligible dormancy after 2,1-insertion has been noted for monocyclopentadienyl

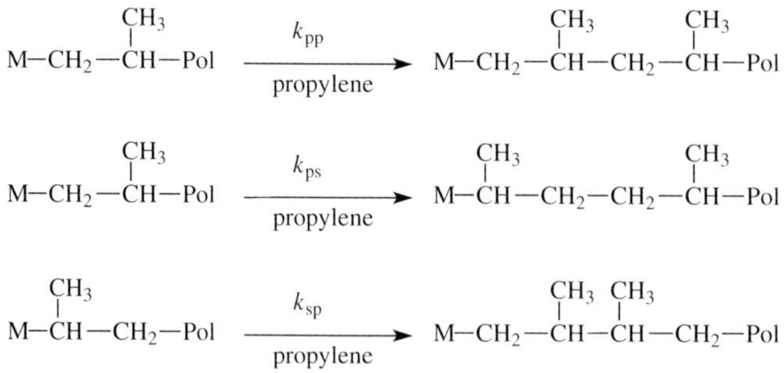

$$
\underset{\substack{\text{CH}_3 \\ \text{M}-\text{CH}_2-\text{CH}-\text{Pol}}}{} \xrightarrow[\text{propylene}]{k_{pp}} \underset{\substack{\text{CH}_3 \quad\quad \text{CH}_3 \\ \text{M}-\text{CH}_2-\text{CH}-\text{CH}_2-\text{CH}-\text{Pol}}}{}
$$

$$
\underset{\substack{\text{CH}_3 \\ \text{M}-\text{CH}_2-\text{CH}-\text{Pol}}}{} \xrightarrow[\text{propylene}]{k_{ps}} \underset{\substack{\text{CH}_3 \quad\quad \text{CH}_3 \\ \text{M}-\text{CH}-\text{CH}_2-\text{CH}_2-\text{CH}-\text{Pol}}}{}
$$

$$
\underset{\substack{\text{CH}_3 \\ \text{M}-\text{CH}-\text{CH}_2-\text{Pol}}}{} \xrightarrow[\text{propylene}]{k_{sp}} \underset{\substack{\text{CH}_3 \ \text{CH}_3 \\ \text{M}-\text{CH}_2-\text{CH}-\text{CH}-\text{CH}_2-\text{Pol}}}{}
$$

**Scheme 5.3** Primary (p) and secondary (s) monomer insertion in propene polymerization.

complexes such as $[Cp^*TiMe_2][MeB(C_6F_5)_3]$ [51]. In contrast to metallocenes, the activity of such complexes is not significantly increased by the addition of hydrogen.

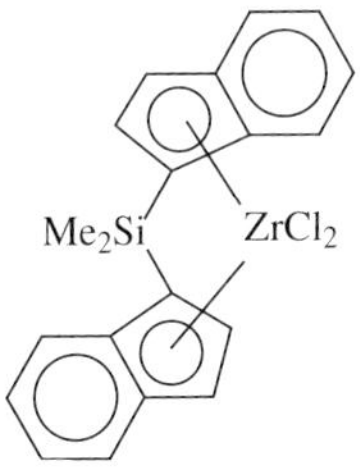

As suggested by Busico, important indicators of high dormancy in propene polymerization are strong catalyst activation upon addition of low amounts of $H_2$ or ethene and a strong tendency for 2,1-units to isomerize to 3,1-units [52]. Dormant site formation as a result of the occasional 2,1-insertion can be considered to be a general phenomenon in propene polymerization with Ziegler–Natta and metallocene catalysts. A further example is the significant increase in propene polymerization activity with the unbridged metallocene $(2\text{-PhInd})_2ZrCl_2/MAO$ observed following the addition of small amounts of ethene or hydrogen [53]. However, an exception to the rule has been reported by Landis *et al.*, who found similar rates for propene insertion into Zr−*n*Bu and Zr−*s*Bu bonds in the active species [*rac*-Et $(Ind)_2ZrBu][MeB(C_6F_5)_3]$ at −80 °C [54]. It was also found that hydrogenolysis was 100 times faster with the secondary zirconium alkyl than with the primary alkyl. Faster reaction with hydrogen at secondary- as opposed to primary-inserted centers has also been found for the non-metallocene catalyst **6** (Bn = benzyl; $R_1$ = cumyl; $R_2$ = methyl) activated with MAO/2,6-di-*tert*-butylphenol [52]. This system did follow the general trend of dormancy after 2,1-propene insertion, with the ratio $k_{sp}/k_{pp}$ estimated to be around 0.03.

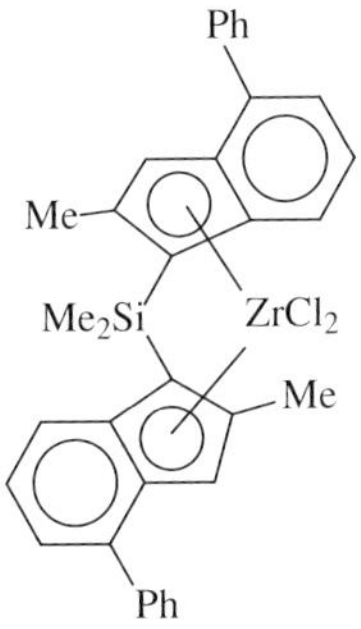

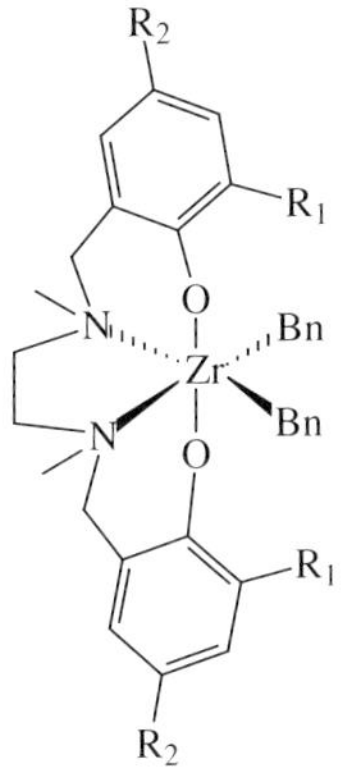

In the absence of hydrogen, a 2,1-inserted unit in propene polymerization can not only isomerize to a 3,1-unit but also lead to chain termination. Mülhaupt and coworkers have observed the formation of 2-butenyl end-groups in polypropylene prepared with MAO-activated *rac*-$Me_2Si(Benz[e]Indenyl)_2ZrCl_2$ (**7**) [55]. The polymer molecular weight was independent of propene concentration, indicating that chain transfer took place via β-H transfer to monomer after 2,1-insertion. The introduction of a 2-methyl substituent as in complex **8** resulted in a lower incidence of monomer-assisted chain transfer and, therefore, higher polymer molecular weight. Addition of hydrogen gave, in both cases, increased activity and fewer 2,1-regioerrors in the polymer chain, as well as significant proportions of *n*-butyl terminated chains resulting from hydrogenolysis after 2,1-insertion. Sacchi and coworkers have investigated polypropylene microstructure as a probe into hydrogen activation with a number of different *ansa*-zirconocenes [56]. Hydrogen activation was observed in all cases, but no clear correlations with the numbers of *n*-butyl terminal groups were found and it was concluded that chain transfer with hydrogen following secondary monomer insertion could not be the only explanation for hydrogen activation with these catalysts. It has now been established that hydrogen activation in metallocene-catalyzed polymerization can also result from reactivation of $\eta^3$-allyl dormant species, discussed below.

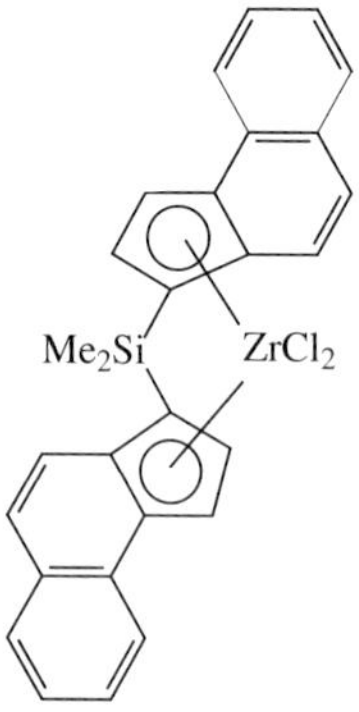

**7** 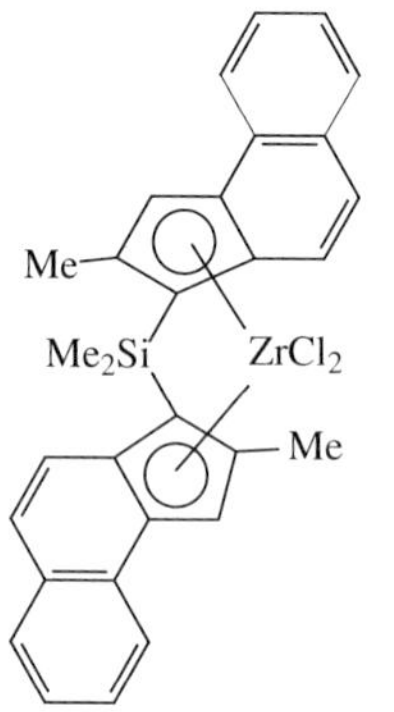                                    **8**

5.3.4
**Effects of $\eta^3$-allylic Species in Propene Polymerization**

An early example of the formation of dormant $\eta^3$-allylic species with a zirconocene was the report by Teuben and coworkers of allylic C—H activation of monomer and isobutene in propene oligomerization, giving the inactive (meth)allyl species $[Cp^*_2Zr (\eta^3\text{-}C_3H_5)]^+$ and $[Cp^*_2Zr(\eta^3\text{-}C_4H_7)]^+$ [57]. Reactivation of the (meth)allyl complexes could be achieved by the addition of hydrogen. Richardson *et al.* have observed the formation of a methallyl species in the gas-phase reaction of $[Cp_2ZrMe]^+$ with propene [58]:

$$[Cp_2ZrMe]^+ + C_3H_6 \rightarrow [Cp_2Zr(\eta^3 - C_4H_7)]^+ + H_2$$

Hydrogen is formed in the above reaction and Karol *et al.* proposed that hydrogen could be formed, along with allyl intermediates, in metallocene-catalyzed olefin polymerization [59]. Subsequent participation of the allyl intermediate in the polymerization process would then lead to side-group unsaturations along the polymer chain. Evolution of hydrogen has indeed been observed in zirconocene-catalyzed propene polymerization and Resconi has proposed that this, and the associated formation of internal vinylidene structure in the polymer, occurs via the type of mechanism depicted in Scheme 5.4 [60]. Theoretical studies by Ziegler and coworkers have indicated that the formation of a dihydrogen allyl complex is a feature that is generally applicable to metallocenes and related catalysts and that the process can be reversed by the addition of excess hydrogen [61–63]. Experimental and theoretical studies by Brintzinger and coworkers have shown that zirconocene–allyl species arising during olefin polymerization can be reactivated either by exchange with an aluminum alkyl or by olefin insertion into the Zr–allyl bond [64]. The rate of olefin insertion into Zr–allyl species was estimated to be at least an order of magnitude lower than insertion into the normal Zr–alkyl species. It was also concluded that reactivation via monomer insertion is slower than that by reaction with hydrogen, which, therefore, provides a further explanation (in addition to chain transfer at 2,1-inserted sites) for the activating effect of hydrogen in propene polymerization with metallocene catalysts.

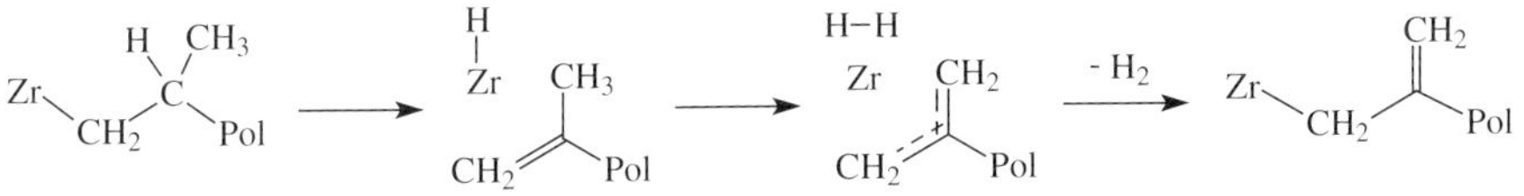

**Scheme 5.4** Formation of "internal" vinylidene units in metallocene-catalyzed propene polymerization.

Dormant $\eta^3$-allyl species that can be reactivated by hydrogen are also formed in the copolymerization of an olefin with a conjugated diene [65, 66]. Ishihara and Shiono obtained very low activity in propene/1,3-butadiene copolymerization with *rac*-Me$_2$Si (2-Me-4-Ph-Ind)$_2$ZrCl$_2$ and MMAO, but found that the activity increased by 3 orders of magnitude in the presence of hydrogen [65]. The polymers prepared in the absence of hydrogen contained olefinic groups resulting from both 1,2- and 1,4-butadiene insertion. Vinyl unsaturation from 1,2-butadiene insertion was also formed in the presence of hydrogen, whereas 1,4-insertion led to saturated $-(CH_2)_4-$ units formed by hydrogenation of dormant $\eta^3$-allyl species.

### 5.3.5
### Chain Epimerization In Propene Polymerization

An interesting phenomenon in 1-alkene polymerization with isospecific ($C_2$-symmetric) *ansa*-metallocenes has been discovered by Busico and Cipullo, who found that polymer isotacticity decreased with decreasing monomer concentration in polymerization [67, 68]. The effect was attributed to an epimerization reaction of the last-inserted monomer unit, which reaction was able to compete with chain propagation at low monomer concentration. The epimerization was proposed to proceed as illustrated in Scheme 5.5.

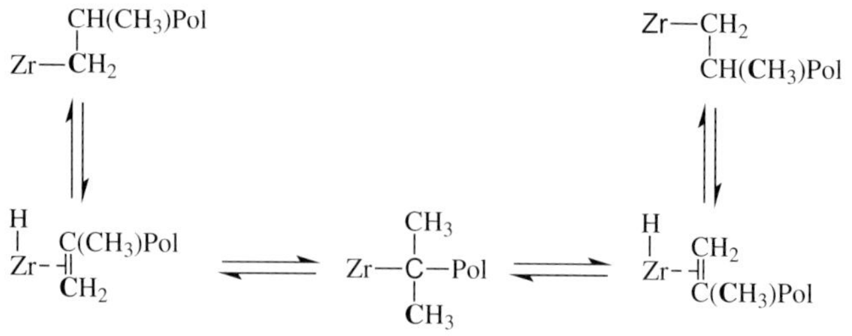

$$\begin{array}{c}CH(CH_3)Pol\\|\\Zr-CH_2\end{array}\qquad\qquad\qquad\begin{array}{c}Zr-CH_2\\|\\CH(CH_3)Pol\end{array}$$

$$\begin{array}{c}H\\|\ \ \ C(CH_3)Pol\\Zr\text{-}\| \\ \ \ CH_2\end{array}\ \rightleftharpoons\ \begin{array}{c}CH_3\\|\\Zr-C-Pol\\|\\CH_3\end{array}\ \rightleftharpoons\ \begin{array}{c}H\\|\ \ \ CH_2\\Zr\text{-}\| \\ \ \ C(CH_3)Pol\end{array}$$

**Scheme 5.5** Epimerization of the last-inserted monomer unit in metallocene-catalyzed propene polymerization.

Experimental evidence for the formation, at low monomer concentration, of stereoerrors by isomerization of the last-inserted monomer unit, rather than errors in the enantiofacial orientation of the inserting olefin, was provided by deuterium labeling studies carried out by Leclerc and Brintzinger [69, 70].

An alternative mechanism to explain chain-end epimerization, based on the reversible formation of a zirconocene allyl dihydrogen complex, has been proposed by Resconi [60].

As indicated in the previous section, this type of allyl intermediate is involved in the formation of internal vinylidene unsaturation, but more recent studies have shown that this mechanism is not involved in chain epimerization [71]. Using doubly labeled propene ($CH_2CD^{13}CH_3$), Yoder and Bercaw were able to prove that chain epimerization occurs as originally proposed by Busico and Cipullo. As shown in Scheme 5.5, this mechanism involves β-H elimination, olefin rotation and insertion to give a tertiary alkyl intermediate, followed by the reverse steps.

## 5.3.6
### Effects of Dormant Site Formation on Polymerization Kinetics

The rate of catalytic olefin polymerization, $R_p$, can be described by the following simple expression (5.2), where [C] represents the active site concentration and [M] is the monomer concentration:

$$R_p = k_p[C][M]^n \tag{5.2}$$

A reaction order $n = 1$ with respect to monomer would normally be expected, but various observations of higher reaction orders ($1 < n < 2$) have been made. For example, Mülhaupt and coworkers found a reaction order of 1.7 in zirconocene-catalyzed propene polymerization and suggested that this might be due to the involvement of propene in an equilibrium between dormant and active catalyst sites [55].

Fait *et al.* have proposed a kinetic model based on a single-center, two-state catalyst [72]:

$$R_p = k_{p,fast}[C_{fast}][M] + k_{p,slow}[C_{slow}][M] \tag{5.3}$$

It was suggested that the fast and slow propagating species could differ in the conformation of the growing chain, a γ-agostic intermediate having higher activity than a β-agostic resting state. An alternative mechanism for higher order in monomer has been put forward by Ystenes, who proposed a mechanism in which monomer insertion is "triggered" by a second monomer molecule [73]. However, this mechanism has been discounted by Busico *et al.*, who were able to show that the competing reactions of chain propagation and chain-end epimerization were both first-order in monomer [74]. Interconversion of catalytic species between a propagating and a "resting" state therefore remains the most plausible explanation for values of $n$ greater than unity in Equation 5.2.

## 5.4
## Late Transition Metal Catalysts

### 5.4.1
### Resting States in Nickel Diimine-Catalyzed Polymerization

The resting state in ethene polymerization catalyzed by cationic α-diimine nickel and palladium complexes is the alkyl ethene complex on the left of Scheme 5.6 [75]. The turnover-limiting step (TLS) is the migratory insertion reaction of the alkyl ethene complex, and chain growth is therefore zero-order in ethene (above a certain ethene pressure). The β-agostic species generated following migratory insertion can undergo a series of β-hydride eliminations and re-additions, resulting in the metal migrating along the polymer chain. This chain-walking mechanism results in polymers with extensive chain branching.

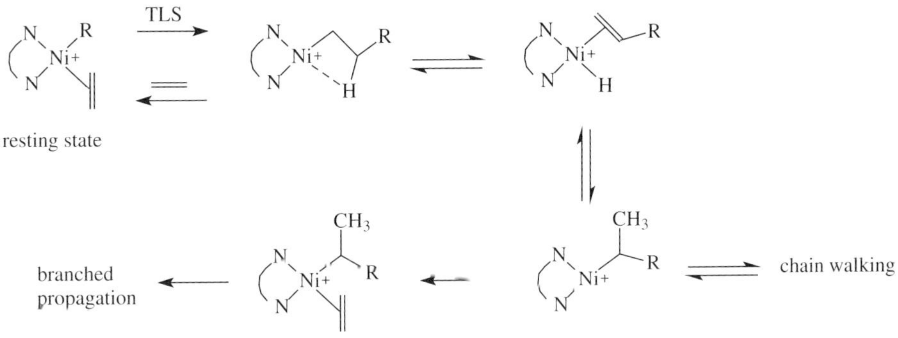

**Scheme 5.6** Resting state and branch formation in ethene polymerization with Ni diimine complexes.

### 5.4.2
### Effects of Hydrogen in Bis(iminopyridyl) Iron-Catalyzed Polymerization

As indicated in this and previous chapters, molecular weight control in catalytic olefin polymerization is generally carried out using hydrogen as the chain transfer agent.

The presence of hydrogen usually leads to significant increase in activity in propene polymerization, as a result of chain transfer at dormant, 2,1-inserted sites, whereas in ethene polymerization decreased activity is often observed on adding hydrogen. However, it has been found that the presence of hydrogen in ethene polymerization using immobilized bis(imino)pyridyl iron catalysts leads not to a decrease but to a surprising increase in activity [76–78]. It was suggested by Zakharov and coworkers that this hydrogen activation effect results from the reactivation via chain transfer of dormant species resulting from 2,1-insertion of vinyl-terminated oligomers into the growing chain, as illustrated in Scheme 5.7 [79, 80]. It has also been proposed that the activating effect of hydrogen in iron-catalyzed polymerization could result from hydrogenolysis of $\eta^3$-allyl species formed after coordination of a vinyl-terminated oligomer (or α-olefin) to the active center and H-transfer from the olefin to the polymer chain, as illustrated in Scheme 5.8 [81]. The formation of the $\eta^3$-allyl complex is accompanied by chain termination, but the addition of $H_2$, leading to hydrogenolysis of the $\eta^3$-allyl complex and reactivation of the active center, does not result in chain transfer. This may explain the fact that the molecular weight of polyethylene synthesized with bis(imino)pyridyl iron catalysts is relatively insensitive to hydrogen pressure unless an α-olefin such as 1-hexene is present, in which case the combined effect of the olefin and hydrogen results in molecular weight lowering [81].

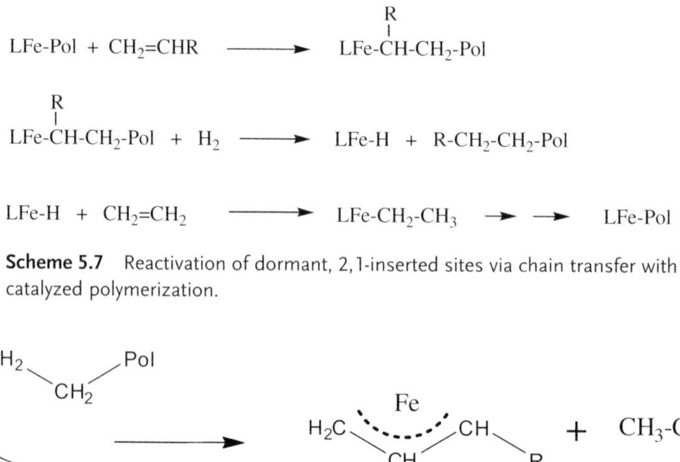

$$LFe\text{-}Pol + CH_2\text{=}CHR \longrightarrow LFe\text{-}\overset{\overset{\displaystyle R}{|}}{C}H\text{-}CH_2\text{-}Pol$$

$$LFe\text{-}\overset{\overset{\displaystyle R}{|}}{C}H\text{-}CH_2\text{-}Pol + H_2 \longrightarrow LFe\text{-}H + R\text{-}CH_2\text{-}CH_2\text{-}Pol$$

$$LFe\text{-}H + CH_2\text{=}CH_2 \longrightarrow LFe\text{-}CH_2\text{-}CH_3 \longrightarrow \longrightarrow LFe\text{-}Pol$$

**Scheme 5.7** Reactivation of dormant, 2,1-inserted sites via chain transfer with hydrogen in iron-catalyzed polymerization.

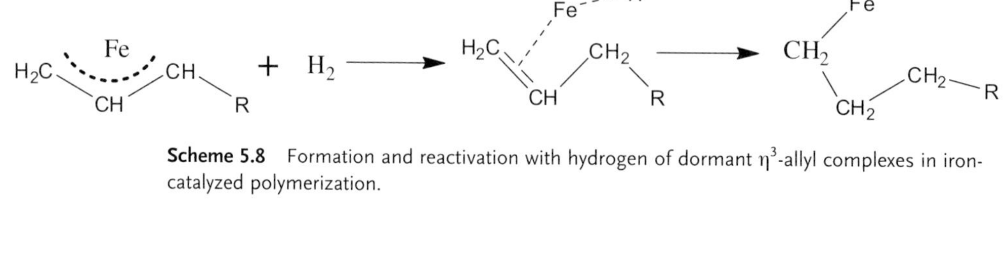

**Scheme 5.8** Formation and reactivation with hydrogen of dormant $\eta^3$-allyl complexes in iron-catalyzed polymerization.

Studies with various bis(imino)pyridyl iron(II) catalysts of type $LFeCl_2$, immobilized on a magnesium chloride support, have revealed that the effects of hydrogen are dependent on the steric bulk of the ligand L [82]. Significant decreases in polyethylene molecular weight with increasing hydrogen pressure were observed with complex **9**, containing relatively bulky ligand substituents. In contrast, the presence of hydrogen in ethene oligomerization with complex **10** gave an overall increase in molecular weight, as a result of greatly decreased formation of low molecular weight polymer and oligomer. Partial deactivation of oligomer-forming active species is, therefore, likely to contribute to the observed increases in the activity of immobilized iron polymerization catalysts in the presence of hydrogen, taking into account the multi-center nature of these catalysts and the proposed hydrogen activation via chain transfer after 2,1-insertion of a vinyl-terminated oligomer into the growing polymer chain and/or via reactivation by hydrogenolysis of dormant $\eta^3$-allyl species formed after coordination of a vinyl-terminated oligomer to the active center.

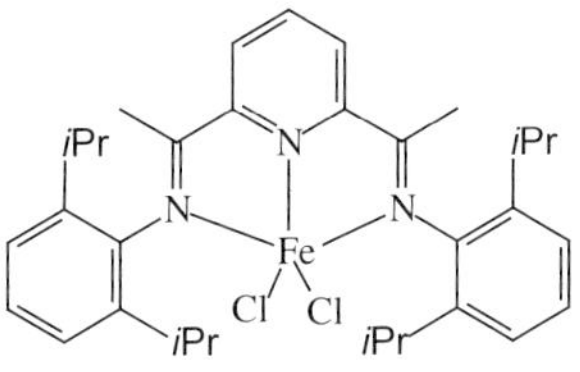

**9** 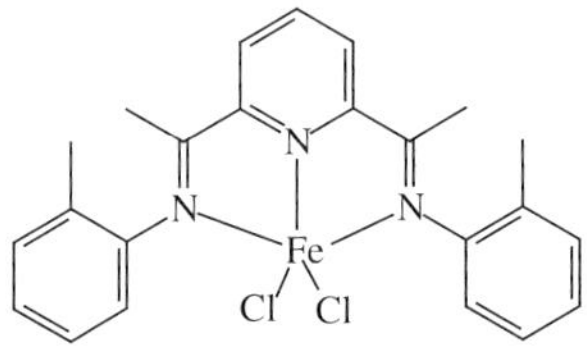                    **10**

## 5.5
## Reversible Chain Transfer in Olefin Polymerization

It has long been known that diethylzinc is an effective chain transfer agent in olefin polymerization with Ziegler–Natta catalysts [83]. More recently, it has been utilized in combination with homogeneous and immobilized early and late transition metal catalysts [84–88]. In zirconocene-catalyzed polymerization, the molecular weight lowering effect of $ZnR_2$ is accompanied by a considerable decrease in activity, which is attributed to the formation of strong heterodinuclear adducts of type $[L_2Zr(\mu\text{-alkyl})(\mu\text{-R})ZnR]^+$, from which the dialkylzinc is not easily displaced by an entering olefin substrate [85]. The stability of these adducts facilitates chain transfer from Zr to Zn.

If chain transfer is reversible, in the sense that the chain is rapidly and reversibly exchanged between the active transition metal center and a non-active main-group metal or other dormant species, the process is referred to as degenerative transfer or coordinative chain transfer polymerization (CCTP). If the rate of chain transfer exchange between active and inactive centers is several times greater than the propagation rate, then both the transition- and main-group metal centers will appear to engage in chain propagation at the same rate. Sita and coworkers have demonstrated degenerative transfer living polymerization when the zirconium amidinate **11** is used in combination with a less than stoichiometric amount of the borate activator

$[PhNMe_2H][B(C_6F_5)_4]$, such that only a fraction of the complex is converted to the active, cationic species [89]. Rapid and reversible chain transfer between active (monoalkyl) and dormant (dialkyl) species as in Equation 5.4 resulted in living polymerization to give a polymer with very narrow polydispersity. Recently, the same group have used the hafnium amidinate **12** together with an excess of $ZnR_2$ (R = Et or $i$Pr) to effect the living polymerization of ethene, $\alpha$-olefins and nonconjugated dienes [90]. Using $ZniPr_2$, it was established that each polyethylene chain contained one isopropyl and one non-branched end group, confirming the participation of zinc as a "surrogate" center for chain growth.

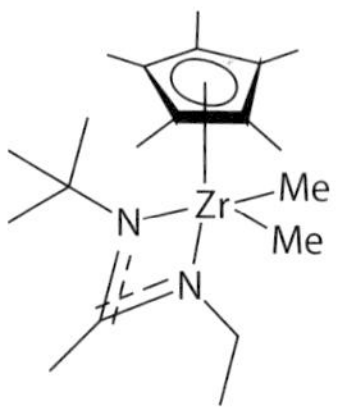

**11** 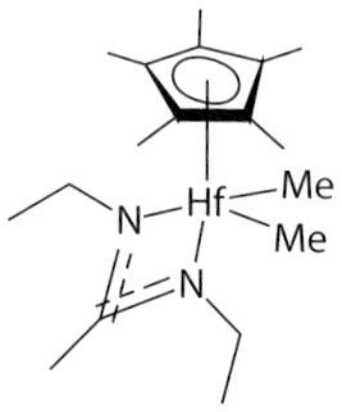 **12**

$$[L_2ZrR]^+ + [L_2ZrR_2] \leftrightarrow [L_2ZrR_2] + [L_2ZrR]^+ \qquad (5.4)$$

Gibson and coworkers have used diethylzinc in combination with a range of different transition metal catalysts for ethene polymerization [88]. Coordinative chain transfer polymerization, generating low molecular weight polyethylene with narrow (Poisson) distribution, was obtained with bis(imino)pyridyl iron complexes in combination with MAO and around 500 equiv of $ZnEt_2$. It was concluded that CCTP was favored by a reasonable match between the bond dissociation energies of both the main-group and the transition metal alkyl species, the key factor being the M$-$C bond energies of the bridging alkyl species. The latter are strongly influenced by the steric environment around the participating metal centers, bulky ligands leading to a weakening of the bridging alkyl bonds and favoring reversible chain transfer. Studies by Mortreux and coworkers have shown that CCTP can also be achieved using a magnesium alkyl (MgEtBu) in combination with rare-earth metallocenes such as $(C_5Me_5)_2NdCl_2Li(OEt)_2$ [91].

The most important development in reversible chain transfer in olefin polymerization has been the recent production by Dow Chemical Company of olefin block copolymers via what is termed chain shuttling polymerization [92, 93]. The concept involves ethene/$\alpha$-olefin (1-octene) copolymerization, in solution at relatively high temperature ($\geq 120\,^\circ C$), using a combination of two transition metal catalysts and a chain shuttling agent such as $ZnEt_2$. One catalyst (**13**, Bn = benzyl) gives low comonomer incorporation, producing a hard, crystalline polyethylene segment, while the other (**14**) gives good comonomer incorporation, producing a soft copolymer segment. The presence of the zinc alkyl results in reversible transfer of the growing chain, via zinc, between the two catalysts to generate a multi-block copolymer.

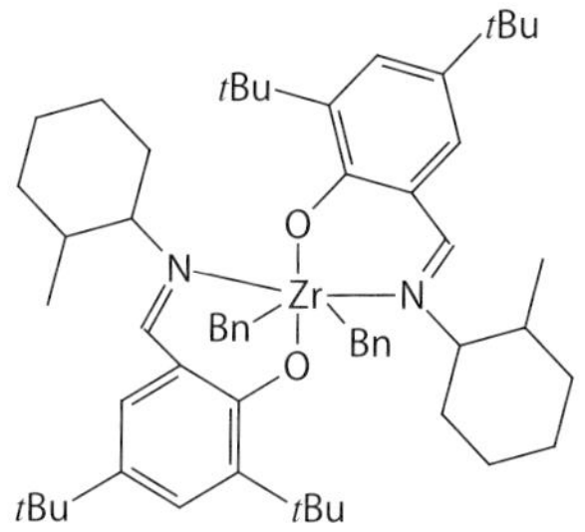

13 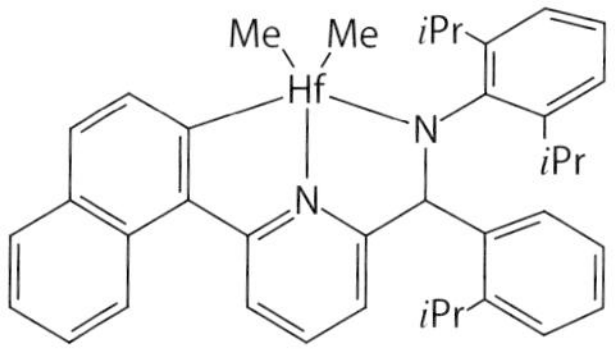                                              14

## 5.6
## Conclusions

The formation of dormant species in catalytic olefin polymerization can have profound effects on both catalyst activity and polymer structure and properties. In propene polymerization with Ziegler–Natta catalysts, an occasional 2,1- rather than the usual 1,2-insertion of the monomer greatly reduces the rate of chain propagation. However, in the presence of hydrogen, chain transfer at dormant, 2,1-inserted centers restores the active species, leading to much higher activities than are obtained in the absence of hydrogen. Even a very low incidence of 2,1-insertion is sufficient to have a profound effect on polypropylene molecular weight and molecular weight distribution. In metallocene-catalyzed polymerization, dormant species can also result from the elimination of hydrogen to form $\eta^3$-allyl species. Other phenomena observed with metallocenes, but not with Ziegler–Natta catalysts, are chain-end isomerization of dormant 2,1-inserted centers to give 3,1-units in the chain and the formation of chain stereoerrors via growing chain epimerization after 1,2-insertion, particularly at low monomer concentration. These isomerizations involve β-H elimination, olefin rotation and reinsertion into the Zr−H bond.

Resting states in olefin polymerization with highly active metallocene-based systems can involve a range of agostic interactions. In systems with lower activity, for example when a strongly coordinating anion or AlMe$_3$ is present, dormant species such as $[Cp_2ZrMe^+...MeB(C_6F_5)_3{}^-]$ or $[Cp_2Zr(\mu\text{-R})(\mu\text{-Me})AlMe_2]^+[X]^-$ hinder effective monomer coordination to the transition metal center. The resting state in polymerization with nickel and palladium diimine complexes is the complex between the metal alkyl and ethene. Monomer insertion results in the formation of β-agostic species which can undergo a series of β-hydride eliminations and re-additions, resulting in chain walking and the formation of chain branching.

A particularly useful exploitation of dormant site formation in olefin polymerization has been the recent discovery of chain shuttling. In this process, rapid and reversible chain transfer between different active transition metal centers and a non-active main group metal, in particular zinc, has resulted in the production of hitherto inaccessible ethene/α-olefin multi-block copolymers comprising alternating crystalline and amorphous chain segments.

## References

1 Kissin, Y.V., Mink, R.I., and Nowlin, T.E. (1999) *J. Polym. Sci., Part A: Polym. Chem.*, **37**, 4255–4272.

2 Kissin, Y.V. and Brandolini, A.J. (1999) *J. Polym. Sci., Part A: Polym. Chem.*, **37**, 4273–4280.

3 Kissin, Y.V., Mink, R.I., Nowlin, T.E., and Brandolini, A.J. (1999) *J. Polym. Sci., Part A: Polym. Chem.*, **37**, 4281–4294.

4 Kissin, Y.V. (2002) *Macromol. Theory Simul.*, **11**, 67–76.

5 Garoff, T., Johansson, S., Pesonen, K., Waldvogel, P., and Lindgren, D. (2002) *Eur. Polym. J.*, **38**, 121–132.

6 Mikenas, T.B., Zakharov, V.A., Echevskaya, L.G., and Matsko, M.A. (2000) *Polimery*, **45**, 349–352.

7 Mikenas, T.B., Zakharov, V.A., Echevskaya, L.G., and Matsko, M.A. (2001) *Macromol. Chem. Phys.*, **202**, 475–481.

8 Matsko, M.A., Bukatov, G.D., Mikenas, T.B., and Zakharov, V.A. (2001) *Macromol. Chem. Phys.*, **202**, 1435–1439.

9 Guastalla, G. and Giannini, U. (1983) *Makromol. Chem., Rapid Commun.*, **4**, 519–527.

10 Busico, V., Cipullo, R., and Corradini, P. (1992) *Makromol. Chem., Rapid Commun.*, **13**, 15–20.

11 Tsutsui, T., Kashiwa, N., and Mizuno, A. (1990) *Makromol. Chem., Rapid Commun.*, **11**, 565–570.

12 Chadwick, J.C., van Kessel, G.M.M., and Sudmeijer, O. (1995) *Macromol. Chem. Phys.*, **196**, 1431–1437.

13 Chadwick, J.C., Morini, G., Albizzati, E., Balbontin, G., Mingozzi, I., Cristofori, A., Sudmeijer, O., and van Kessel, G.M.M. (1996) *Macromol. Chem. Phys.*, **197**, 2501–2510.

14 Busico, V., Cipullo, R., Polzone, C., Talarico, G., and Chadwick, J.C. (2003) *Macromolecules*, **36**, 2616–2622.

15 Busico, V., Chadwick, J.C., Cipullo, R., Ronca, S., and Talarico, G. (2004) *Macromolecules*, **37**, 7437–7443.

16 Chadwick, J.C., van der Burgt, F.P.T.J., Rastogi, S., Busico, V., Cipullo, R., Talarico, G., and Heere, J.J.R. (2004) *Macromolecules*, **37**, 9722–9727.

17 Moscardi, G., Piemontesi, F., and Resconi, L. (1999) *Organometallics*, **18**, 5264–5275.

18 Randall, J.C., Ruff, C.J., Vizzini, J.C., Speca, A.N., and Burkhardt, T.J. (1999) *Metalorganic Catalysts for Synthesis and Polymerization* (ed W. Kaminsky), Springer, Berlin, pp. 601–615.

19 Chadwick, J.C., Heere, J.J.R., and Sudmeijer, O. (2000) *Macromol. Chem. Phys.*, **201**, 1846–1852.

20 Kissin, Y.V. and Rishina, L.A. (2002) *J. Polym. Sci., Part A: Polym. Chem.*, **40**, 1353–1365.

21 Bochmann, M. (1996) *J. Chem. Soc., Dalton Trans.*, 255–270.

22 Jia, L., Yang, X., Stern, C.L., and Marks, T.J. (1997) *Organometallics*, **16**, 842–857.

23 Beck, S., Prosenc, M.-H., and Brintzinger, H.H. (1998) *J. Mol. Catal. A: Chem.*, **128**, 41–52.

24 Schaper, F., Geyer, A., and Brintzinger, H.H. (2002) *Organometallics*, **21**, 473–483.

25 Busico, V., Cipullo, R., and Esposito, V. (1999) *Macromol. Rapid Commun.*, **20**, 116–121.

26 Beck, S., Lieber, S., Schaper, F., Geyer, A., and Brintzinger, H.H. (2001) *J. Am. Chem. Soc.*, **123**, 1483–1489.

27 Song, F., Cannon, R.D., and Bochmann, M. (2003) *J. Am. Chem. Soc.*, **125**, 7641–7653.

28 Liu, Z., Somsook, E., White, C.B., Rosaaen, K.A., and Landis, C.R. (2001) *J. Am. Chem. Soc.*, **123**, 11193–11207.

29 Landis, C., Rosaaen, K.A., and Sillars, D.R. (2003) *J. Am. Chem. Soc.*, **125**, 1710–1711.

30 Bochmann, M. (2004) *J. Organometal. Chem.*, **689**, 3982–3998.

31  Bochmann, M., Cannon, R.D., and Song, F. (2006) *Kinet. Catal.*, **47**, 160–169.

32  Song, F., Lancaster, S.J., Cannon, R.D., Schormann, M., Humphrey, S.M., Zuccaccia, C., Macchioni, A., and Bochmann, M. (2005) *Organometallics*, **24**, 1315–1328.

33  Chen, M.-C., Roberts, J.A.S., and Marks, T.J. (2004) *J. Am. Chem. Soc.*, **126**, 4605–4625.

34  Coevoet, D., Cramail, H., and Deffieux, A. (1996) *Macromol. Chem. Phys.*, **197**, 855–867.

35  Forlini, F., Fan, Z.-Q., Tritto, I., Locatelli, P., and Sacchi, M.C. (1997) *Macromol. Chem. Phys.*, **198**, 2397–2408.

36  Forlini, F., Princi, E., Tritto, I., Sacchi, M.C., and Piemontesi, F. (2002) *Macromol. Chem. Phys.*, **203**, 645–652.

37  Sacchi, M.C., Forlini, F., Losio, S., Tritto, I., and Locatelli, P. (2003) *Macromol. Symp.*, **193**, 45–56.

38  Bryliakov, K.P., Talsi, E.P., Voskoboynikov, A.Z., Lancaster, S.J., and Bochmann, M. (2008) *Organometallics*, **27**, 6333–6342.

39  Busico, V., Cipullo, R., Pellecchia, R., Talarico, G., and Razavi, A. (2009) *Macromolecules*, **42**, 1789–1791.

40  Busico, V., Cipullo, R., Cutillo, F., Friederichs, N., Ronca, S., and Wang, B. (2003) *J. Am. Chem. Soc.*, **125**, 12402–12403.

41  Stapleton, R.A., Galan, B.R., Collins, S., Simons, R.S., Garrison, J.C., and Youngs, W.J. (2003) *J. Am. Chem. Soc.*, **125**, 9246–9247.

42  Busico, V., Cipullo, R., and Corradini, P. (1993) *Macromol. Chem., Rapid Commun.*, **14**, 97–103.

43  Grassi, A., Zambelli, A., Resconi, L., Albizzati, E., and Mazzocchi, R. (1988) *Macromolecules*, **21**, 617–622.

44  Busico, V., Cipullo, R., Chadwick, J.C., Modder, J.F., and Sudmeijer, O. (1994) *Macromolecules*, **27**, 7538–7543.

45  Resconi, L., Fait, A., Piemontesi, F., Colonnesi, M., Rychlicki, H., and Zeigler, R. (1995) *Macromolecules*, **28**, 6667–6676.

46  Pilmé, J., Busico, V., Cossi, M., and Talarico, G. (2007) *J. Organometal. Chem.*, **692**, 4227–4236.

47  Yano, A., Yamada, S., and Akimoto, A. (1999) *Macromol. Chem. Phys.*, **200**, 1356–1362.

48  Busico, V., Cipullo, R., Chadwick, J.C., and Borriello, A. (1995) *Macromol. Rapid Commun.*, **16**, 269–274.

49  Busico, V., Cipullo, R., and Ronca, S. (2002) *Macromolecules*, **35**, 1537–1542.

50  Ewen, J.A., Elder, M.J., Jones, R.L., Rheingold, A.L., Liable-Sands, L.M., and Sommer, R.D. (2001) *J. Am. Chem. Soc.*, **123**, 4763–4773.

51  Ewart, S.W., Parent, M.A., and Baird, M.C. (1999) *J. Polym. Sci., Part A: Polym. Chem.*, **37**, 4386–4389.

52  Busico, V., Cipullo, R., Romanelli, V., Ronca, S., and Togrue, M. (2005) *J. Am. Chem. Soc.*, **127**, 1608–1609.

53  Lin, S., Kravchenko, R., and Waymouth, R.M. (2000) *J. Mol. Catal. A: Chem.*, **158**, 423–427.

54  Landis, C.R., Sillars, D.R., and Batterton, J.M. (2004) *J. Am. Chem. Soc.*, **126**, 8890–8891.

55  Jüngling, S., Mülhaupt, R., Stehling, U., Brintzinger, H.-H., Fischer, D., and Langhauser, F. (1995) *J. Polym. Sci., Part A: Polym. Chem.*, **33**, 1305–1317.

56  Carvill, A., Tritto, I., Locatelli, P., and Sacchi, M.C. (1997) *Macromolecules*, **30**, 7056–7062.

57  Eshuis, J.J.W., Tan, Y.Y., Meetsma, A., Teuben, J.H., Renkema, J., and Evens, G.G. (1992) *Organometallics*, **11**, 362–369.

58  Richardson, D.E., Alameddin, N.G., Ryan, M.F., Hayes, T., Eyler, J.R., and Siedle, A.R. (1996) *J. Am. Chem. Soc.*, **118**, 11244–11253.

59  Karol, F.J., Kao, S.-C., Wasserman, E.P., and Brady, R.C. (1997) *New J. Chem.*, **21**, 797–805.

60  Resconi, L. (1999) *J. Mol. Catal. A: Chem.*, **146**, 167–178.

61  Margl, P.M., Woo, T.K., Blöchl, P.E., and Ziegler, T. (1998) *J. Am. Chem. Soc.*, **120**, 2174–2175.

62  Margl, P.M., Woo, T.K., and Ziegler, T. (1998) *Organometallics*, **17**, 4997–5002.

63  Zhu, C. and Ziegler, T. (2003) *Inorg. Chim. Acta*, **345**, 1–7.

**64** Lieber, S., Prosenc, M.-H., and Brintzinger, H.-H. (2000) *Organometallics*, **19**, 377–387.

**65** Ishihara, T. and Shiono, T. (2005) *J. Am. Chem. Soc.*, **127**, 5774–5775.

**66** Niu, H. and Dong, J.-Y. (2007) *Polymer*, **48**, 1533–1540.

**67** Busico, V. and Cipullo, R. (1994) *J. Am. Chem. Soc.*, **116**, 9329–9330.

**68** Busico, V. and Cipullo, R. (1995) *J. Organometal. Chem.*, **497**, 113–118.

**69** Leclerc, M.K. and Brintzinger, H.H. (1995) *J. Am. Chem. Soc.*, **117**, 1651–1652.

**70** Leclerc, M.K. and Brintzinger, H.H. (1996) *J. Am. Chem. Soc.*, **118**, 9024–9032.

**71** Yoder, J.C. and Bercaw, J.E. (2002) *J. Am. Chem. Soc.*, **124**, 2548–2555.

**72** Fait, A., Resconi, L., Guerra, G., and Corradini, P. (1999) *Macromolecules*, **32**, 2104–2109.

**73** Ystenes, M. (1991) *J. Catal.*, **129**, 383–401.

**74** Busico, V., Cipullo, R., Cutillo, F., and Vacatello, M. (2002) *Macromolecules*, **35**, 349–354.

**75** Ittel, S.D., Johnson, L.K., and Brookhart, M. (2000) *Chem. Rev.*, **100**, 1169–1203.

**76** Huang, R., Liu, D., Wang, S., and Mao, B. (2004) *Macromol. Chem. Phys.*, **205**, 966–972.

**77** Mikenas, T.B., Zakharov, V.A., Echevskaya, L.G., and Matsko, M.A. (2005) *J. Polym. Sci., Part A: Polym. Chem.*, **43**, 2128–2133.

**78** Semikolenova, N.V., Zakharov, V.A., Paukshtis, E.A., and Danilova, I.G. (2005) *Top. Catal.*, **32**, 77–82.

**79** Zakharov, V.A., Semikolenova, N.V., Mikenas, T.B., Barabanov, A.A., Bukatov, G.D., Echevskaya, L.G., and Mats'ko, M.A. (2006) *Kinet. Catal.*, **47**, 303–312.

**80** Barabanov, A.A., Bukatov, G.D., Zakharov, V.A., Semikolenova, N.V., Mikenas, T.B., Echevskaja, L.G., and Matsko, M.A. (2006) *Macromol. Chem. Phys.*, **207**, 1368–1375.

**81** Mikenas, T.B., Zakharov, V.A., Echevskaya, L.G., and Matsko, M.A. (2007) *J. Polym. Chem., Part A: Polym. Chem.*, **45**, 5057–5066.

**82** Huang, R., Koning, C.E., and Chadwick, J.C. (2007) *J. Polym. Chem., Part A: Polym. Chem.*, **45**, 4054–4061.

**83** Boor, J. Jr. (1979) *Ziegler-Natta Catalysts and Polymerizations*, Academic, San Diego CA.

**84** Kim, J.D. and Soares, J.B.P. (1999) *Macromol. Rapid Commun.*, **20**, 347–350.

**85** Ní Bhriain, N., Brintzinger, H.-H., Ruchatz, D., and Fink, G. (2005) *Macromolecules*, **38**, 2056–2063.

**86** Britovsek, G.J.P., Cohen, S.A., Gibson, V.C., Maddox, P.J., and van Meurs, M. (2002) *Angew. Chem. Int. Ed.*, **41**, 489–491.

**87** Britovsek, G.J.P., Cohen, S.A., Gibson, V.C., and van Meurs, M. (2004) *J. Am. Chem. Soc.*, **126**, 10701–10712.

**88** van Meurs, M., Britovsek, G.J.P., Gibson, V.C., and Cohen, S.A. (2005) *J. Am. Chem. Soc.*, **127**, 9913–9923.

**89** Zhang, Y., Keaton, R.J., and Sita, L.R. (2003) *J. Am. Chem. Soc.*, **125**, 9062–9069.

**90** Zhang, W., Wei, J., and Sita, L.R. (2008) *Macromolecules*, **41**, 7829–7833.

**91** Chenal, T., Olonde, X., Pelletier, J.-F., Bujadoux, K., and Mortreux, A. (2007) *Polymer*, **48**, 1844–1856.

**92** Arriola, D.J., Carnahan, E.M., Hustad, P.D., Kuhlman, R.L., and Wenzel, T.T. (2006) *Science*, **312**, 714–719.

**93** Hustad, P.D., Kuhlman, R.L., Arriola, D.J., Carnahan, E.M., and Wenzel, T.T. (2007) *Macromolecules*, **40**, 7061–7064.

# 6
# Transition Metal Catalyzed Olefin Oligomerization

## 6.1
## Introduction

The oligomerization of ethene to produce linear 1-alkenes represents one of the most important industrial applications for homogeneous catalysis. Olefins such as 1-butene, 1-hexene and 1-octene are used as comonomers in the production of linear low-density polyethylene. Olefins in the $C_6$–$C_{10}$ range are used as starting materials for plasticizers, while $C_{10}$–$C_{20}$ 1-alkenes are required for the manufacture of surfactants and lubricant oil additives. Important and well-established industrial processes for the production of 1-alkenes are aluminum-catalyzed ethene oligomerization and the nickel-based Shell Higher Olefins Process (SHOP). Aspects of the latter process are described in Section 6.6. The aluminum-catalyzed production of 1-alkenes is based on Ziegler's *Aufbaureaktion*, involving a series of olefin insertions into an Al—C bond, using an alkyl such as $AlEt_3$ [1]. As indicated in Scheme 6.1, β-H transfer generates a 1-alkene and an Al—H species, which reacts with ethene to (re) generate Al—Et, leading to further oligomeric chain growth.

Early studies on olefin oligomerization, reviewed in 1991–1992, also included a wide range of Ziegler–Natta systems based on titanium and zirconium [2, 3]. Examples of such systems are $TiCl_4$/$EtAlCl_2$ and $Ti(OR')_4$/$AlR_3$. The latter system is utilized in the Alphabutol process for the dimerization of ethene to 1-butene [4, 5]. More recently, attention has been paid to the application of well-defined single-center complexes for ethene oligomerization. Early transition metal catalysts described in this chapter include zirconium complexes used for olefin oligomerization, half-sandwich titanium complexes which trimerize ethene to 1-hexene, and chromium catalysts for ethene trimerization or tetramerization. Of these, the chromium-based systems have received a particularly large amount of attention in the past decade. Late transition metal catalysts for ethene oligomerization comprise, in addition to the nickel-based SHOP systems, bis(imino)pyridyl iron complexes, which are also covered here. The final section in this chapter describes tandem catalyst systems for the synthesis of linear low-density polyethylene using ethene as sole monomer source. Such systems involve the use of an ethene polymerization catalyst in

*Homogeneous Catalysts: Activity – Stability – Deactivation*, First Edition. Piet W.N.M. van Leeuwen and John C. Chadwick.
© 2011 Wiley-VCH Verlag GmbH & Co. KGaA. Published 2011 by Wiley-VCH Verlag GmbH & Co. KGaA.

$$Al\text{--}Et \quad + \quad n\ CH_2{=}CH_2 \quad \longrightarrow \quad Al\text{--}(C_2H_4)_n\ Et$$

$$Al\text{--}(C_2H_4)_n\ Et \quad \longrightarrow \quad Al\text{--}H \quad + \quad CH_2{=}CH\text{--}(C_2H_4)_{n-1}\ Et$$

$$Al\text{--}H \quad + \quad CH_2{=}CH_2 \quad \rightleftharpoons \quad Al\text{--}Et$$

**Scheme 6.1** Aluminum-catalyzed oligomerization of ethene.

combination with, for example, a catalyst which is able to trimerize ethene to 1-hexene, generating the comonomer *in situ*.

## 6.2
## Zirconium Catalysts

Metallocene complexes for the dimerization or oligomerization of ethene, propene and higher 1-alkenes have been reviewed by Janiak [6]. Most of this chapter concerns ethene oligomerization, but metallocenes can, in contrast to many other oligomerization catalysts, also be used for propene oligomerization [7]. The nature of chain termination in propene oligomerization and polymerization with zirconocene/MAO systems depends on the cyclopentadienyl ring substitution. $Cp_2ZrCl_2$ and indeed most metallocenes, give vinylidene end-groups, resulting from $\beta$-H transfer from the growing chain to the monomer. However, highly substituted metallocenes such as $Cp^*_2ZrCl_2$ ($Cp^* = C_5Me_5$) give vinyl-terminated polymers and oligomers formed by $\beta$-$CH_3$ elimination from the growing chain [8]:

$$Cp_2^*Zr^+\text{--}CH_2\text{--}CH(CH_3)\text{--}Pol \rightarrow Cp_2^*Zr^+\text{--}CH_3 + CH_2{=}CH\text{--}Pol$$

Termination by $\beta$-$CH_3$ transfer was first reported by Teuben and coworkers, who obtained 4-methyl-1-pentene and 4,6-dimethyl-1-heptene and higher vinyl-terminated oligomers on contacting propene with $[Cp^*_2ZrMe(THT)]^+[BPh_4]^-$ (THT = tetrahydrothiophene) [9]. Under similar conditions, ethene gave relatively high molecular weight polymer. Further investigation of propene oligomerization with this system and its hafnium analogue revealed that catalyst deactivation took place via the formation of allyl and 2-methallyl complexes, as described in Chapter 5.3.4 [10].

Zirconocene-catalyzed homo- and co-oligomerization of ethene and propene has been investigated by Ciardelli and coworkers [11]. With $Cp_2ZrMe_2/MAO$, increasing the temperature from 20 to 80 °C in ethene polymerization led to an order of magnitude decrease in polymer molecular weight. The incorporation of $AlEt_3$ as a third catalyst component decreased both the catalyst activity and the polymer molecular weight. In the case of propene, $Cp_2ZrMe_2/MAO$ gave liquid oligomers. Oligomers in the $C_6$–$C_{30}$ range were also obtained in ethene/propene co-oligomerization. Chain-end analysis indicated that chain transfer took place predominantly after propene insertion.

The ethoxyboratabenzene complex **1** has been shown to be an effective catalyst for ethene oligomerization when activated by MAO, giving almost exclusively linear 1-alkenes [12]. In contrast, the aminoboratabenzene complex **2** gave polyethylene and the phenyl-substituted complex **3** gave mainly 1-alkenes and 2-alkyl-1-alkenes [13, 14].

The 2-alkyl-1-alkenes resulted from insertion of a 1-alkene into the growing oligomer chain, followed by β-H elimination. The different product distributions obtained with complexes 1–3 were attributed to differences in electron density at the metal site.

A heterogeneous catalyst system for propene oligomerization has been reported by Jacobs and coworkers, who immobilized a borate activator by reacting $B(C_6F_5)_3$ and dimethylaniline with an MCM-41 support [15]. Contact with $Cp_2ZrMe_2$ then generated active species of type $[Si-O-B(C_6F_5)_3]^-[Cp_2ZrMe]^+$. Reaction with propene was carried out at the relatively high temperature of 90 °C to avoid polymerization, resulting in a Schulz–Flory distribution of oligomers in which the main dimeric product was 2-methyl-1-pentene. Oligomerization of γ-branched 1-alkenes using $Me_2Si(Cp)_2ZrCl_2$/MAO has been investigated by Longo and coworkers, who observed selective dimerization in the case of vinylcyclohexane, whereas less hindered monomers such as 3-methyl-1-butene gave higher oligomers and propene gave low molecular weight polymer [16].

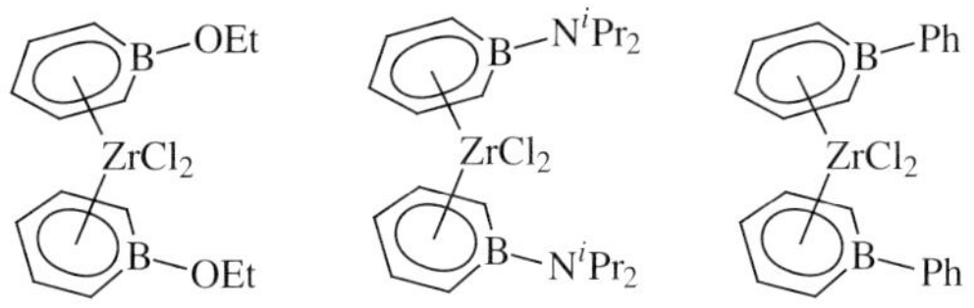

| 1 | 2 | 3 |

Bis(amido) zirconium complexes such as 4, used in combination with a borate activator, have been found to be effective in ethene oligomerization under mild conditions [17]. A lower catalyst decay rate, leading to higher productivity, was obtained when a branched aluminum alkyl such as tris(2,4,4-trimethylpentyl)aluminum or its aluminoxane derivative was incorporated as an additional component in the catalyst system [18].

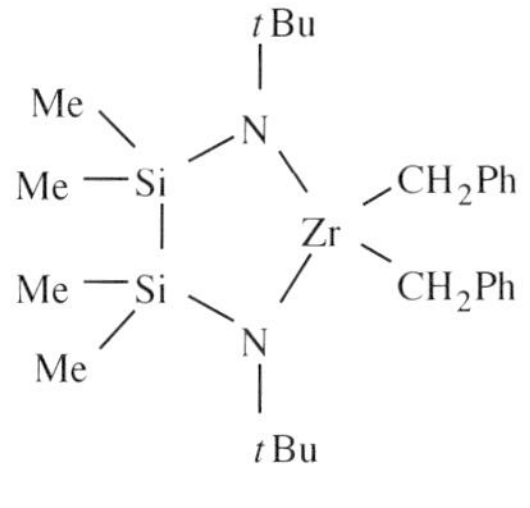

4

## 6.3
## Titanium Catalysts

An interesting transformation of an ethene polymerization catalyst to a catalyst for the trimerization of ethene to 1-hexene, making use of a hemilabile ancillary ligand, has been reported by Deckers et al. [19]. This transformation was achieved by changing the ligand substituent R in $(\eta^5\text{-}C_5H_4CMe_2R)TiCl_3$ from a methyl to a phenyl group. In combination with MAO, complex **5** gave 95–98% selectivity in ethene trimerization, producing a $C_6$ fraction containing >99% 1-hexene. A $C_{10}$ fraction was also obtained, comprising cotrimers of ethene and 1-hexene. The thermal stability of the catalyst was modest and increasing the temperature in the range 30–80 °C led to significantly decreased productivity as a result of catalyst deactivation. Increased catalyst stability was observed when a second $CMe_2Ph$ substituent was introduced, as in complex **6** [20]. This catalyst was less active than **5**, but in a 2 h run gave higher productivity as a result of greatly improved stability. It was also observed that **5**/MAO degraded less rapidly in toluene than in an octane/toluene mixture, indicating catalyst stabilization by transient toluene coordination. Different stabilities were also observed for complexes **7** and **8** A lower stability of **7**/MAO relative to **8**/MAO was suggested to be due to the electron-donating *tert*-butyl group, increasing the electron density on Ti and making it less prone to stabilization by arene coordination. It was also noted that deactivation by cyclometalation of a cyclopentadienyl ring substituent would be more likely for $CMe_3$ than for $SiMe_3$, by analogy with results obtained by Marks and coworkers with cationic zirconocene species [21]. The most active complex for ethene trimerization was found to be **9**, containing methyl substitution in the aryl ring in addition to the trimethylsilyl substituent in the cyclopentadienyl ring [20]. Deckers and Hessen have also studied the stability of titanium tribenzyl complexes with cyclopentadienyl-arene ligands [22]. Thermolysis of $(C_5H_4CMe_2Ph)Ti(CH_2Ph)_3$ in solution at 50 °C led to *ortho*-cyclometalation of the pendant aryl group, forming complex **10**. The cationic dibenzyl species $[(C_5H_4CMe_2Ph)Ti(CH_2Ph)_2]^+$ decomposed similarly at ambient temperature, whereas the dimethyl species $[(C_5H_4CMe_2Ph)TiMe_2]^+$ was more stable, indicating that displacement of the coordinated arene is required for cyclometalation.

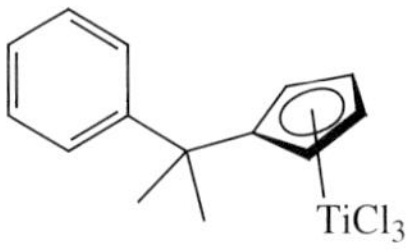

**5**

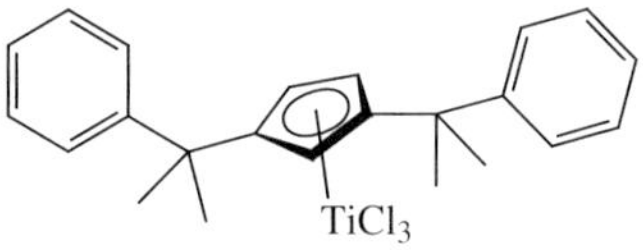

**6**

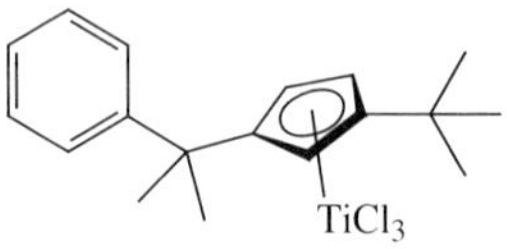

**7**

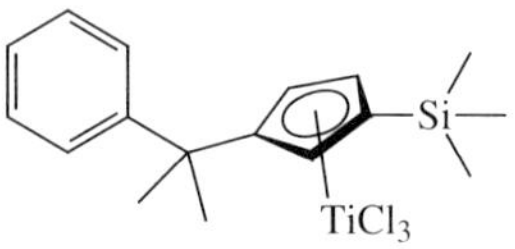

**8**

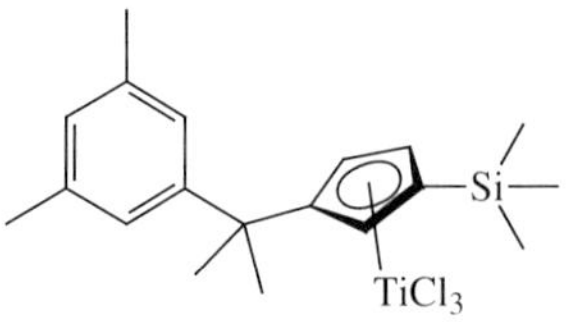

**9**

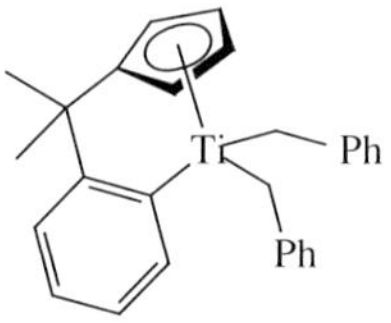

**10**

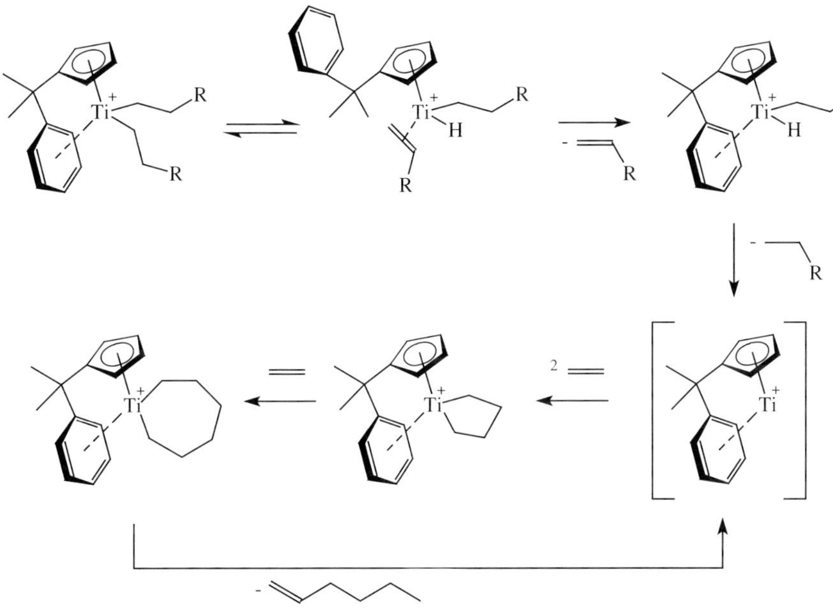

**Scheme 6.2** Ethene trimerization catalyzed by $(\eta^5\text{-}C_5H_4CMe_2Ph)TiCl_3/MAO$.

The trimerization of ethene to 1-hexene in these systems proceeds via metallacycle intermediates, as shown in Scheme 6.2 [19]. The initially generated species $[(C_5H_4CMe_2Ph)TiMe_2]^+$ first undergoes multiple ethene insertions to give $[(C_5H_4CMe_2Ph)Ti(CH_2CH_2R)_2]^+$, which is in equilibrium (through β-H elimination) with hydride–olefin species. Displacement of the olefin by the pendant aryl moiety is then followed by reductive elimination to give a Ti(II) species, which coordinates two molecules of ethene to give a titana(IV)cyclopentane. Insertion of a further ethene unit gives a titana(IV)cycloheptane. Theoretical studies revealed that the liberation of 1-hexene from the titanacycloheptane arises from direct, intramolecular $C_\beta$ to $C_\alpha$, hydrogen transfer [23–25].

The driving force for trimerization rather than polymerization in the above systems is the hemilabile character of the arene-Cp ligand [26, 27]. Whereas under homogeneous conditions $(C_5H_4CMe_2Ph)TiCl_3$ is a trimerization catalyst, immobilization on a MgCl$_2$-based support has been found to result in the exclusive formation of polyethylene [28]. The transformation from trimerization to polymerization suggests that coordination of the pendant aryl ring to the metal center, an essential feature in the trimerization mechanism, is no longer operative after immobilization on MgCl$_2$. Even under homogeneous conditions, 2–5 wt% of polyethylene is produced as a side product. Its formation is caused by at least two different species, including partially alkylated species from the reaction of $(C_5H_4CMe_2Ph)TiCl_3$ with MAO and species formed by degradation of the ligand system via loss of the aryl-bearing substituent [29]. Polymer formation, particularly at elevated temperatures, has also been observed in ethene trimerization with the half-sandwich titanium complex **11**, bearing a pendant thienyl group [30]. At low temperature (0 °C), **11**/ MAO gave 1-hexene with a selectivity >95%, whereas the related complex **12** gave mainly polyethylene, indicating that 1-hexene formation resulted from an $\eta^1$-S rather than an $\eta^5$ coordination of the pendant ligand.

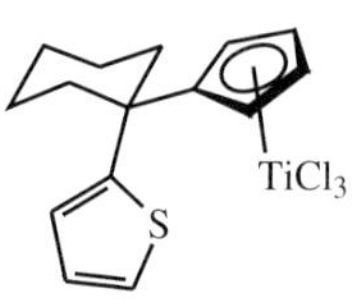

**11** 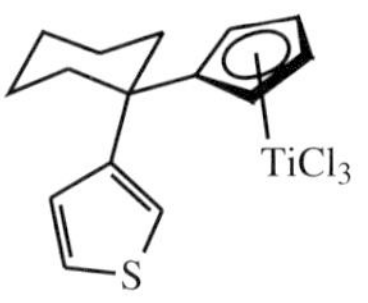              **12** 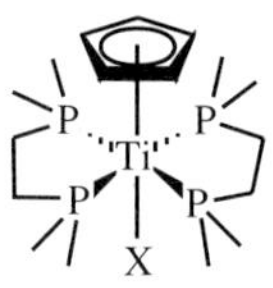              **13**

Monocyclopentadienyl titanium(II) complexes of type **13** (X = Cl, H or Me) have been found to be active dimerization catalysts when contacted with ethene in the absence of any cocatalyst [31]. The product distribution obtained, comprising mainly 1-butene, 2-ethyl-1-butene and 3-methyl-1-pentene, provided strong evidence for an olefin coupling mechanism involving metallacyclopentane intermediates. It was concluded that the active species were Ti(II) centers, and that the oligomerization of ethene to dimers and trimers in titanium-based catalysts is diagnostic of the presence of the titanium(II) oxidation state. On the basis of a similar product distribution, it was proposed that the catalytically active species in the Ti(OR')$_4$/ AlR$_3$-based Alphabutol process for ethene dimerization are Ti(II) centers, whereas in Ziegler–Natta olefin polymerization catalysts the active species are in a higher oxidation state.

## 6.4
### Tantalum Catalysts

A "ligand-free" catalyst system for the selective trimerization of ethene to 1-hexene, comprising TaCl$_5$ and an alkylating agent such as SnMe$_4$ or ZnMe$_2$, has been reported by Sen and coworkers [32]. The trimerization mechanism involves first the formation of a tantalum(III) species, which reacts with two molecules of ethene to form a Ta(V) metallacycle. Insertion of a third molecule of ethene gives the metallacycloheptane, after which reductive elimination of 1-hexene regenerates Ta(III). The precursor to the active species was found to be TaMe$_2$Cl$_3$. *In situ* reduction of TaCl$_5$ to TaCl$_3$ has also been achieved as shown in Scheme 6.3, resulting in an active system for ethene trimerization [33]. A selectivity of 98.5% for 1-hexene formation was obtained.

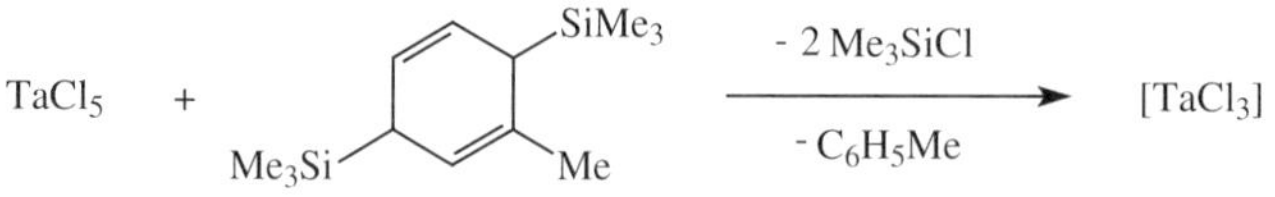

**Scheme 6.3**   Salt-free reduction of TaCl$_5$.

## 6.5
## Chromium Catalysts

### 6.5.1
### Chromium-catalyzed Trimerization

The chromium-catalyzed trimerization of ethene to 1-hexene has received consider-able industrial interest and, in 2003, a plant for the manufacture of 1-hexene via ethene trimerization was brought on stream by Chevron Phillips. Advances in selective ethene trimerization up to 2004 have been reviewed by Morgan and coworkers, who described the use of pyrrolyl ligands in trimerization catalyst development by Phillips [34]. A particularly active catalyst system described by Phillips comprised chromium(III) 2-ethylhexanoate, 2,5-dimethylpyrrole, $AlEt_3$ and $AlEt_2Cl$ [35]. There is relatively little public knowledge concerning the fundamental aspects of such catalyst systems, but it has been found that catalytic activity and selectivity is strongly dependent on organochloro components used as a halogen source together with $Cr(2\text{-ethylhexanoate})_3$, 2,5-dimethylpyrrole and $AlEt_3$ [36]. Luo et al. obtained high activity and selectivity using 2-fluoro-6-chloro-$\alpha,\alpha,\alpha$-trichlorotoluene and proposed various possibilities for coordination of the halide to the metal center [37]. Trimeriza-tion of ethene to 1-hexene using chromium catalysts takes place via a metallacycle mechanism, as illustrated in Scheme 6.4 [38]. A theoretical study by Janse van Rensburg et al. was based on the presence of Cr(II) and Cr(IV) active species in the catalytic cycle and indicated hemilabile coordination of the pyrrole ring [39].

In the past decade, new chromium catalysts for both trimerization and tetramer-ization have been discovered [40]. At BP, it was found that a catalyst system comprising $CrCl_3(THF)_3$ and the ligand $Ar_2PN(Me)PAr_2$ (Ar = 2-methoxyphenyl), activated with MAO, produced 1-hexene with an overall selectivity of around 90% and at a productivity exceeding $10^6$ g (g Cr)$^{-1}$ h$^{-1}$ [41]. Reaction was carried out in toluene for 1 h at 80 °C and 20 bar. The productivity was found to be second-order with respect to ethene pressure, consistent with the metallacycle mechanism. The catalyst was also very stable, no deactivation being observed over the run time. In addition to 1-hexene, $C_{10}$ cotrimers of ethene and 1-hexene were formed. It was proposed that the pendant methoxy group plays an important role in stabilizing the coordinatively unsaturated intermediates formed during the catalytic cycle. Studies by Labinger, Bercaw and coworkers provided support for the metallacycle mechanism and showed that the 2-methoxy substituted PNP ligand can display $\kappa^2$-(P,P) and fac-$\kappa^3$-(P,P,O)

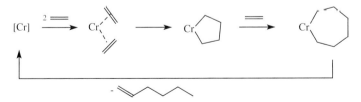

**Scheme 6.4** Chromium-catalyzed trimerization of ethene.

coordination modes [42–44]. A coordination number of six was observed for the Cr (III) center, such as in complex **14** (Ar = 2-methoxyphenyl), obtained from the reaction of the PNP ligand with $CrCl_3(THF)_3$. Activation of such complexes using MAO or borate activators generated ethene trimerization catalysts in which it was proposed that the active species was a cationic (PNP)-chromium species that shuttles between Cr(III) and Cr(I), with one or more methoxy groups stabilizing the various intermediates along the catalytic cycle [45]. In these studies, carried out at 1 bar ethene pressure, a catalyst initiation period was followed by decreasing ethene consumption, apparently due to a first-order catalyst deactivation.

**14**          **15**          **16**

McGuinness *et al.* have obtained high activity and selectivity towards 1-hexene with MAO-activated Cr(III) complexes of tridentate ligands, such as complex **15** (R = Et) [46]. The catalyst behavior was sensitive to the reaction conditions. At 100 °C, significant catalyst deactivation was observed within 30 min, but greater catalyst stability was obtained at 80 °C. At 50 °C, the catalyst activity dropped significantly and a high proportion of polymer was formed. Cr(III)–SNS complexes such as **16** (R = *n*-butyl or *n*-decyl) have also been found to be efficient catalysts for ethene trimeriza-tion, in combination with MAO [47]. A perceived advantage of such complexes was cheaper synthesis of the SNS ligands, relative to the PNP analogues. In these tridentate ligands, the N–H functionality was found to be essential for high activity and selectivity; replacement of N–H by N–Me or N–benzyl led to decreased activity and greatly increased polymer formation [48]. It was suggested that deprotonation to give an anionic ligand could be occurring with complexes such as **15** and **16**, but this was not supported by other studies [49].

The role of the metal oxidation state in ethene trimerization with chromium–SNS complexes has been investigated by Gambarotta, Duchateau and coworkers [49]. Reaction of [CySCH₂CH₂N(H)CH₂CH₂SCy]CrCl₃ with AlMe₃ gave the trivalent dimer **17**, whereas with isobutylaluminoxane the divalent complex **18** was formed. Complex **17** was found to be thermally robust, but decomposition was accelerated by the presence of excess MAO or AlMe₃. Similarities in the ethene trimerization behavior of **17** and **18**, in the presence of MAO, suggested that these complexes were precursors to the same catalytically active species, most likely containing chromium in the divalent state. It was also concluded that cationization of the metal center was an important factor influencing both the catalytic activity and the retention of the ligand system, which was not found to be deprotonated by the cocatalyst. Further insight into the oxidation state was obtained when $CrCl_2(THF)_2$ was treated with either AlEtCl₂ or AlEt₃ in the presence of the SNS ligand [50]. In the first case, the

Cr(II) complex **19** was formed. However, treatment with AlEt$_3$ gave the Cr(III) complex **17**, the formation of which implies a disproportionation reaction and the concomitant generation of species with a valence lower that 2. The AlEt$_3$-induced re-oxidation of Cr(II) to give an active and selective Cr(III) catalyst precursor therefore indicated that, contrary to the previous suggestion, selective trimerization may result from a trivalent catalyst.

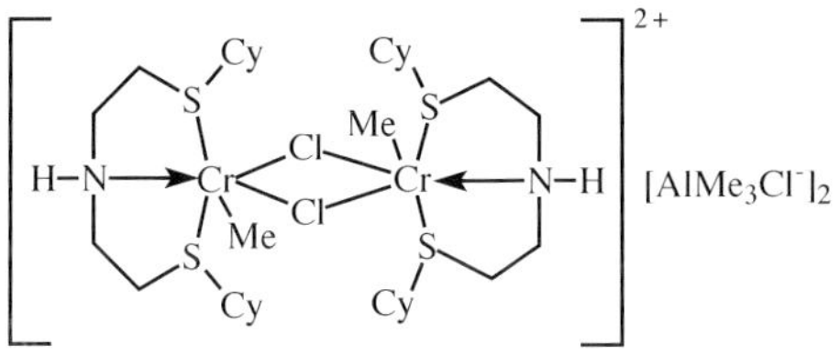

**17**

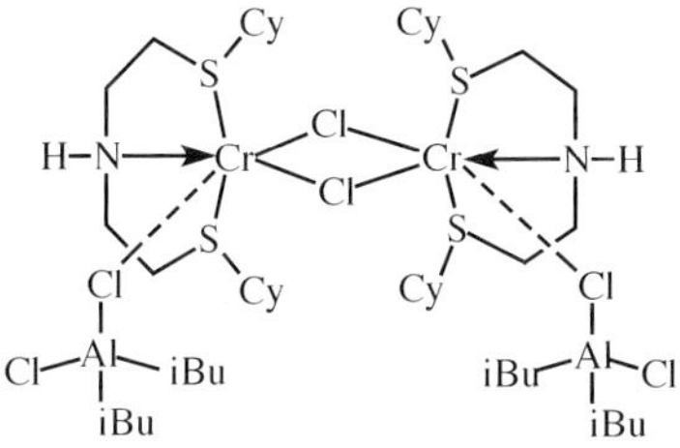

**18**

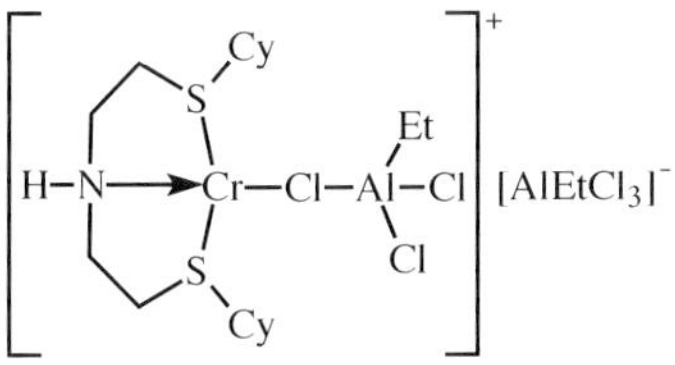

**19**

Additional studies revealed that selective trimerization of ethene to 1-hexene could also be obtained after MAO activation of the trivalent complex **20** (R = phenyl or cyclohexyl), whereas divalent species gave a statistical distribution of oligomers [51]. The presence of the pyridine ring stabilized the individual oxidation states and the fact that this ligand does not contain an N−H function indicates that the presence of the latter is not a general requirement for high selectivity in SNS-based systems.

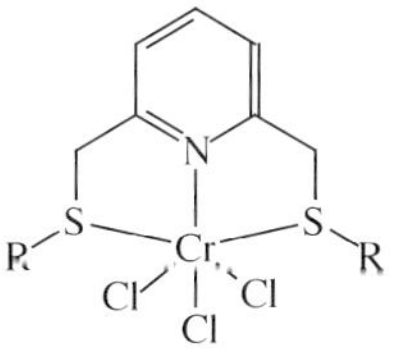

**20**

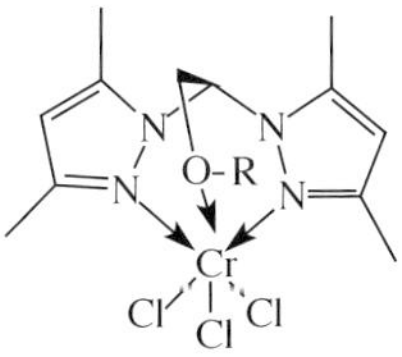

**21**

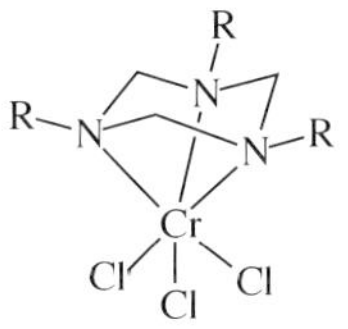

**22**

Chromium(III) complexes with [NON] and [NSN] heteroscorpionate ligands, such as **21**, have also been found to be highly selective ethene trimerization catalysts [52]. It was observed, however, that extending the reaction time from 20 to 60 min at 80 °C led to a decrease in activity, indicating a limited catalyst lifetime. Recent studies with MAO-activated [NNN] pyrazolyl complexes have indicated that residual AlMe$_3$ present in MAO played a key role in both metal reduction and Cr−Cl activation [53].

It was suggested that the importance of adventitious $AlMe_3$ in MAO-activated ethene oligomerization processes may have been underestimated.

Selective trimerization of 1-alkenes such as 1-hexene has been obtained with the 1,3,5-triazacyclohexane complex **22** (R = *n*-octyl or *n*-dodecyl), activated by MAO [54]. With ethene, however, these complexes give polymerization rather than trimerization [55]. Activation of **22** with $[PhNHMe_2][B(C_6F_5)_4]$ and $AliBu_3$ gave 1-hexene trimerization activities similar to those obtained with MAO and showed that reduction to Cr(I) can occur in these systems [56]. Evidence was obtained for decomposition via transfer of the triazacyclohexane ligand from Cr to Al in solution in toluene, generating $[(triazacyclohexane)AliBu_2][B(C_6F_5)_4]$ and $[(arene)_2Cr][B(C_6F_5)_4]$.

## 6.5.2
## Chromium-catalyzed Tetramerization of Ethene

An important advance in ethene oligomerization took place in 2004, when researchers at Sasol reported chromium-catalyzed ethene tetramerization, producing 1-octene with high selectivity [57]. It had previously been considered that tetramerization would be improbable, requiring discrimination between 7- and 9-membered metallacycle rings. The catalyst was formed by contacting $CrCl_3(THF)_3$ with a PNP ligand similar to that used in the trimerization complex **14**, but without the methoxy substituents. With MAO as activator and at a pressure of 45 bar at 45 °C, around 70 wt % of 1-octene was obtained, along with 15–20 wt% 1-hexene. Activities were higher under these conditions than at 30 bar and 65 °C, which was attributed to reduced catalyst deactivation at the lower temperature, as well as the higher ethene pressure [58]. The switch from trimerization to tetramerization on removing the *ortho*-methoxy aryl substituents in $Ar_2PN(R)PAr_2$ was proposed to be related to – OMe coordination to the metal center, retarding coordination and insertion of ethene into the 7-membered metallacycle. Steric bulk was also found to be an important factor; whereas Ar = phenyl gave predominantly 1-octene, Ar = 2-ethylphenyl gave 1-hexene with 93% selectivity [59].

Along with 1-hexene, side products in ethene tetramerization are methylcyclopentane and methenecyclopentane, formed in a 1:1 ratio [60, 61]. The production of cyclic side products is influenced by the substituent attached to the nitrogen atom in the PNP ligand, replacement of a methyl by an isopropyl, cyclohexyl or 2-alkylcyclohexyl group giving greater selectivity for 1-hexene within the $C_6$ fraction [57, 62]. Budzelaar has proposed that the formation of methylcyclopentane and methylcyclopentene takes place as illustrated in Scheme 6.5 [63].

A kinetic study of ethene tetramerization with the system $Cr(acac)_3/Ph_2PN(iPr)$ $PPh_2/MAO$ indicated an overall reaction order of 1.6 in ethene concentration [64]. The catalyst deactivation rate increased with temperature in the range 35–60 °C. Subsequent studies carried out in a continuous tube reactor revealed a higher reaction order in ethene for 1-octene than for 1-hexene formation, increased ethene pressure leading to greater selectivity for 1-octene [65]. In another study, carried out at 50 °C with the system $CrCl_3(THF)_3/Ph_2PN(iPr)PPh_2/MAO$, similar deactivation rates were observed for ligand/Cr molar ratios >1.0, but at a ligand/Cr ratio of

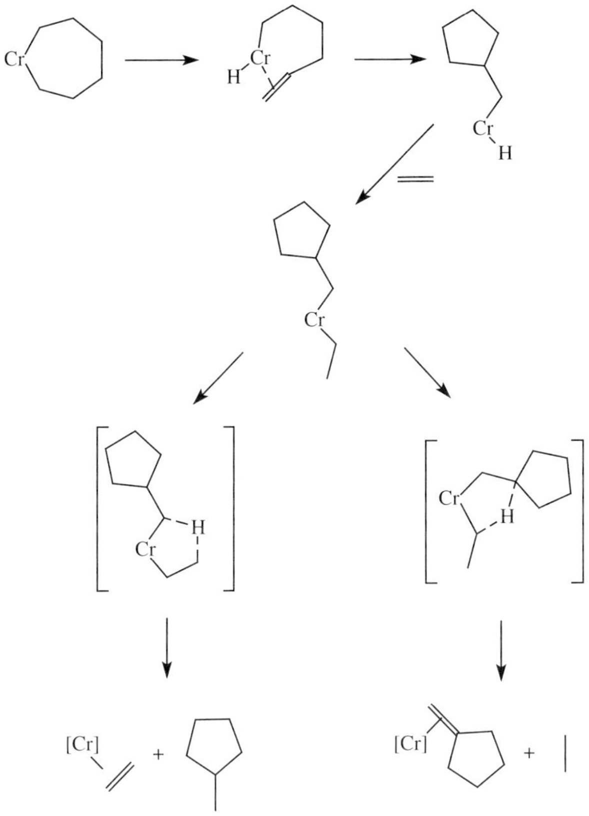

**Scheme 6.5** Formation of cyclic products in Cr-catalyzed tri-/tetramerization of ethene.

0.5 a long induction period was observed, followed by an accelerating rate [66]. This sub-stoichiometric ligand/Cr ratio also resulted in a shift to a Schulz–Flory product distribution and the formation of odd-numbered $\alpha$-olefins as by-products.

Relatively stable catalysts for ethylene tri- and tetramerization have been reported by Bercaw and coworkers, who reacted $CrCl_3(THF)_3$ with ligands such as $Ph_2PN(R)$ $PPh_2$, where R represents an ether-containing group such as $-CH_2(o\text{-}OCH_3)$ $C_6H_4$ [61]. Catalyst activity remained nearly constant for 2 h at 25 °C, whereas with systems containing the ligand $(o\text{-}MeO\text{-}C_6H_4)_2PN(Me)P(o\text{-}MeO\text{-}C_6H_4)_2$ catalyst stability could not be maintained reproducibly over 20 min.

Following the above studies in which MAO was used as cocatalyst, activation with B $(C_6F_5)_3/AlR_3$ and $[Ph_3C][B(C_6F_5)_4]/AlR_3$ was investigated [67]. Selectivities similar to those in MAO activation were obtained, but rapid catalyst deactivation and significant polyethylene formation was observed. The rate of deactivation increased with increasing $AlEt_3$ concentration, indicating degradation of the borane/borate through alkyl exchange with $AlR_3$, as described in Chapter 2.2.3 for metallocene-catalyzed olefin polymerization. The use of excess $[Ph_3C][B(C_6F_5)_4]$ resulted in loss of the PNP ligand from Cr and a shift from tetramerization to Schulz–Flory oligomerization. In a search for a more robust counterion, it was found that $[Ph_3C][Al\{OC(CF_3)_3\}_4]$, used in combination with $AlEt_3$, gave a much more stable and longer lived catalyst [68]. The $[Al\{OC(CF_3)_3\}_4]^-$ anion is reported to be one of the most weakly coordinating and robust anions known [69]. High selectivity for 1-octene was obtained, whereas the use of $Al(OC_6F_5)_3$ resulted in a shift towards trimerization, as a result of stronger anion coordination increasing the steric hindrance around the metal center.

Additional evidence for a catalytic cycle involving formally Cr(I) and Cr(III) active species in ethene tetramerization was obtained by the synthesis of the cationic Cr(I) aluminate complex $[\{Ph_2PN(iPr)PPh_2\}Cr(CO)_4][Al\{OC(CF_3)_3\}_4]$, which in combination with $AlEt_3$ gave $C_6$ and $C_8$ selectivities similar to those obtained with a Cr(III)/MAO system [70]. However, different redox couples, such as Cr(II)/Cr(IV), have also been considered as possible oxidation states in chromium-catalyzed tri- and tetramerization. The Cr(II) complex **23**, prepared either by reaction of $[(PNP)CrCl_3]_2$ $(PNP = Ph_2PN(cyclohexyl)PPh_2)$ with $AlMe_3$ or by reaction of $CrCl_2(THF)_2$ with PNP in the presence of $AlMe_3$, was active in ethene tetramerization when used with MAO [71]. The presence of a second PNP ligand in the metal coordination sphere and cationization by an Al cocatalyst were apparently central to the stability of Cr(II) derivatives of the PNP ligand.

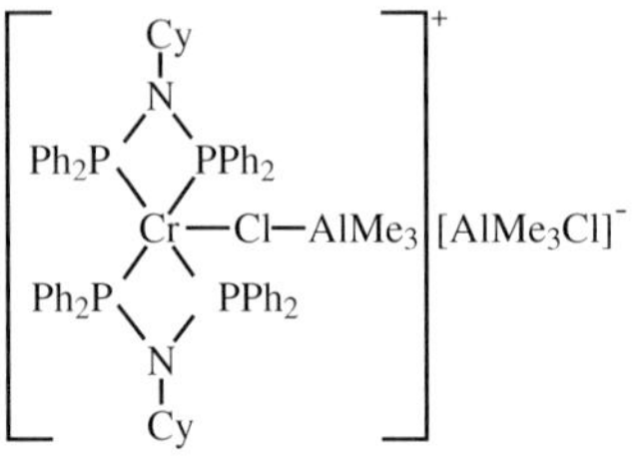

**23**

## 6.5.3
### Chromium-Catalyzed Oligomerization

In addition to the catalysts for ethene tri- and tetramerization, described above, several chromium-based systems for the preparation of higher oligomers have been described. Gibson and coworkers have reported Cr(III) complexes of CNC-pincer carbene ligands, such as **24**, which with MAO gave exceptionally active catalysts for ethene oligomerization, producing a Schulz–Flory distribution of 1-alkenes [72]. The catalyst activity decreased with time, and deactivation was particularly rapid above $50\,°C$. Complex **24** has also been used for the oligomerization of propene and higher $\alpha$-olefins; metallacycle intermediates gave mainly head-to-tail dimers with vinylidene unsaturation [73]. A metallacycle mechanism in chromium-catalyzed ethene oligomerization was unequivocally demonstrated with complex **25**, again using MAO as activator [74]. GC–MS analysis of oligomers prepared using a mixture of $C_2H_4$ and $C_2D_4$ showed only isotopomers with 0, 4, 8... deuteriums, consistent with chain propagation via large-ring metallacycles. Further studies by McGuinness *et al.* demonstrated this mechanism for oligomerization with complex **24**, but found that a change to a bidentate ligand structure, as in **26**, resulted in a change from a metallacycle to a linear chain growth mechanism and greatly decreased activity [75]. It was concluded that only chromium catalysts which support a metallacycle mechanism promote oligomerization with high activity.

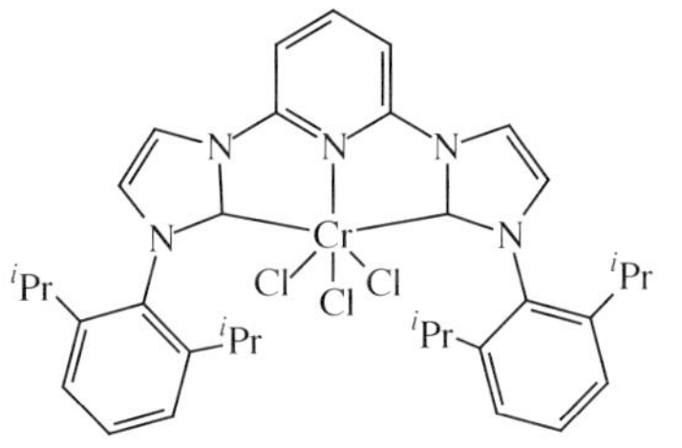

24 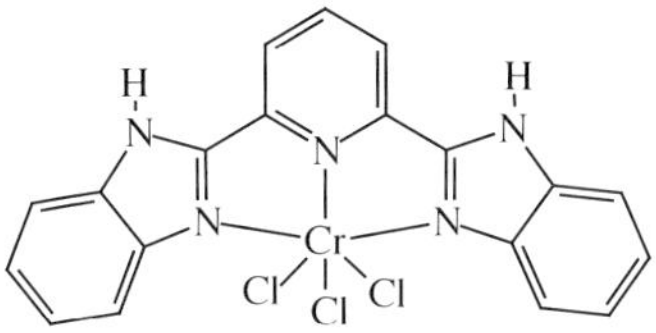                                           25

Small *et al.* have investigated ethene oligomerization using bis(imino)pyridyl chromium complexes, activated with MMAO [76]. Complex **27** dimerized ethene to 1-butene, while replacement of the methyl substituents by *tert*-butyl or the introduction of a second ortho-substituent resulted in the formation of a mixture of wax and low molecular weight polymer. These catalysts exhibited high activity up to at least 100 °C.

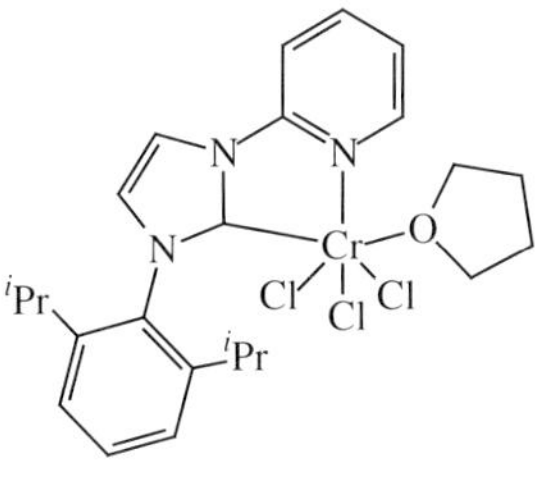

26 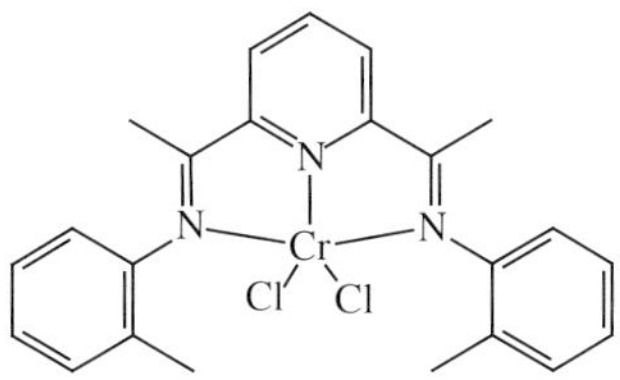                                           27

Ethene oligomerization activities with various pyrollide complexes have been found to be dependent on the metal oxidation state [77]. Using MAO as activator, activities were six times higher with a divalent than with a trivalent chromium complex. An unexpected cocatalyst effect was also observed, replacement of MAO by $iBu_2AlOAliBu_2$ leading to polymerization rather than oligomer formation.

Several examples of reversible chain transfer, leading to a living catalytic system and narrow (Poisson) product distribution, have been reported for chromium-catalyzed oligomerization. This process is analogous to coordinative chain transfer polymerization, described in Chapter 5.5. Bazan and coworkers observed rapid transmetalation during ethene oligomerization when $Cp^*CrMe_2(PMe_3)$ (**28**) was activated with MAO or with $B(C_6F_5)_3$ and $AlMe_3$ or $AlEt_3$ [78, 79]. Chain transfer between active and dormant (alkyl-bridged) species was proposed, as illustrated in Scheme 6.6. No deactivation was observed in experiments carried out at ambient

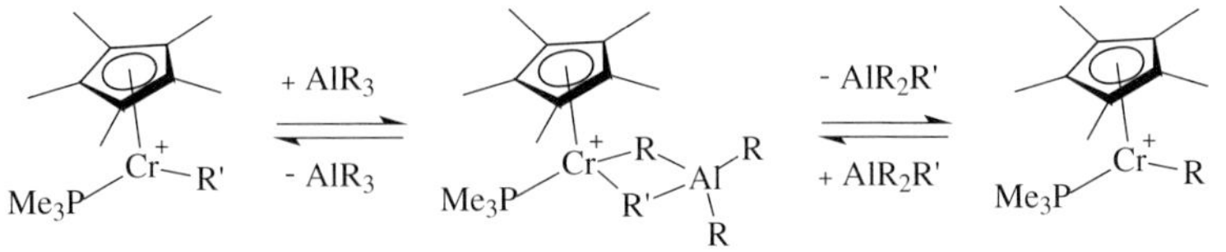

**Scheme 6.6** Active and dormant species in chromium-catalyzed coordinative chain transfer oligomerization.

temperature. Ganesan and Gabbai reported living oligomerization at ambient temperature using complex **29** in combination with AlEt$_3$ [80].

      **28**                                **29**

## 6.5.4
### Single-component Chromium Catalysts

The catalysts described in the previous sections comprise a chromium complex used in combination with a cocatalyst such as MAO. However, examples of single-component chromium catalysts have recently been reported by Gambarotta, Duchateau and coworkers [81, 82]. Treatment of Cr(tBuNPNtBu)$_2$ with AlMe$_3$ gave the divalent chromium complex **30** (Scheme 6.7). Exposure of this complex to ethene yielded polyethylene with moderate activity [81]. In contrast, a statistical distribution of ethene oligomers was formed when **30** was activated with MAO. It was suggested that active species formed from complexes such as **30**, which might survive even in the presence of a large excess of activator, might be responsible for the polymer formation commonly observed during ethene oligomerization.

    Reaction of Cr(tBuNPNtBu)$_2$ with different stoichiometric ratios of AliBu$_3$ was found to give complexes **31** and **32** (Scheme 6.8) [82]. Both complexes are potent ethene polymerization catalysts, active in the absence of any cocatalyst. The similar

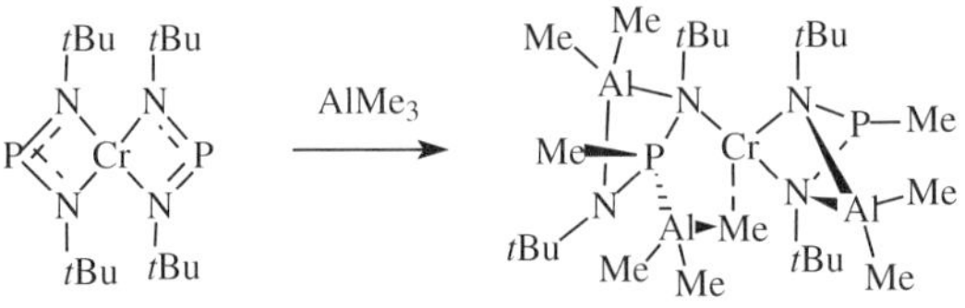

                                          **30**

**Scheme 6.7** Reaction of Cr(tBuNPNtBu)$_2$ with AlMe$_3$.

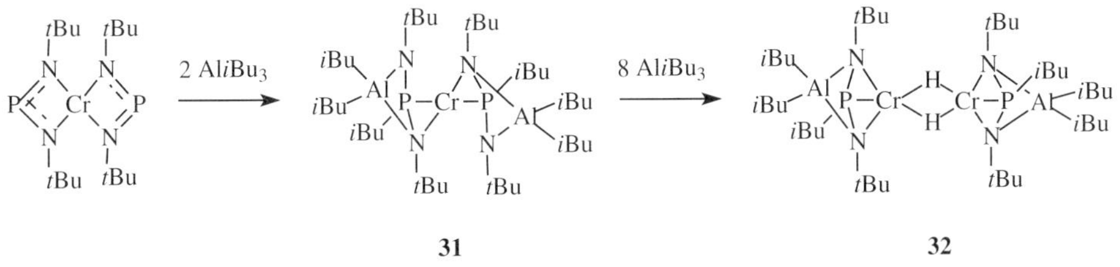

**Scheme 6.8** Reaction of Cr(tBuNPNtBu)$_2$ with Al$i$Bu$_3$.

activities and polymer molecular weights obtained with these complexes indicated that complex **31** is most likely converted to **32** under the polymerization conditions. Activation of these complexes with a large excess of Al$i$Bu$_3$ resulted in a switch in selectivity from polymerization to trimerization, possibly due to the formation of coordinatively unsaturated Cr(I) species.

Partial replacement of the *t*Bu groups in Cr(*t*BuNPN*t*Bu)$_2$ with an aryl group (Ar = 2,6-$i$Pr$_2$C$_6$H$_3$), followed by reaction with 4 equiv of AlMe$_3$, generated complex **33** (Scheme 6.9) [83]. This was shown to be a single-component trimerization catalyst, producing exclusively 1-hexene at 50 °C. Extending the reaction time did not lead to increased productivity, indicating rapid catalyst deactivation (in less than 30 min). Increasing the temperature to 80 °C led to loss of selectivity and polymer formation. The selectivity of complex **33** also changed in the presence of a cocatalyst. Treatment with MAO resulted in greatly increased activity and a statistical distribution of oligomers, whereas $i$Bu$_2$AlOAl$i$Bu$_2$ gave polyethylene.

A single-component catalyst for ethene trimerization to 1-hexene was obtained via the reaction of Cr(2-ethylhexanoate)$_3$ with 2,3,4,5-tetrahydrocarbazole, AlEt$_3$ and AlEt$_2$Cl [84]. The reaction product of this Phillips-type trimerization catalyst synthesis was the Cr(I) complex **34**. This complex had negligible activity when exposed to ethene in toluene, but was active in methylcyclohexane, producing 1-hexene with only traces of higher oligomers. The poisoning effect of toluene in Phillips catalytic systems was interpreted as illustrated in Scheme 6.10, assuming the dissociation of one ligand to vacate the necessary ethene coordination sites and blocking of these sites by toluene coordination.

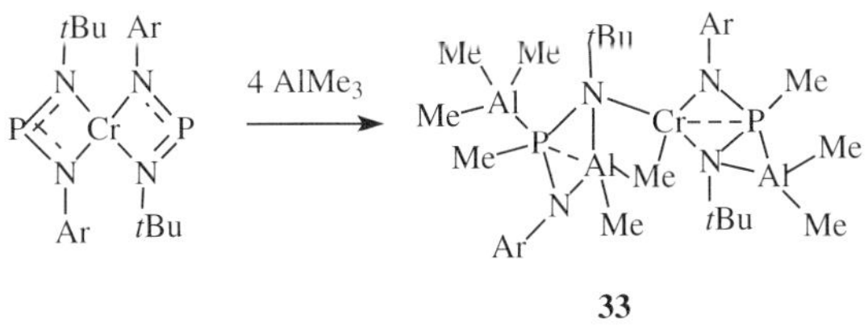

**Scheme 6.9** Reaction of Cr(ArNPN*t*Bu)$_2$ with AlMe$_3$.

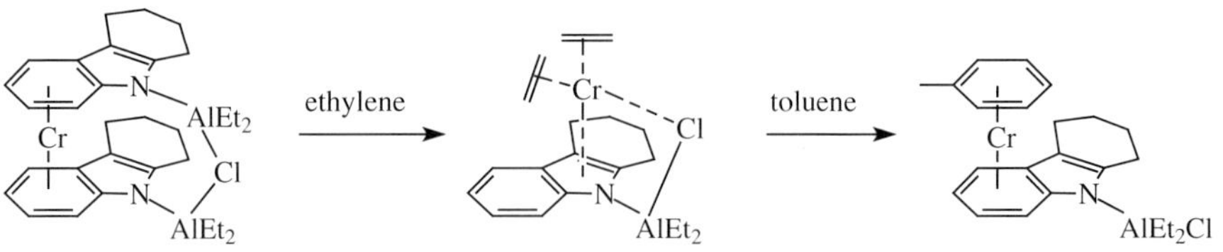

**Scheme 6.10** Catalyst deactivation in the presence of toluene.

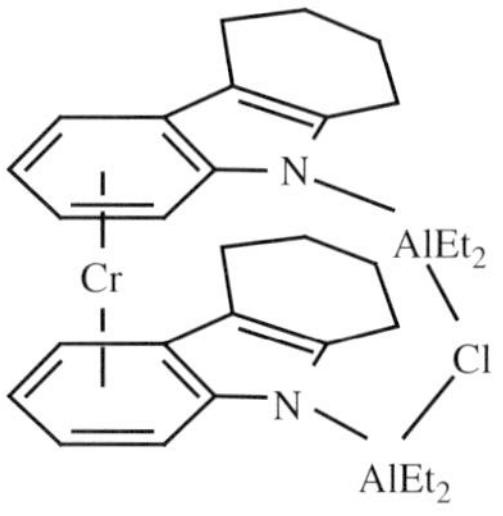

**34**

On the basis of the above studies, a link between the metal oxidation state and the catalytic behavior was proposed, with Cr(III) leading to nonselective oligomerization, Cr(II) to polymerization and Cr(I) to selective trimerization, although it was noted that it is too early to generalize these findings [84].

## 6.6
## Nickel Catalysts

Nickel is a widely used metal in the field of industrial homogeneous catalysis [85]. One of the largest applications is in the Shell Higher Olefins Process (SHOP), in which ethene is oligomerized to higher 1-alkenes. Nickel complexes active in the SHOP process contain a (P^O) chelate ligand, which controls the catalytic activity and selectivity. The active catalyst is a nickel hydride complex, formed for example by insertion of ethene into a Ni–phenyl group followed by elimination of styrene, as shown in Scheme 6.11.

Ethene oligomerization takes place via coordination and migratory insertion of ethene to generate Ni-alkyl species. As the alkyl group chain is assumed to be more strongly coordinated in the position trans to the oxygen atom, an isomerization step

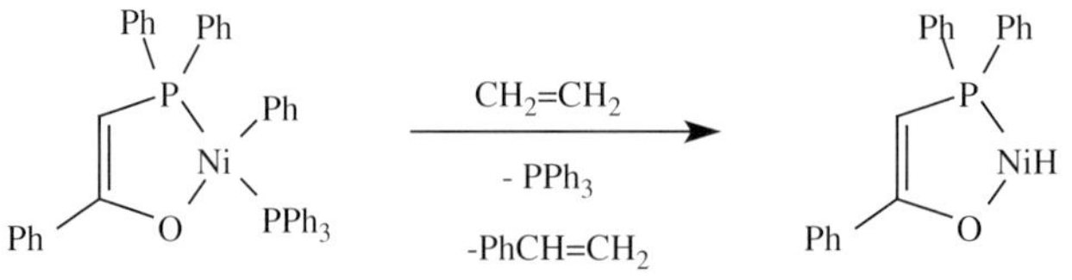

**Scheme 6.11** Activation of a SHOP catalyst for ethene oligomerization.

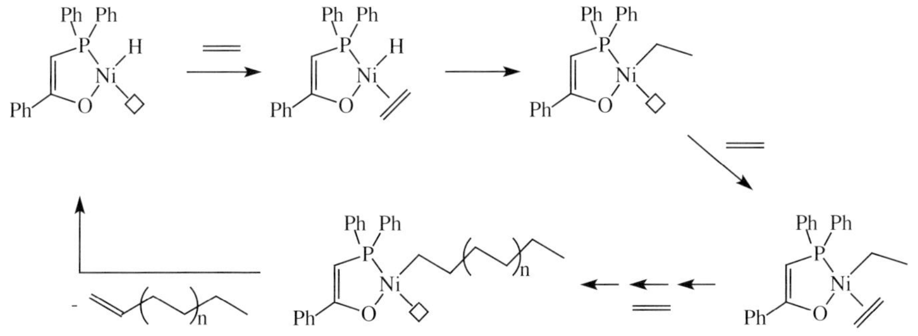

**Scheme 6.12** Catalytic cycle for ethene oligomerization with a SHOP catalyst.

prior to the following insertion will lead to the catalytic cycle illustrated in Scheme 6.12 [86]. A Schulz–Flory distribution of oligomers is obtained.

The balance between chain propagation and chain termination is dependent on the ligand L in complex **35**. Replacement of $PPh_3$ by the more strongly coordinating $PMe_3$ hinders ethene coordination, giving decreased activity and low molecular weight oligomers. On the other hand, a weakly coordinating ligand such as pyridine, or abstraction of the $PPh_3$ ligand by a phosphine sponge such as bis(cyclooctadiene) nickel, results in the formation of polyethylene. Bulkier ligands than $PPh_3$, for example $P(o\text{-Tol})_3$, are less coordinating and give higher molecular weight oligomers and polymers [87]. Productivity also decreases, as the stability of the active species is dependent on the effective coordination of the ligand.

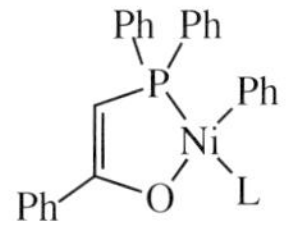

**35** 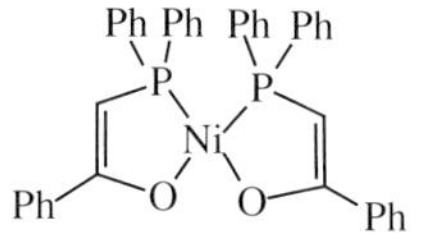                               **36**

Deactivation of SHOP catalysts can take place via the formation of a bis-chelate complex such as **36** [88, 89]. Increased catalyst activity can be obtained in the presence of an aluminum alkyl and it has been proposed that reactivation of the catalyst can take place as shown in Scheme 6.13 [86]. This reactivation is, however, accompanied by a significant drop in selectivity for 1-alkene formation. Binuclear deactivation to form complexes such as **36** can be avoided by catalyst immobilization and various supports, including silica, have been used for the synthesis of heterogeneous

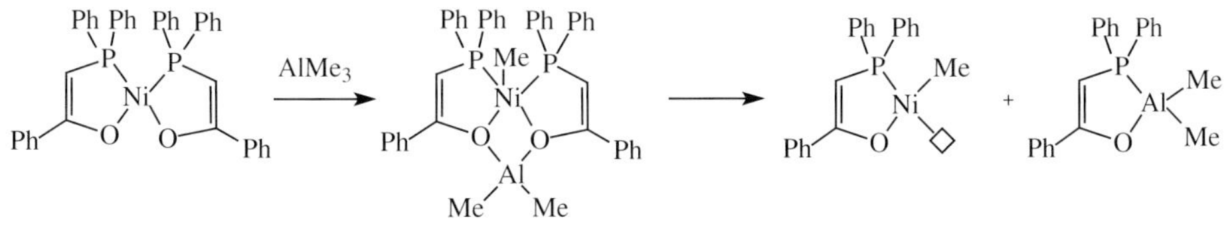

**Scheme 6.13** Aluminum alkyl-mediated reactivation of a SHOP catalyst.

catalysts [90]. Effective site isolation, preventing the irreversible formation of the inactive bis-chelate complex, has been achieved using a dendritic P,O ligand [91]. The dendritic catalyst was far more active than its parent complex and it was suggested that the increased propensity of the dendrimer ligand to form mono-ligated species might be applied to other transition metal catalysts for which the formation of bis-ligated or bimetallic complexes plays a role in catalyst deactivation.

## 6.7
## Iron Catalysts

Bis(imino)pyridyl iron complexes of type $\{2,6\text{-}[ArN=C(Me)]_2C_5H_3N\}FeCl_2$ used in ethene polymerization have been discussed in Chapter 3. The polymer molecular weight obtained with such complexes is dependent on the steric bulk of the aryl ligand; high molecular weight is obtained with Ar = 2,6-diisopropylphenyl. On the other hand, the presence of only a single ortho-substituent in the aryl rings, such as in complex **37**, leads to the formation of low molecular weight oligomers rather than polymers [92–95]. Under homogeneous conditions, using MAO as activator, high activities can be obtained in ethene oligomerization with **37** and related complexes. However, as described in Chapter 3, bis(imino)pyridyl iron complexes deactivate relatively rapidly during the course of polymerization or oligomerization, particularly at elevated temperatures. After 1 h at 50 °C, oligomerization activities were found to be 10–20% of the initial activity [94]. It has also been noted that the use of Al$i$Bu$_3$ as a scavenger in ethene oligomerization with **37**/MAO can lead to further deactivation, due to the lack of sufficient steric bulk in the *ortho*-aryl position of the ligand [94].

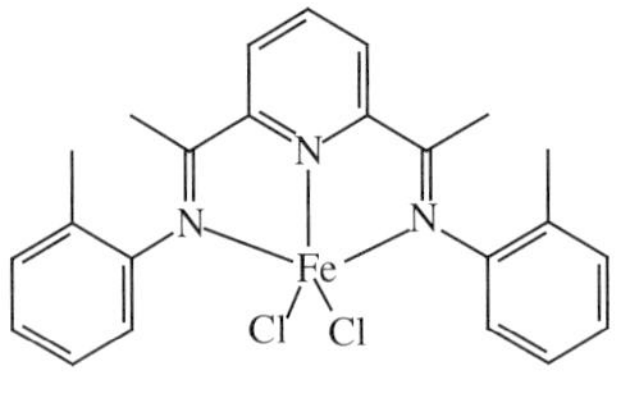

**37**

The product molecular weight in iron-catalyzed oligomerization and polymerization is dependent not only on the presence and steric bulk of *ortho*-aryl substituents, but also on electronic effects. Electron-withdrawing substituents, such as F and Cl, have been shown to give low molecular weight polymers and oligomers [96]. However, the introduction of halogen substituents can also lead to decreased activity. In contrast, high activity can be obtained by introducing an alkyl or aryl substituent at one or both meta-positions in the aryl rings in bis(imino)pyridyl iron complexes [94, 97]. It has been reported that, at high temperature (100–120 °C), the lifetime of complex **37** in MMAO-activated ethene oligomerization was only 3 min [97]. In contrast, complex **38**, in which one of the imino aryl groups has two bulky *meta*-aryl substituents, had a lifetime of 20 min and gave a more linear Schulz–Flory oligomer distribution. The increased thermal stability of the catalyst was attributed to the remote steric protection provided by the *meta*-aryl groups. Improved high-temperature stability has also been obtained with bis(imino)pyridyl complexes such as **39**, containing a boryl group [98].

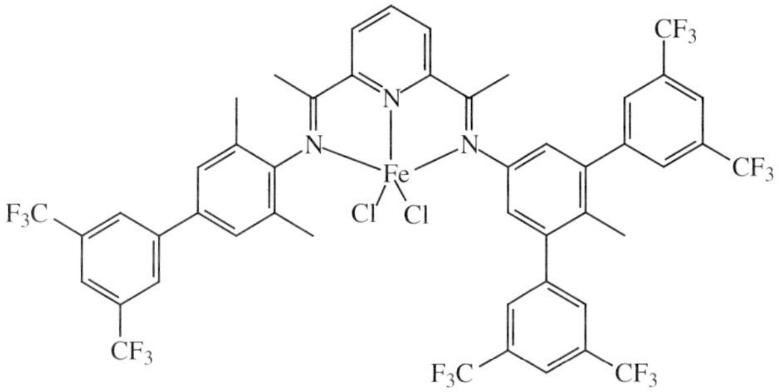

**38**

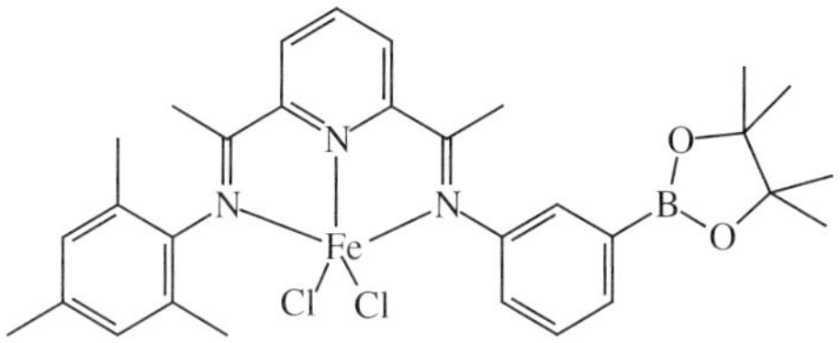

**39**

The importance of steric bulk in relation to the stability of active species has also been noted for complexes of type **40** [99]. Higher activity in MAO-activated ethene oligomerization was obtained when the substituent R was isopropyl rather than methyl or ethyl and this effect was attributed to the steric protection provided by the 2,6-diisopropyl substitution in the aryl ring.

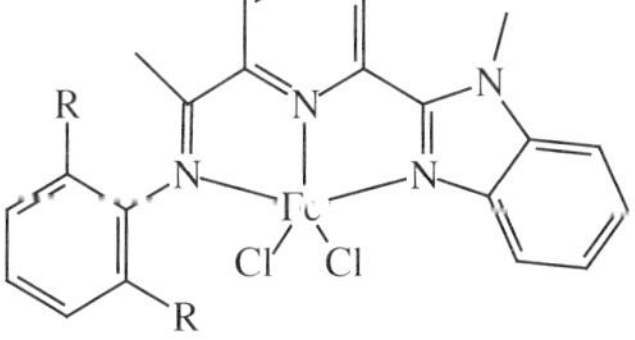

**40**

Deactivation of iron catalysts can be greatly reduced or even eliminated by immobilization on a support, as indicated in Chapter 4. It has been found that immobilization of **37** on a magnesium chloride support resulted in the retention of high activity, using AlEt$_3$ as activator [100]. The complex remained predominantly an oligomerization catalyst, generating oligomers and polymer in an approximately 9:1

weight proportion. Interestingly, the presence of hydrogen led to a large reduction in oligomers formation and a slight increase in polymer formation, indicating deactivation of oligomer-forming species. Very little chain transfer with hydrogen took place and in both cases the product comprised predominantly vinyl-terminated chains.

## 6.8
### Tandem Catalysis involving Oligomerization and Polymerization

The concept of linear low-density polyethylene (LLDPE) synthesis from a single monomer (ethene) via tandem catalysis dates from the 1980s. Beach and Kissin described a dual-functional catalyst system comprising a dimerization catalyst such as $Ti(OR)_4$–$AlEt_3$ and a Ziegler–Natta catalyst based on $MgCl_2/TiCl_4$ [101, 102]. The product is a short-chain branched polyethylene formed via Ziegler–Natta copolymerization of ethene with the 1-butene produced by the dimerization catalyst. A modified Phillips catalyst, containing two different chromium species, giving oligomerization and polymerization respectively, has also been reported [103], as has the combination of a Phillips-type catalyst with an ylid-nickel oligomerization catalyst [104].

More recently, the availability of a wide range of single-center oligomerization and polymerization catalysts has led to the investigation of new possibilities for tandem catalysis involving the *in situ* generation of 1-butene, 1-hexene and vinyl-terminated oligomers and their copolymerization with ethene [105]. Examples include the combination of a bis(imino)pyridyl iron oligomerization catalyst and a zirconocene [106, 107], and systems comprising a nickel dimerization catalyst and a constrained geometry catalyst [108, 109]. Titanium and chromium trimerization catalysts, as well as a cobalt oligomerization catalyst, have also been used in combination with a constrained geometry catalyst [110–112].

The abovementioned studies have all been carried out under homogeneous conditions, typically in solution in toluene. Relatively little has been reported on tandem catalysis using immobilized single-center catalysts, other than studies involving separate addition of an oligomerization catalyst and an immobilized polymerization catalyst to the reactor [113, 114]. An investigation of the effects of co-immobilization of an oligomerization and a polymerization catalyst on the same support would seem to be worthwhile, given the fact that close proximity of different catalytic species on a support should enhance their tandem interaction. For example, the 1-hexene generated by a trimerization catalyst should, if in the vicinity of a polymerization catalyst, be readily incorporated into the polymer chain. The importance of the close proximity of different catalyst species in tandem catalysis has been demonstrated in homogeneous polymerization by Marks and coworkers, who used a binuclear borate activator to spatially confine two different Zr- and Ti-based constrained geometry species via tight ion pairing [115]. Furthermore, Przybyla and Fink reported evidence for chain transfer between different zirconocenes on silica when the two catalysts were simultaneously immobilized on the support, but not when the catalysts were immobilized separately [116]. A cooperative effect of two metal centers has been reported by Osakada and coworkers, who synthesized a bimetallic complex comprising a Zr polymerization center and a Ni oligomerization center and found efficient incorporation of oligomers into the polymer chain [117].

## 6.9
## Conclusions

In the past decade, considerable advances have been made in the field of ethene oligomerization, following the discovery of highly active chromium catalysts for ethene trimerization and tetramerization. These catalysts produce 1-hexene and 1-octene with high selectivity, as opposed to the statistical (Schulz–Flory) product distribution of $\alpha$-olefins produced in aluminum- and nickel-catalyzed processes for ethene oligomerization. Chromium-catalyzed ethene tri- and tetramerization is typically carried out with MAO or a simple aluminum alkyl as activator and takes place via a metallacycle mechanism in which the relative stabilities of 7- and 9-membered rings with respect to olefin elimination are dependent on the nature of the ligand. PNP ligands of type $Ar_2PN(R)PAr_2$ give predominantly 1-hexene when Ar = 2-methoxyphenyl, as opposed to 1-octene when Ar = phenyl. Selective trimerization is also obtained using tridentate ligands such as $R_2PCH_2CH_2NHCH_2CH_2PR_2$ and $RSCH_2CH_2NHCH_2CH_2SR$.

High activity and selectivity in the trimerization of ethene to 1-hexene has also been achieved with titanium-based complexes such as $(C_5H_4CMe_2Ph)TiCl_3$, again using MAO as activator. The driving force for trimerization rather than polymerization is hemilabile coordination of the aryl group to the metal center. This system is less resistant to deactivation than the chromium-based systems. Productivity drops with increasing temperature in the range 30–80 °C, but the catalyst stability can be increased by the introduction of a second $CMe_2Ph$ substituent into the cyclopentadienyl ring and by the use of an aromatic rather than an aliphatic solvent.

Deactivation in nickel-catalyzed oligomerization, such as in the SHOP process, can take place via the formation of an inactive bis-($P^\wedge O$) chelate complex. Binuclear deactivation to form such complexes can, however, be avoided by embedding the ($P^\wedge O$)Ni complex in a dendritic framework, thereby enhancing the catalyst productivity.

Bis(imino)pyridyl iron complexes with relatively little steric bulk in the ligand give low molecular weight ethene oligomers rather than polyethylene. These complexes have high activity but deactivate rapidly under homogeneous conditions, particularly at elevated temperatures. Improved catalyst stability can be obtained by increasing the steric bulk of the ligand and, in some cases, this can be achieved without transformation from an oligomerization to a polymerization system.

Various tandem catalyst systems involving the combination of an oligomerization and a polymerization catalyst for the production of LLDPE using ethene as sole monomer source have been reported, but most studies of tandem catalysis have been carried out under homogeneous conditions. A disadvantage here is that different deactivation rates of the two catalyst components would lead to a shift in product composition during the reaction. Investigation of heterogeneous tandem systems would be of interest, in view of the much greater catalyst stability frequently obtained after catalyst immobilization and also the possibility that the tandem interaction of two different catalysts might be enhanced by close proximity on a support.

## References

1 Ziegler, K., Gellert, H.G., Kühlhorn, H., Martin, H., Meyer, K., Nagel, K., Sauer, H., and Zosel, K. (1952) *Angew. Chem.*, **64**, 323–329.

2 Skupinska, J. (1991) *Chem. Rev.*, **91**, 613–648.

3 Al-Jarallah, A.M., Anabtawi, J.A., Siddiqui, M.A.B., Aitani, A.M., and Al-Sa'doun, A.W. (1992) *Catal. Today*, **14**, 1–121.

4 Commereuc, D., Chauvin, Y., Gaillard, J., Leonard, J., and Andrews, J. (1984) *Hydrocarb. Proc., Int. Ed.*, **63**, 118–120.

5 Al-Sa'doun, A.W. (1993) *Appl. Catal. A: Gen.*, **105**, 1–40.

6 Janiak, C. (2006) *Coord. Chem. Rev.*, **250**, 66–94.

7 Hungenberg, K.-D., Kerth, J., Langhauser, F., Müller, H.-J., and Müller, P. (1995) *Angew. Makromol. Chem.*, **227**, 159–177.

8 Resconi, L., Piemontesi, F., Franciscono, G., Abis, L., and Fioriani, T. (1992) *J. Am. Chem. Soc.*, **114**, 1025–1032.

9 Eshuis, J.J.W., Tan, Y.Y., Teuben, J.H., and Renkema, J. (1990) *J. Mol. Catal.*, **62**, 277–287.

10 Eshuis, J.J.W., Tan, Y.Y., Meetsma, A., Teuben, J.H., Renkema, J., and Evens, G.G. (1992) *Organometallics*, **11**, 362–369.

11 Michelotti, M., Altomare, A., Ciardelli, F., and Ferrarini, P. (1996) *Polymer*, **37**, 5011–5016.

12 Rogers, J.S., Bazan, G.C., and Sperry, C.K. (1997) *J. Am. Chem. Soc.*, **119**, 9305–9306.

13 Bazan, G.C. and Rodriguez, G. (1996) *J. Am. Chem. Soc.*, **118**, 2291–2292.

14 Bazan, G.C., Rodriguez, G., Ashe, A.J. III, Al-Ahmad, S., and Kampf, J.W. (1997) *Organometallics*, **16**, 2492–2494.

15 Kwanten, M., Carrière, B.A.M., Grobet, P.J., and Jacobs, P.A. (2003) *Chem. Commun.*, 1508–1509.

16 Boccia, A.C., Costabile, C., Pragliola, S., and Longo, P. (2004) *Macromol. Chem. Phys.*, **205**, 1320–1326.

17 Shell Internationale Research Maatschappij, invs.: Horton, A.D. and de With, J. (1999) Int. Patent WO 96/27439; (1996) *Chem. Abstr.*, **125**, 276896.

18 Montell Technology Company, invs.: Horton, A.D., Ruisch, B.J., von Hebel, K., and Deuling, H.H. (1999) Int. Patent WO 99/52631;(1999) *Chem. Abstr.*, **131**, 299819.

19 Deckers, P.J.W., Hessen, B., and Teuben, J.H. (2001) *Angew. Chem. Int. Ed.*, **40**, 2516–2519.

20 Deckers, P.J.W., Hessen, B., and Teuben, J.H. (2002) *Organometallics*, **21**, 5122–5135.

21 Yang, X., Stern, C.L., and Marks, T.J. (1994) *J. Am. Chem. Soc.*, **116**, 10015–10031.

22 Deckers, P.J.W. and Hessen, B. (2002) *Organometallics*, **21**, 5564–5575.

23 Blok, A.N.J., Budzelaar, P.H.M., and Gal, A.W. (2003) *Organometallics*, **22**, 2564–2570.

24 de Bruin, T.J.M., Magna, L., Raybaud, P., and Toulhoat, H. (2003) *Organometallics*, **22**, 3404–3413.

25 Tobisch, S. and Ziegler, T. (2003) *Organometallics*, **22**, 5392–5405.

26 Hessen, B. (2004) *J. Mol. Catal. A: Chem.*, **213**, 129–135.

27 de Bruin, T., Raybaud, P., and Toulhoat, H. (2008) *Organometallics*, **27**, 4864–4872.

28 Severn, J.R. and Chadwick, J.C. (2004) *Macromol. Chem. Phys.*, **205**, 1987–1994.

29 Hagen, H., Kretschmer, W.P., van Buren, F.R., Hessen, B., and van Oeffelen, D.A. (2006) *J. Mol. Catal. A: Chem.*, **248**, 237–247.

30 Huang, J., Wu, T., and Qian, Y. (2003) *Chem. Commun.*, 2816–2817.

31 You, Y. and Girolami, G.S. (2008) *Organometallics*, **27**, 3172–3180.

32 Andes, C., Harkins, S.B., Murtuza, S., Oyler, K., and Sen, A. (2001) *J. Am. Chem. Soc.*, **123**, 7423–7424.

33 Arteaga-Müller, R., Tsurugi, H., Saito, T., Yanagawa, M., Oda, S., and Mashima, K. (2009) *J. Am. Chem. Soc.*, **131**, 5370–5371.

34 Dixon, J.T., Green, M.J., Hess, F.M., and Morgan, D.H. (2004) *J. Organometal. Chem.*, **689**, 3641–3668.

35 Phillips Petroleum Company, invs.: Freeman, J.W., Buster, J.L., and Knudsen, R.D. (1999) US Patent 5856257;(1999) *Chem. Abstr.*, **130**, 95984.

36 Yang, Y., Kim, H., Lee, J., Paik, H., and Jang, H.G. (2000) *Appl. Catal. A: Gen.*, **193**, 29–38.

37 Luo, H.-K., Li, D.-G., and Li, S. (2004) *J. Mol. Catal. A: Chem.*, **221**, 9–17.

38 Briggs, J.R. (1989) *J. Chem. Soc., Chem. Commun.*, 674–675.

39 Janse van Rensburg, W., Grové, C., Steynberg, J.P., Stark, K.B., Huyser, J.J., and Steynberg, P.J. (2004) *Organometallics*, **23**, 1207–1222.

40 Wass, D.F. (2007) *Dalton Trans.*, 816–819.

41 Carter, A., Cohen, S.A., Cooley, N.A., Murphy, A., Scutt, J., and Wass, D.F. (2002) *Chem. Commun.*, 858–859.

42 Agapie, T., Schofer, S.J., Labinger, J.A., and Bercaw, J.E. (2004) *J. Am. Chem. Soc.*, **126**, 1304–1305.

43 Agapie, T., Day, M.W., Henling, L.M., Labinger, J.A., and Bercaw, J.E. (2006) *Organometallics*, **25**, 2733–2742.

44 Agapie, T., Labinger, J.A., and Bercaw, J.E. (2007) *J. Am. Chem. Soc.*, **129**, 14281–14295.

45 Schofer, S.J., Day, M.W., Henling, L.M., Labinger, J.A., and Bercaw, J.E. (2006) *Organometallics*, **25**, 2743–2749.

46 McGuinness, D.S., Wasserscheid, P., Keim, W., Hu, C., Englert, U., Dixon, J.T., and Grove, C. (2003) *Chem. Commun.*, 334–335.

47 McGuinness, D.S., Wasserscheid, P., Keim, W., Morgan, D., Dixon, J.T., Bollmann, A., Maumela, H., Hess, F., and Englert, U. (2003) *J. Am. Chem. Soc.*, **125**, 5272–5273.

48 McGuinness, D.S., Wasserscheid, P., Morgan, D.H., and Dixon, J.T. (2005) *Organometallics*, **24**, 552–556.

49 Jabri, A., Temple, C., Crewdson, P., Gambarotta, S., Korobkov, I., and Duchateau, R. (2006) *J. Am. Chem. Soc.*, **128**, 9238–9247.

50 Temple, C., Jabri, A., Crewdson, P., Gambarotta, S., Korobkov, I., and Duchateau, R. (2006) *Angew. Chem. Int. Ed.*, **45**, 7050–7053.

51 Temple, C., Gambarotta, S., Korobkov, I., and Duchateau, R. (2007) *Organometallics*, **26**, 4598–4603.

52 Zhang, J., Braunstein, P., and Hor, T.S.A. (2008) *Organometallics*, **27**, 4277–4279.

53 Zhang, J., Li, A., and Hor, T.S.A. (2009) *Organometallics*, **28**, 2935–2937.

54 Köhn, R.D., Haufe, M., Kociok-Köhn, G., Grimm, S., Wasserscheid, P., and Keim, W. (2000) *Angew. Chem. Int. Ed.*, **39**, 4337–4339.

55 Köhn, R.D., Haufe, M., Mihan, S., and Lilge, D. (2000) *Chem. Commun.*, 1927–1928.

56 Köhn, R.D., Smith, D., Mahon, M.F., Prinz, M., Mihan, S., and Kociok-Köhn, G. (2003) *J. Organometal. Chem.*, **683**, 200–208.

57 Bollmann, A., Blann, K., Dixon, J.T., Hess, F.M., Killian, E., Maumela, H., McGuinness, D.S., Morgan, D.H., Neveling, A., Otto, S., Overett, M., Slawin, A.M.Z., Wasserscheid, P., and Kuhlmann, S. (2004) *J. Am. Chem. Soc.*, **126**, 14712–14713.

58 Overett, M.J., Blann, K., Bollmann, A., Dixon, J.T., Hess, F., Killian, E., Maumela, H., Morgan, D.H., Neveling, A., and Otto, S. (2005) *Chem. Commun.*, 622–624.

59 Blann, K., Bollmann, A., Dixon, J.T., Hess, F.M., Killian, E., Maumela, H., Morgan, D.H., Neveling, A., Otto, S., and Overett, M.J. (2005) *Chem. Commun.*, 620–621.

60 Overett, M.J., Blann, K., Bollmann, A., Dixon, J.T., Haasbroek, D., Killian, E.,

Maumela, H., McGuinness, D.S., and Morgan, D.H. (2005) *J. Am. Chem. Soc.*, **127**, 10723–10730.

61 Elowe, P.R., McCann, C., Pringel, P.G., Spitzmesser, S.K., and Bercaw, J.E. (2006) *Organometallics*, **25**, 5255–5260.

62 Kuhlmann, S., Blann, K., Bollmann, A., Dixon, J.T., Killian, E., Maumela, M.C., Maumela, H., Morgan, D.H., Prétorius, M., Taccardi, N., and Wasserscheid, P. (2007) *J. Catal.*, **245**, 279–284.

63 Budzelaar, P.H.M. (2009) *Can. J. Chem.*, **87**, 832–837.

64 Walsh, R., Morgan, D.H., Bollmann, A., and Dixon, J.T. (2006) *Appl. Catal. A: Gen.*, **306**, 184–191.

65 Kuhlmann, S., Paetz, C., Hägele, C., Blann, K., Walsh, R., Dixon, J.T., Scholz, J., Haumann, M., and Wasserscheid, P. (2009) *J. Catal.*, **262**, 83–91.

66 Wöhl, A., Müller, W., Peulecke, N., Müller, B.H., Peitz, S., Heller, D., and Rosenthal, U. (2009) *J. Mol. Catal. A: Chem.*, **297**, 1–8.

67 McGuinness, D.S., Overett, M., Tooze, R.P., Blann, K., Bollmann, A., Dixon, J.T., and Slawin, A.M.Z. (2007) *Organometallics*, **26**, 1108–1111.

68 McGuinness, D.S., Rucklidge, A.J., Tooze, R.P., and Slawin, A.M.Z. (2007) *Organometallics*, **26**, 2561–2569.

69 Krossing, I. and Raabe, I. (2004) *Angew. Chem. Int. Ed.*, **43**, 2066–2090.

70 Rucklidge, A.J., McGuinness, D.S., Tooze, R.P., Slawin, A.M.Z., Pelletier, J.D.A., Hanton, M.J., and Webb, P.B. (2007) *Organometallics*, **26**, 2782–2787.

71 Jabri, A., Crewdson, P., Gambarotta, S., Korobkov, I., and Duchateau, R. (2006) *Organometallics*, **25**, 715–718.

72 McGuinness, D.S., Gibson, V.C., Wass, D.F., and Steed, J.W. (2003) *J. Am. Chem. Soc.*, **125**, 12716–12717.

73 McGuinness, D.S. (2009) *Organometallics*, **28**, 244–248.

74 Tomov, A.K., Chirinos, J.J., Jones, D.J., Long, R.J., and Gibson, V.C. (2005) *J. Am. Chem. Soc.*, **127**, 10166–10167.

75 McGuinness, D.S., Suttil, J.A., Gardiner, M.G., and Davies, N.W. (2008) *Organometallics*, **27**, 4238–4247.

76 Small, B.L., Carney, M.J., Holman, D.M., O'Rourke, C.E., and Halfen, J.A. (2004) *Macromolecules*, **37**, 4375–4386.

77 Crewdson, P., Gambarotta, S., Djoman, M.-C., Korobkov, I., and Duchateau, R. (2005) *Organometallics*, **24**, 5214–5216.

78 Rogers, J.S. and Bazan, G.C. (2000) *Chem. Commun.*, 1209–1210.

79 Bazan, G.C., Rogers, J.S., and Fang, C.C. (2001) *Organometallics*, **20**, 2059–2064.

80 Ganesan, M. and Gabbai, F.P. (2004) *Organometallics*, **23**, 4608–4613.

81 Albahily, K., Koç, E., Al-Baldawi, D., Savard, D., Gambarotta, S., Burchell, T.J., and Duchateau, R. (2008) *Angew. Chem. Int. Ed.*, **47**, 5816–5819.

82 Albahily, K., Al-Baldawi, D., Gambarotta, S., Koç, E., and Duchateau, R. (2008) *Organometallics*, **27**, 5843–5947.

83 Albahily, K., Al-Baldawi, D., Gambarotta, S., Duchateau, R., Koç, E., and Burchell, T.J. (2008) *Organometallics*, **27**, 5708–5711.

84 Jabri, A., Mason, C.B., Sim, Y., Gambarotta, S., Burchell, T.J., and Duchateau, R. (2008) *Angew. Chem. Int. Ed.*, **47**, 9717–9721.

85 Keim, W. (1990) *Angew. Chem. Int. Ed.*, **29**, 235–244.

86 Kuhn, P., Sémeril, D., Matt, D., Chetcuti, M.J., and Lutz, P. (2007) *Dalton Trans.*, 515–528.

87 Heinicke, J., Köhler, M., Peulecke, N., Kindermann, M.K., Keim, W., and Fink, G. (2004) *J. Catal.*, **225**, 16–23.

88 Klabunde, U. and Ittel, S.D. (1987) *J. Mol. Catal.*, **41**, 123–134.

89 Klabunde, U., Mülhaupt, R., Herskovitz, T., Janowicz, A.H., Calabrese, J., and Ittel, S.D. (1987) *J. Polym. Sci., Part A: Polym. Chem.*, **25**, 1989–2003.

90 Nesterov, G.A., Zakharov, V.A., Fink, G., and Fenzl, W. (1991) *J. Mol. Catal.*, **66**, 367–372.

91 Müller, C., Ackerman, L.J., Reek, J.N.H., Kamer, P.C.J., and van Leeuwen, P.W.N.M. (2004) *J. Am. Chem. Soc.*, **126**, 14960–14963.

92 Small, B.L. and Brookhart, M. (1998) *J. Am. Chem. Soc.*, **120**, 7143–7144.

93 Britovsek, G.J.P., Gibson, V.C., Kimberley, B.S., Maddox, P.J., McTavish, S.J., Solan, G.A., White, A.J.P., and

Williams, D.J. (1998) *Chem. Commun.*, 849–850.

94 Britovsek, G.J.P., Mastroianni, S., Solan, G.A., Baugh, S.P.D., Redshaw, C., Gibson, V.C., White, A.J.P., Williams, D.J., and Elsegood, M.R.J. (2000) *Chem. Eur. J.*, **6**, 2221–2231.

95 Bianchini, C., Giambastiani, G., Guerrero Rios, I., Mantovani, G., Meli, A., and Segarra, A.M. (2006) *Coord. Chem. Rev.*, **250**, 1391–1418.

96 Chen, Y., Chen, R., Qian, C., Dong, X., and Sun, J. (2003) *Organometallics*, **22**, 4312–4321.

97 Ionkin, A.S., Marshall, W.J., Adelman, D.J., Bobik Fones, B., Fish, B.M., and Schiffhauer, M.F. (2006) *Organometallics*, **25**, 2978–2992.

98 Ionkin, A.S., Marshall, W.J., Adelman, D.J., Bobik Fones, B., Fish, B.M., and Schiffhauer, M.F. (2008) *Organometallics*, **27**, 1902–1911.

99 Sun, W.-H., Hao, P., Zhang, S., Shi, Q., Zuo, W., and Tang, X. (2007) *Organometallics*, **26**, 2720–2734.

100 Huang, R., Koning, C.E., and Chadwick, J.C. (2007) *J. Polym. Sci., Part A: Polym. Chem.*, **45**, 4054–4061.

101 Beach, D.L. and Kissin, Y.V. (1984) *J. Polym. Sci., Part A: Polym. Chem.*, **22**, 3027–3042.

102 Kissin, Y.V. and Beach, D.L. (1986) *J. Polym. Sci., Part A: Polym. Chem.*, **24**, 1069–1084.

103 Benham, E.A., Smith, P.D., and McDaniel, M.P. (1988) *Polym. Eng. Sci.*, **28**, 1469–1472.

104 Ostoja Starzewski, K.A., Witte, J., Reichert, K.H., and Vasilou, G. (1988) *Transition Metals and Organometallics as Catalysts for Olefin Polymerization* (eds W. Kaminsky and H. Sinn), Springer-Verlag, Berlin, pp. 349–360.

105 Komon, Z.J.A. and Bazan, G.C. (2001) *Macromol. Rapid Commun.*, **22**, 467–478.

106 Quijada, R., Rojas, R., Bazan, G., Komon, Z.J.A., Mauler, R.S., and Galland, G.B. (2001) *Macromolecules*, **34**, 2411–2417.

107 Zhang, Z., Cui, N., Lu, Y., Ke, Y., and Hu, Y. (2005) *J. Polym. Sci., Part A: Polym. Chem.*, **43**, 984–993.

108 Komon, Z.J.A., Bu, X., and Bazan, G.C. (2000) *J. Am. Chem. Soc.*, **122**, 1830–1831.

109 Komon, Z.J.A., Diamond, G.M., Leclerc, M.K., Murphy, V., Okazaki, M., and Bazan, G.C. (2002) *J. Am. Chem. Soc.*, **124**, 15280–15285.

110 Ye, Z., AlObaidi, F., and Zhu, S. (2004) *Macromol. Rapid Commun.*, **25**, 647–652.

111 De Wet-Roos, D. and Dixon, J.T. (2004) *Macromolecules*, **37**, 9314–9320.

112 Bianchini, C., Frediani, M., Giambastiani, G., Kaminsky, W., Meli, A., and Passaglia, E. (2005) *Macromol. Rapid Commun.*, **26**, 1218–1223.

113 Musikabhumma, K., Spaniol, T.P., and Okuda, J. (2003) *J. Polym. Sci., Part A: Polym. Chem.*, **41**, 528–544.

114 Zhang, Z., Guo, C., Cui, N., Ke, Y., and Hu, Y. (2004) *J. Appl. Polym. Sci.*, **94**, 1690–1696.

115 Abramo, G.P., Li, L., and Marks, T.J. (2002) *J. Am. Chem. Soc.*, **124**, 13966–13967.

116 Przybyla, C. and Fink, G. (1999) *Acta. Polym.*, **50**, 77–83.

117 Kuwabara, J., Takeuchi, D., and Osakada, K. (2006) *Chem. Commun.*, 3815–3817.

# 7
# Asymmetric Hydrogenation

## 7.1
## Introduction

Asymmetric hydrogenation is one of the success stories of homogeneous catalysis. Hydrogenation that leads to non-chiral products proceeds equally well, but since there are many highly selective heterogeneous catalysts available there is no need to go through the cumbersome process of catalyst and product separation as may be the case for homogeneous catalysts. Modification of heterogeneous metal catalysts with chiral molecules, especially cinchona alkaloids on Pd and Pt and tartaric acid on Ni, to induce asymmetric hydrogenation has also been successful and many examples have been reported, but few if any practical applications have been revealed [1]. A search in SciFinder on "asymmetric hydrogenation" gives more than 6000 hits and thus it is clear that a complete coverage of this reaction cannot be presented in one chapter. Narrowing down on "deactivation" gives only 34 publications, of which several are not relevant and other search criteria (inhibition, incubation, and so on) were needed to expand the set. An edited volume by De Vries and Elsevier [2] has been published on the topic of homogeneous asymmetric hydrogenation and the reader is referred to this source for complete information. It contains a chapter on inhibition and deactivation processes by Heller, De Vries and De Vries, which has been an excellent, additional source of information for the present chapter [3].

Many studies have been published on the initial formation of the active catalyst from the precursor applied (incubation), especially diene complexes of Rh(I) (Section 7.2). The ligands used are very diverse in rhodium catalyzed hydrogenation, ranging from strongly electron-donating alkylphosphines to electron-withdrawing phosphites. As a result the mechanistic sequence of the elementary steps may change from one system to the other, even between alkyl- and aryl-substituted phosphines. Likewise, the resting state of the catalyst can be different and the routes towards them from the different precursors can also vary. Inhibition can occur by substrate, product, polar additives or impurities that coordinate to the metal as ligands more strongly than the substrate, coordinating anions, or even polar solvents such as nitriles and aromatic solvents (Section 7.3). Another mechanism of deactivation is the formation of metal dimers or oligomers, a common feature in homogeneous

*Homogeneous Catalysts: Activity – Stability – Deactivation*, First Edition. Piet W.N.M. van Leeuwen and John C. Chadwick.
© 2011 Wiley-VCH Verlag GmbH & Co. KGaA. Published 2011 by Wiley-VCH Verlag GmbH & Co. KGaA.

catalysis resulting from our attempts to make unsaturated metal species (Section 7.4). This problem might be attacked by site-isolation, but examples are scarce.

Most of these reactions are reversible and their effect on the overall kinetics depends on the equilibrium constants. Inhibition by impurities is an extreme case. The common poisons in catalytic systems for the conversion of simple alkenes (polymerization, hydroformylation) are conjugated dienes, alkynes, enones, hydroperoxides and the like, but these types of impurities are less abundant in fine chemicals, the substrates for asymmetric hydrogenation. In addition, the substrate to catalyst ratio is usually much lower in fine chemistry than in bulk chemistry, and, furthermore, the noble metal catalysts are less sensitive to heteroatoms in impurities than ETM catalysts. Irreversible deactivation can be caused by formation of inactive metal species (e.g., "over-reduction") or ligand decomposition. We would have expected that cleavage of the P—C bond in phosphines would be a frequent mechanism of decomposition. For instance, palladium acetate in the presence of hydrogen will cleave the P—C bond in triphenylphosphine giving inactive phosphido-bridged palladium dimers and oligomers (Section 1.4.3), but this does not seem to be an important reaction for the hydrogenation systems based on rhodium, ruthenium and iridium (Section 7.5). In hydrogen transfer catalysts based on rhodium and iridium and imine-based ligands (including robust species such as bipyridine and phenanthroline) especially the very fast catalysts often have a short lifetime, but nothing is known about their decomposition pathway, as the interest in this is small, after having achieved one million turnovers in one hour!

Section 7.6 will show a few examples of inhibition by the product and in Section 7.7 the formation of metal as a mechanism of deactivation will be presented.

Kinetic studies are indispensable to establish the importance and character of catalyst deactivation. For development of an industrial process the data are important as well. Using dynamic experiments and modeling, the kinetics of a homogeneous hydrogenation (Scheme 7.1) with cationic rhodium-PyrPhos {[Rh(PyrPhos)(COD)] BF$_4$} was investigated by Greiner and Ternbach [4]. A set of three batch experiments only allowed discrimination between six models (Michaelis–Menten kinetics combined with substrate inhibition, product inhibition, and deactivation). In this case also, evidence for catalyst deactivation was gained.

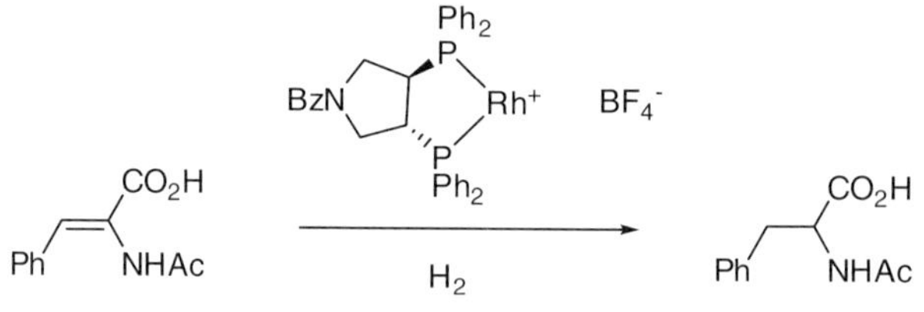

**Scheme 7.1** Hydrogenation with cationic rhodium-PyrPhos.

Heller and coworkers [5] developed a more sophisticated model that allows variation of the ee obtained, as they inserted several diastereomeric intermediates and the equilibria between them in their reaction scheme. In the case of isobaric conditions the hydrogen consumption can be described by an equation analogous to simple Michaelis–Menten-type kinetics. The interpretation of the resulting constants is briefly demonstrated for a few practical examples. The temperature dependence of the enantiomeric ratio for the resulting products, as for example for a selection process, is also considered for the investigated models.

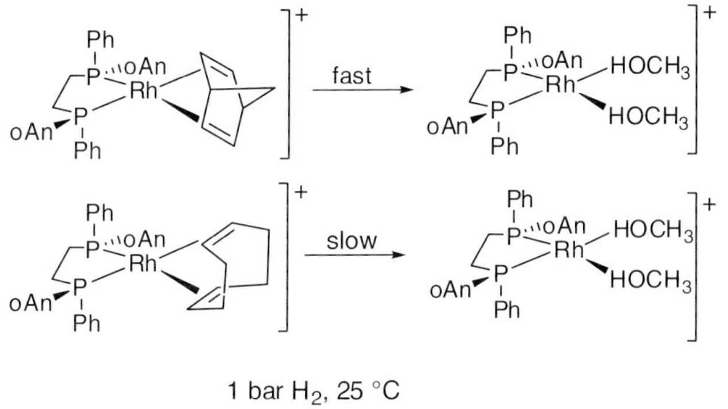

1 bar H$_2$, 25 °C

**Scheme 7.2** Slow and fast conversion of precursor diene complexes in active catalyst (RR-dipamp, o-An = ortho-anisyl).

Blackmond discusses the kinetic aspects of nonlinear effects in asymmetric catalysis [6], to which we will return in Section 7.8, where we discuss selective activation and deactivation (selective poisoning) of enantioselective catalysts.

## 7.2
### Incubation by Dienes in Rhodium Diene Precursors

Common precursors for rhodium hydrogenation of alkenes are cationic rhodium diene complexes of the general formula (diene)$_2$Rh(WCA), in which WCA stands for weakly coordinating anion (such as BF$_4$, PF$_6$, BArF). The first diene is readily replaced by the phosphine ligand(s) added forming (diene)(L$_2$)Rh$^+$, in which L$_2$ is either a bidentate ligand or two monodentate ones. Heller discovered [7] that the type of diene remaining in the diphosphine complex made an enormous difference to the incubation observed, as the replacement of the second diene by the actual substrate in most instances is very slow compared to the hydrogenation of the substrate. The diene of the precursor has to be removed by hydrogenation and the rate of this reaction depends enormously on the particular diene (Scheme 7.2). Surprisingly, this effect was already known for many years (since 1977), although with much less quantification, for rhodium complexes of dinitrogen ligands (bipy, phen) from the work of Mestroni and coworkers [8, 9]. They reported that 1,5-hexadiene-based catalysts showed a much shorter incubation time than 2,5-norbornadiene (NBD) complexes, which in turn were much faster than 1,5-cyclooctadiene (COD) complexes. COD readily displaced 1,5-hexadiene in the rhodium and iridium complexes. In neutral media the rhodium complexes catalyze the hydrogenation of alkenes and in basic alcohols ketones are also hydrogenated. The fastest catalyst was the one containing 4,7-Me$_2$-phenanthroline, the most sterically hindered one. Alkene hydrogenation is slower than that of phosphine complexes, but ketone hydrogenation is faster when diimine ligands are used. An excess of ligand gave the best results, showing that the coordination is not all that strong ([Rh] = 1–8 mM). Also bis-ligand complexes are formed, and these may be more stable resting states than other complexes that might form.

The slow reaction of certain dienes with several Rh precursors was already noted by Heldal and Frankel [10] in their studies on the hydrogenation of oligo-ene alkanoate esters, for example, methyl sorbate (methyl (*E,E*)-hexa-2,4-dienoate), at low pressure, 1 bar. With RhCl(PPh$_3$)$_3$, constant ratios between *trans*-2- and *trans*-3-hexenoate indicated approximately the same activation energy for 1,2-addition of H$_2$ on the Δ4 double bond of the sorbate and for 1,4-addition. The complexes RhCl(CO)(PPh$_3$)$_3$ and [Rh(NBD)(diphos)] + PF$_6^-$ (diphos = diphosphine) were inactive in the hydrogenation of methyl sorbate, but they catalyzed the hydrogenation of methyl linoleate. Catalyst inhibition was apparently caused by stronger complexation with sorbate (conjugated diene) than with the diene of methyl linoleate.

The observation concerning the difference between COD and NBD was found, by Heller and coworkers, to be very general [11]. In asymmetric hydrogenation with the use of rhodium complexes the latter two complexes are the most common ones as they are commercially available. Complexes of monoalkenes may be by far better precursors, but their shelf-life is much shorter. The *in situ* preparation of the ligand complex leaves one molecule of the diene in solution, which for most substrates will coordinate more rapidly and strongly and thus both diene molecules must be hydrogenated (to the monoalkenes) before the catalyst starts the hydrogenation of the substrate. The ambiguity of the induction period can be circumvented by using methanol complexes of the cationic diphosphine rhodium complex, and also by using higher temperatures and pressures. The differences in rates of hydrogenation between COD and NBD were studied at 25 °C and 1 bar of dihydrogen.

Within the group of complexes of either COD or NBD the rates of diene hydrogenation may differ by three orders of magnitude, depending on the bidentate phosphorus ligand used. For the same ligand, the ratio of the rates for COD and NBD can differ from only a factor of 5 to as much as 3000. These differences are not understood, although Heller notes that the complexes containing larger rings for the rhodium-ligand moieties show large distortions from planarity in their crystal structures of the COD complexes, and that these complexes also give the slowest hydrogenation of COD. NBD is smaller and less prone to distortions. One might speculate that rotation of the COD diene is more hindered than that of NBD when the oxidative addition of dihydrogen takes place.

The effectiveness of cyclooctadiene and norbornadiene precatalysts of the type [Rh(DuPHOS)(diolefin)]BF$_4$ in catalytic asymmetric hydrogenation of various prochiral olefins was studied by Cobley and coworkers [12]. At higher concentrations they confirmed the differences in NBD and COD precursors, but at low concentrations, as required for an economic use of such complexes in an industrial application, they found, for both systems, that incubation was nearly absent. One way to demonstrate this was the use of equal amounts of NBD and COD catalysts having ligands of the opposite DuPHOS chirality, and in these mixed catalyst systems no significant ees were observed, indicating that NBD and COD gave equally active catalysts under these conditions (e.g., 0.06 mM of Rh complex in MeOH, 5 bar $H_2$, 25 °C, and a reactive substrate such as dimethyl itaconate). Thus, identical reaction profiles were observed for either precatalyst, but an important determinant of reaction rate was the efficiency of hydrogen introduction to the mixture. In addition, the effectiveness of a catalyst loading of 50 000/1 for the hydrogenation of di-Me itaconate with [Rh(DuPHOS)(COD)]$BF_4$ demonstrates that this precatalyst can be put to highly economical use [13].

A kinetic investigation of the selective hydrogenation of 2,5-norbornadiene to norbornene catalyzed by [Rh(NBD)(PPh$_3$)$_2$]BF$_4$ at room temperature was carried out by Esteruelas and coworkers [14]. The reaction was found to be independent of the substrate concentration, while it is first order in catalyst and hydrogen pressure. Furthermore, the addition of triphenylphosphine inhibits the reaction. Moreover, it was observed that, under hydrogen atmosphere and at low temperature, the precursor was in equilibrium with the dihydride *cis-trans*-[RhH$_2$(NBD)(PPh$_3$)$_2$]BF$_4$, and that the addition of tolylphosphine complexes led to exchange of phosphines. On the basis of these observations and other spectroscopic results, they concluded that the hydrogenation of 2,5-norbornadiene to norbornene proceeds in this case by the five-coordinate dihydride [RhH$_2$(NBD)(PPh$_3$)]$^+$, which is formed by oxidative addition of molecular hydrogen to both a bisphosphine complex and a tri-coordinate monophosphine complex [Rh(NBD)(PPh$_3$)]$^+$, depending on the concentration of free phosphine in the catalytic solution.

Another example in which the presence of COD slowed down the reaction is the reductive amination of aldehydes with secondary amines and $H_2$ by rhodium diphosphine complexes containing COD [15]. Many intermediates and half-products were identified, and several of these intermediates were successfully hydrogenated independently under these conditions, but a detailed analysis could not be given. Alcohols were also formed as product via direct hydrogenation of the aldehydes. In several cases it was shown that pre-hydrogenation of (diphosphine)Rh(COD)$^+$ in order to remove COD was advantageous for the rate of reaction.

## 7.3
### Inhibition by Substrates, Solvents, Polar Additives, and Impurities

### 7.3.1
### Inhibition by Substrates: Iridium

When the substrate is a diene this may also cause inhibition or show a slow hydrogenation reaction. Burgess [16] studied the hydrogenation of 2,3-diphenylbutadiene at 25 °C and 1 bar of dihydrogen using an NHC variant **1** of the Pfaltz's catalyst [17] for alkene hydrogenation **2**, in its turn a bidentate, chiral variant of Crabtree's catalyst **3** [18]. As described before, in a slow reaction COD has to be removed first from the iridium complex [19]. Then the reaction occurs in two stages:

first hydrogenation to the monoene, and then a twice as fast reaction produces 2,3-diphenylbutane.

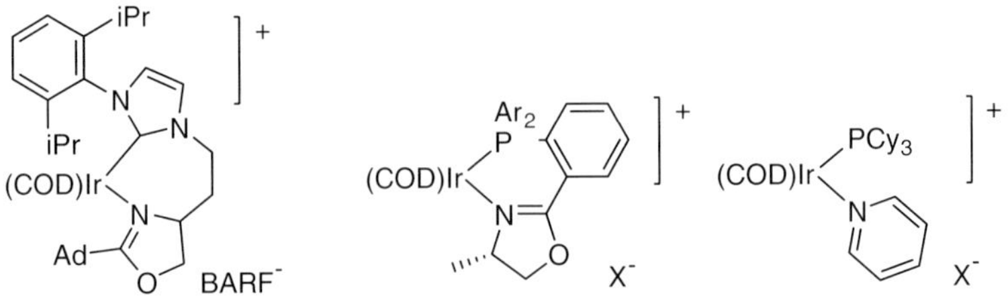

The first step shows poor stereoselectivity, but the second step gives high stereoselectivity and thus mainly the meso product (76%) and one of the rac isomers are obtained (Scheme 7.3).

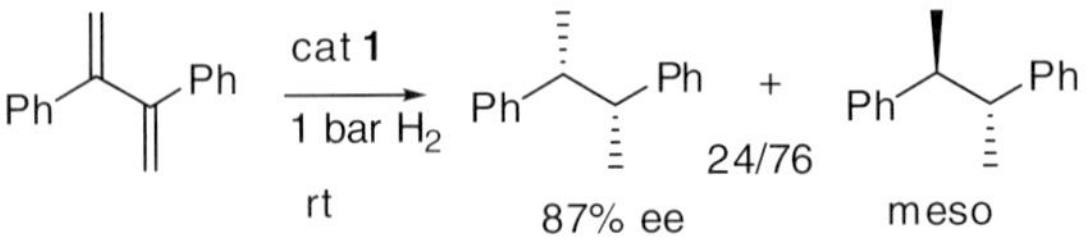

**Scheme 7.3** Hydrogenation of 2,3-diphenylbutadiene by **1**.

The authors do not propose allylic intermediates as dormant states, as is known for hydroformylation or polymerization of monoenes containing dienes as impurities. Instead they propose, in analogy with Pfaltz's results, that a fast hydrogenation may involve iridium(V) species containing several hydrides, formed by concomitant insertion of the alkene into the Ir−H bond and oxidative addition of another dihydrogen molecule. When a diene is present in the complex together with the bidentate ligand and two hydrides, the 18-electron complex thus obtained is saturated and the proposed reaction cannot take place.

### 7.3.2
### Inhibition by Substrates, Additives: Rhodium

The first example was discussed in Section 7.2 where we mentioned the unsuccessful hydrogenation of sorbate esters due to the formation of stable adducts with the conjugated diene of methyl sorbate (methyl (*E,E*)-hexa-2,4-dienoate).

Heller and coworkers reported on the inhibition of hydrogenation by itaconic acid, a substrate used by many authors, preferably though as its dimethyl ester [20]. The catalyst used was (DIPAMP)rhodium(methanol)$_2$ tetrafluoroborate. The solution of the active catalyst is orange, but it turns colorless when the hydrogenation stops. The colorless product was a rhodium(III) species containing the coordinated substrate as an alkyl species, the structure of which was confirmed by X-ray analysis. Its formation was explained by oxidative addition of the second carboxylic acid group after the first one was deprotonated (Scheme 7.4).

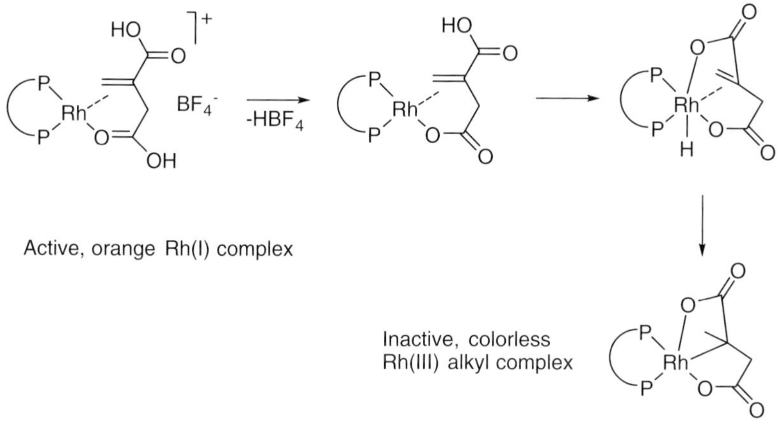

Active, orange Rh(I) complex

Inactive, colorless
Rh(III) alkyl complex

**Scheme 7.4** Formation of inactive Rh(III) species via oxidative addition of carboxylic acid.

This rhodium(III) species can neither return to rhodium(I) nor react with $H_2$ and the activity stops completely. Reductive elimination of the substrate would lead to highly strained lactones and this does not happen. In other instances in which oxidative addition of protic species occurs, a base might either prevent this or return the species by base-aided reductive elimination back to Rh(I).

Prior to this work, Brown studied complexes formed between a variety of diphosphine (DIPAMP, DIOP) rhodium complexes solvated by methanol and several unsaturated acids and acid amides [21]. They found that many complexes formed and that the structure of such diastereomeric adducts bore no relation to the ultimate ee obtained in the catalytic hydrogenation reaction. For instance, (R,R)-[DIPAMP]bis(methanol)rhodium cation and 2-methylenesuccinic acid and its methyl esters gave a variety of complexes, including tridentate species where both carboxyl groups and olefin were concomitantly bound. They also noted that acids yielded more efficient hydrogenation in the presence of triethylamine, which we now understand thanks to the work by Heller cited above.

Itaconic acid was hydrogenated also by Takahashi and Ichiwa using BCPM **4** and BPPM **5** ligands [22]. They found that higher pressures of $H_2$ and the addition of triethylamine would enhance the reaction rate and dramatically increase the ee obtained at ambient temperatures. It can be imagined that here also base prevents the oxidative addition of carboxylic acid to Rh(I).

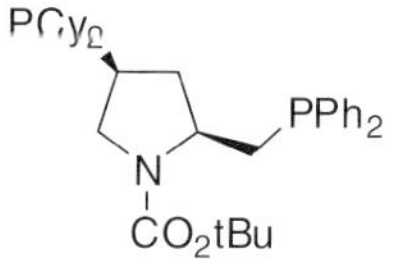

BCPM

**4**

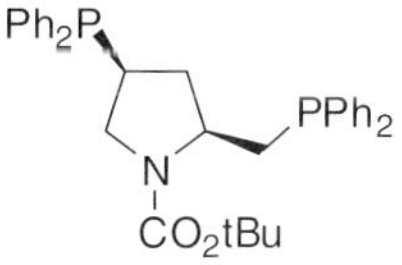

BPPM

**5**

Either the substrate or the solvent can be a source of CO, and, if this is the case, a carbonyl rhodium complex forms which is inactive as a hydrogenation catalyst. It is sufficient to present one example, from Chaudhari and coworkers, who found that hydrogenation of allyl alcohol under mild conditions with Wilkinson's catalyst produced $RhCl(CO)(PPh_3)_2$ [23]. The complex precipitated from the ethanol solvent used. In their case the carbonyl species was isolated and characterized at the end of the reaction. It is assumed that alcohols are dehydrogenated to give aldehydes, and that the formed aldehydes oxidatively add to rhodium(I) and are decarbonylated, as was reported by Kollar [24].

The well known paper by Landis and Halpern reporting the "major/minor" phenomenon in asymmetric hydrogenation, also contains an interesting item on "poisoning" by the substrate dihydrogen [25]. They reported on the hydrogenation of methyl (Z)-α-acetamidocinnamate (mac), catalyzed by $[Rh(dipamp)]^+$, and on the thermodynamics and kinetics of the formation of the two diastereomeric [Rh(dipamp)(mac)]$^+$ adducts, and the oxidative addition of $H_2$ to these complexes, the rate-determining step (Scheme 7.5). It is concluded that the predominant product enantiomer, (S)-N-acetylphenylalanine methyl ester, is derived from the minor (less stable) [Rh(dipamp)(mac]$^+$ adduct by virtue of its much higher reactivity toward $H_2$. Less well known is that they also reported on the influence of temperature and $H_2$ pressure on the rate and enantioselectivity. The inverse dependence of the optical yield on the $H_2$ partial pressure is due to trapping of the [Rh(dipamp)(mac)]$^+$ adducts by reaction with $H_2$, thus inhibiting the interconversion of the diasteromeric intermediates. Thus, this is an effect of reactant concentration on the enantioselectivity. This effect can be offset by increasing the temperature.

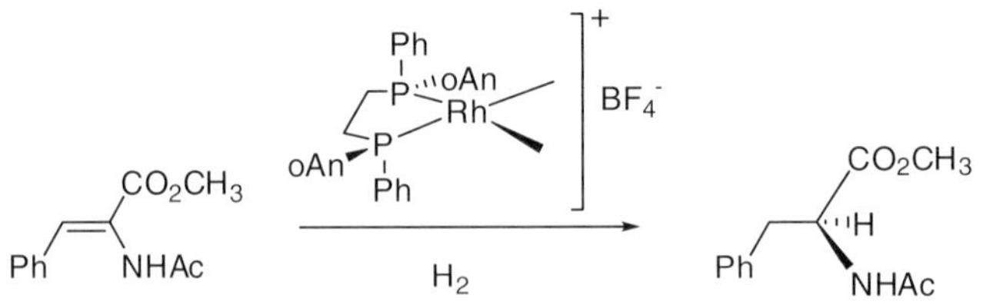

Scheme 7.5 RR-dipamp Rh(I) giving the main product, (S)-N-acetylphenylalanine methyl ester (o-An = ortho-anisyl).

Heller and coworkers reported on the inhibiting influence of aromatic solvents on the activity of asymmetric hydrogenations [26]. It was found that both weakly and strongly coordinating prochiral olefins formed stable $\eta^6$-arene Rh complexes in MeOH–arene solvent mixtures, thus blocking the catalytic activity.

Deactivation by α-methylstyrene was noted during the homogeneous hydrogenation of α-methylstyrene with Wilkinson's catalyst, $RhCl(PPh_3)_3$ at 325 K [27]. The reaction is faster than that for styrene hydrogenation with the same catalyst and a first order in $H_2$ was reported. The deactivation products were not studied.

In an example of PHIP-NMR spectroscopic detection of hydrogenation intermediates, Bargon and coworkers showed that styrene as the substrate remained initially coordinated to the rhodium catalyst through the arene ring and then detached from the cationic rhodium(I) catalyst in a subsequent slow reaction step that influenced the overall rate of the hydrogenation [28].

Heller, Boerner and coworkers found a pressure-dependent, highly enantioselective hydrogenation of unsaturated β-amino acid precursors using Me- or Et-DuPHOS [29] in polar solvents (Scheme 7.6). Much higher rates were found in polar solvents than in toluene. The ees obtained for Z- and E-3-(acetylamino)-2-butenoate methyl esters **6** depended on the polar solvent used (alcohols, THF, ees 12–35% at high pressure, ~40 bar), but it was observed that the rate of hydrogenation of the Z-isomer is dramatically accelerated in polar solvents. Also for the Z-isomer, a decreasing $H_2$ pressure resulted in a dramatic increase in the enantioselectivity. The ee obtained for the E-isomer was the same at 1 bar and at 35 bar (97%). Both isomers led to the same enantiomer in excess. Mixtures could be hydrogenated with the same catalyst system giving ees >90% at 1 bar. The pressure effect was explained with reference to the work described above by Landis and Halpern [25], according to the major–minor concept of Halpern and Brown by pressure-dependent disturbance of the pre-equilibrium to the minor substrate complex [30]. Apparently the E- and Z-isomers are affected differently (the overall hydrogenation of the E-isomer is much slower and perhaps "equilibration" between the manifolds is reached).

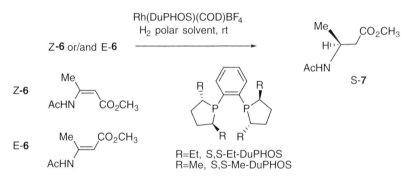

**Scheme 7.6** Hydrogenation of Z-**6** and E-**6** both leading to S-**7**.

Impurities in the feed may retard or hamper catalysis, although given the relatively high catalyst loading in the laboratory this phenomenon is less common here than in bulk chemical processes, especially those using "polymer grade" alkenes. During the development of industrial applications the deleterious effect of impurities often shows up. Hydrogenation of 2-methylenesuccinamic acid monoamide, $HO_2CC(:CH_2)CH_2C(O)NH_2$ required larger amounts of catalyst (Rh Et-DuPHOS) and longer reaction times when the starting material was purified with hydrochloric acid, which led to residues of chloride in the substrate [31], although the enantiomeric excess of the product was unaffected. Turnover numbers as high as 100 000 were achieved when chloride was avoided by using $\{[(S,S)\text{-Et-DuPHOS}]Rh(COD)\}^+ BF_4^-$ as the precatalyst for hydrogenation of 2-methylenesuccinamic acid (MeOH, 140 bar

$H_2$, 45 °C), and (R)-2-methylsuccinamic acid was obtained in 96% ee. The solid isolated contained less than 1 ppm rhodium.

In this context we also mention the rhodium-catalyzed, enantioselective isomerization step in the synthesis of menthol, although this is not a hydrogenation reaction. It concerns the Takasago process for the commercial production of (−)menthol from myrcene. The catalyst used is a rhodium complex of BINAP [32]. The synthesis of menthol is given in the reaction scheme, Scheme 7.7. The key reaction [33] is the enantioselective isomerization of the allylamine to the asymmetric enamine. It is proposed that this reaction proceeds via an allylic intermediate.

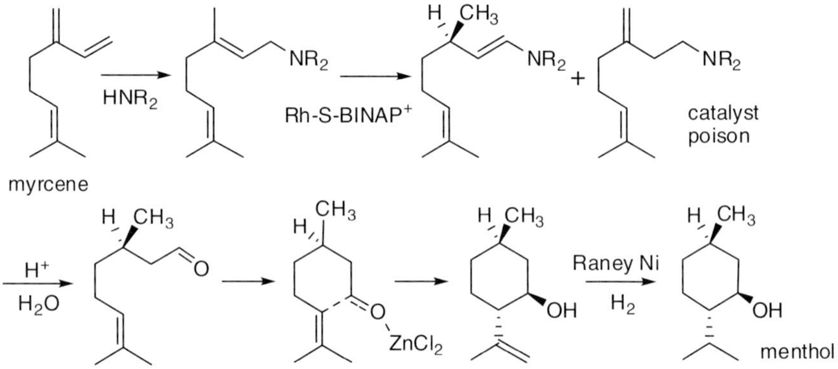

Scheme 7.7   The Takasago process for (−)menthol showing the amine catalyst poison.

This is the only step that needs to be steered to the correct enantiomer, since the other two stereocenters are produced in the desired stereochemistry with the route depicted. After the enantioselective isomerization the enamine is hydrolyzed. A Lewis acid-catalyzed ring closure gives the menthol skeleton. In the six-membered ring of the intermediate all substituents can occupy an equatorial position and thus this intermediate is strongly favored. In a subsequent step the isopropenyl group is hydrogenated over a heterogeneous Raney nickel catalyst.

Both the ligand, S-BINAP, and rhodium are rather expensive and the turnover per mole of catalyst should be high, >50 000 [34]. Without pre-treatment of the allylamine the TON is only 100. When the substrate was treated with Vitride (sodium bis(2-methoxyethoxy)aluminum hydride) the TON was increased to 1000. Removal of an amine isomer (catalyst poison, see Scheme 7.7) was essential (TON 8000). A 10-fold increase was obtained by using a ligand to rhodium ratio of 2 instead of 1. Recovery with small losses eventually gave turnover numbers of 400 000. The method was commercialized in 1984. The main part of the annual production of 11 800 tonnes (1998) [35] is still obtained from the natural source, *Méntha arvensis*. This may also contain the related pulegone, which is a poison and its level in food is kept below 20 ppm. In the last decade the cultivation of *Méntha arvensis* has increased enormously, especially in India, and with an annual production of 20 000 tons a$^{-1}$ the proportion of synthetic menthol has probably seen its share decreasing.

While added halides often show a negative effect, Zhang and coworkers reported a case in which the addition of potassium bromide and lutidine accelerated the reaction and enhanced the enantioselectivity [36]. [Rh(cod)Cl]$_2$ catalyzed the asymmetric hydrogenation of simple ketones in the presence of PennPhos ligands (R = Me, iPr, **8**). For example, [Rh(cod)Cl]$_2$ MePennPhos catalyzed the hydrogenation of PhC(O)Me to give (S)-PhCHMeOH with 97% yield and 95% ee. Interestingly, aliphatic ketones also gave high ees. Both 2,6-lutidine and KBr accelerated the reaction and enhanced the enantioselectivity; no explanation was given.

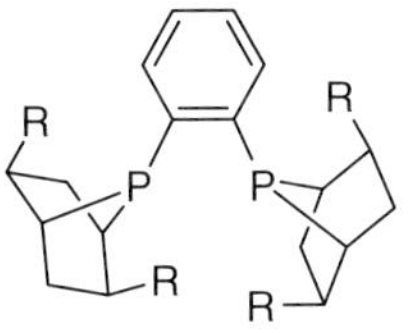

R = Me, i-Pr

**8**

The chiral rhodium complex obtained from a reaction of pentamethylcyclopentadienylrhodium chloride dimer and 1R,2S-1-amino-2-indanol provides a superior catalyst for the rapid, high-yielding asymmetric transfer hydrogenation of acetophenone with 2-propanol to produce (R)- and (S)-(1)-phenylethanol. The effects of various reaction parameters, such as reaction temperature, catalyst and substrate concentration, gaseous environment, and acetone concentration on conversion and enantioselectivity were investigated [37]. The results indicate that the catalyst can be deactivated by high temperature and air atmosphere. When the base (sodium isopropoxide) was added before the ketone substrate, the catalyst was much less active, presumably, it was speculated, because inactive dimers form [38]. The addition of acetone at the beginning of the reaction reduces the reaction rate, but not the enantioselectivity. Thus, acetone is an inhibitor. At prolonged reaction times the enantioselectivity decreases as a result of racemization via the equilibration reaction.

## 7.3.3
### Inhibition by Substrates: Ruthenium

Hydrogenation of allyl alcohol with RuCl$_2$(PPh$_3$)$_3$ as the precursor led to the formation of RuCl$_2$(CO)(PPh$_3$)$_2$, similar to the case of the rhodium-based Wilkinson catalyst in the hydrogenation of allyl alcohol (Section 7.3.2) [39]. Ruthenium complexes are also well-known for their ability to decarbonylate aldehydes and to catalyze hydrogen transfer from alcohols to unsaturated substrates [40].

Garrou prepared the trifluoroacetate (tfa) complexes Ru(tfa)$_2$(CO)(PPh$_3$)$_2$ and Ru(tfa)$_2$(CO)(PPh$_3$)(diphos) to be used as hydrogenation and dehydrogenation catalysts of ketones and alcohols [41]. They found that the complexes containing PPh$_3$ only gave more readily the inactive dicarbonyl species RuCl$_2$(CO)$_2$(PPh$_3$)$_2$ than the various complexes containing, in part, diphosphines, which were also more active in the hydrogenation reactions.

Internal alkynes can be hydrogenated to E-alkenes, as was shown by Bargon and coworkers by [Cp*Ru(alkene)]$^+$ complexes [42]. There are not many complexes that will give *trans*-alkenes; palladium(0) diimine complexes [43] for instance always lead to *cis*-alkenes. The reaction (Ru) was studied by *in situ* PHIP (para-hydrogen induced polarization)-NMR spectroscopy. With this method the initially formed products can be identified and characterized, even at very low concentrations and low conversions. A mechanism involving a binuclear complex **9** is proposed to explain the formation of the (*E*)-alkenes. The catalyst is not active in combination with terminal alkynes, possibly due to the formation of a vinylidene complex, often observed for ruthenium complexes.

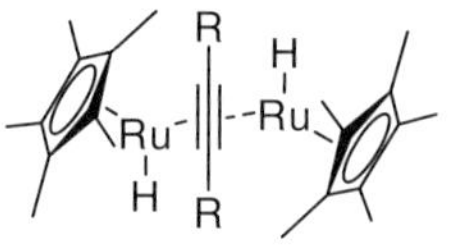

**9**

While CO may decrease the activity of the ruthenium hydrogenation catalyst, or inhibit hydrogenation completely when dicarbonyl species are formed, Fogg and coworkers [44] reported that in complexes $RuHCl(H_2)(PCy_3)(L)$ and $RuHCl(CO)$ $(PCy_3)(L)$ (L = IMes, 1,3-dimesitylimidazol-2-ylidene) the latter are more stable and give higher turnovers than the former. The carbon monoxide $\pi$-acceptor reduces the electron density somewhat on the metal containing strong $\sigma$-donor ligands and apparently this stabilizes the complex. The greater stability compensates for the loss of activity caused by the lower propensity of ligand loss, most likely $PCy_3$.

Yi and coworkers studied the activation of catalysts $RuHCl(PCy_3)_2(CO)$ by strong acids and found that a drastic acceleration took place when $HBF_4$ was added [45]. One phosphine ligand is protonated, forming the phosphonium salt, and a highly active, unsaturated 14-electron ruthenium species remains (Scheme 7.8). The active complex slowly decomposes in benzene to form a tetrameric complex $[Ru_4Cl_7(P-Cy_3)_4(CO)_4]BF_4$ containing bridging chlorides, six chlorides bridging two ruthenium atoms (two double and two single bridges) on the edges of the square arrangement of $Ru_4$ and one $\mu_4$ chlorine in the middle, bridging four ruthenium atoms. This is typical of the highly basic phosphine, as with triphenyl phosphine as the ligand both remain coordinated to the metal and instead an inactive dimer is formed [46] (Scheme 7.8) and $H_2$ is lost.

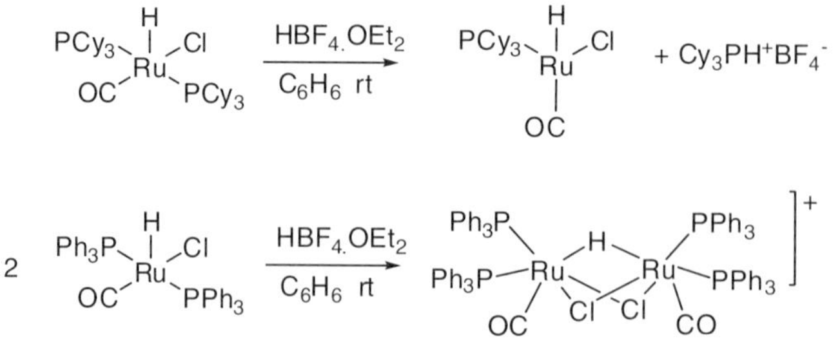

**Scheme 7.8** Protonation of Ru complexes.

Carpentier and coworkers reported substrate inhibition in an asymmetric hydrogen transfer hydrogenation reaction of functionalized ketones using catalysts of the type β-amino alcohol)(arene)ruthenium(II) [47]. The structure of the catalyst was systematically screened using a wide variety of $[(\eta^6\text{-arene})RuCl_2]_2$ complexes and β-amino alcohols $R^1CH(OH)CHR^2NHR^3$. The scope of the reaction was studied for many substrates. The catalyst precursor $[\{\eta^6\text{-p-cymene}\}[\eta^2\text{-N,O-(L)}]RuCl]$ (**10**), the 16-electron true catalyst $[\{\eta^6\text{-p-cymene}\}\{\eta^2\text{-N,O-(L1-)}\}Ru]$, and the hydride $[\{\eta^6\text{-p-cymene}\}\{\eta^2\text{-N,O-(L)}\}RuH]$ have been isolated and characterized. This allowed confirmation of the mechanistic pathway hitherto envisaged on the basis of computational studies [48]. Here, we will only be concerned about the inhibition processes of the active Ru species during the course of the reaction. Deactivation of the catalytic species by β-dicarbonyl substrates was found via formation of inactive (β-diketonato) Ru(II) complexes **11**, representing an intrinsic limitation of the Ru-catalyzed transfer hydrogenation process (Scheme 7.9).

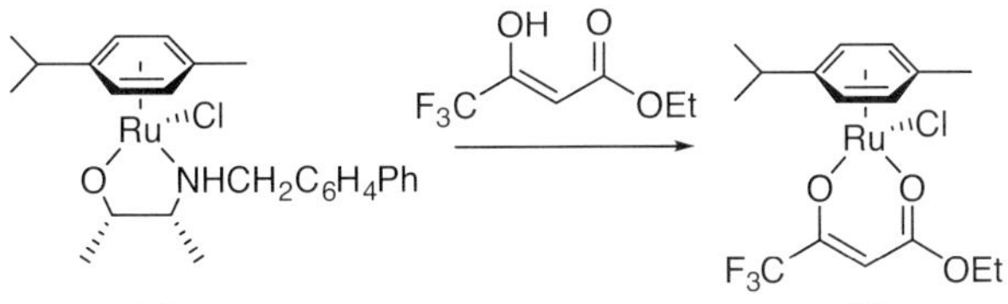

**Scheme 7.9** Deactivation of catalyst **10** by reaction with β-ketoesters.

The presence of water in hydrogen transfer catalysts of rhodium, iridium, and ruthenium is often detrimental to the catalysis. In most catalysts a strong base must be added and sodium hydroxide is often effective, showing that small amounts of water do not always inhibit the reaction. The order in which base and substrate (acetophenone, mostly) are added is also critical, and often the base should be added first to generate the active species, probably a hydride formed via β-elimination from a metal alkoxide intermediate [49]. Formic acid can be used as the hydrogen donor, which has the advantage compared to isopropanol that the equilibrium lies far to the alcohol product (and $CO_2$) side and no loss of enantioselectivity occurs at prolonged reaction times. Since hydrogen transfer requires neutral to basic conditions, usually the azeotrope formic acid/NEt3 is used or mixtures of formic acid and sodium formate. The Noyori–Ikariya Ru-Ts-dpen **12** turned out to be active and enantioselective for hydrogen transfer in aqueous systems [50]. Turnover frequencies of around $100\,h^{-1}$ at $40\,°C$ were obtained in homogeneous systems while the immobilized ones were somewhat slower. An interesting study by Xiao showed that at low pH (<5) the reaction is inhibited. Much faster, more enantioselective, and more productive catalysis was attained in water under slightly basic conditions (pH 5–8) [51]. At low pH values, the reaction was much less efficient and appears to operate through a different mechanism. It was suggested that under neutral and basic conditions the reaction proceeds via the concerted Noyori mechanism [52], while at low pH the classic insertion mechanism is operative.

**12**  Ru–(R,R)-Ts-dpen

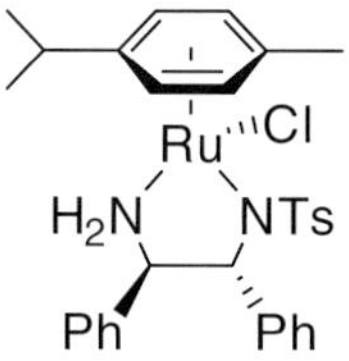

A polymer-supported Gao–Noyori [53] catalyst (**13**) was found to be inhibited by traces of water, showing that water resistance depends strongly on the nature of the catalyst. The ligand is a $P_2N_2$ ligand that coordinates in the square plane in the ruthenium dichloride precursor. The amine-containing ligand performed much better than the imine intermediate. Liese and coworkers found that the supported catalyst prepared by copolymerization of the vinylbenzene complex with methylhydrosiloxane-dimethylsiloxane copolymer and $(MeOCH_2CH_2O)_3SiCH:CH_2$ [54] can be retained by ultrafiltration membranes. They found that this catalyst in continuously operated membrane reactors required a continuous isopropoxide dosage in order to compensate deactivation caused by water residues in the feed stream.

**13**

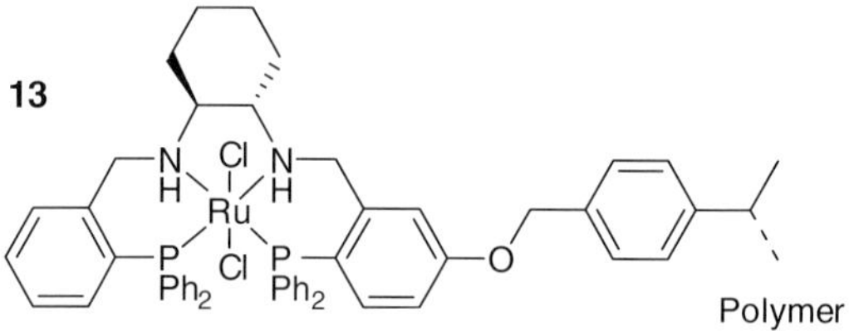

Polymer

Continuous addition of base was not needed for an immobilized Noyori catalyst made from ligand **14** and cymene ruthenium dichloride in the hydrogen transfer reaction from isopropanol to acetophenone (ee ~90%). Once the catalyst was activated by KOtBu at room temperature it could be used without loss of activity for one week at room temperature [55]. In this case the catalyst on silica is much more stable than the homogeneous one. The increased stability was tentatively ascribed to site isolation effects, preventing dimer formation.

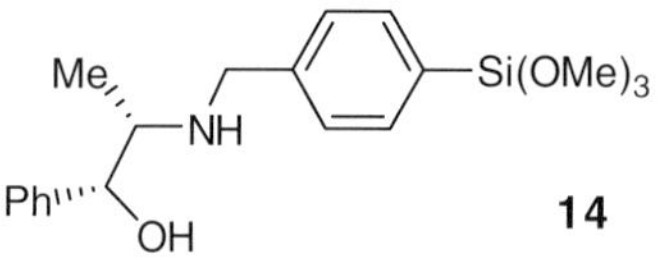

**14**

## 7.4
## Inhibition by Formation of Bridged Species

When dimer formation becomes important one might attempt to destabilize the dimer relative to the monomer. For instance making the ligand very bulky might prevent dimer formation. Models show that the ligand size must be substantially increased to arrive at the desired effect. Another approach is so-called "site isolation" as was described by Grubbs [56]. His well-known example concerns a titanocene catalyst that is used as a hydrogenation catalyst. The intermediate titanium hydride is converted almost completely to a dimer rendering the catalyst with a low activity (Scheme 7.10). Immobilization of the catalyst on a resin support prevents dimerization and an active catalyst is obtained. One Cp ring is connected to a cross-linked polystyrene and its anion is reacted with CpTiCl$_3$ to give the active catalyst.

**Scheme 7.10** Fulvalene titanium dimer formation from Cp$_2$Ti.

### 7.4.1
### Inhibition by Formation of Bridged Species: Iridium

As early as 1977 Crabtree *et al.* [57] described the formation of dimeric species containing bridging hydrides of the formula [Ir$_2$($\mu$-H)$_3$H$_2$L$_4$]PF$_6$ when working with their iridium catalysts with the formula [Ir(cod)Lpy]PF$_6$ (L = phosphine). The reaction takes place in the absence of substrate and is irreversible and the catalysis cannot be restarted. Sometimes the inhibition started before all substrate had been consumed. The reaction worked best in dichloromethane. When ethanol was used as the solvent, it was noted that the solvent competed with the alkene substrate for iridium coordination.

Crocker stabilized the monomeric iridium catalysts [Ir(COD)(PPh$_3$)$_2$]$^+$ and [Ir(COD)(NCMe)(PCy$_3$)]$^+$ via intercalation in montmorillonite clay [58] via an ion-exchange procedure. Although the initial activity of the intercalated catalysts for the hydrogenation of cyclohexene was generally 50–80% of that of their homogeneous analogs, the intercalated catalysts remained active for longer periods, indicating that dimer- and trimer-forming reactions, responsible for deactivation of the homogeneous catalysts, are partially suppressed upon intercalation.

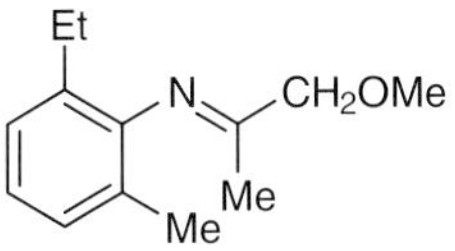

**15** 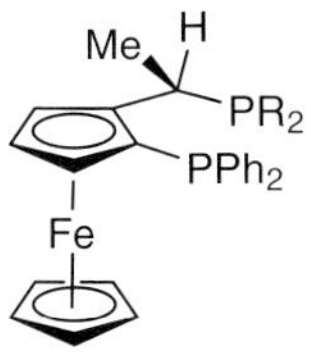                    **16**

The above examples seem to suggest that dimer or trimer formation in iridium catalysts can hardly be avoided, but that is not true either and we finish this section by mentioning a highly successful hydrogenation reaction, namely the asymmetric hydrogenation of the imine precursor for S-Metolachlor, **15**. Spindler and co-workers [59] reported that a new class of Ir ferrocenyldiphosphine **16** complexes turned out to be stable and, in the presence of HOAc and iodide, gave extraordinarily active and productive catalysts. **15** was hydrogenated at a $H_2$ pressure of 80 bars and a temperature of 50 °C with a substrate/catalyst ratio of 500 000 within 12 h, and 1 000 000 turnovers were achieved in 30 h. Optical yields were 79% only, but for this particular herbicide this was considered to be acceptable. Metolachlor is the chloroacetyl amide of amine **15**. It is used as a weedkiller in corn and soya plantations.

Thus, while Josiphos had solved the stability problem for the hydrogenation of the metolachlor precursor, other imine substrates requiring the use of Ir catalysts with DIOP, BDPP, and BPPM, chiraphos, and norphos ligands still suffered from rapid decomposition via dimer and trimer formation. Blaser and coworkers found two solutions, one via immobilization and the second with the use of labile, protecting metal hydrides [60]. The best results were obtained with complexes immobilized on silica gel and with bimetallic W–Ir and Mo–Ir complexes. In most cases, enantiomeric excesses as well as initial rates were only slightly affected while the stability of the catalysts was enhanced significantly. It was reasoned that the use of another metal hydride forming heterodimers with the iridium hydride species in a reversible manner, might prevent irreversible homodimerization. The co-hydrides selected were $Cp_2WH_2$ and $Cp_2MoH_2$, which were known, from the work by Venanzi [61], to dimerize with cationic iridium hydride species. Metalation on one of the Cp rings by iridium may occur (Scheme 7.11).

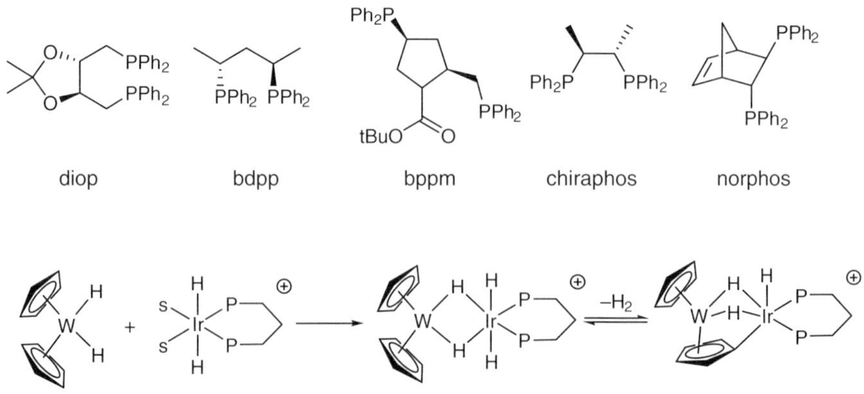

diop        bdpp        bppm        chiraphos        norphos

**Scheme 7.11** Ligands used and alternative, reversible hetero-dimer formation for iridium hydrogenation catalysts.

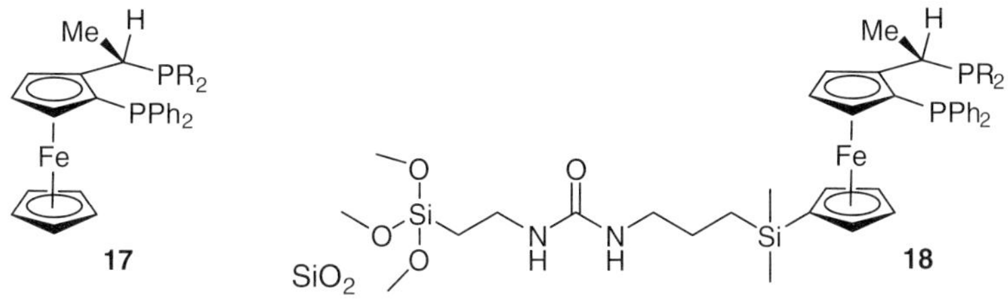

R = 3,5-xylyl

In an attempt to further improve the actual catalyst of the metolachlor process, Xyliphos, **17**, the Solvias research group immobilized this ligand on silica and polystyrene, via well established methods [62]. A variety of alkoxysilane and isocyanate-substituted linkers was investigated. The best heterogeneous catalyst **18** (0.04 mmol (g $SiO_2$)$^{-1}$, ca. 5 wt%) exhibited TONs >100 000 and TOFs up to 20 000 h$^{-1}$, the best values so far for immobilized catalysts. The immobilized catalysts gave similar enantioselectivities but lower activities than the homogeneous analogues. Surprisingly, in the present case the heterogeneous catalysts gave higher deactivation rates than the homogeneous ones. These negative effects were explained by the higher local catalyst concentrations on the support surface leading to an increased tendency to deactivation by irreversible dimer formation, the opposite of what was aimed at. Lower loadings did not give better results. An advantage of the system is the easy and efficient separation of these catalysts by filtration. The authors concluded though that industrial application was not attractive.

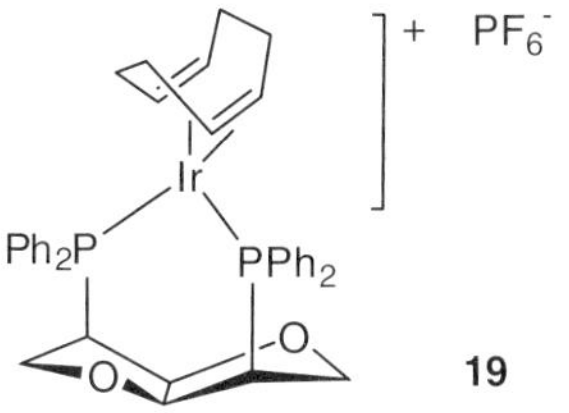

Complexes of the type [Ir(ddppm)(COD)]X **19** were prepared and tested in the asymmetric hydrogenation of a range of imine substrates [63]. Contrary to known iridium catalysts [64], the ddppm complexes formed efficient catalysts under an atmospheric hydrogen pressure, whereas at higher pressures the catalytic activity of the system was drastically reduced. Depending on the reaction conditions, N-arylimines, Ar'N=CMeAr, were hydrogenated to the corresponding secondary amines in high yields and enantioselectivities (80–94% ee). In contrast to the BF$_4$$^-$ and PF$_6$$^-$ complexes, coordinating anions such as chloride did not form active Ir-ddppm hydrogenation catalysts. The cationic Ir-ddppm hydrogenation system performed well in chlorinated solvents, whereas coordinating solvents such as THF and methanol deactivated the system. When the catalyst was pretreated with H$_2$ before the imine substrate was added, an inactive system was also obtained. Dimeric and trimeric Ir(III) polyhydride complexes were formed from the reaction of [Ir(ddppm)(COD)]PF$_6$ with molecular hydrogen at 1 bar and were found to inhibit catalytic activity (Scheme 7.12). It was thought that as long as imine is present that can coordinate to the cationic, monomeric iridium complex, no dimer or trimer is formed.

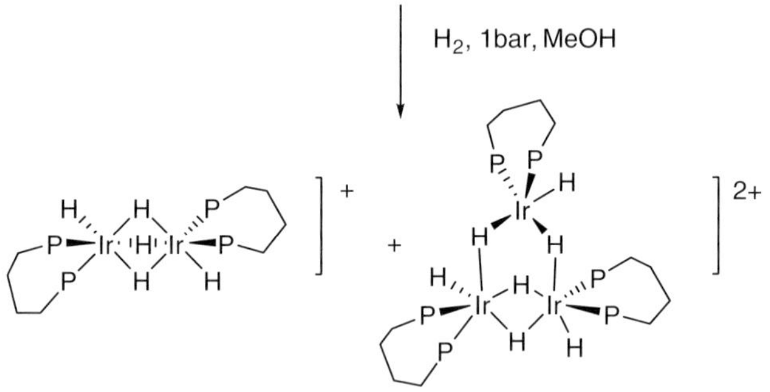

**Scheme 7.12** Trimer and dimer formation from **19** in the absence of substrate.

The Crabtree catalyst [Ir(Py)(PCy$_3$)(COD)]PF$_6$ is one of the few catalysts that will hydrogenate tri- or tetra-substituted alkenes. The reaction with highly substituted alkenes is slow and deactivation takes place during the reaction, usually caused by hydrido dimer and trimer formation. Pflatz's Ir PHOX catalysts, used for asymmetric hydrogenation of alkenes, were much faster and more resistant to deactivation. Thus Pfaltz and coworkers reasoned that achiral PHOX ligands with a suitable steric bulk might be suitable for hydrogenation of these substrates [65]. Four iridium complexes with achiral phosphino-oxazoline (PHOX) ligands were readily prepared. The air-stable complexes with tetrakis[3,5-bis(trifluoromethyl)phenyl]borate (BArF) as counterion showed high reactivity in the hydrogenation of a range of tri- and tetra-substituted alkenes. Excellent results were obtained with an iridium complex derived from a dicyclohexylphosphino-oxazoline ligand bearing no additional substituents in the oxazoline ring. With several substrates, which gave only low conversion with the Crabtree catalyst, full conversion was observed. The productivity of the Crabtree catalyst could be strongly increased by replacing the hexafluorophosphate anion with BArF.

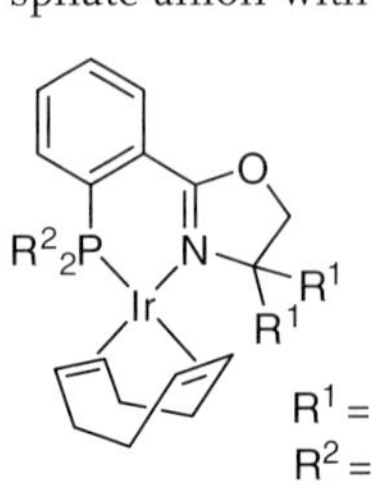

R$^1$ = H, Me
R$^2$ = Ph, Cy

The anion effect had been studied before by Pfaltz for the iridium-PHOX catalysts [66]. In the asymmetric hydrogenation of unfunctionalized olefins with cationic iridium-PHOX catalysts, the reaction kinetics and, as a consequence, catalyst activity and productivity, depend heavily on the counterion. A strong decrease in the reaction rate was observed in the series $[Al\{OC(CF_3)_3\}_4]^- > BArF^- > [B(C_6F_5)_4]^- > PF_6^- > BF_4^- > CF_3SO_3^-$. With the first two anions, high rates, turnover frequencies (TOF $> 5000\,h^{-1}$ at $4\,°C$), and turnover numbers of 2000–5000 are routinely achieved. The hexafluorophosphate salt reacts at lower rates, although they are still respectable; in addition, this salt suffers from deactivation during the reaction and extreme water-sensitivity, especially at low catalyst loading. Catalysts containing the first three counterions do not lose activity during the reaction and remain active, even after all the substrate is consumed. They were much less sensitive to moisture and, in general, rigorous exclusion of water and oxygen was not necessary.

## 7.4.2
### Inhibition by Formation of Bridged Species: Rhodium

Dimer formation in rhodium hydrogenation catalysts is a well known phenomenon, as in the early papers of Wilkinson their formation was already proposed [67]. In order to make a vacant site in Wilkinson's catalyst a triphenylphosphine must dissociate and this species can coordinate to an alkene, a solvent molecule, or it may dimerize to form $[RhCl(PPh_3)_2]_2$. At that time many questions remained unresolved due to the absence of sensitive $^{31}$P-NMR spectroscopy.

Dimerization was also studied by Czakova and Capka by hydrogenation of alkenes catalyzed by homogeneous Rh(I) complexes prepared *in situ* from μ,μ'-dichlorobis[di (alkene)rhodium] and phosphines of the type $RPPh_2$ [$R = (CH_2)_n(OEt)_3$, $n = 1$–6, $R = CH_2SiMe_{3-m}(OEt)_m$, $m = 1$–3] and by their heterogeneous analogues anchored to silica. Conditions used were 1.1 bar of $H_2$ and 37–67 $°C$ [68]. The hydrogenations catalyzed by catalysts of both types were first order in the alkenes. The deactivation of the rhodium catalysts was due to the dimerization of catalytically active species. Interestingly, in this case dimerization takes place also on the surface of the support and depends on the length of the spacer group separating the diphenylphosphino group from the surface. Apparently a lower surface coverage is needed to obtain efficient site isolation.

Ducket, Newell, and Eisenberg studied Wilkinson's catalyst more recently and observed new intermediates in hydrogenation catalyzed by Wilkinson's catalyst using parahydrogen-induced polarization (PHIP) [69]. The use of PHIP enabled the detection of previously unobserved dihydride species in the reaction system. Specifically, the binuclear complexes $H_2Rh(PPh_3)_2(\mu-Cl)_2Rh(PPh_3)$(alkene) for *cis-* and *trans-*2-hexene, 2-pentene, methyl methacrylate, styrene, and substituted styrenes were detected and characterized. While readily observable during hydrogenation catalysis under 1–2 bar of para-enriched hydrogen, $H_2Rh(PPh_3)_2(\mu-Cl)_2Rh(PPh_3)$(alkene) is not directly in the catalytic cycle; its appearance suggests reduced activity of the system through the formation of less active binuclear species. A second new species observed with styrene, *p*-chlorostyrene, and *p*-methylstyrene as substrates is the mononuclear complex $RhH_2(alkene)(PPh_3)_2(Cl)$, which appears to be an intermediate in hydrogenation catalysis. From PHIP spectra, it is possible to assign this species as having *cis*-phosphines as well as *cis*-hydrides. Their results also confirm that small amounts of free phosphine inhibit the formation of binuclear complexes, thereby enhancing the catalytic activity of Wilkinson's complex.

Later studies showed that, in addition to the binuclear dihydride $[Rh(H)_2(PPh_3)_2(\mu\text{-}Cl)_2Rh(PPh_3)_2]$, the tetrahydride complex $[Rh(H)_2(PPh_3)_2(\mu\text{-}Cl)]_2$ was also readily formed (Scheme 7.13) [70]. While magnetization transfer from free $H_2$ into both the hydride resonances of the tetrahydride and $[Rh(H)_2Cl(PPh_3)_3]$ is observable, neither transfer into $[Rh(H)_2(PPh_3)_2(\mu\text{-}Cl)_2Rh(PPh_3)_2]$ nor transfer between the two binuclear complexes is seen. Consequently $[Rh(H)_2(PPh_3)_2(\mu\text{-}Cl)]_2$ and $[Rh(H)_2(PPh_3)_2(\mu\text{-}Cl)_2Rh(PPh_3)_2]$ are not connected on the NMR time-scale by simple elimination or addition of $H_2$. The rapid exchange of free $H_2$ into the tetrahydride proceeds via reversible halide bridge rupture and the formation of $[Rh(H)_2(PPh_3)_2(\mu\text{-}Cl)RhCl(H)_2(PPh_3)_2]$. When these reactions were studied in dichloromethane, the formation of the solvent complex $[Rh(H)_2(PPh_3)_2(\mu\text{-}Cl)_2Rh(CD_2Cl_2)(PPh_3)]$ and the deactivation products $[Rh(Cl)(H)(PPh_3)(\mu\text{-}Cl)(\mu\text{-}H)RhCl(Cl)(H)(PPh_3)_2]$ and $[Rh(Cl)(H)(CD_2Cl_2)(PPh_3)(\mu\text{-}Cl)(\mu\text{-}H)Rh(Cl)(H)(PPh_3)_2]$ was indicated. In the presence of an alkene and parahydrogen, signals corresponding to binuclear alkene complexes mentioned above were detected. These complexes undergo intramolecular hydride interchange in a process that is independent of the concentration of styrene and catalyst. and involves halide bridge rupture, followed by rotation about the remaining Rh–Cl bridge, and bridge re-establishment. This process is facilitated by electron-rich alkenes. Magnetization transfer from the hydride ligands of these complexes into the alkyl group of the hydrogenation product is also observed. Hydrogenation is proposed to proceed via binuclear complex fragmentation and trapping of the resultant intermediate $[RhCl(H)_2(PPh_3)_2]$ by the alkene. Studies on a number of

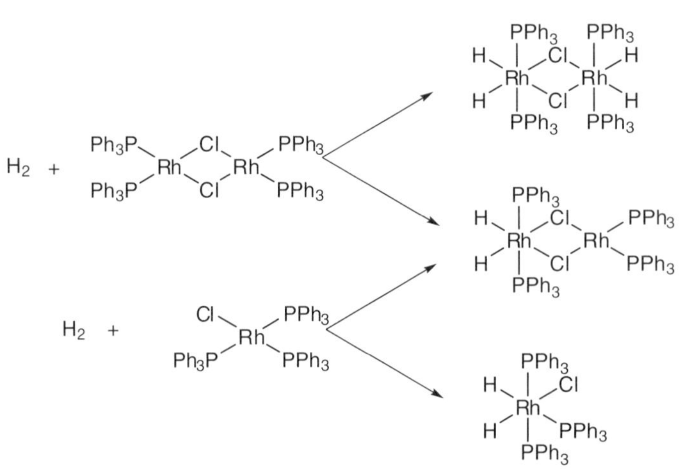

**Scheme 7.13** Dihydrogen addition products of rhodium hydrogenation precursors.

other binuclear dihydride complexes derived from PMe$_3$ as the ligand reveal that such species are able to play a similar role in hydrogenation catalysis.

A study of the analogous iodide complexes [RhI(PPh$_3$)$_2$]$_2$ and [RhI(PPh$_3$)$_3$] revealed the presence of the analogous complexes, including the binuclear alkene-dihydride products. The catalyst is more active, but the higher initial activity of these precursors is offset by the formation of the trirhodium phosphide bridged deactivation product, [{(H)(PPh$_3$)Rh($\mu$-H)($\mu$-I)($\mu$-PPh$_2$)Rh(H)(PPh$_3$)}($\mu$-I)$_2$Rh(H)$_2$(PPh$_3$)$_2$] containing a phosphido bridge, and thus ligand decomposition occurs here as well.

Rhodium phosphine complexes containing carborane monoanions were synthesized and studied as hydrogenation catalysts by Rifat *et al.* [71]. Addition of Ag[closo-CB$_{11}$H$_{12}$] to [(PPh$_3$)$_2$RhCl]$_2$ afforded the new exopolyhedrally coordinated complex [(PPh$_3$)$_2$Rh(closo-CB$_{11}$H$_{12}$)], which was fully characterized. The use of the less nucleophilic [closo-CB$_{11}$H$_6$Br$_6$]$^-$ anion afforded the arene-bridged dimer [(PPh$_3$)(PPh$_2$-$\eta^6$-C$_6$H$_5$)Rh]$_2$[closo-CB$_{11}$H$_6$Br$_6$]$_2$. With the precursor complexes [(PPh$_3$)$_2$Rh(nbd)][closoborane] as starting materials, treatment with H$_2$ afforded the complexes in higher purity and yield. The norbornadiene complexes have been evaluated as internal alkene hydrogenation catalysts using the substrates cyclohexene, 1-methyl-cyclohexene, and 2,3-dimethylbut-2-ene under mild conditions. These new catalysts have also been compared with [(PPh$_3$)$_2$Rh(nbd)][BF$_4$] and Crabtree's catalyst, [(py)(PCy$_3$)Ir(cod)][PF$_6$]. The counterion had a clear effect and the hexabromoclosoborane was significantly better than the other two, matching Crabtree's catalyst in hydrogenation efficiency. Only for 2,3-dimethylbut-2-ene hydrogenation was Crabtree's found to be better. Nevertheless, these results are excellent for a rhodium complex, which have traditionally been considered as ineffective catalysts for the hydrogenation of internal alkenes under mild conditions. The deactivation product in the catalytic cycle was reported to be [(PPh$_3$)$_2$HRh($\mu$-Cl)$_2$($\mu$-H)RhH(PPh$_3$)$_2$][CB$_{11}$H$_{12}$].

Heller and coworkers reported the formation of trinuclear hydroxy rhodium complexes as dormant states of rhodium hydrogenation catalysts [72]. Various trinuclear rhodium complexes of the type [Rh$_3$(PP)$_3$($\mu_3$-OH)$_x$($\mu_3$-OMe)$_{2-x}$]BF$_4$ (where PP = Me-DuPhos, dipamp, dppp, dppe; different ligands and $\mu$-bridging anions) were presented, which were formed upon addition of bases such as NEt$_3$ to solvated complexes [Rh(PP)(solvent)$_2$]BF$_4$. Their *in situ* formation resulting from basic additives or basic prochiral olefins (without addition of another base) can cause deactivation of the asymmetric hydrogenation. The effect can be reversed by means of acidic additives.

A related trimerization was observed in the BINAP rhodium catalyst used for the asymmetric isomerization of a myrcene-derived allylic alcohol, the key step for the production of menthol in the Takasago process [73]. The addition of aqueous ammonia to an acetone solution of [Rh(BINAP)(MeOH)$_2$]ClO$_4$ gave [{Rh(BINAP)}$_3$(OH)$_2$]ClO$_4$ [74]. The two hydroxy groups are triply bridging on each side of an approximately regular Rh$_3$ triangle. Each rhodium atom is coordinated in a square plane by two phosphorus atoms of BINAP and two hydroxy oxygen atoms. This species was also found to be responsible for the deactivation of the catalysis of 1,3-H migration for allylamine, a model for the myrcene-derived allylic amine.

Monophosphonite complexes (Scheme 7.14) are hydrogenation catalysts which, in the absence of substrate or in the presence of weakly coordinating substrates, form dimers via arene binding together with bis-solvento species [75]. In the presence of hydrogen and substrate a hydride intermediate was observed by NMR spectroscopy, showing that the hydrogenation reaction follows the normal reaction scheme. Such binding of ligand arenes has been reported more often [76] and this serves only as an example.

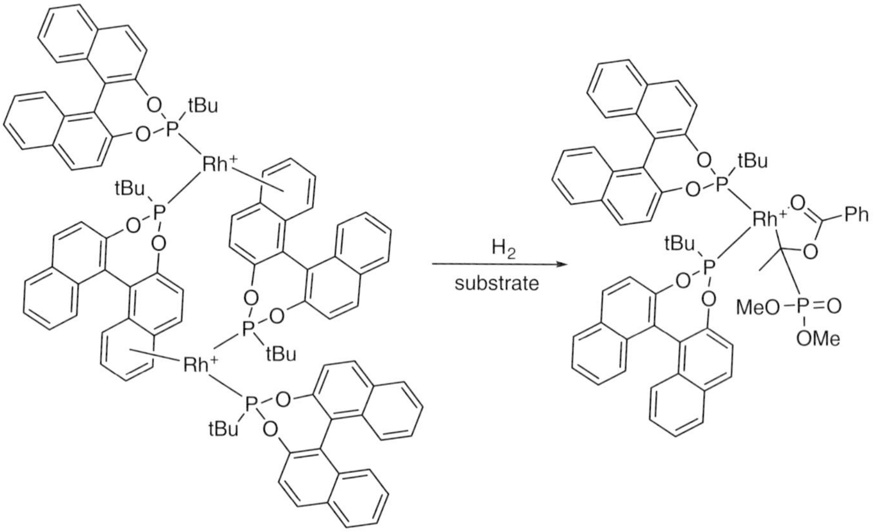

**Scheme 7.14** Formation of hydrogenation catalyst from phosphonite dimer.

## 7.5
## Inhibition by Ligand Decomposition

As mentioned in Section 1.4.3 the cleavage of P—C bonds in the presence of transition metals and dihydrogen is a common reaction, but surprisingly in relation to hydrogenation catalysis few data were found.

In Section 7.4.2 it was mentioned that complexes $[RhI(PPh_3)_2]_2$ and $[RhI(PPh_3)_3]$ revealed the formation of the trirhodium phosphido bridged deactivation product, $[\{(H)(PPh_3)Rh(\mu\text{-}H)(\mu\text{-}I)(\mu\text{-}PPh_2)Rh(H)(PPh_3)\}(\mu\text{-}I)_2Rh(H)_2(PPh_3)_2]$ containing a phosphido bridge, the first example of ligand decomposition via P—C bond cleavage [70]. While this is a very common reaction, in relation to catalytically active hydrogenation compounds hardly any reports have been found.

Nindakova *et al.* studied the transformations of the chiral diphosphine rhodium catalyst $[(1,5\text{-}COD)Rh(-)\text{-}R,R\text{-}DIOP]^+ \ CF_3SO_3^-$ under conditions of hydrogenation [77]. They reported on the products obtained in its reactions with $H_2$, base ($NEt_3$), and solvents in the absence of a substrate, as studied by $^1H$ and $^{31}P$ NMR spectroscopy. In addition to dimeric complexes mentioned above they found benzene arising from the reaction of ligand with $H_2$, caused by destruction of the diphosphine ligand. Most likely the mechanism involves P—C bond cleavage.

Ruthenium acetate BINAP complexes are very effective asymmetric hydrogenation catalysts [78], as studied by Noyori and Takaya. They have been used for the hydrogenation of a wide variety of functionalized alkenes and ketones. It was found that ruthenium in diphosphine complexes may easily lead to P–C cleavage reactions in the ligands employed when acids are added to the complex. The reaction was first discovered for MeO-Biphep ligands [79]. The findings by Pregosin on BINAP, ruthenium and tetrafluoroborate are highly relevant to catalysis. The use of inert anions such as $BF_4^-$ would not arouse much suspicion, but his work shows that fluorodiphenylphosphines may form in such catalysts, even at very low temperatures (Scheme 7.15) [80]. Water and acetate may also function as the nucleophile. The reaction might involve phosphorane intermediates, as observed in the work reported by Grushin for BINAP(O) palladium compounds [81]. When triflic acid containing small amounts of water is used the phosphine product formed is diphenylphosphine oxide. Interestingly, the P–C bond of the naphthyl group is cleaved in both instances and not that of a phenyl group. The reaction is preceded by coordination of the $\eta^6$ arene of the backbone to ruthenium, which occurs in the absence of other suitable ligands [82].

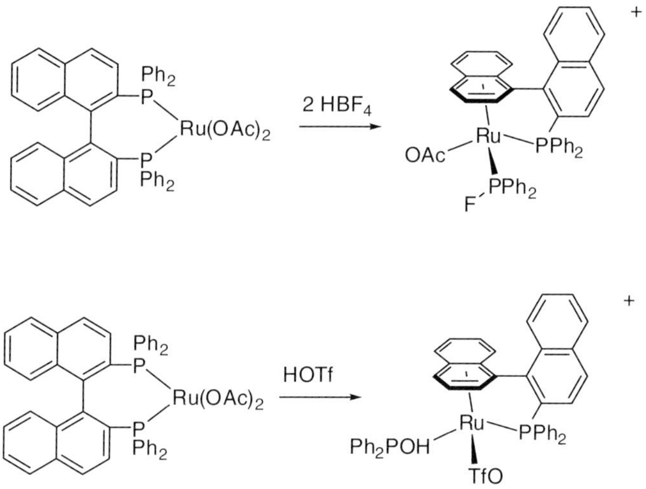

**Scheme 7.15** Ruthenium catalyzed P–C bond cleavage in BINAP.

## 7.6
## Inhibition by the Product

### 7.6.1
### Inhibition by the Product: Rhodium

The product will be an efficient inhibitor when it coordinates more strongly to the catalyst than the substrate, in the same way as polar solvent molecules. The most popular substrates similar to the L-DOPA precursors do not lead to products that may cause inhibition, as the substrates coordinate to rhodium as bidentates with an amido function and the alkene function, and the products lack the alkene function. Thus, after hydrogenation of the alkene function, they are much weaker ligands than the substrate. Imine hydrogenation results in amines as the product and one might expect that they would be inhibitors of the reaction, as the amines may coordinate more strongly to the metal than the starting imines. James and coworkers [83] found that hydrogenations of PhN=CHPh and PhCH$_2$N=CHPh are catalyzed by the precursor [Rh(COD)(PPh$_3$)$_2$]PF$_6$ and indeed product inhibition was observed. However, the amine product PhNHCH$_2$Ph poisons the catalyst by coordination to Rh through an arene moiety, while the other amine product, (PhCH$_2$)$_2$NH, forms a labile N-bonded species that does not poison the catalyst system.

## 7.6.2
## Ruthenium

Wills and coworkers studied the transfer hydrogenation to a wide range of substituted cyclic and acyclic α,β-unsaturated ketones catalyzed by ruthenium and rhodium diamine and amino alcohol complexes [84]. Cyclic α,β-unsaturated ketones appear to be more suitable substrates for the synthesis of enantiomerically pure allylic alcohols than do acyclic unsaturated ketones. In spite of the similarity with other, active substrates, the ketone precursors for the alcohols **21a** and **21b** showed no conversion with catalyst **22**. This was assigned to either substrate inhibition or product inhibition. Catalyst **23** did not undergo inhibition with these substrates or products.

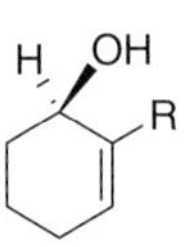

**21a** R = OBn
**21b** R = NH$_2$CO$_2$Me
**22**
**23** 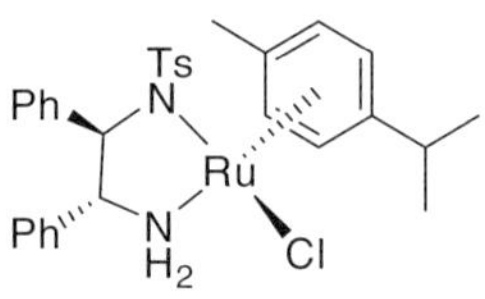

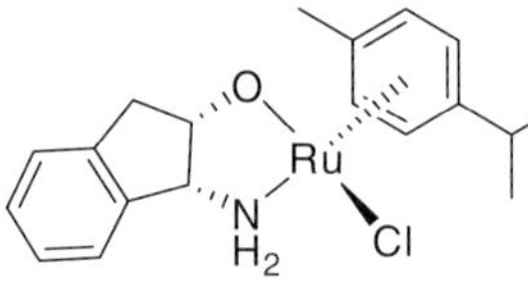

Several α-amino- and α-alkoxy-substituted ketones underwent asymmetric transfer hydrogenation successfully, with high yields and enantioselectivity, when the appropriate protecting group was used [85]. Exceptions were α-chloro- and α-methoxyacetophenone for which either product or educt inhibition was proposed.

Whole cell bioconversions for α-chloroacetophenone and β-chloropropiophenone to the corresponding chiral alcohols have been achieved, see for example Ref. 86 and references therein, using whole cells of the white-rot fungus *Merulius tremellosus*. For the latter the main product was the dechlorinated alcohol (Scheme 7.16); several fungi and bacteria show dechlorination activity. The Noyori type catalysts similar to **22** showed no activity for transfer hydrogenation with these substrates and it was suggested that either the substrate or the product give oxidative addition to the metal, the onset to dechlorination, leading to chloride poisoning of the catalyst.

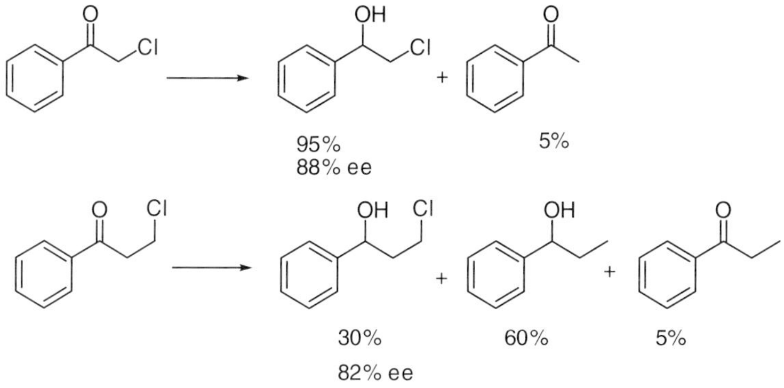

95%
88% ee

5%

30%
82% ee

60%

5%

Scheme 7.16  Products of whole-cell reduction of chloroketones.

## 7.7
## Inhibition by Metal Formation; Heterogeneous Catalysis by Metals

The majority of the hydrogenation and transfer hydrogenation catalysts reviewed so far are truly homogeneous catalysts and participation of bulk metal is excluded and metal precipitation is not a common deactivation mechanism when mild conditions are used for the metals we have focused on in this chapter. The enantioselectivities obtained are a proof that the chiral (di)phosphines, P−N ligands, or diimine ligands are coordinated to the metal as the chiral inducers, but they do not prove the homogeneous character of the catalyst, as many heterogeneous catalysts surface-modified with chiral modifiers are known [87]. The two most common modifiers are tartaric acid (on Ni especially) and cinchona-type molecules (on Pt and Pd), rather than the typical ligands used in the studies reviewed here and high ees have been reported for these catalysts, especially when ketones and ketoesters are used as substrates [88].

Under forcing conditions of high pressure of $H_2$ and high temperatures the formation of metal is more common. Formation of bulk metal at high catalyst loadings can be easily observed, but low loadings accompanied by the formation of nanosized particles has been an issue of much debate over the last 30 years. Nanoparticles may be as small as 1 nm (the size of a large ligand!) and they can be stabilized by anions, polymers, surfactant molecules, ligands, dendrimers, and so on. Several means to discover the participation of metal particles have been proposed (filtration, addition of Hg, addition of $CS_2$, recording TEM spectra, and so on), but, as has been clearly underpinned and explained in recent years by Finke and coworkers, this task is far from easy. There is not one single experiment that can distinguish between homogeneous and heterogeneous catalysts, as had become clear over the years. Widegren and Finke proposed a package of measurements to be undertaken in order to determine the true nature of the catalyst [89]. The package includes (i) catalyst isolation and identification, (ii) kinetic studies (incubation, reproducibility), (iii) quantitative poisoning and recovery experiments, and (iv) mechanistic studies showing that the identity of the catalyst is consistent with all the data.

For alkene, ketone, imine, ester, and so on hydrogenation, one has always readily accepted the homogeneous nature of the proposed catalyst, but hydrogenation of monoaromatics has always been subject to debate as regards the nature of the catalyst, as usually harsh conditions were required ($>50$ bar, $>80\,°C$), while the metals used –rhodium and ruthenium– are highly active as metallic catalysts for this reaction. The Nb(V) and Ta(V) hydrido aryloxide complexes developed by Rothwell and coworkers form an exception as these must be truly homogeneous catalysts, among other things because thermodynamically niobium and tantalum metal cannot be obtained under these conditions from hydrogen and the metal aryloxides [90]. Moreover, the catalysts show a peculiar preference for intramolecular hydrogenation of the aryloxide ligands.

Widegren, Bennett and Finke studied benzene hydrogenation with the use of Ru(II)($\eta^6$-C$_6$Me$_6$)(OAc)$_2$ as the precatalyst at 60 bar H$_2$, $100\,°C$ [91], one of the oldest "homogeneous" benzene hydrogenation catalysts, the definitive identification of which had been pending for many years.

The key observations that led to the conclusion that the true catalyst consists of bulk Ru metal particles, and not a homogeneous metal complex or a soluble nanocluster, were as follows. First, the catalytic benzene hydrogenation reaction showed the typical sigmoidal kinetics reported before for metal(0) formation from homogeneous precursors [92]. The bulk Ru metal formed had sufficient activity to account for all the observed activity. Furthermore, the filtrate from the product solution was inactive until bulk metal was formed. Also, the addition of Hg(0), a known heterogeneous catalyst poison, completely inhibits further catalysis. Lastly, TEM fails to detect nanoclusters and thus it was concluded that the bulk metal is the actual hydrogenation catalyst.

A tris-ruthenium cluster [Ru$_3$($\mu_2$-H)$_3$($\eta^6$-C$_6$H$_6$($\eta^6$-C$_6$Me$_6$)$_2$($\mu_3$-O)]$^+$ was reported by Süss-Fink as a very active benzene hydrogenation catalyst at 60 bar H$_2$, $110\,°C$ with 1000s of turnovers per hour [93]. The clusters contain a Ru$_3$ core (24), bridged with three hydrides, or two hydrides and a hydroxy group, and capped with an oxide ion, yielding a 1+ charge. Two ruthenium atoms are $\eta^6$ bonded to two Me$_6$C$_6$ molecules and the third one contains a $\eta^6$ benzene molecule. In this case the mercury test did not negatively affect the hydrogenation activity. It was thought that perhaps the $\eta^6$ coordination of benzene to the cluster might mimic a metal surface thus rendering this unusual activity. The trimer was generated *in situ* from the dimer reported in Chapter 1, Scheme 1.17, and a cationic monomeric precursor.

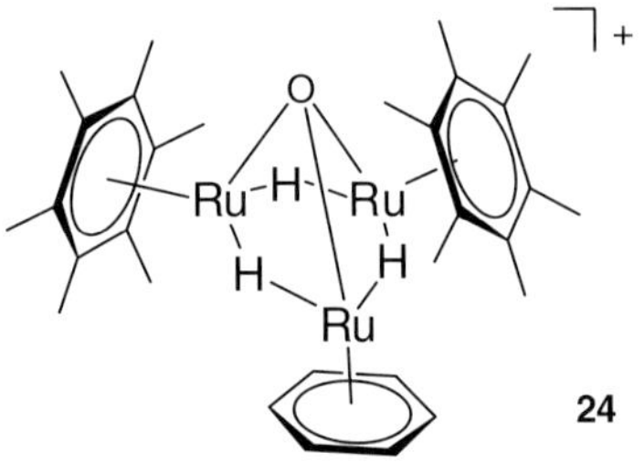

24

In a joint effort of the Süss-Fink and Finke groups it was shown eventually that the "supramolecular catalysis concept" thought to be associated with the homogeneous arene hydrogenation was not correct [94], although the supramolecular, host–guest complexation of benzene to the trimeric cluster was unequivocally demonstrated, including by X-ray. Thus, the $\eta^6$ $\kappa^3$ trimetallic ligated benzene case stands out as a special one compared to the other monometallic catalysts proposed, and it was worth investigating in much detail. The methodology employed for the analysis of the catalysis was the same as that mentioned above. The results provided a compelling case that $[Ru_3(\mu^2\text{-}H)_3(\eta^6\text{-}C_6H_6)(\eta^6\text{-}C_6Me_6)_2(\mu^3\text{-}O)]^+$, is *not* the true benzene hydrogenation catalyst, as was previously believed; instead, all evidence was consistent with trace Ru(0) derived from the homogeneous precursor under the reaction conditions as the true, active catalyst. An important conclusion of this work is that one should never rely on just one of the following measurements alone, such as recovery of the catalyst precursor after reaction, product selectivities, mercury poisoning, TEM studies – the observed clusters can be side products or form during the sample preparation or in the electron beam.

Finney and Finke conducted similar studies on Pt(1,5-COD)Cl$_2$ and Pt(1,5-COD) (CH$_3$)$_2$ as hydrogenation catalysts to answer the question of "Is it homogeneous or heterogeneous catalysis" [95]. The complexes were used for the hydrogenation of alkenes. Using product studies, kinetic evidence, and mercury poisoning experiments, it was shown that Pt(1,5-COD)Cl$_2$ is a precatalyst and must be reduced to Pt(0) nanoclusters and bulk metal as the true hydrogenation catalyst. An investigation of the related complex Pt(1,5-COD)(CH$_3$)$_2$ revealed that this complex does not form a hydrogenation catalyst by itself under H$_2$, in agreement with the literature. Kinetic and mercury poisoning evidence confirm that Pt(1,5-COD)(CH$_3$)$_2$, too, forms a Pt(0) heterogeneous catalyst if other metals (Ir, Pt) are used as seeds to initiate the reduction of Pt(II).

Another longstanding question on the catalyst nature in the hydrogenation of benzene and cyclohexene concerns rhodium complexes of the structure $[Rh(\eta^5\text{-}C_5Me_5)Cl_2]_2$, first reported by Maitlis and coworkers [96]. Over the years the catalyst was extensively studied and there were indications that the reported precursor may not be the actual catalyst. For complete references on this see the recent report by the groups of Maitlis and Finke who studied in depth this reaction of the rhodium dimeric catalyst [97]. The true *benzene* hydrogenation catalyst derived from $[Rh(\eta^5\text{-}C_5Me_5)Cl_2]_2$ was found to be a nanoparticle heterogeneous catalyst. This answered a 25 year-old question and corrected the previous belief that the *benzene* hydrogenation catalyst was a monometallic, homogeneous catalyst. The evidence came from the kinetics, the mercury test, the formation of the metal product, the vigorous conditions, and so on. The true *cyclohexene* hydrogenation catalyst, though, derived from the same precursor was reported to be a homogeneous catalyst, under the mild conditions employed herein, as originally postulated and as concluded by Collman and coworkers [98]. The cyclohexene hydrogenation catalyst was obtained at the milder conditions of 22 °C and 3.7 bar H$_2$ and was a non-nanocluster, homogeneous catalyst, most likely the previously identified complex, $[Rh(\eta^5\text{-}C_5Me_5)H_2(solvent)]$ [99].

Thus the methodology employed has the ability to identify both heterogeneous and homogeneous catalysts from the same catalyst precursor.

Manners reported the hydrogenation of cyclohexene using $Me_2NH-BH_3$ as the hydrogen donor and rhodium colloids as the (heterogeneous) catalyst [100]. Subsequently, they looked at the dehydrocoupling of $Me_2NH-BH_3$ and $Ph_2PH-BH_3$ using several rhodium precursors such as $[\{Rh(1,5\text{-cod})(\mu\text{-Cl})\}_2]$, $Rh/Al_2O_3$, Rh-colloid/ $[Oct_4N]Cl$, and $[Rh(1,5\text{-cod})_2]OTf$. The dihydrocoupling of $Me_2NH-BH_3$ turned out to be heterogeneous, while the dehydrocoupling of $Ph_2PH-BH_3$ (to give $Ph_2PH-BH_2-PPh_2-BH_3$) was a homogeneously catalyzed reaction, even when $Rh/Al_2O_3$ was used as the precursor. This shows that the interaction of phosphorus with rhodium is indeed very different from that of dimethylamine, while the interaction with borane is reverse in strength.

## 7.8
### Selective Activation and Deactivation of Enantiomeric Catalysts

The concept is simple; one takes a non-enantiopure catalyst, adds a chiral ligand that coordinates selectively to one of the enantiomers, and in doing so one enantiomer is deactivated or activated and we are left with an enantiopure catalyst giving optimal enantioselectivity in the catalytic reaction. Since there are fewer catalyst centers available the rate will go down, but the selectivity may increase. For a recent review see Ref. [101], as here we will present only a few examples. This approach is closely related to non-linear effects in asymmetric catalysis. Clearly, if less than stoichiometric amounts are added we end up with intermediate enantioselectivities [102]. The first application of the principle of "deactivating" one enantiomer of a racemic catalyst by an enantiomerically pure agent is probably due to Brown and coworkers [103]. Actually, their procedure was slightly different and called *in situ resolution*. In their experiment an iridium complex prepared from enantiopure (*R*)-menthyl-(*Z*)-R-benzamidocinnamate, was reacted with 2 equiv of racemic CHIRAPHOS to selectively bind only the (*S,S*)-enantiomer of the diphosphine. Subsequent addition of Rh(I) then allowed the remaining free (*R,R*)-CHIRAPHOS to form a Rh complex that was used to selectively hydrogenate methyl (*Z*)-R-benzamidocinnamate.

Mikami and coworkers discuss the possible ways of obtaining more or less enantioselective catalysts by mixing racemic (diphosphine) complexes **25** with either chiral deactivators, or chiral activators such as *S,S*-DPEN, **26** [104, 105]. For instance the racemic mixture of BINAP–Ru(II) complexes reacts with 0.5 equiv of enantiopure 3,3'-dimethyl-2,2'-diamino-1,1'-binaphthyl (*R*-DM-DABN) **27** to give selectively 0.5 mole of enantiopure, diastereomeric (*R*-DM-DABN)-*R*-BINAP-RuCl$_2$ catalyst and 0.5 mole of the remaining enantiomer of *S*-BINAP-Ru, an inactive BINAP-Ru complex; even the presence of an excess of the chiral diamine gave the same result. The lower catalytic activity of (*R*-DM-DABN)-*R*-BINAP-Ru stems from the electron delocalization from the Ru center to the diamine moiety in contrast to the BINAP-Ru(II)/DPEN complex where the highest electron densities are localized in the Ru-N region. Then a second amine, for example, *S,S*-**26**, is added to form the active complex from the remaining *S*-BINAP-*S,S*-DPEN-Ru complex. The mixture was successfully used in asymmetric ketone hydrogenation.

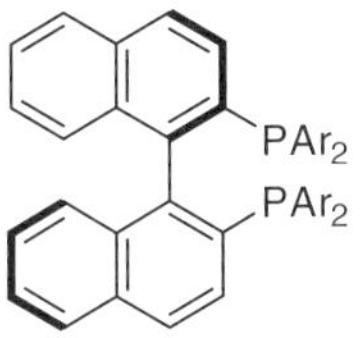

R-TolBINAP, R=4-MeC$_6$H$_4$     S,S-DPEN     R-DM-DABN

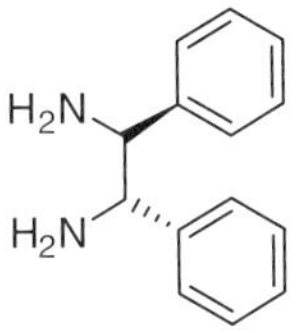

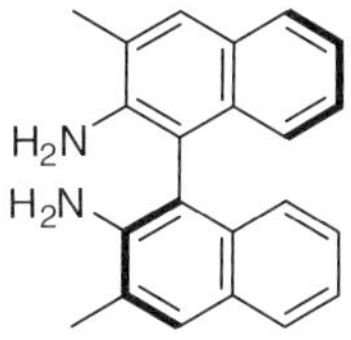

**25**        **26**        **27**

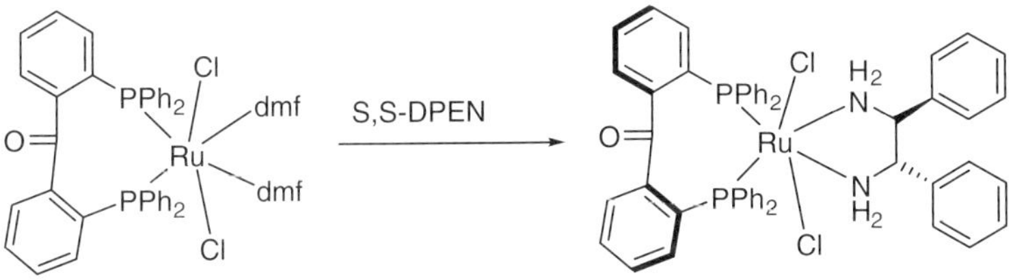

**Scheme 7.17**   Formation of one atropoisomer of diphosphine via complexation of *S,S*-DPEN.

Instead of a non-enantiopure diphosphine one can also use a non-chiral diphosphine that assumes a chiral if racemic structure upon complex formation, and control the diasteromeric excess by the chiral diamine added [106]. To this end Mikami used 2,2'-bisdiphenylphosphinobenzophenone, RuCl$_2$ and (1*S*,2*S*)-(−)-1,2-diphenylethylenediamine (*S,S*)-DPEN) as the chiral activator (Scheme 7.17). DPBP is a so-called tropos ligand, similar to 2,2'-bisdiphenylphosphinobiphenyl, to be discussed below. Only one diastereoisomer was formed (containing one preferred atropoisomer of DPBP) and ees as high as 99% were achieved with this non-chiral diphosphine. A comparative study showed that this system for all substrates tested was more selective than BINAP! Hydrogenation of the ketone group of the ligand was negligible.

Leitner and coworkers studied racemic BINAP and a chiral proline-based ionic liquid (cIL) in the asymmetric hydrogenation of dimethyl itaconate to see whether the chiral cIL would influence the outcome [107]. The reaction led to identical ee values as enantiopure BINAP. The enantio-differentiation results primarily from a diastereomeric interaction of the BINAP-Rh complex and the proline ester moiety. Thus, in this case, one enantiomeric rhodium complex is selectively deactivated.

When tropos ligands are used, such as BIPHEP, the outcome is more complicated, but surprisingly moderate ees could still be obtained [108]. Free BIPHEP will rapidly racemize from one to the other enantiomeric conformer, but in a complex the ring system formed with rhodium prohibits such a fast exchange (Scheme 7.18). If we compare the system with the example of rac-BINAP, the cIL forms preferentially a complex with one of the enantiomeric forms of the complex and the remaining enantiomer can act as an enantioselective catalyst. It resembles the case of Scheme 7.17, apart from the fact that now the adduct is the inactive species. In the

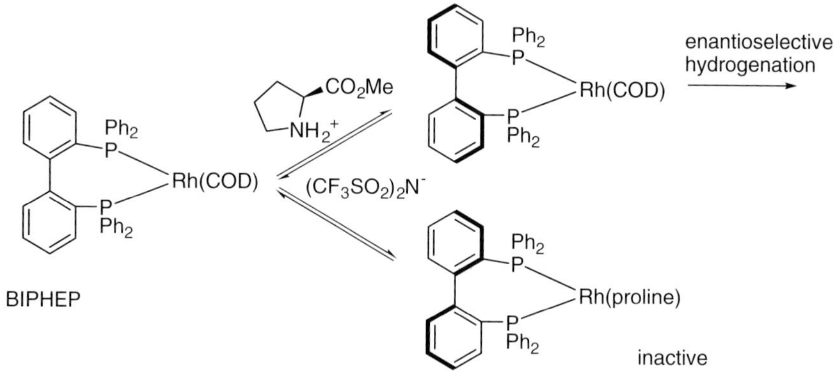

**Scheme 7.18** Deactivation of one atropoisomeric complex by a chiral ionic liquid component.

longer term, though, all complex may convert to the enantiomer that forms the most stable adduct with the solvent and catalysis will come to a halt.

## 7.9
## Conclusions

One of the hydrogenation reactions studied in most detail as regards inhibition or deactivation is the rhodium(I)-catalyzed hydrogenation using Rh(cod) or Rh(nbd) complexes as precursors. In particular, the work by Heller and coworkers has shown that, under mild conditions, the norbornadiene precursors show a much faster onset of the reaction. At higher temperatures and pressures the difference in incubation times may vanish, but it would be good practice to use norbornadiene precursors. A plausible explanation is that the more bulky cyclooctadiene is more slowly hydrogenated, because the oxidative addition of $H_2$ requires rotation of the diene, which is slower for the more bulky diene.

Inhibition by impurities or polar solvents seems to be rare, as the alkene substrate is a relatively strongly bound ligand for (diphosphine)Rh complexes. In the absence of substrate many groups will coordinate to cationic Rh(I), polar ones, and arenes, and so on. More strongly bound unsaturated impurities such as enones and alkynes might inhibit the hydrogenation of alkenes, but under laboratory conditions of low substrate to metal ratios this normally does not show up. The same holds for iridium and ruthenium.

Known substrates that inhibit catalysis are carboxylic acids, giving oxidative addition to rhodium(I), and bifunctional substrates that may form stable salts with ruthenium.

Since hydrogenation requires the creation of a free coordination site on the metal, the resulting MX species may dimerize to give $M-\mu X_2-M$ species rather than the desired alkene adducts. When X is a halide the dimers may become suitable resting states and precursors. If they are hydrides, metal hydrides, or hydroxy groups then dimer or trimer formation may retard the reaction considerably. Especially for iridium hydrides dimer and trimer formation has often been found to be irreversible, but also good solutions have been reported. In the presence of substrate the reaction is much less pronounced.

Product inhibition is not so common; the alkane product, also when it carries more functional groups, is a less strongly bound ligand than its alkene precursor. In transfer hydrogenation there are a few examples for which it is not known whether the substrate or the product is the inhibitor.

Ligand decomposition via P—C cleavage during catalysis seems to be an uncommon reaction for the typical hydrogenation metals Rh, Ru, and Ir, unless it has remained unnoticed. For Pt and Pd more reports can be found. Pregosin reported a fluorine–carbon exchange in BINAP Ru complexes, which surprisingly turned out to be a facile reaction occurring at low temperatures, the source of the fluorine nucleophile being tetrafluoroborate.

Forcing conditions under a hydrogen atmosphere may lead easily to metal formation, be it nanoparticles, bulk metal, or a film on the reactor wall. It has been realized that the determination of the nature of the true catalysts in such systems requires an extensive study and cannot be decided on the basis of one simple experiment. A thorough approach to solve this question has been developed by Finke and coworkers and it was used, *inter alia*, to solve the longstanding questions on arene hydrogenation by Cp\*-Rh (Maitlis) and Cp\*-Ru (Süss-Fink) complexes, which both were proven, in joint efforts, to contain heterogeneous species causing the catalytic hydrogenation.

Selective activation or deactivation of catalyst complexes by chiral, enantiopure additives can be used successfully to enhance the enantioselective performance of racemic or non-enantiopure catalysts, or even achiral catalysts possessing tropos type properties, as was shown by Faller and Mikami. The action differs for each metal; for Rh the active species is usually a diphos-Rh species, and the diphos-diamine-Rh complex may take away one of the enantiomers to make the way clear for the remaining enantiomer for catalysis. For Ru the active species involves $RuN_2P_2$ coordination, and the deactivated one may be either $RuP_2$, or a $RuN_2P$ species with a different amine, that is not active. Clearly, optimal use of metal and ligand is obtained when enantiopure ligands are used, however neat the tricks might be. An achiral tropos ligand, used with a chiral amine only, is a winner, of course.

## References

1 Studer, M., Blaser, H.-U., and Exner, C. (2003) *Adv. Synth. Catal.*, **345**, 45–65.

2 de Vries, J.G. and Elsevier, C.J. (eds) (2006) *The Handbook of Homogeneous Hydrogenation*, Wiley-VCH Verlag GmbH, Weinheim.

3 Heller, D., De Vries, A.H.M., and de Vries, J.G. (2006) in *The Handbook of Homogeneous Hydrogenation*, J.G. de Vries, and C.J. Elsevier, (eds), Wiley-VCH Verlag GmbH, Weinheim, Ch. 11, pp. 1481–1514.

4 Greiner, L. and Ternbach, M.T. (2004) *Adv. Synth. Catal.*, **346**, 1392–1396.

5 Heller, D., Thede, R., and Haberland, D. (1997) *J. Mol. Catal. A: Chem.*, **115**, 273–281.

6 Blackmond, D.G. (2000) *Acc. Chem. Res.*, **33**, 402–411.

7 (a) Heller, D., Kortus, K., and Selke, R. (1995) *Liebigs Ann.*, **3**, 575–581; (b) Heller, D., Borns, S., Baumann, W., and Selke, R. (1996) *Chem. Ber.*, **129**, 85–89.

8 We are indebted to Dr. B. de Bruin (University of Amsterdam) for pointing out Reference 9.

9 Mestroni, G., Zassinovich, G., and Camus, A. (1977) *J. Organomet. Chem.*, **140**, 63–72.

10 Heldal, J.A. and Frankel, E.N. (1985) *J. Am. Oil Chem. Soc.*, **62**, 1117–1120.

11 (a) Heller, D., Kortus, K., and Selke, R. (1995) *Liebigs Ann.*, 575–581; (b) Heller, D., Borns, S., Baumann, W., and Selke, R. (1996) *Chem. Ber.*, **129**, 85–89; (c) Baumann, W., Mansel, S., Heller, D., and Borns, S. (1997) *Magn. Reson. Chem.*, **35**, 701; (d) Drexler, H.-J., Baumann, W., Spannenberg, A., Fischer, C., and Heller, D. (2001) *J. Organomet. Chem.*, **621**, 89–102; (e) Börner, A. and Heller, D. (2001) *Tetrahedron Lett.*, **42**, 223–225; (f) Heller, D., Drexler, H.-J., You, J., Baumann, W., Drauz, K., Krimmer, H.-P., and Börner, A. (2002) *Chem. Eur. J.*, **8**, 5196; (g) Drexler, H.-J., You, J., Zhang, S., Fischer, C., Baumann, W., Spannenberg, A., and Heller, D. (2003) *Org. Proc. Res. Dev.*, **7**, 355; (h) Braun, W., Salzer, A., Drexler, H.-J., Spannenberg, A., and Heller, D. (2003) *Dalton Trans.*, 1606; (i) Preetz, A., Drexler, H.-J., Fischer, C., Dai, Z., Boerner, A., Baumann, W., Spannenberg, A., Thede, R., and Heller, D. (2008) *Chem. Eur. J.*, **14**, 1445–1451.

12 Cobley, C.J., Lennon, I.C., McCague, R., Ramsden, J.A., and Zanotti-Gerosa, A. (2001) *Tetrahedron Lett.*, **42**, 7481–7483.

13 Cobley, C.J., Lennon, I.C., McCague, R., Ramsden, J.A., and Zanotti-Gerosa, A. (2003) *Chemical Industries (Dekker)*, **89** (Catalysis of Organic Reactions), 329–339.

14 Esteruelas, M.A., Herrero, J., Martin, M., Oro, M.L.A., and Real, V.M. (2000) *J. Organomet. Chem.*, **599**, 178–184.

15 Tararov, V.I., Kadyrov, R., Riermeier, T.H., and Borner, A. (2002) *Adv. Synth. Catal.*, **344**, 200–208.

16 Cui, X. and Burgess, K. (2003) *J. Am. Chem. Soc.*, **125**, 14212–14213.

17 Lightfoot, A., Schnider, P., and Pfaltz, A. (1998) *Angew. Chem., Int. Ed.*, **37**, 2897–2899.

18 Crabtree, R. (1979) *Acc. Chem. Res.*, **12**, 331–337.

19 Perry, M.C., Cui, X., Powell, M.T., Hou, D.-R., Reibenspies, J.H., and Burgess, K. (2003) *J. Am. Chem. Soc.*, **125**, 113–123.

20 Schmidt, T., Drexler, H.-J., Sun, J., Dai, Z., Baumann, W., Preetz, A., and Heller, D. (2009) *Adv. Synth. Catal.*, **351**, 750–754.

21 Brown, J.M. and Parker, D. (1982) *J. Org. Chem.*, **47**, 2722–2730.

22 Takahashi, H. and Achiwa, K. (1987) *Chem. Lett.*, 1921–1922.

23 Wadkar, J.G. and Chaudhari, R.V. (1983) *J. Mol. Catal.*, **22**, 103–116.

24 Kollar, L., Toros, S., Heil, B., and Marko, L. (1980) *J. Organometal. Chem.*, **192**, 253–256.

25 Landis, C.R. and Halpern, J. (1987) *J. Am. Chem. Soc.*, **109**, 1746–1754.

26 Heller, D., Drexler, H.-J., Spannenberg, A., Heller, B., You, J., and Baumann, W. (2002) *Angew. Chem. Int. Ed.*, **41**, 777–780.

27 Fu, C.C. and McCoy, B.J. (1988) *Ind. Eng. Chem. Res.*, **27**, 233–237.

28 Giernoth, R., Huebler, P., and Bargon, J. (1998) *Angew. Chem., Int. Ed.*, **37**, 2473–2475.

29 Heller, D., Holz, J., Drexler, H.-J., Lang, J., Drauz, K., Krimmer, H.-P., and Boerner, A. (2001) *J. Org. Chem.*, **66**, 6816–6817.

30 (a) Chan, A.S.C., Pluth, J.J., and Halpern, J. (1980) *J. Am. Chem. Soc.*, **102**, 5952; (b) Brown, J.M. and Chaloner, P.A. (1980) *J. Chem. Soc., Chem.Commun.*, 344.

31 Cobley, C.J., Lennon, I.C., Praquin, C., Zanotti-Gerosa, A., Appell, R.B., Goralski, C.T., and Sutterer, A.C. (2003) *Org. Proc. Res. Dev.*, **7**, 407–411.

32 Miyashita, A., Yasuda, A., Takaya, H., Toriumi, K., Ito, T., Souchi, T., and Noyori, R.J. (1980) *J. Am. Chem. Soc.*, **102**, 7932.

33 Tani, K., Yamagata, Y., Tatsuno, Y., Yamagata, Y., Tomita, K., Agutagawa, S.,

Kumobayashi, H., and Otsuka, S. (1985) *Angew. Chem. Int. Ed. Engl.*, **24**, 217.

34 Akutagawa, S. (1992) Chapter 16, in *Chirality in Industry* (eds A.N. Collins, G.N. Sheldrake, and J. Crosby), John Wiley & Sons, Inc., New York.

35 Clark, G.S. (1998) *Menthol. Perfumer Flavorist*, **23**, 33.

36 Jiang, Q., Jiang, Y., Xiao, D., Cao, P., and Zhang, X. (1998) *Angew. Chem., Int. Ed*, **37**, 1100–1103.

37 Sun, X., Manos, G., Blacker, J., Martin, J., and Gavriilidis, A. (2004) *Org. Proc. Res. Dev.*, **8**, 909–914.

38 Gladiali, S., Pinna, L., Delogu, G., Martin, S., Zassinovich, G., and Mestroni, G. (1990) *Tetrahedron: Asymmetry*, **1**, 635–648.

39 Patil, S.R., Sen, D.N., and Chaudhari, R.V. (1983) *J. Mol. Catal.*, **19**, 233–241.

40 (a) Bolton, P.D., Grellier, M., Vautravers, N., Vendier, L., and Sabo-Etienne, S. (2008) *Organometallics*, **27**, 5088–5509. (b) Chaudret, B., Cole-Hamilton, D.J., Nohr, R.S., and Wilkinson, G. (1977) *J. Chem. Soc., Dalton Trans.*, 1546–1557; (c) Ikariya, T. and Blacker, A.J. (2007) *Acc. Chem. Res.*, **40**, 1300–1308.

41 Jung, C.W. and Garrou, P.E. (1982) *Organometallics*, **1**, 658–666.

42 Schleyer, D., Niessen, H.G., and Bargon, J. (2001) *New J. Chem.*, **25**, 423–426.

43 van Laren, M.W. and Elsevier, C.J. (1999) *Angew. Chem., Int. Ed.*, **38**, 3715–3717.

44 Beach, N.J., Blacquiere, J.M., Drouin, S.D., and Fogg, D.E. (2009) *Organometallics*, **28**, 441–447.

45 Yi, C.S., Lee, D.W., He, Z., Rheingold, A.L., Lam, K.-C., and Concolino, T.E. (2000) *Organometallics*, **19**, 2909–2915.

46 Sanchez-Delgado, R.A., Valencia, N., Marquez-Silva, R.-L., Andriollo, A., and Medina, M. (1986) *Inorg. Chem.*, **25**, 1106–1111.

47 (a) Everaere, K., Mortreux, A., Bulliard, M., Brussee, J., van der Gen, A., Nowogrocki, G., and Carpentier, J.-F. (2001) *Eur. J. Org. Chem.*, 275–291; (b) Everaere, K., Mortreux, A., and Carpentier, J.-F. (2003) *Adv. Synt. Catal.*, **345**, 67–77.

48 (a) Noyori, R. and Hashiguchi, S. (1997) *Acc. Chem. Res.*, **30**, 97–102; (b) Petra, D.G.I., Reek, J.N.H., Handgraaf, J.-W., Meijer, E.J., Dierkes, P., Kamer, P.C.J., Brussee, J., Schoemaker, H.E., and van Leeuwen, P.W.M.N. (2000) *Chem. Eur. J.*, **6**, 2818–2829.

49 (a) Zassinovich, G., Mestroni, G., and Gladiali, S. (1992) *Chem. Rev.*, **92**, 1051–1069; (b) Gladiali, S. and Alberico, E. (2006) *Chem. Soc. Rev.*, **35**, 226–236.

50 (a) Liu, P.N., Deng, J.G., Tu, Y.Q., and Wang, S.H. (2004) *Chem. Commun.*, 2070–2071; (b) Himeda, Y., Onozawa-Komatsuzaki, N., Sugihara, H., Arakawa, H., and Kasuga, K. (2003) *J. Mol. Catal. A*, **195**, 95–100; (c) Li, X.G., Wu, X.F., Chen, W.P., Hancock, F.E., King, F., and Xiao, J. (2004) *Org. Lett.*, **6**, 3321–3324.

51 Wu, X., Li, X., King, F., and Xiao, J. (2005) *Angew. Chem. Int. Ed.*, **44**, 3407–3411.

52 (a) Noyori, R., Yamakawa, M., and Hashiguchi, S. (2001) *J. Org. Chem.*, **66**, 7931–7944; (b) Alonso, D.A., Brandt, P., Nordin, S.J.M., and Andersson, P.G. (1999) *J. Am. Chem. Soc.*, **121**, 9580–9588; (c) Petra, D.G.I., Reek, J.N.H., Handgraaf, J.W., Meijer, E.J., Dierkers, P., Kamer, P.C.J., Brussee, J., Schoemaker, H.E., and van Leeuwen, P.W.N.M. (2000) *Chem. Eur. J.*, **6**, 2818–2829.

53 Gao, J.-X., Ikariya, T., and Noyori, R. (1996) *Organometallics*, **15**, 1087–1089.

54 Laue, S., Greiner, L., Woltinger, J., and Liese, A. (2001) *Adv. Synth. Catal.*, **343**, 711–720.

55 Sandee, A.J., Petra, D.G.I., Reek, J.N.H., Kamer, P.C.J., and van Leeuwen, P.W.N.M. (2001) *Chem. Eur. J.*, **7**, 1202–1208.

56 (a) Bonds, W.D., Brubaker, C.H., Chandrasekaran, E.S., Gibsons, C., Grubbs, R.H., and Kroll, L.C. (1975) *J. Am. Chem. Soc.*, **97**, 2128; (b) Grubbs, R.H., Gibbons, C., Kroll, L.C., Bonds, W.D., and Brubaker, C.H. (1973) *J. Am. Chem. Soc.*, **95**, 2373–2375.

57 (a) Crabtree, R.H., Felkin, H., and Morris, G.E. (1977) *J. Organomet. Chem.*, **141**, 205–215; (b)Crabtree, R.H., Chodosh, D.F., Quirk, J.M., Felkin, H., Khan-Fillebeen, T., and Morris, G.E. (1979) *Fundamental Research in Homogeneous Catalysis*, vol. 3 (ed. M. Tsutsui), Plenum Press, New York, pp. 475–485.

**58** Crocker, M. and Herold, R.H.M. (1993) *Catal. Lett.*, **18**, 243–251.

**59** Spindler, F., Pugin, B., Jalett, H.-P., Buser, H.-P., Pittelkow, U., and Blaser, H.-U. (1996) *Chemical Industries (Dekker)*, **68** (Catalysis of Organic Reactions), 153–166.

**60** Blaser, H.-U., Pugin, B., Spindler, F., and Togni, A. (2002) *Compt. Rend. Chim.*, **5**, 379–385.

**61** Albinati, A., Togni, A., and Venanzi, L.M. (1986) *Organometallics*, **5**, 1785.

**62** Pugin, B., Landert, H., Spindler, F., and Blaser, H.-U. (2002) *Adv. Synth. Catal.*, **344**, 974–979.

**63** Dervisi, A., Carcedo, C., and Ooi, L. (2006) *Adv. Synth. Catal.*, **348**, 175–183.

**64** Schnider, P., Koch, G., Prétôt, R., Wang, G., Bohnen, F.M., Krüger, C., and Pfaltz, A. (1997) *Chem. Eur. J.*, **3**, 887–892.

**65** Wustenberg, B. and Pfaltz, A. (2008) *Adv. Synth. Catal.*, **350**, 174–178.

**66** Smidt, S.P., Zimmermann, N., Studer, M., and Pfaltz, A. (2004) *Chem. Eur. J.*, **10**, 4685–4693.

**67** Jardine, F.H., Osborn, J.A., Wilkinson, G., and Young, J.F. (1965) *Chem. Ind. (London)*, 560;(1966) *J. Chem. Soc. (A)*, 1711.

**68** Czakova, M. and Capka, M. (1981) *J. Mol. Catal.*, **11**, 313–322.

**69** Duckett, S.B., Newell, C.L., and Eisenberg, R. (1994) *J. Am. Chem. Soc.*, **116**, 10548–10556.

**70** Colebrooke, S.A., Duckett, S.B., Lohman, J.A.B., and Eisenberg, R. (2004) *Chem. Eur. J.*, **10**, 2459–2474.

**71** Rifat, A., Patmore, N.J., Mahon, M.F., and Weller, A.S. (2002) *Organometallics*, **21**, 2856–2865.

**72** Preetz, A., Baumann, W., Drexler, H.-J., Fischer, C., Sun, J., Spannenberg, A., Zimmer, O., Hell, W., and Heller, D. (2008) *Chem. Asian J.*, **3**, 1979–1982.

**73** Tani, K., Yamagata, Y., Tatsuno, Y., Yamagata, Y., Tomita, K., Agutagawa, S., Kumobayashi, H., and Otsuka, S. (1985) *Angew. Chem. Int. Ed. Engl.*, **24**, 217.

**74** Yamagata, T., Tani, K., Tatsuno, Y., and Saito, T. (1988) *J. Chem. Soc., Chem. Commun.*, 466–468.

**75** Gridnev, I.D., Fan, C., and Pringle, P.G. (2007) *Chem. Commun.*, 1319–1321.

**76** Faller, J.W., Mazzieri, M.R., Nguyen, J.T., Parr, J., and Tokunaga, M. (1994) *Pure Appl. Chem.*, **66**, 1463.

**77** Nindakova, L.O. and Shainyan, B.A. (2001) *Russ. Chem. Bul.*, **50**, 1855–1859.

**78** Noyori, R. and Takaya, H. (1990) *Acc. Chem. Res.*, **23**, 345–350.

**79** den Reijer, C.J., Ruegger, H., and Pregosin, P.S. (1998) *Organometallics*, **17**, 5213–5215.

**80** (a) Geldbach, T.J. and Pregosin, P.S. (2002) *Eur. J. Inorg. Chem.*, 1907; (b) Geldbach, T.J., Pregosin, P.S., and Albinati, A. (2003) *Organometallics*, **22**, 1443; (c) den Reijer, C.J., Dotta, P., Pregosin, P.S., and Albinati, A. (2001) *Can. J. Chem.*, **79**, 693; (d) Geldbach, T.J. and Pregosin, P.S. (2001) *Organometallics*, **20**, 2990–2997.

**81** Marshall, W.J. and Grushin, V.V. (2003) *Organometallics*, **22**, 555–562.

**82** Feiken, N., Pregosin, P.S., and Trabesinger, G. (1997) *Organometallics*, **16**, 3735–3736.

**83** Marcazzan, P., Patrick, B.O., and James, B.R. (2003) *Organometallics*, **22**, 1177–1179.

**84** Peach, P., Cross, D.J., Kenny, J.A., Mann, I., Houson, I., Campbell, L., Walsgrove, T., and Wills, M. (2006) *Tetrahedron*, **62**, 1864–1876.

**85** Kenny, J.A., Palmer, M.J., Smith, A.R.C., Walsgrove, T., and Wills, M. (1999) *Synlett.*, 1615–1617.

**86** Hage, A., Petra, D.G.I., Field, J.A., Schipper, D., Wijnberg, J.B.P.A., Kamer, P.C.J., Reek, J.N.H., van Leeuwen, P.W.N.M., Wever, R., and Schoemaker, H.E. (2001) *Tetrahedron: Asym.*, **12**, 1025–1034.

**87** Klabunovskii, E., Smith, G.V., and Zsigmond, A. (2006) Heterogeneous enantioselective hydrogenation, theory and practice, in *Catalysis by Metal Complexes*, vol. 31 (eds B.R. James and P.W.N.M. van Leeuwen), Springer, Dordrecht, the Netherlands.

**88** Studer, M., Blaser, H.-U., and Exner, C. (2003) *Adv. Synth. Catal.*, **345**, 45–65.

**89** Widegren, J.A. and Finke, R.G. (2003) *J. Mol. Catal. A: Chem.*, **198**, 317–341.

**90** Rothwell, I.P. (1997) *Chem. Commun.*, 1331.

91 Widegren, J.A., Bennett, M.A., and Finke, R.G. (2003) *J. Am. Chem. Soc.*, **125**, 10301–10310.

92 (a) Watzky, M.A. and Finke, R.G. (1997) *J. Am. Chem. Soc.*, **119**, 10382; (b) Widegren, J.A., Aiken, J.D. III, Ozkar, S., and Finke, R.G. (2001) *Chem. Mater.*, **13**, 312.

93 Suss-Fink, G., Faure, M., and Ward, T.R. (2002) *Angew. Chem., Int. Ed.*, **41**, 99.

94 Hagen, C.M., Vieille-Petit, L., Laurenczy, G., Süss-Fink, G., and Finke, R.G. (2005) *Organometallics*, **24**, 1819–1831.

95 Finney, E.E. and Finke, R.G. (2006) *Inorg. Chim. Acta*, **359**, 2879–2887.

96 Hamlin, J.E., Hirai, K., Millan, A., and Maitlis, P.M. (1980) *J. Mol. Catal.*, **7**, 543.

97 Hagen, C.M., Widegren, J.A., Maitlis, P.M., and Finke, R.G. (2005) *J. Am. Chem. Soc.*, **127**, 4423–4432.

98 Collman, J.P., Kosydar, K.M., Bressan, M., Lamanna, W., and Garrett, T. (1984) *J. Am. Chem. Soc.*, **106**, 2569.

99 Gill, D.S., White, C., and Maitlis, P.M. (1978) *J. Chem. Soc., Dalton Trans.*, 617.

100 (a) Jaska, C.A. and Manners, I. (2004) *J. Am. Chem. Soc.*, **126**, 1334; (b) Jaska, C.A. and Manners, I. (2004) *J. Am Chem. Soc.*, **126**, 9776.

101 Faller, J.W., Lavoie, A.R., and Parr, J. (2003) *Chem. Rev.*, **103**, 3345–3367.

102 Blackmond, D.G. (2000) *Acc. Chem. Res.*, **33**, 402–411.

103 Alcock, N.W., Brown, J.M., and Maddox, P.J. (1986) *J. Chem. Soc., Chem. Commun.*, 1532.

104 Mikami, K., Korenaga, T., Ohkuma, T., and Noyori, R. (2000) *Angew. Chem., Int. Ed.*, **39**, 3707–3710.

105 Mikami, K., Korenaga, T., Yusa, Y., and Yamanaka, M. (2003) *Adv. Synt. Catal.*, **345**, 246–254.

106 Mikami, K., Wakabayashi, K., and Aikawa, K. (2006) *Org. Lett.*, **8**, 1517–1519.

107 Chen, D., Schmitkamp, M., Francio, G., Klankermayer, J., and Leitner, W. (2008) *Angew. Chem., Int. Ed.*, **47**, 7339–7341.

108 Schmitkamp, M., Chen, D., Leitner, W., Klankermayer, J., and Francio, G. (2007) *Chem. Commun.*, 4012–4014.

# 8
# Carbonylation Reactions

## 8.1
## Introduction

The development of carbonylation reactions took off in the late 1930s and early 1940s with the inventions of Roelen and Reppe who both focused on the activity of first row transition metals such as cobalt for hydroformylation and methanol carbonylation, and nickel for alkene carbonylations [1]. The initial achievements in terms of activity and selectivity were rather modest by today's standards, but at the time they were revolutionary. Their applications were retarded by the second world war, but soon thereafter both cobalt hydroformylation and methanol carbonylation were applied industrially. For these ligand-free systems the obvious decomposition reaction was metal deposition and the way to avoid that was to apply higher CO pressures, lower temperatures, or, in the case of certain applications of nickel, to increase the amount of acid that could regenerate Ni(II) via oxidative addition. As the catalytic cycle starts with the loss of one coordinated CO ligand to make room for a substrate, an alkene in this instance, the processes are operated at conditions where metal precipitation may readily occur, because a minimal CO pressure gives the highest rates. Phosphorus ligands may be used to prevent metal cluster formation, but unsaturated metal complexes may also cause breakdown of the ligands.

The copolymerization of ethene and carbon monoxide to give polyketones was also discovered by Reppe [2] and related to this are hydroxy- and methoxycarbonylation of alkenes. In this chapter these three major carbonylation reactions will be discussed: hydroformylation (Sections 8.2 and 8.3), alkene/CO reactions (Section 8.4), and carbonylation (Section 8.5). For all reactions the introduction of second row transition metals (Rh and Pd) led to major breakthroughs in the rate and selectivity of the reaction. For hydroformylation and alkene/CO reactions the introduction of phosphorus ligands gave enormous improvements in rate and selectivity, including stereoselectivity, while methanol carbonylation can be accelerated by phosphine ligands facilitating the oxidative addition of methyl iodide to Rh(I), but this is only of a temporary nature as the phosphines eventually will end

*Homogeneous Catalysts: Activity – Stability – Deactivation*, First Edition. Piet W.N.M. van Leeuwen and John C. Chadwick.
© 2011 Wiley-VCH Verlag GmbH & Co. KGaA. Published 2011 by Wiley-VCH Verlag GmbH & Co. KGaA.

up as their oxides or phosphonium salts and thus one loses the enhanced rates. Cobalt phosphine-modified hydroformylation catalysts are more selective, but slower than the unmodified catalysts.

For the older reactions, such as hydroformylation and carbonylation, not all that many explicit studies on catalyst decomposition have been published, while for recent developments such as homogeneous coordination polymerization and metathesis many more studies have appeared. Cross-coupling, also an area advanced by major developments in the last decade, is an exception to this as one will see in Chapter 9, as here also the explicit attention to deactivation is relatively low given the huge amount of contributions. As yet the most popular approach to find better catalysts in cross-coupling catalysis is to screen more ligands, bases, and conditions, rather than concentrate on decomposition pathways.

In hydroformylation and carbonylation the issues of selectivity and turnover have been solved, it would seem, without the need to go into much detail on the reasons why certain catalysts did not perform well. One part of the explanation may be that the area is relatively old and several industrial applications use unmodified catalysts (i.e., no phosphines are used). Secondly, common ligands are triphenylphosphine or its sulfonated congener tppts, which are very cheap and, in spite of improved performance of more modern ligands that give a better performance at higher ligand costs, their replacement has a high barrier. Thirdly, carbonylation reactions require high pressures and suitable equipment is not at hand in most laboratories and in fine chemical production units. Thus carbonylation steps are not a routine step in organic synthesis and their use is restricted to large scale products, with a few exceptions such as the synthesis of Ibuprofen in its racemic form via the Hoechst route. Therefore the number of academic studies on carbonylation catalyst decomposition is low. No doubt study of catalyst stability has been conducted in industry, but publications are scarce.

## 8.2
### Cobalt-Catalyzed Hydroformylation

The first cobalt-catalyzed hydroformylation processes used unmodified hydrido-cobalt carbonyls as the catalyst. It is well-known that the hydroformylation of alkenes was discovered accidentally by Roelen in the late 1930s while he was studying the Fischer–Tropsch reaction with a heterogeneous cobalt catalyst. Roelen studied whether alkenes were intermediates in the "Aufbau" process of syn-gas (from coal, Germany 1938) to fuel by recycling the ethene formed [3]. He found that alkenes were converted to aldehydes or alcohols containing one more carbon atom (115 °C, >100 bar). It took almost a decade (1945) before the reaction was taken further, but now it was the conversion of petrochemical hydrocarbons, being the key feedstock since the 1950s, into oxygenates. It was discovered that the reaction was not catalyzed by the supported cobalt but by $HCo(CO)4$ which was formed in the liquid state at sufficiently high pressure. The chemistry of the cobalt catalysts was recently reviewed [4]. The catalyst can be recovered by thermal decomposition to metal and filtered off, or, more commonly, the catalyst can be oxidized and separated via an extraction with water, or the catalyst can be recycled without decomposition, as in the Shell and Kuhlmann processes [5].

The oxidation process developed by BASF is still being applied and improved [6]. The effluent of the hydroformylation reactor, containing catalyst, aldehyde and alkene, is treated with water, acid, and oxygen and the cobalt salts are removed in the water layer by phase separation. After concentration they are reduced by syn-gas and the hydrocarbon-soluble $HCo(CO)_4$ formed is extracted with the alkene [7] into the feed stream to the reactor.

A more elegant way of recycling the catalyst is that of the Produits Chimiques Ugine Kuhlmann process (now Exxon). The process is still in use, and being improved [8]. In this process the hydroformylation is done in one organic phase consisting of alkene and aldehyde. The reactor is often a loop reactor or a reactor with an external loop to facilitate heat transfer. A liquid/liquid separation of product and catalyst is carried out in separate vessels after the reaction has taken place. The reaction mixture is sent to a gas separator and from there to a counter current washing tower in which the effluent is treated with aqueous $Na_2CO_3$. The acidic $HCo(CO)_4$ is transformed into the water-soluble conjugate base $NaCo(CO)_4$. The product is scrubbed with water to remove traces of base. The oxo-crude goes to the distillation unit. $Co_2(CO)_8$ is not extracted this way, unless it disproportionates under the extraction conditions into $Co^{2+}$ salts and tetracarbonylcobaltate anions.

The basic solution in water containing $NaCo(CO)_4$ is treated with sulfuric acid in the presence of syn-gas and $HCo(CO)_4$ is regenerated. This can be extracted from water into the substrate alkene. The catalyst is returned to the reactor dissolved in the alkene. Although the salts are produced in "catalytic" amounts, today one would prefer to avoid salt production even at this level. As we will see later for rhodium catalysts better solutions have been found.

Another catalyst being recycled without decomposition is Shell's ligand-modified cobalt catalyst. Cobalt-catalyzed hydroformylation was one of the first industrial processes to make use of phosphine ligands to obtain a more stable and selective catalyst for the hydroformylation of detergent range ($C_{11-12}$) alkenes [9], the SHF process (Shell hydroformylation process, 1970). The SHF process leads to more hydrogenation, both of the substrate (to alkane, up to 10%, a real loss) and the product (to the alcohol, the desired product anyway). The ligand used is a trialkylphosphine, which for years was described in the literature as $Bu_3P$, but most likely this would have been too volatile; nowadays it is accepted that the ligand is a phobane (a mixture of 1,5- and 1,4-cyclooctanediyl phosphine) derivative carrying a long $C_{20}$ alkyl chain [10] (see Figure 8.1) The ring structure renders the phosphine less basic than an acyclic phosphine, but still the basic properties retard CO dissociation and higher temperatures (180 vs. 140 °C) are needed for the Shell catalyst than for the unmodified one, while lower pressures can be applied (80 vs. 200–300 bar). The Shell catalyst is much slower but also more stable than the unmodified one; therefore the SHF catalyst can be recycled after separation by distillation [11].

Unconverted alkene, aldehyde, and alcohol are distilled off under reduced pressure in a wiped-film evaporator and the bottom of the distillation vessel is recycled to

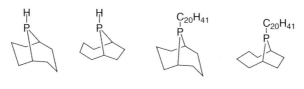

phobane mixture

**Figure 8.1** Phobane and Shell's ligand mixture used in hydroformylation with cobalt.

the reactor with a bleed stream to remove heavy ends (aldol condensation products) continuously. The bleed stream is sent to an incinerator. There are few reports, in general, on P—C carbon bond cleavage as a mechanism for phosphine decomposition and perhaps this is not a route for catalyst decomposition in the present system. Alkylphosphines are prone to oxidation and any oxygenate (water, aldehyde, carbon dioxide) can oxidize an alkylphosphine to phosphine oxide, as this reaction will be thermodynamically allowed and cobalt might function as the catalyst. Ingress of molecular oxygen is a more likely cause of oxidation. In the acidic medium of palladium carbonylation catalysis, quaternization of alkylphosphine ligands may take place, but under the circumstances of cobalt hydroformylation this is not expected.

Sasol developed a set of ligands similar to phobane, the Lim ligands [12]. These ligands are based on the addition of $PH_3$ to limonene (the R-enantiomer). A mixture of two diastereoisomeric compounds is obtained due to the two configurations of the methyl group at the C-4 position (Scheme 8.1). The Lim-H compounds obtained can be derivatized at the phosphorus atom via radical additions of alkenes or substitution reactions of their conjugate bases formed with electrophiles.

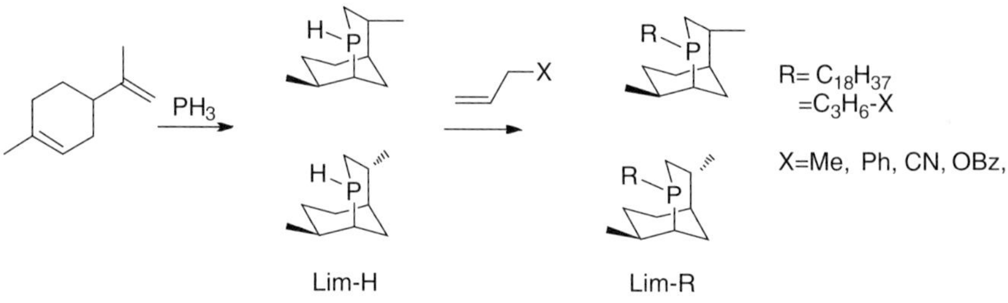

R= $C_{18}H_{37}$
=$C_3H_6$-X

X=Me, Ph, CN, OBz,

Lim-H          Lim-R

**Scheme 8.1** Sasol's Lim ligands.

The Shell phobane derivatives (Figure 8.1) contain two secondary hydrocarbyl groups and are therefore relatively bulky, comparable to $(i\text{-}Pr)_2P\text{-}n$-alkyl ligands. The Lim ligands contain two primary alkyl groups, but the branching at the other positions may compensate for this and they may still behave as bulky ligands. Electronically, the alkyl groups are strong donors, but it is expected that the steric strain caused by the ring systems will raise the $\chi$-value of the ligands (i.e., make them better $\pi$-acceptors). The differences observed in oxidation and quaternization of the two phobane isomers and the two Lim diastereoisomers show that the different ring structures lead to slightly different acid–base properties for each pair.

Small electronic effects do have an effect on the l:b ratio, but as the authors remarked, this is not an intrinsic ligand effect on the reaction; it is a measure of the amount of phosphine-free catalyst **5** that is present in the equilibrium. Thus, the weaker donor ligands bind less strongly to cobalt, give more phosphine-free catalyst, and thus produce more branched aldehyde at a higher rate. At lower temperatures the phobane ligands form slightly more of the phosphine hydride complex than the Lim ligands, but at 170 °C the equilibria between dimers, hydrides, and phosphine-free complexes are the same for both ligands under identical conditions [12].

Iron carbonyls are a potential contaminant in hydroformylation reactors under these conditions (70 bar, 170 °C) and the effects thereof were also studied by Dwyer and coworkers using high-pressure $^{31}$P NMR spectroscopy [12]. Thermodynamically, $Fe(CO)_4$(Lim) complexes were found to be more stable than the cobalt complexes present under hydroformylation conditions, but below 120 °C $Fe(CO)_5$ only reacted with free excess phosphine in the system and phosphine transfer from cobalt to iron did not take place in the timeframe studied. At 170 °C all ligand was transferred from cobalt to iron within 2 h. Thus, accumulation of iron carbonyl in these reactor systems has to be prevented as otherwise cobalt will be deprived of ligand. Ligand consumption (by iron or oxidation) will be noticed by an increase in rate and a lowering of the l:b ratio.

## 8.3
## Rhodium-Catalyzed Hydroformylation

### 8.3.1
### Introduction of Rhodium-Catalyzed Hydroformylation

In the patent by Slaugh concerning the SHF process rhodium was also mentioned as a potential active catalyst, but since the focus was on alkylphosphines the results were not very exciting (much later Cole-Hamilton and coworkers showed how alkylphosphines could be used efficiently with rhodium catalysts to give alcohols in protic solvents [13]). In the mid-1960s the work of Wilkinson showed that *aryl*phosphines should be used for rhodium and that even under mild conditions active catalysts can be obtained [14]. Soon thereafter Pruett and Smith introduced phosphites as modifying ligands for rhodium-catalyzed hydroformylation [15]. The Wilkinson papers were a striking breakthrough, because he and his coworkers achieved hydroformylation at room temperature and ambient pressure, which was unthought of until then, given the process conditions of cobalt catalysts!

The first rhodium-catalyzed, ligand-modified process came on stream in 1974 (Celanese) and was soon followed by UCC (Union Carbide Corporation, 1976) and in MCC (Mitsubishi Chemical Corporation, 1978); all these processes used triphenylphosphine as the ligand. The UCC process (now Dow) has been licensed to many other users and is often referred to as the LPO process (low-pressure oxo process). Not only are rhodium catalysts much faster – which is translated into milder reaction conditions – but also their feedstock utilization (or atom economy) is much higher than that of cobalt catalysts. For example, the cobalt-alkylphosphine catalyst may give as much as 10% of alkane as a by-product. Since the mid-1970s the rhodium catalysts started to replace the cobalt catalysts in propene and butene hydroformylation. For detergent alcohol production though, even today, the cobalt systems are still in use, because there is no good alternative yet for the hydroformylation of internal higher alkenes to mainly linear products. Palladium catalysts have been very near to replacing cobalt as drop-in catalysts in existing cobalt plants [16], but this has been postponed, it seems.

The Ruhrchemie/Rhône-Poulenc process utilizing a two-phase system containing water-soluble rhodium-tppts in one phase and the product butanal in the organic phase has been in operation since 1984 by Ruhrchemie (or Celanese, nowadays) and this represents the third generation hydroformylation process. For lower alkenes (the process for butene came on stream in 1995) this is the most economic process; ligand decomposition is less, separation of product and catalyst is extremely facile, and heat exchange is very well taken care of (the hydroformylation reaction is exothermic and the heat generated is used for the separation of the isomeric products by distillation).

Phosphites as ligands in rhodium-catalyzed hydroformylation, as introduced by Pruett and Smith [15], came into focus again after the work by van Leeuwen and coworkers, who discovered the peculiar effect of bulky monophosphites giving very high rates [17]. Bryant and coworkers at Union Carbide have expanded this work, first by making more stable bulky monophosphites [18]. Later they focused on diphosphites, which have become very popular since then for their high regioselectivity and, with the use of chiral ligands, for the enantioselectivities obtained [19]. There is only one relatively small commercial application of "bulky monophosphite" by Kuraray for the hydroformylation of 3-methylbut-3-en-1-ol [20]. Much research has been devoted to diphosphites in the last decades aiming at a variety of applications [21]. Phosphite decomposition has been studied extensively by UCC and enormous improvements have been made in chemical stability by changing the substitution pattern of the phosphites.

In the following we will look at catalyst formation and incubation, and the decomposition reactions that may take place with several ligands. As mentioned before, there are only very few detailed studies and often one can only guess what might have been the problem. First, we will summarize a few characteristic ligand effects. Rhodium hydroformylation in the absence of decomposition reactions or reactive impurities is a well controlled and predictable reaction. A recent DFT study by Jensen and coworkers explained the basic phenomena of ligand effects observed over the years [22].

The main species observed under hydroformylation conditions are shown in Figure 8.2 [23]. Complexes **1** and **2** are stable complexes that can be isolated, **1** for triphenylphosphine (tpp), and **2** for small phosphites, for example. Under standard laboratory hydroformylation conditions (10–20 bar syn gas, 70–100 °C, 1 mM Rh) triphenylphosphine will be replaced from **1** and species **3** and **4** will form. A 10–20-fold excess of tpp is needed under these conditions to form the active species **3**, which exists as two isomers having tpp in equatorial–equatorial or apical–equatorial positions. Low hydrogen pressures lead to formation of dimer **5**, or, if the

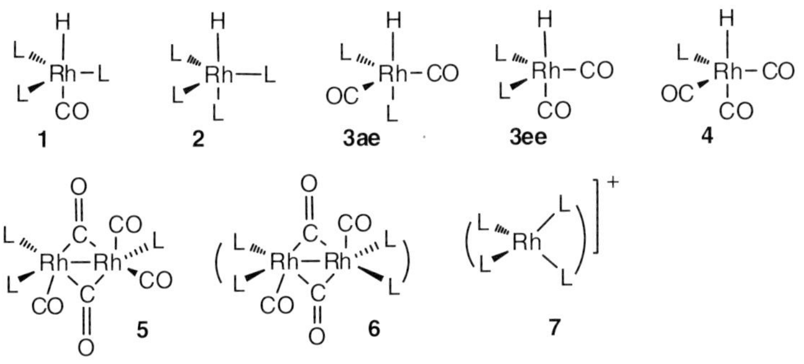

**Figure 8.2** The most common rhodium complexes under hydroformylation conditions.

CO pressure is also low, to formation of a related dimeric complex containing fewer CO molecules, as was already discovered by Wilkinson and coworkers in the 1960s [24]. It is known as the so-called orange dimer **5**, generated from HRh (PPh$_3$)$_3$CO; CO deficient complexes are dark red. Since then dimers have been reported by several authors [25 and references therein]. Initially Wilkinson observed that under sub-ambient pressures of H$_2$ and CO the reaction was first order in H$_2$; actually, this was due to the predominant formation of **5**, and higher pressures shifted the equilibrium from **5** to the side of **1–4**, and not to a rate-limiting step involving H$_2$ as has often been quoted – after a premature suggestion Wilkinson proved convincingly that this is not the case. Work-up of hydroformylation solutions often leads to formation of dimers. It seems likely that in a liquid recycle of a continuous reactor rhodium will occur as such a dimeric species. Since the reaction with hydrogen is reversible it presents a means to recycle rhodium. Dimer formation is not restricted to phosphines as phosphites behave similarly.

Diphosphines will form the desired complexes **3** in a higher proportion than monophosphines, which always give rise to mixtures. Diphosphines and diphosphites also require an excess of ligand at 1 mM of Rh and CO pressures >5–10 bar, although a 2–3-fold excess may suffice. Depending on the bite angle of the diphosphine more or less of **3ee** or **3ae** will be present in the equilibrium. Isomer **3ee** is the preferred one when high linearities are desired [26], for example, BISBI and Xantphos-type ligands will do so. The dimers formed in this instance at low H$_2$ pressure have structure **6**. Certain diphosphines, such as Nixantphos [27] and bis (dibenzophospholyl)-Xantphos [28], have a high propensity to form such dimers, even at H$_2$ pressures that normally do not show dimer formation, which is not yet understood. It was observed that in ionic liquids the same ligands showed no dimer formation at the same hydrogen pressure [29]. In NMR experiments, for which higher concentrations are applied than those in catalysis, dimer formation is more prominent. For example, for rhodium carbonyl hydride complexes of aminophosphine phosphinite ligands mainly dimer was observed at moderate H$_2$ pressures [30].

Other diphosphines show formation of ionic complexes such as **7** (the anionic counterpart does not contain phosphorus and has not been identified) [31]. Very bulky ligands are expected to give mono P-ligand complexes **4**, as phosphites do, *vide infra*, but for bulky phosphines this seems a rather subtle equilibrium and usually the phosphines fall off completely, rather than giving type **4** complexes [32].

For phosphites the same equilibria were observed and many structures **3** have been identified by NMR and IR spectroscopy. Structure **3ae** was observed for Binaphos, a ligand producing high ees in styrene hydroformylation, while wide-bite-angle phosphites having a C3–4 bridge, give mainly **3ee**, complexes that give high linearity in alkene hydroformylation.

In most catalysts, dissociation of CO and coordination of alkene occur in the rate equation (Scheme 8.2, [33]), the rate-determining step being either alkene coordination or insertion, and thus electron-withdrawing ligands accelerate hydroformylation [34], because they facilitate both CO dissociation and alkene coordination, while migratory insertion is often not affected or accelerated by electron-withdrawing ligands. The kinetic equation involves a positive order in alkene concentration and a negative order in CO and/or phosphorus ligand concentration, but it does not contain dihydrogen pressure. Bulky phosphites were shown to give mono-phosphite complexes **4** under CO pressure. These are fast hydroformylation catalysts for electronic and steric reasons [35]. In this peculiar case hydrogenolysis of rhodium acyl with dihydrogen is rate determining; the first reactions of the cycle are all accelerated but the reaction with $H_2$, most likely involving an oxidative addition, may be retarded. In this case the rate is independent of alkene concentration and first order in dihydrogen pressure.

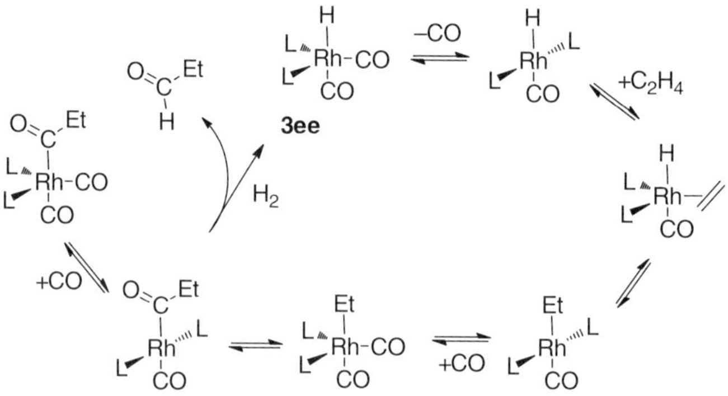

**Scheme 8.2** Hydroformylation mechanism for ethene according to Heck and Breslow, Co replaced by Rh.

Electron-poor ligands also accelerate isomerization of alkenes via β-hydride elimination of the alkylrhodium intermediates; low CO pressures have the same effect. Electron-rich alkylphosphines react via a different mechanism, as will be outlined later. Electron-rich ligands such as NHC ligands are not expected to give fast catalysts, if the Heck–Breslow mechanism is operative; ligand dissociation and alkene coordination are slow. Oxidative addition of dihydrogen is predicted to be faster, but this is not the rate-determining step and certainly not in this instance.

8.3.2
## Catalyst Formation

As only a few hydrides of structure **1** are sufficiently stable to be isolated and used as catalysts (or their precursors), usually catalysts are formed *in situ* from a rhodium salt and ligand under syn gas pressure. It was found that aryl phosphite complexes having structures **1** or **2** form easily metallated species upon standing and warming and van Leeuwen and coworkers suggested that these should also be good catalyst precursors for hydroformylation [36], as was indeed observed [37].

A variety of dicarbonyl rhodium and 1,5-cyclooctadiene rhodium complexes have been used as precursor. The reaction equation is shown in Scheme 8.3. In order of decreasing suitability the anions used are methoxide > acetylacetonate > thiolate > acetate > chloride. Rhodium methoxide under hydroformylation conditions will give methanol and rhodium hydride species, but thermal decomposition in the absence of hydrogen will also give rhodium hydride via β-hydride elimination together with formaldehyde. Its use is not widespread because of its low stability and the lack of commercial availability. A comparative example showing higher rates for methoxide than acac precursors was reported by Jacobs [38].

$$(L'RhX)_n + H_2 + 2L + 2CO \rightleftharpoons L_2(CO)_2RhH + L' + HX$$

L' = 2CO, 1,5-COD
X = MeO, RS, acac, AcO, Cl
$n$ = 1,2

**Scheme 8.3** Transformation of precursors into catalysts.

Slightly better than acetylacetonate rhodium dicarbonyl is its dipivaloylmethanate analogue, which has a longer shelf-life in solution [39]. At temperatures below 80 °C the incubation time of acac precursors can be substantial, especially for phosphite ligands. It is recommended to have the hydride formation take place at 80 °C before alkene is added, if one is interested in the initial reaction rate. At 40 °C, 9 bar of syn gas, incubation times up to 10 h were noted by *in situ* IR spectroscopy [40].

The dimers of rhodium thiolates, introduced by Kalck as hydroformylation catalysts [41], have long been thought to be a new type of catalyst [42], deviating from the classic Wilkinson catalyst, acting in its dimeric form. Davis and coworkers showed in kinetic studies that the rhodium thiolates used were active as the monomeric species [43]. Diéguez, van Leeuwen and coworkers proved by *in situ* IR studies that the thiolates are precursors and that under the reaction conditions monomeric rhodium hydrides **1–3** are formed; the activity of the system runs parallel with the amount of hydride formed according to the equilibrium shown in Scheme 8.3 [25, 44]. The position of the equilibrium depends strongly on the structure of the thiolates, the phosphine ligands added, and the conditions.

Gao and Angelici reported that rhodium immobilized as dimeric species on silica via thiolate linkages formed active catalysts, provided that phosphine was present in the solution. The catalyst was recovered unchanged [45]. We tend to conclude that the active catalyst forms in solution by partial leaching from the surface – TOFs are modest – and that upon release of the pressure rhodium thiolates on silica are formed again. Often in these systems a positive response to higher $CO/H_2$ pressures is observed, as one would expect if the equilibrium in Scheme 8.3 plays a role. In general, thiols, thioethers, and even dithioethers are weak ligands towards rhodium hydride species and will be displaced by CO or phosphines, bidentate thiolate ligands being the best choice for obtaining stable rhodium thiolates [46]. Studies continue until today to prove or disprove the involvement of dirhodium dithiolates as catalysts or "active precursors", as they are coined [42].

It is worth mentioning that not all sulfur-based ligands are displaced by carbon monoxide at elevated pressures; thiourea and its derivatives have been known as ligands in palladium catalysts used for carbonylation chemistry for more than 40 years [47]. More recently they have been used in cross-coupling catalysis (Pd), hydrogen transfer reactions (Rh), Pauson Khand reactions (Pd), and hydroformylation (Rh), apart from their extensive use as organic catalysts [48]. Chiral thioureas gave moderate but distinct ees in the rhodium-catalyzed hydroformylation of styrene in the absence of other chiral ligands, showing that also under rhodium hydroformylation conditions thioureas are strong enough ligands that are not displaced by CO [49].

Initially rhodium acetates were often used as the precursors, but at higher concentrations of carboxylic acids the formation of rhodium hydrides is not complete [50].

Rhodium chloride complexes of CO and 1,5-cod have been used extensively as precursors for hydroformylation, but in the absence of base the rhodium chloride versus rhodium hydride equilibrium lies on the chloride side. Therefore, since the 1960s, an amine, such as triethylamine, was added to scavenge HCl. An excess of $NEt_3$ does not harm the reaction as neither amines nor other nitrogen ligands will coordinate to the apolar rhodium hydride catalysts. Phosphine and phosphite ligands may contain residues of HCl or its salts from the synthesis and these impurities can be present in deleterious quantities when large excesses of ligand are used with respect to rhodium. Formation of rhodium chloride salts can, in these instances, also be prevented by addition of $EtN_3$. Ionic rhodium complexes function as isomerization and hydrogenation catalysts, and these side-reactions may point to their presence. Sometimes their formation is promoted deliberately and tandem hydroformylation–hydrogenation reactions are performed, or the catalyst can be switched from one stage to the other, as was shown to be reversible for an immobilized Nixantphos catalyst [51].

The non-coordinative behavior of amines towards rhodium hydrides opens possibilities for other uses of the amino groups. In Xantphos containing phenyl groups substituted by dimethylaminomethyl groups the amino groups do not participate in rhodium complexation and they can be used to extract the complexes as the ammonium salts formed by protonation from the product/educt mixture [52]. Andrieu and coworkers studied $\alpha$, $\beta$, and $\gamma$ amino-phosphines as hemilabile ligands in hydroformylation reactions [53]. Even in the rhodium chloride precursors amino groups did not coordinate to rhodium under CO pressure. Due to the presence of the amino group the chloride precursors were active without addition of amines under circumstances where PPh$_3$ showed no activity. α-Aminophosphine gave the most active catalyst, which was explained by a concerted heterolytic cleavage of dihydrogen facilitated by the proximity of the amine function. In the presence of NEt$_3$ the rate enhancement was partly retained and it was speculated that the amine might also play a role in the hydrogenolysis step of the rhodiumacyl intermediate.

Numerous examples have been reported concerning the use of rhodium perchlorate or tetrafluoroborate complexes as catalyst precursor in the presence of the desired phosphine, phosphite, and so on. The use of these salts, as their 1,5-cod or norbornadiene rhodium complexes, is rather general for hydrogenation reactions, which usually proceed better when "cationic" complexes are used rather than neutral rhodium chloride species. In view of the above equilibria outlined for carboxylates and chlorides it is surprising that the cationic complexes give such fast catalysts [54]. If the reaction equation were indeed that of Scheme 8.3, HBF$_4$ would form and it is not likely that this represents the equilibrium situation. Probably the acid decomposes or a phosphonium salt is formed with the excess of phosphine present, thus facilitating the formation of rhodium hydrides.

## 8.3.3
### Incubation by Impurities: Dormant Sites

Irreversible decomposition of ligands (and thus catalysts) will be discussed later, while here we are only concerned with temporary "poisoning" of the catalyst. Dienes and alkynes are well-known poisons for many alkene processes. In polyolefin manufacture they must be carefully removed as they deactivate the catalyst completely. Insertion of conjugated dienes is even much more rapid than insertion of ethene and propene. The resulting π-allyl species are inactive as catalysts.

In rhodium-catalyzed hydroformylation the effect is less drastic and often remains unobserved, but diene impurities can obscure, for instance, the kinetics of alkene hydroformylation [55]. Because the effect is often only temporary, we summarize it here under "dormant sites". Hydroformylation of conjugated alkadienes is much slower than that of alkenes, but alkadienes are more reactive than alkenes toward rhodium hydrides [56, 57]. Butadiene can be hydroformylated by diphosphine rhodium complexes but the TOF is only a few hundred per hour at 120 °C, while 1-octene would show a TOF of several tens of thousands per hour; in part this may be due to the volatility of butadiene. Stable π-allyl complexes are formed that undergo a very slow insertion of carbon monoxide (Scheme 8.4). The resting state of the catalyst for dienes is a π-allyl species and hardly any rhodium hydride is available for alkene

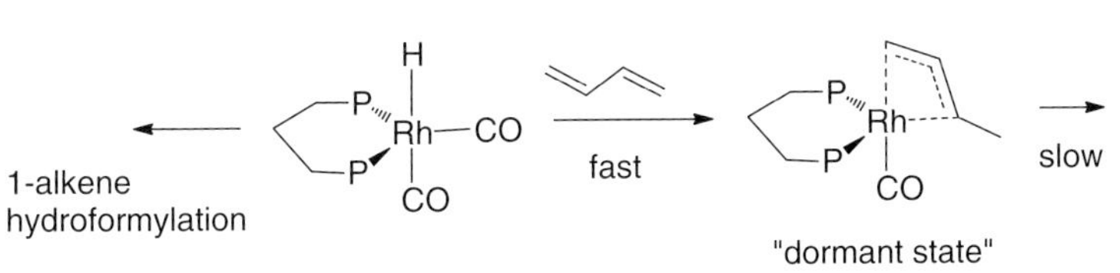

Scheme 8.4 Dormant site formed from dienes.

hydroformylation. Thus, alkadienes must be thoroughly removed, as described by Garland [58]; this was especially important to obtain reliable results in his kinetic studies. Thus 1,3- and 1,2-diene impurities in 1-alkenes retard, if not inhibit, the hydroformylation of alkenes [59]. Partial or temporary inhibition by 1,3-dienes was also observed for unmodified catalysts by Liu and Garland [60].

The hydroformylation of butadiene with the use of dppe as the ligand for rhodium is unique in that it gives only pentanal as the product [56]. In all other cases a mixture of unsaturated aldehydes is obtained. This is also true for isoprene, as was reported by Gusovskaya and coworkers [61]. They found an unselective reaction to unsaturated aldehydes for all ligands tested. Among the monodentate phosphines $Cy_3P$ gave the fastest catalyst. The rate was increased by higher pressures of both $H_2$ and CO, and higher ligand concentrations. The kinetics was not well understood.

Hydroformylation of endocyclic double bonds in *para*-menthenic terpenes containing non-conjugated dienes under mild conditions gave a sluggish and incomplete formation of aldehydes which was assigned to the formation of unreactive π-allylic intermediates by Gusevskaya and co-workers [62]. Fast and complete reactions were obtained when "bulky" phosphite ($o$-tBu-$C_6H_4$) was used [63], although the selectivity remained low.

Often the kinetics of rhodium-catalyzed hydroformylation of 1-alkenes does not follow the expected first order dependence on the alkene concentration and the minus one order in CO pressure [64], which may well be related to changes in the catalyst composition due to phosphine loss.

Severe or slight incubation has also been observed. It is known [65] that not only 1,3-alkadienes but also impurities such as enones and terminal alkynes may be the cause of such behavior. Detailed studies were not known until recently, although several workers in the field took precautions to make sure that the alkene substrates were not contaminated with enones, dienes or alkynes. In an *in situ* IR study combined with high-pressure NMR studies Walczuk and coworkers have shown how this temporary inhibition takes place and what intermediates are formed [66]. *In situ* IR clearly showed the disappearance of rhodium hydride catalyst when these "poisons" were added and new, dormant species were observed. Subsequently, the dormant states were slowly depopulated, and after all impurities were converted the hydroformylation of 1-octene started, while simultaneously the hydride resting state of the catalyst was recovered. Enones gave rise to stable alkoxycarbonyl species **8** (Scheme 8.5) that were slowly converted to ketones after deinsertion of CO. The degree of inhibition depends both on the structure of the inhibitor and on the ligand.

Oxygenates and other unsaturated impurities are present, in particular, in alkene feedstocks derived from a Fischer–Tropsch product mixture [65]. Triphenylphosphine shows considerable incubation with enones, alkynes, and dienes present in such a feed. In a continuous process this translates into substantial amounts of catalysts tied up in a dormant state. This was found to be strongly dependent on the phosphine used and addition of bidentate phosphines such as Xantphos or DPEphos to the $PPh_3$ system was found to reduce the incubation in a batch reaction considerably.

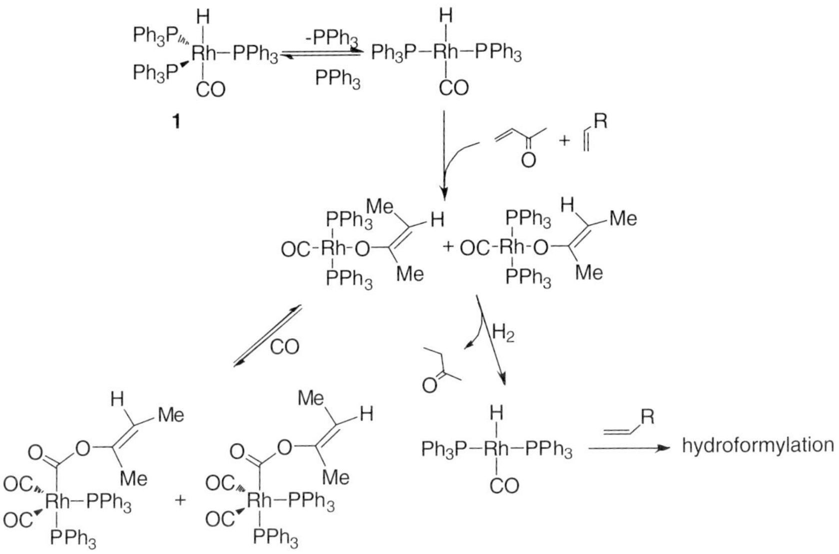

**Scheme 8.5** Resting states formed with enone and rhodium hydride.

Hydrogenation of enones under these conditions, using syn gas, was exploited by Scheuermann and Jaekel [67]. Using Chiraphos as the ligand they achieved ees up to 90% in the hydrogenation of substituted cyclohexenones and pentenones; isophorone gave the highest ee. Interestingly, unlike in hydrogenation catalyzed by cationic rhodium complexes, which operate via the classic dihydride mechanism, the two hydrogen atoms incorporated in the product do not stem from the same $H_2$ molecule, as the catalyst is a monohydride of type **3ae**. In solution they identified, in addition to **3ae**, dimers **5** (carrying two CO molecules less than **5**), and ionic species **7** having acac as the anion. No hydroformylation took place at 60 °C.

The intermediate rhodium enolates in Scheme 8.5 are intriguing compounds as they were used as catalysts for the Tischchenko reaction to convert benzaldehyde into benzyl benzoate [68], notably at room temperature with a TOF of 120 molecules of benzaldehyde per hour. A variety of alkyl and aryl aldehydes were found to react this way. Apparently this reaction does not take place in the presence of CO; as the catalytic species the highly unsaturated complex $(PPh_3)_2RhH$ was proposed.

Inhibition by alkynes can take place in more ways than one. First a direct complexation may occur as was observed by Liu and Garland for unmodified catalysts (Figure 8.3) [69]. Twenty different alkynes were tested. They tentatively proposed that the species formed are substituted dirhodium carbonyl species, viz. $Rh_2(CO)_6\{\mu\text{-}\eta^1\text{-}(CO\text{-}HC_2R)\}$ for terminal alkynes and $Rh_2(CO)_6\{\mu\text{-}\eta^1\text{-}(CO\text{-}R^1C_2R^2)\}$ and $Rh_2(CO)_6\{\mu\text{-}\eta^2\text{-}(R^1C_2R^2)\}$ for disubstituted alkynes. In the case of terminal alkynes, the dirhodium-alkyne complexes undergo rapid CO insertion under 20 bar of CO. For disubstituted alkynes, CO insertion is not complete and an equilibrium was observed between the bridged alkyne species and the insertion product. In both

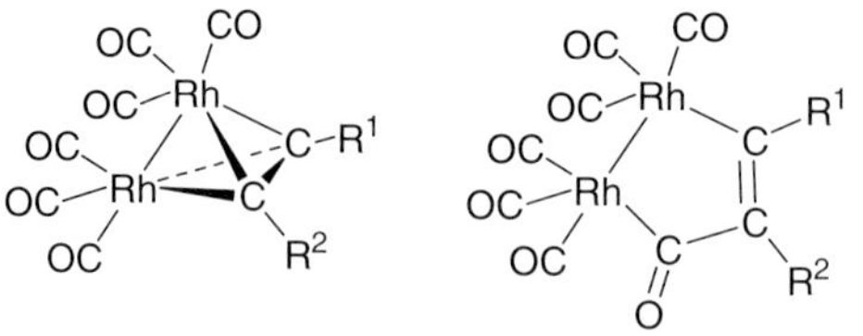

**Figure 8.3** Structures proposed by Garland for alkynes and rhodium carbonyl complexes.

cases the final alkyne complexes were stable under CO, even in the presence of molecular $H_2$. This was thought to be the primary reason why trace quantities of alkynes are able to poison the catalytic alkene hydroformylation reaction in unmodified systems.

When phosphine-modified catalysts are used, hydroformylation takes place, albeit at a low rate. Probably an enone is first formed as the product that presents the poisoning effect mentioned above. As its formation is slow and the catalyst can slowly hydrogenate the enone to aldehyde, the reaction is not completely blocked.

Ligand metallation was mentioned briefly in Section 8.3.2. At low pressure of $H_2$ the metallated form of a catalyst can also be a dormant state and it is dormant because usually the active hydride can be recovered via reaction with hydrogen [36, 37]. Aryl phosphites often show metallation; work-up of rhodium-phosphite catalyst solutions after hydroformylation often shows partial formation of metallated species, especially when bulky phosphites are used [70]. Alkane elimination may also lead to the metallated complex. The reaction is reversible for rhodium and $H_2$ and thus the metallated species could function as a stabilized form of rhodium during a catalyst recycle. Many metallated phosphite complexes have been reported (see Scheme 8.6 [36a]).

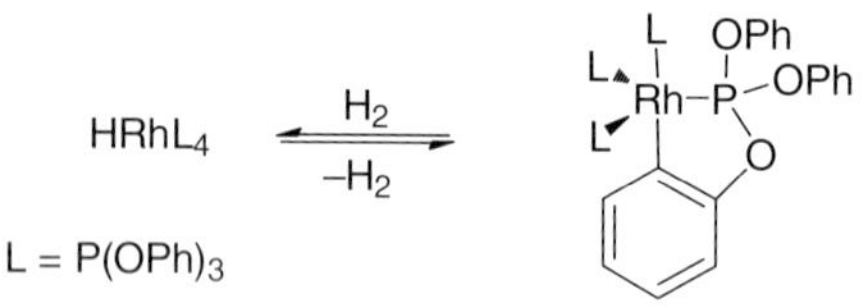

**Scheme 8.6** Reversible metallation in rhodium phosphite complexes.

Rosales and coworkers were interested in the hydroformylation of the alkenes in a naphtha stream without purification in order to improve the fuel value of this feed for automotive use. Since the crude of importance was from Venezuela it was rich in sulfur compounds. Therefore they were interested in the effect of such impurities on the hydroformylation reaction at 120 °C and a rather low pressure of 3 bar. They studied the influence of thiophene, benzo[b]thiophene, and dibenzo[b,d]thiophene on the hydroformylation of 1-hexene catalyzed by rhodium complexes RhH$(CO)_4$, RhH$(CO)_2(PPh_3)_2$ and RhH$(CO)_2(dppe)$. Even at this low pressure they found no effect of the sulfur compounds on the rate of reaction at impurity loadings between 100–1000 ppm, in accord with the statement in Section 8.3.2 where the influence of sulfur compounds resulting from precursors was discussed.

8.3.4
**Decomposition of Phosphines**

Phosphine ligands are prone to oxidation to give phosphine oxides. The latter cannot be reduced by $H_2$ or CO, a desirable reaction also for the recycling of $PPh_3P=O$ in Wittig and Mitsunobu reactions [71], and one may conclude that even water and carbon dioxide may oxidize phosphines to the corresponding oxides. Transition metals may catalyze these reactions and examples are known for Pd, Rh and $CO_2$ [72]. For example, $(Ph_3P)_3RhCl$ and $[(cyclooctene)RhCl]_2$ were found to catalyze the oxidation of phosphines by carbon dioxide in refluxing decalin. The rate of oxidation increases in the order $PPh_3 < PBuPh_2 < PEt_3$ [73].

Water can be the source of oxygen, while the metal (e.g., $Pd^{2+}$) is the oxidant (see Section 1.4.2), or water itself is the oxidant [74]. More often water reacts in a different fashion causing P–C cleavage, *vide infra*. To avoid oxidation oxygen and hydroperoxides have to be thoroughly removed from the reagents and solvents before starting hydroformylation. Purification of the alkene feed is often neglected. Since the alkene may be present in a thousand-fold excess of the ligand, careful removal of hydroperoxides in the alkene is an absolute must. Hydroperoxides are the ideal reagents for oxidizing phosphines. Percolation over neutral alumina is usually sufficient to remove the hydroperoxides before hydroformylation. Treatment over sodium metal or sodium–potassium alloy on a support will remove hydroperoxides in addition to 1-alkynes and 1,2 or 1,3-dienes that may also influence the catalyst performance. Distillation from sodium to remove dienes and alkynes may give isomerization of 1-alkenes as an undesirable side-reaction. For most functionalized alkenes this drastic treatment cannot be applied. Removal of 1,3-dienes can be done efficiently by treatment with maleic anhydride [58].

Prolonged hydroformylation of propene using $RhH(CO)(PPh_3)_3$ as the catalyst leads to a variety of decomposition products, such as benzene, benzaldehyde, propyldiphenylphosphine, and various rhodium cluster compounds containing diphenylphosphido moieties (Scheme 8.7) [75]. Thermal decomposition of RhH $(CO)(PPh_3)_3$, in the absence of $H_2$ and CO, leads to a stable cluster shown in Scheme 8.7 containing $\mu_2$-$PPh_2$ fragments [76]. Presumably clusters of this type also form eventually in LPO hydroformylation plants. Recovery of rhodium from these inert clusters is a tedious operation. Reaction of the cluster mixture with reactive organic halides such as allyl chloride has been described to give allyldiphenylphosphine and rhodium chloride, which can be easily extracted into a water layer [77].

Propyldiphenylphosphine formed in the reactor is more basic and less bulky than $PPh_3$, and thus will coordinate more strongly to rhodium than $PPh_3$. Because of its higher basicity CO dissociation from the new complex is slower and as a result hydroformylation will be slower. Propyldiphenylphosphine therefore has to be removed from the reactor in the recycle loop; this can be carried out with an aqueous

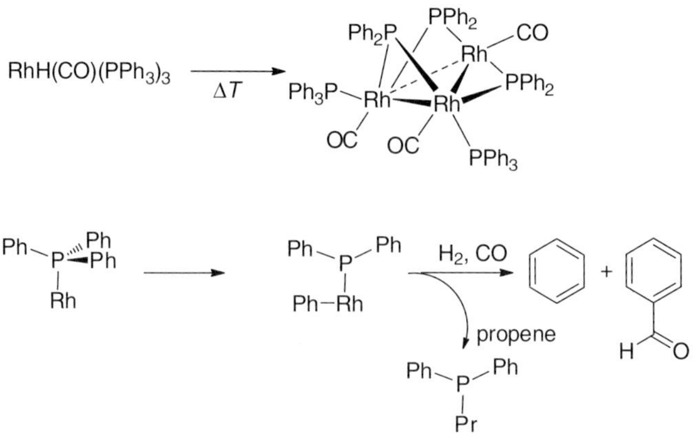

**Scheme 8.7** Decomposition of RhH(CO)(PPh$_3$)$_3$ during hydroformylation.

solution of an appropriate acid to form the corresponding phosphonium salt while PPh$_3$ remains in the organic solution.

Benzene and benzaldehyde by-products were also observed in the Ruhrchemie–Rhône Poulenc (RC–RP) process using the trisulfonated analogue of triphenylphosphine [78], but the decomposition was reported to be much slower for tppts than for PPh$_3$. This is an accidentally favorable aspect of the RC–RP process, which cannot be easily explained. One might have thought that in water phosphine decomposition might be even faster than in organic media, especially via the route involving nucleophilic attack at the coordinated, activated phosphorus atom, thus initiating the second mode of decomposition observed for methoxyrhodium intermediates (*vide infra*).

Aldehydes might lead to rapid decomposition of triarylphosphines, but so far this has only been observed for formaldehyde. A catalytic decomposition of triphenylphosphine was reported [79] in a reaction involving rhodium carbonyls, formaldehyde, water, and carbon monoxide. Several hundreds of moles of phosphine can be decomposed this way per mole of rhodium per hour! The reactions that may be involved are shown in Scheme 8.8. The key step is the nucleophilic attack of the

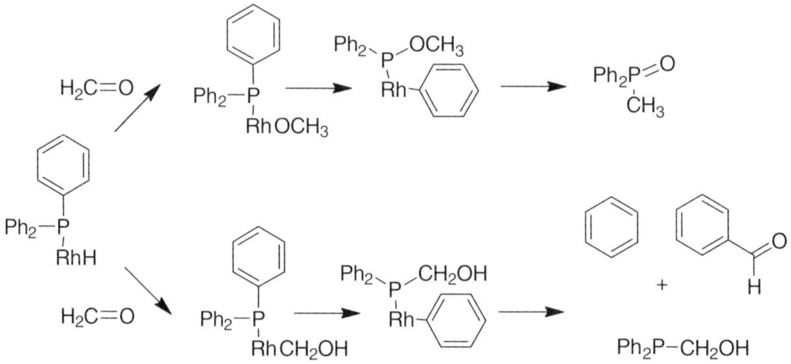

**Scheme 8.8** Decomposition of triphenylphosphine via nucleophilic attack.

methoxy group or the hydroxymethyl group at phosphorus. Related to this chemistry is the hydroformylation of formaldehyde to give glycolaldehyde, and further hydrogenation to ethylene glycol, which would be an attractive route from syn-gas to ethylene glycol, circumventing the use of ethene derived from naphtha. The reaction was indeed accomplished by Chan and coworkers using rhodium arylphosphine complexes as the catalysts [80]. Obviously, phosphine decomposition is a major problem to be solved before formaldehyde hydroformylation can be applied commercially. Formation of methoxy species can perhaps be disfavored by choosing the correct ligand, but the products indicate that hydroxymethyl groups are also involved in the phosphine decomposition and they are an inevitable intermediate in the hydroformylation of formaldehyde to glycolaldehyde.

The bimetallic catalyst developed by Stanley and coworkers [81] shows an interesting inhibition by $PPh_3$. The bimetallic catalyst (Scheme 8.9) shows a high linear to branched product ratio (l:b = 25, TOF = 1200 m m$^{-1}$ h$^{-1}$) under its normal operating conditions, 6 bar, 90 °C, but addition of only 1 mol of $PPh_3$ per dimer reduces the rate to half of its initial value and the l:b ratio to 3. Further addition of $PPh_3$ blocks the catalyst completely [82], without completely decomposing the bimetallic structure since the normal $PPh_3$ catalysis was not observed either.

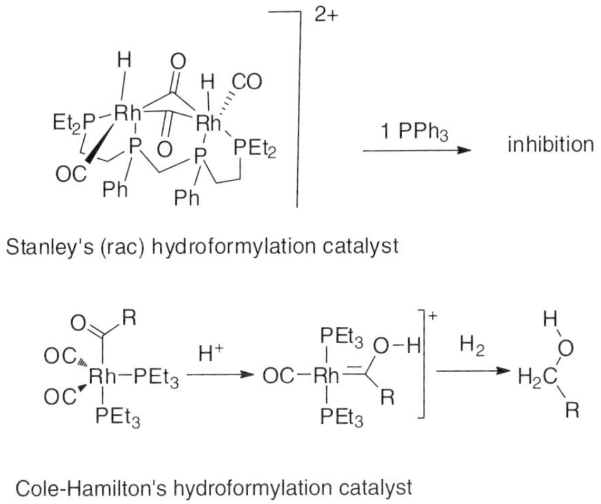

Stanley's (rac) hydroformylation catalyst

Cole-Hamilton's hydroformylation catalyst

**Scheme 8.9** $PPh_3$ inhibition in Stanley's bimetallic catalyst and Cole-Hamilton's catalyst giving alcohols.

Unfortunately there were always mixtures of complexes observed by $^{31}$P NMR spectroscopy, but most likely the active species has a dimeric nature and in view of its behavior and performance, the alkyl nature of the ligand, and so on, it differs considerably from the classic $L_4RhH$ catalysts.

The catalysts based on alkyl monophosphines developed by Cole-Hamilton and coworkers resemble more the classic catalysts, but for a few of them, especially $Et_3P$, the products obtained are alcohols and not aldehydes, when alcohols are used as the solvent [13]. For ethene turnover frequencies as high as 54 000 h$^{-1}$ were obtained. As in the case of cobalt more severe conditions are needed (120 °C, 40 bar) than those required for PPh$_3$ or phosphites. The considerations for phosphine decomposition for cobalt (Section 8.2) are also relevant here. It was proven that alcohols are the primary product and thus a different mechanism must be operative for these catalysts. The cycle may have the same starting hydride as small changes in the phosphine or the use of aprotic solvents make it return to the "normal" aldehyde product. Et$_3$P gives bis-ligand rhodium acyl species, which are highly electron-rich and the oxygen can be protonated, in protic solvents (see Scheme 8.12). More bulky ligands give mono-phosphine rhodium acyl intermediates and these are not protonated and form aldehyde via the normal route [83].

The groups of Krause, Reinius, and Pakkanen made a study of ortho-substituted triphenylphosphine derivatives in the hydroformylation of 1-alkenes and methyl methacrylate. Krause and Reinius used 2-thiomethylphenyldiphenylphosphine as the ligand in rhodium-catalyzed methyl methacrylate hydroformylation [84]. The new catalyst had a significant effect on the conversion, but only a minor effect on the selectivity compared to PPh$_3$. At higher temperatures significant decomposition took place; donating groups on PPh$_3$ enhance decomposition, most likely towards phosphido complexes [85]. The effect of ortho-alkyl substitution of PPh$_3$ on catalyst performance in propene hydroformylation was extensively studied by Krause, Reinius, and coworkers [86]. *Ortho*-tolyl and *ortho*-ethylphenyl substitution led to higher branched aldehyde formation, while the reaction was slightly slower. Apart from this, the behavior was very similar to that of PPh$_3$, including the decomposition. Pakkanen and coworkers studied the effect of heteroatom ortho-substituents (S, N, O) on propene and 1-hexene hydroformylation [87]. In general conversions were low, and 1-hexene gave substantial isomerization.

As mentioned before in Section 8.3.1, still larger arylphosphine ligands lose the competition with CO even under these mild conditions and tend to fall off [32].

A rather bulky phosphole ligand is 1,3,5-triphenylphosphole, which was studied by Neibecker and coworkers. First, they noticed that a large excess did not change the selectivity or rate of the hydroformylation reaction of styrene [88], which is an indication of the relative bulkiness of the ligand, a behavior comparable to that of bulky phosphites (see Section 8.3.4). Catalyst identification showed that two isomeric species of formula **3** were present, of which one was formulated as the bis-equatorial bis-phosphine isomer, and the spectrum of the other isomer was interpreted as a species containing an equatorial hydride, the first of its kind [89]. The authors overlooked the possibility, reported for several phosphites [90], that the two **3ea** isomers may exchange extremely rapidly without passing through **3ee** via a hydride shift from one face of the tetrahedron to another, in between the two CO groups. Note that such a movement across an edge containing a phosphorus ligand at one of the vertices would be much slower. In styrene hydroformylation kinetics were observed that contained the initial substrate concentration – higher concentrations slowing the reaction – but incubation effects caused by impurities were not considered [91]. Interestingly, at higher conversions product inhibition by aldehyde was observed in spite of the mild conditions.

Rafter and coworkers studied the hydroformylation of 1-octene using 2-arylphe-nyldiphenylphosphines [32], which are successful ligands in cross-coupling reactions using Pd, as explored by Buchwald and coworkers. The ligands are bulky in an asymmetric way, in that they block a coordination site cis to the phosphorus donor atom; this led to highly efficient Pd catalysts. In rhodium hydroformylation catalysts they turned out to be very weak ligands and under hydroformylation conditions ligand-free rhodium carbonyls were obtained. The so-called bowl-shaped phosphines, for example, tris(2,2″,6,6″-tetramethyl-*m*-terphenyl-5′-yl)phosphine, as studied by Tsuji in catalysis [92] form bis-ligand complexes even under hydroformylation conditions, although the ligands are very bulky. In contrast with the Buchwald type ligands they are bulky having $C_3$ symmetry rather than having the bulk exposed in one direction. As a result they show a $PPh_3$-like complexation and hydroformylation behavior.

Bianchini and coworkers studied a range of dppf-like ligands in the hydroformylation of 1-hexene: $1,1'$-$(PPh_2)_2$-ferrocene, dppf, $1,1'$-$(PPh_2)_2$-ruthenocene, dppr, $1,1'$-$(PPh_2)_2$-osmocene, dppo, $1,1'$-$(-PPh_2)_2(Me)_8$ferrocene, dppomf, $1,1'$- (*o*-isopropylphenyl$_2$P)$_2$-ferrocene, *o*-*i*Pr-dppf [93]. Under hydroformylation conditions five-coordinate hydride(dicarbonyl) complexes $RhH(CO)_2(P–P)$ were formed that exist in solution as two rapidly equilibrating isomers **3ee** and **3ae**. Replacing Fe in dppf by Ru or Os has hardly any effect on the product distribution, in spite of the expected bite angle effect. The reaction of the *o*-*i*Pr-dppf precursor with syngas at 60 °C gave a trigonal-bipyramidal dicarbonyl complex with a dative Fe–Rh bond, while the dppomf complex decomposed to various CO-containing rhodium complexes. HP-IR spectroscopy allowed one to distinguish the ee and ea geometric isomers of the hydride(dicarbonyl) resting states with dppf, dppr, and dppo. *o*-*i*Pr-dppf formed a stable dicarbonyl complex, while the dppomf dicarbonyl was unstable under hydroformylation conditions, converting into phosphine-free carbonyl Rh compounds. Thus, also here, a large steric bulk prohibits the formation of the diphosphine catalyst.

## 8.3.5
### Decomposition of Phosphites

The use of phosphites as ligands in rhodium-catalyzed hydroformylation was reported in the late 1960s [15] by researchers at Union Carbide. After the discovery by Shell of the accelerating effect of bulky phosphites [17], Union Carbide reinitiated research on phosphites in the late 1980s, which led to an extensive follow-up by many academic and industrial groups. Aryl phosphites were the ligands of choice for the commercial production of adiponitrile via the butadiene hydrocyanation process, but phosphites have not been applied industrially on a large scale for hydroformylation and only small-scale applications hve been reported. Phosphites are easier to synthesize and less prone to oxidation than phosphines. They are much cheaper than most phosphines and a wide variety can be obtained commercially as they are used as anti-oxidants, for instance, in polypropylene. Disadvantages of the use of phosphites as ligands include several decomposition reactions: hydrolysis, alcoholysis, trans-esterification, and the Arbuzov rearrangement, which is catalyzed by acids and metal salts. Aryl phosphites do not undergo Arbuzov rearrangements and for that reason they are often preferred.

One example of an Arbuzov reaction during hydroformylation under relatively mild conditions was reported by Perez-Torrente, Pardey and coworkers [94]. They used thiolate precursors, which continue to be of interest. In this instance *gem*-dithiolate dirhodium complexes were used together with triphenyl phosphite, trimethyl phosphite and $Ph_3P$ as the modifying ligands (see Scheme 8.10). Without phosphorus ligands no activity was observed and even in the presence of P-ligands activities are rather modest, with TOFs of a few hundred while other precursors under the same conditions and with the same ligands would give TOFs of the order of 5000 ($PPh_3$) to 10 000 (($PhO)_3P$) [57]. Spectroscopic studies under pressure (HP NMR and HP IR) evidenced the formation of hydrido mononuclear species under catalytic conditions that are responsible for the observed catalytic activity. Apparently, the *gem*-dithiolate complexes are relatively stable and only in part converted to complexes of type **3** under their standard conditions (7 bar, 80 °C). When $(MeO)_3P$ was used as the ligand the authors could identify a dimethyl phosphite monomeric complex by NMR spectroscopy, shown in Scheme 8.10. Its formation involves an Arbuzov-like rearrangement catalyzed by Rh; the fate of the methyl moiety was not reported. If the mechanism were similar to that of the rhodium-catalyzed reaction reported in Chapter 1 (Scheme 1.36), the co-product would be methane in this instance (i.e., oxidative addition of MeX in **1** or **2**, followed by reductive elimination of MeH).

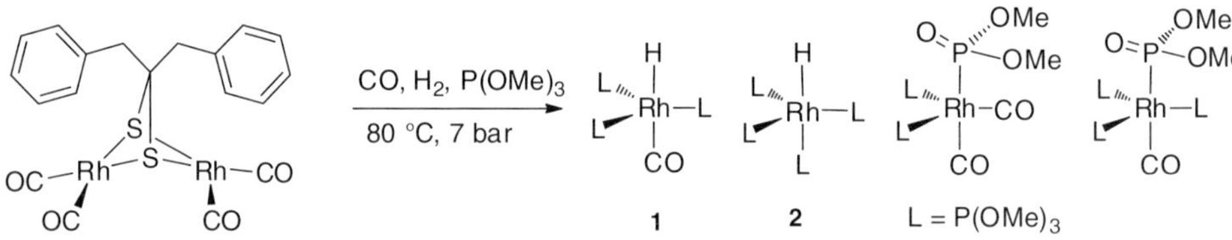

**Scheme 8.10** Gem-dithiolates as precursor and the formation of dimethyl phosphite rhodium.

Phosphites are not very sensitive towards oxidation, but, nevertheless, in a continuous process in which oxygen ingress may occur in the low-pressure distillation before the catalyst is recycled, oxidation was noted by Borman and Gelling [95]. The wide bite-angle, bulky diphosphites are used in excess (*vide infra*) and the free ligand can be oxidized. Since these ligands are costly their oxidation should be prevented, which was done in this example by adding the cheap and more easily oxidized $PPh_3$ as a scavenger. In addition, the excess $PPh_3$ may stabilize rhodium hydride species in the absence of CO during the recycle by forming a stable diphosphite-triphenylphosphine rhodium hydride monocarbonyl complex, instead of a labile diphosphite-rhodium hydride dicarbonyl complex.

Bryant and coworkers have studied the decomposition of phosphites extensively [96]. Stability involves thermal stability, hydrolysis, alcoholysis, and stability toward aldehydes. For typical phosphites see Figure 8.4. The precise structure has an enormous influence on the stability.

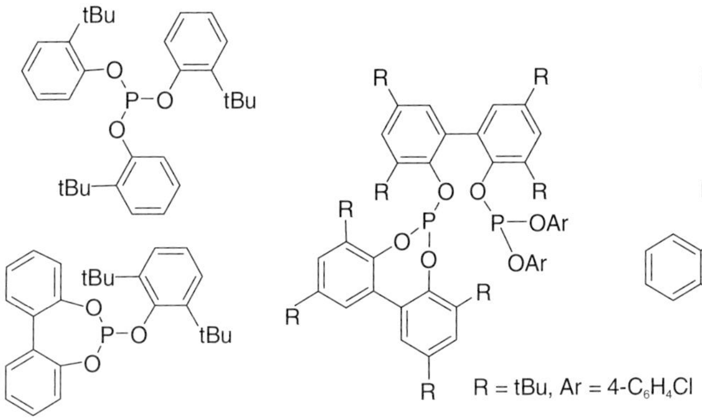

**Figure 8.4** Typical bulky monophosphites and diphosphites.

By thorough exclusion of moisture one can prevent hydrolysis of phosphites in the laboratory reactor. In a continuous operation, under severe conditions, traces of water may form via aldol condensation of the aldehyde product. Weak and strong acids and strong bases catalyze the reaction. The reactivity for individual phosphites spans many orders of magnitude. Dihydrocarbyl phosphites formed react with aldehydes to give α-hydroxyalkylphosphonates (see Scheme 8.11).

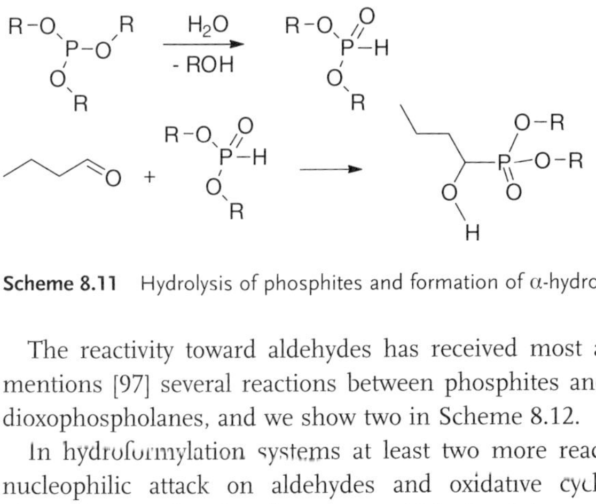

**Scheme 8.11** Hydrolysis of phosphites and formation of α-hydroxyalkylphosphonate.

The reactivity toward aldehydes has received most attention. Older literature mentions [97] several reactions between phosphites and aldehydes, for example, dioxophospholanes, and we show two in Scheme 8.12.

In hydroformylation systems at least two more reactions may occur, namely nucleophilic attack on aldehydes and oxidative cyclizations with aldehydes. The addition of a phosphite to an aldehyde giving a phosphonate is the most important reaction [98]. The reaction is catalyzed by acid and since the product is acidic, the reaction is autocatalytic. Furthermore, acids catalyze hydrolysis and alcoholysis and, therefore, the remedy proposed is continuous removal of the phosphonate over a basic resin (Amberlyst A-21). The examples in the patents illustrate that very stable systems can be obtained when the acidic decomposition products are continuously removed. Babin and Billig reported another effective way of removing the (presumed) hydroxyalkylphosphonates by adding cyclohexene oxide

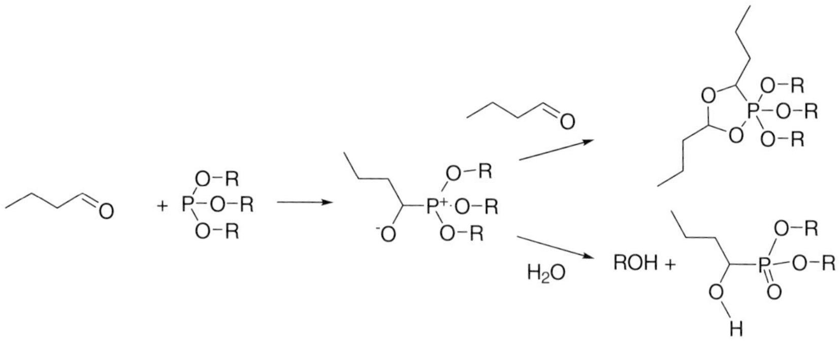

**Scheme 8.12** Reactions of phosphites and aldehydes.

to the hydroformylation mixture [99]. Presumably the α-hydroxyalkylphosphonate reacts with the oxirane producing an alcohol that is much less acidic.

The detailed structure of monophosphites was shown to be decisive for their stability. The thermal decomposition of a few monophosphites with aldehydes is illustrated in Figure 8.5.

Mitsubishi reported on the continuous removal of water from a diphosphite rhodium catalyst used in propene hydroformylation from the gas phase, while the dried gases were recycled to the reactor [100].

Chiral diphosphites were used for enantioselective hydroformylation of styrene and derivatives [101]. Stability and decomposition were not studied, but it was noted by Buisman and coworkers that the most stable, and equatorial–equatorially coordinating ligands gave the highest ees [102].

Sulfonated phosphites and their use as ligands in biphasic aqueous hydroformylation have been reported, but, in view of their instability in water, biphasic fluorophase hydroformylation seems a better option. An example shows that high rates and high l/b ratios can be obtained with the use of triphenyl phosphite equipped with fluoroalkyl tails and that the catalyst can be recycled efficiently [103].

Other electron-withdrawing ligands mimicking the electronic properties of phosphites usually show the same instability issues as do phosphites [104]. A few examples have been collected in Figure 8.6, ligands **9–11**. Electron-withdrawing ligands show, in general, high l/b ratios, but **9–11** decompose relatively quickly.

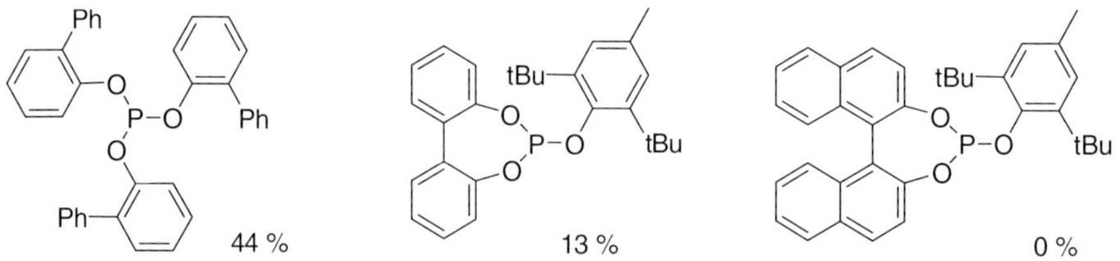

**Figure 8.5** Reactivity of various phosphites toward C5-aldehyde. Percentage decomposition after 23 h at 160 °C.

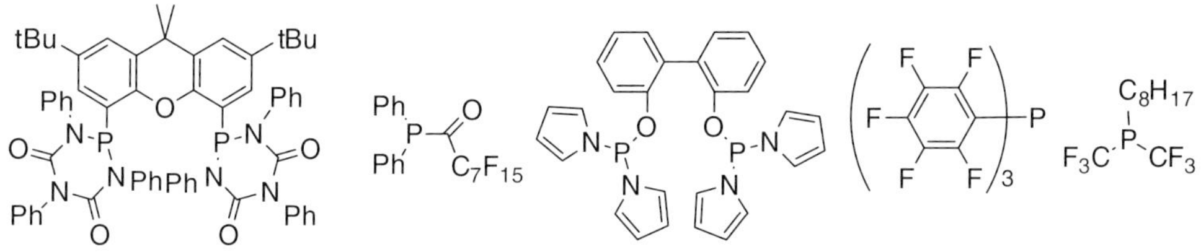

**Figure 8.6** Ligands with electron-withdrawing groups (references 107a–d).

Ligand **11** gave excellent results (TOF $100\,000\,\text{mol}\,\text{mol}^{-1}\,\text{h}^{-1}$, l/b $= 100$, at $80\,^{\circ}\text{C}$, 20 bar), one of the best reported, but its life-time is only a few hours. Ligand **12** showed low activity, even at $100\,^{\circ}\text{C}$, 20 bar. Surprisingly, ligand **13**, like **12** having a $\chi$-value equal to phosphites, only shows hydrogenation activity under standard hydroformylation conditions.

An important contribution to the field of enantioselective hydroformylation was presented by Takaya, Nozaki and coworkers with the discovery of BINAPHOS. The hydroformylation of a wide variety of olefins with rhodium(I) complexes containing this chiral ligand provided throughout high regio- and enantio-selectivity under mild conditions [105]. Several other chiral bidentate phosphorus ligands have emerged as efficient chiral modifiers for Rh(I)-catalysts in the asymmetric hydro-formylation of styrene [102b], and high stereoselectivities were obtained in the asymmetric hydroformylation of vinyl acetate using a chiral Rh(I)-catalyst with a bis (diazophospholidine) ligand [106]. With styrene this last catalyst provides the branched aldehyde in low yield and in racemic form. More recently, new ligands were launched that perform well for a broad range of substrates [107]. In spite of these advances, no practical application of asymmetric hydroformylation to the synthesis of chiral fine chemicals has been reported in the technical literature to date. Perhaps the total turnover that can be achieved for BINAPHOS is not all that high. Solinas and coworkers reported that at prolonged reaction times the ee of the styrene and aryloxy-substituted ethylenes hydroformylation product dropped sharply (substrate to catalyst ratio 2000/1, $60\,^{\circ}\text{C}$, 80 bar, ee 40%, 40 h). For an industrial application this ratio must probably be one or two orders of magnitude higher. The decomposition was attributed to a reaction of the aldehyde product with the phosphite moiety of BINAPHOS, via a reaction as shown in Scheme 8.12 producing an α-hydroxyalkylphosphonate, but a definite proof was not pursued. When the reaction was stopped at 50% conversion, 20 h, the ee was still high (90%). Zn was added for the reduction of Rh(III) chloride to monovalent Rh as in the absence of Zn no hydroformylation took place.

## 8.3.6
## Decomposition of NHCs

The boom in the use of NHCs as ligands took off after the publications by Bertrand and Arduengo concerning isolable NHC compounds, which made the synthesis of

NHC complexes a lot easier [108]. In addition, the bulky ligands applied yielded much more stable complexes than those known hitherto. NHCs take part as a ligand in several excellent catalysts for cross-coupling reactions [109] and butadiene dimerization [110], both using Pd as the metal, and of course Ru-catalyzed metathesis (Chapter 10). Rhodium complexes were reported by Lappert and coworkers in the early 1970s [111] and their main synthetic route involved the reaction of metal complexes with electron-rich, tetraamine-substituted alkenes. Some of their samples were already tested as hydroformylation catalysts, but results in those days were rather disappointing.

Among other catalytic reactions, hydroformylation with Rh–NHC complexes was recently reviewed by Praetorius and Crudden [112]. Many references therein used rhodium chloride complexes as catalyst precursors, which in phosphite and phosphine complexes do not lead to hydroformylation activity in the absence of a base such as $NEt_3$, which is used for the elimination of HCl. The high activities observed in several mixed complexes of phosphines and NHC might be due to phosphine hydride complexes generated via HCl elimination with the use of an NHC as the base. Examples of fast catalysts in the absence of $PPh_3$ (100 °C, 50 bar) can be found in references [113] with rates up to 3500 mol mol$^{-1}$ h$^{-1}$, albeit that the activity for isomerization points to the formation of carbonyl rhodium catalysts. The catalyst precursor in this study is a rhodium halide NHC complex and thus, without addition of a base, no hydride formation would be expected, unless the NHC functions as a base. Interestingly, in this study the most electron-poor NHCs gave the fastest catalysts, but this could also be due to the precursor instability rather than an intrinsic change in activity. Unlike phosphine and phosphite-based catalysts, the steric bulk in NHC-based catalysts did not change the product distribution or rate.

It was shown by HP NMR spectroscopy that the Rh—C(NHC) bond in part survives hydroformylation conditions, but it was stated that this was no definite proof for the catalytic activity for hydroformylation of these species [114]. Catalytic activity in most cases was only observed in the presence of phosphines. Square-planar bis-NHC rhodium carbonyl hydrides have been prepared, but they were not tested in hydroformylation [115]. A phosphine-free mono-NHC rhodium dicarbonyl acetate gave moderate hydroformylation activity, which could be boosted by the addition of 1 equiv of $PPh_3$ [116].

Until today there are still doubts whether NHC complexes of rhodium lead to hydroformylation activity. In the following a few examples have been collected to underscore this. Laï and coworkers synthesized a Rh complex containing a chiral NHC and hydroformylation of styrene with this complex gave results similar to phosphorus-ligand-free carbonyl catalysts, that is, a low branched/linear ratio and negligible ee (80 °C, 12 bar) [117]. Addition of $PPh_3$ to this system gave a higher conversion to branched aldehyde but also low ee.

Rhodium(III) cyclopropenylidene complexes of the type [RhCl$_3$(PPh$_3$)$_2$(2,3-di(aryl) cyclopropenylidene)] were synthesized via oxidative addition of 1,1-dichloro-2,3-diarylcyclopropene fragments to rhodium(I) precursors by Wass and coworkers [118]. Hydroformylation of 1-hexene (90 °C, 20 bar) with these complexes led to catalysis results which were strongly suggestive of decomposition of the carbene complex.

The rhodium(I) complexes (1,5-cod)Rh(BIAN-SIMes)Cl and (1,5- cod)Rh(BIAN-SIPr)Cl were synthesized by Dastgir and co-workers [119]. The catalytic activities for the rhodium(I) complexes were evaluated for the hydroformylation of 1-octene at 100 °C and 20–55 bar. The low l/b ratios (often <1) suggest that catalysis is due to carbonyl-only species, unless this selectivity was also an effect of NHCs (Scheme 8.13; in the presence of ligands L also complexes **1**, **2**, and so on may form). For none of these systems was *in situ* IR reported as a way to prove the presence of modified rhodium carbonyl hydride complexes.

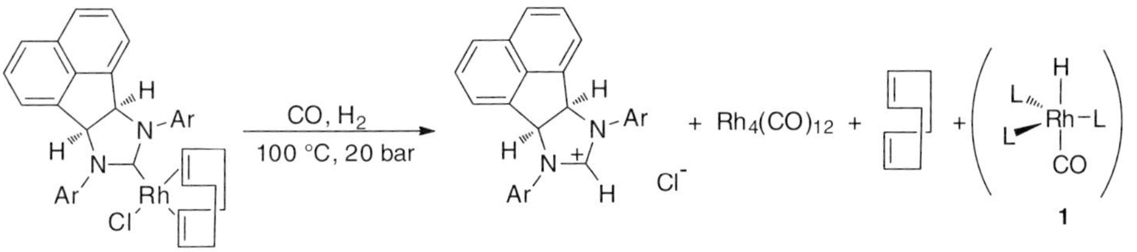

**Scheme 8.13** Formation of rhodium carbonyls from NHC complexes.

Rhodium(I) carbene complexes of the type [Rh(NHC)(cod)(halide)] with P(OPh)$_3$ as a modifying ligand for 1-hexene hydroformylation under mild conditions (80 °C, 10 bar) were also studied by Trzeciak and coworkers [120]. The l/b ratios found were higher for this catalyst than for that not containing NHC in the precursor, but the rates were not as high as one can obtain with pure triphenyl phosphite systems, in the absence of halides [57]. It was assumed that under the conditions of the catalytic process, rhodium(I) carbene complexes, [Rh(NHC)(P(OPh)$_3$)$_2$X], reacted with H$_2$/CO, giving a catalytically active rhodium(I) hydrido complex containing an *N*-heterocyclic carbene ligand, although for triphenyl phosphite no evidence was presented and only the presence of [HRh(CO)(P(OPh)$_3$)$_3$] and [Rh(NHC)(P(OPh)$_3$)(CO)X] complexes was confirmed. When P(OCH$_2$CF$_3$)$_3$ was used as the modifying ligand with [Rh(NHC)(cod)Br] as the precursor, formation of two hydride species HRhL$_4$ and HRhL$_3$(CO) was evidenced, while a residual hydride signal was assigned to a species containing one NHC and two phosphite ligands; direct proof for a Rh−C (NHC) bond could not be obtained. This ligand gave even higher l/b ratios, but these were also reported previously for P(OCH$_2$CF$_3$)$_3$, in the absence of NHC [121].

The present brief survey of NHC complexes in rhodium-catalyzed hydroformylation will be concluded by describing the work of Veige and coworkers [122]. They considered that in hydrogenation (and other organometallic catalytic reactions) "the Achilles heel is the tendency for NHC reductive elimination to the imidazolium salt [NHC-H]$^+$X$^-$", as reported by Cavell and coworkers [123], more especially for Pd (Chapter 1, Scheme 1.42, and Scheme 8.14).

With the aim to stabilize NHC complexes and to reduce the tendency to imidazolium salt elimination, they decided to synthesize chelating bisNHCs, and, better still, chiral ones, in order to find proof of ligand participation in case ees would be obtained. Thus they synthesized a chiral ethanoanthracene ligand bearing two NHC

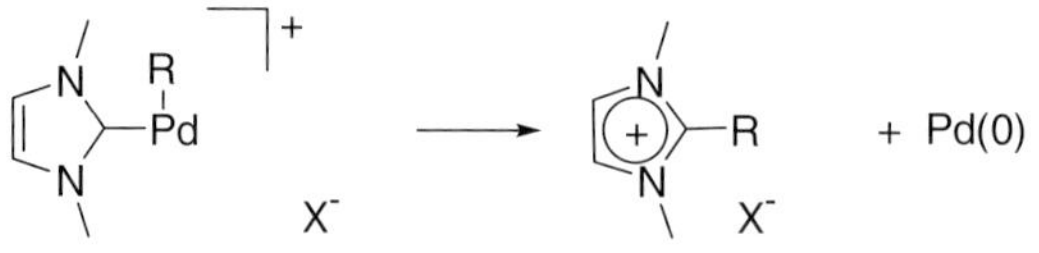

**Scheme 8.14** Reductive elimination of an NHC-alkyl salt (R=alkyl, hydride).

moieties. Hydroformylation of styrene was conducted at 50 °C and 30–100 bar. The gas uptake curves had a clear sigmoidal character and the ees were very low. Different ligands gave very similar results and after the reaction imidazolium salts were identified in the NMR spectra. It was concluded that rhodium carbonyl species not containing NHCs were responsible for the catalysis.

With few exceptions [13, 81], rhodium catalysts for hydroformylation contain weak electron donors or electron acceptors as phosphorus ligands, while cobalt catalysts perform best with strong σ-donors. Van Rensburg and coworkers argued therefore that NHCs might be more suitable ligands for Co-catalyzed hydroformylation than for Rh catalysis [124]. They reported the synthesis and structure of the first cobalt carbonyl carbene dimer $Co_2(CO)_6(IMes)_2$. Using this compound at 170 °C and 60 bar of syn gas (2 : 1) no hydroformylation activity was observed. Instead direct observation of the elimination of imidazolium salts with $Co(CO)_4^-$ as the counter anion was found (see Scheme 8.14 for Pd).

## 8.3.7
### Two-Phase Hydroformylation

There are several ways for two-phase hydroformylation (three-phase, including the gas phase!) to occur, *viz.* aqueous/organic phase, ionic liquid/organic, fluoro phase/ organic, and immobilized catalyst as a solid phase.

From a commercial point of view two-phase (aqueous/organic) hydroformylation is very important as it was reported that this is the most economic process for lower alkenes (propene and 1-butene) [125]. Butenes may be applied as a mixture, because, using the present catalyst and conditions, only the fraction of 1-butene is converted. An alkene recycle is not convenient as soon 2-butene would be the major alkene present; the solution applied is to send the rest of the stream of alkenes to a second reactor containing a cobalt catalyst. Also, one could use the remaining butenes for another process, such as butylation of aromatics. This two-phase hydroformylation process was discovered by Rhone Poulenc [126] and commercialized by Ruhrchemie in 1984 [127]. The basic principle is the use of two liquid phases in both reactor and separator, one phase being the crude product and the other phase containing Rh catalyst and excess ligand, thus allowing efficient catalyst/product separation. The second, polar phase in this instance is water, and water-soluble ligands are used. The most successful ligand (used in the Ruhrchemie/Rhone Poulenc process) is triphenylphosphine tri-meta-sulfonate (tppts). Tppts synthesis via triphenylphosphine sulfonation [128] is not straightforward, but has been optimized using purification via extraction with amines [129] and minimizing phosphine oxidation using water-free $H_2SO_4/H_3BO_3$ [130]. In the two-phase hydroformylation reactor, apolar propene diffuses into the water phase and apolar aldehydes and by-products leave the water phase again; the major component in the reactor is the aqueous catalyst phase and the organic, unproductive portion is kept to a minimum (6 : 1). In this way water becomes the continuous phase, which gives an overall high Rh concentration in the reactors and a high space–time velocity. The extraction takes place in the reactor, while the two phases are strongly mixed by vigorous agitation, forming a yellowish milky liquid. Note that in this process the apolar heavy ends leave the reactor and separator in the organic stream, where they are separated from the product by distillation and the heavy ends remaining can be sent to an incinerator. In distillative processes heavy ends accumulation constitutes a major problem, as they either remain in the reactor (stripping process), or they are recycled with the bottom of the distillation unit together with the catalyst. In principle, this extraction system is an "open" system, that is, it would allow Rh loss by simple solubility in the organic phase. Separation of the two phases is extremely fast, as was shown in laboratory experiments, and it is also highly efficient as the losses of rhodium in the organic phase are extremely low (1 ppb) [127].

Figure 8.7 shows a schematic view of the process [125, 131]. The reactor is mechanically stirred, to maximize mass transfer for both gas/liquid- and liquid/liquid phase transfer. In spite of this, the reaction still seems to be mass-transfer limited and restricted to the liquid/liquid interface [43], which may be caused by high catalyst activity and temperature combined with the low solubility of propene. The reactor cooling is integrated with the butanal distillation reboiler. The temperature is 120 °C, the pressure is circa 50 bar ($H_2$/CO ratio of 1.0). The relatively high

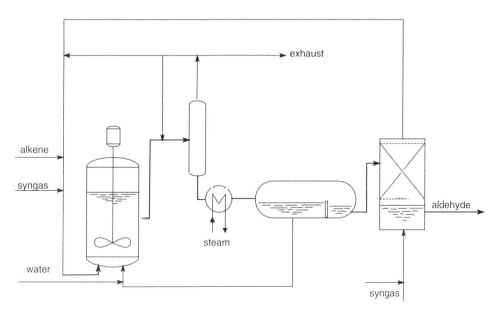

**Figure 8.7** The Ruhrchemie–Rhone Poulenc two-phase process.

temperature of 120 °C seems to be required because of the limited solubility of propene in water, the desire for energy integration, and the need for high propene conversion per pass. This higher temperature might have an impact on catalyst stability, although a high ligand/Rh ratio is applied. The water phase contains circa 300 ppm Rh at a tppts/Rh molar ratio of 50–100. Catalyst stability is higher than that of the Union Carbide process (now Dow), but it should be mentioned that tppts is added continuously, as the ratio should be kept at 50 or higher to ensure catalyst stability. The organic decomposition products of tppts were not identified and apparently no phosphido-rhodium species are formed as in the one-phase distillative process. It is surprising that the catalyst in this process is more stable than that in the Dow process since water seems, in general, a major source of phosphine degradation. There are no comparative data on phosphine decomposition, but the important thing is that the RCRP catalyst stays intact. CO pressure should be maintained relatively high, because its solubility in water is around 10 times lower than that in organic media and at lower pressures catalyst decomposition takes place.

Heavy ends are minimized by pH control (pH 5.5–6.2); the presence of some $CO_2$ (1–3%) in the syn gas seems to be beneficial to minimize heavy ends [132]. Water is slightly soluble in butanal and water lost that way has to be replaced continuously. Minor side-products may also involve enones resulting from aldol condensation and, since heavy end formation is lower and the organic species are taken to the organic phase, formation of dormant species from enones is also less in this process than in one-phase processes (see Scheme 8.5).

The major advantages of this process are [126]: (i) the high level of heat integration, (ii) simple separation of catalyst and product/heavy ends, (iii) better selectivities (99% aldehydes, 95% linearity; no alkene hydrogenation) and (iv) lower sensitivity to some poisons (the preference of some poisons for the organic product layer). The higher product linearity seems to be hardly an advantage in view of the value and market size for isobutanal derivatives. The higher reactor pressure and the heat integration might increase the capital expenditure somewhat.

Higher alkenes cannot be hydroformylated in aqueous biphasic media as the solubility of alkenes decreases drastically with higher carbon numbers in the chain and already at 1-octene the activity of the tppts system is practically zero. Many additives, extended ligands forming micelles or vesicles raising alkene solubility [133], mixed solvent systems [134], and so on have been investigated, but often separation turned out to be much less effective than that of the RCRP system.

Another approach involves one-phase catalysis followed by two-phase separation as discussed in the introduction (Section 8.2, the Kuhlmann process for Co). An acid–base switch, or base–acid switch to remove the catalyst from the organic phase into water increases the chances of catalyst decomposition or inhibition by the acidic group during the one-phase catalytic step, as was shown by Buhling and coworkers [50b]. They use *meta-* and *para-*carboxylic acid functionalized triarylphosphines, which can be removed from the organic solvent after the reaction by treatment with an aqueous basic solution together with the rhodium catalyst. Unfortunately, the large excess of carboxylic acid gives large amounts of inactive rhodium carboxylates under mild conditions (80 °C, 20 bar syn gas, Scheme 8.15, method A). The reverse acid–base couple (method B) leads to one-phase hydroformylation without inhibition, but in their protonated form in water the *p*-aminoarylphosphines are susceptible to P–C cleavage if the aromatic ring is activated as a result of protonation of nitrogen. Separation of the charge on the amine from the aromatic ring by a methylene group solved this problem effectively, method C [52]. As mentioned in Section 8.2 the production of salts for each cycle is less desirable nowadays, even though only "catalytic amounts" are produced. Scheme 8.15 summarizes an example of each method.

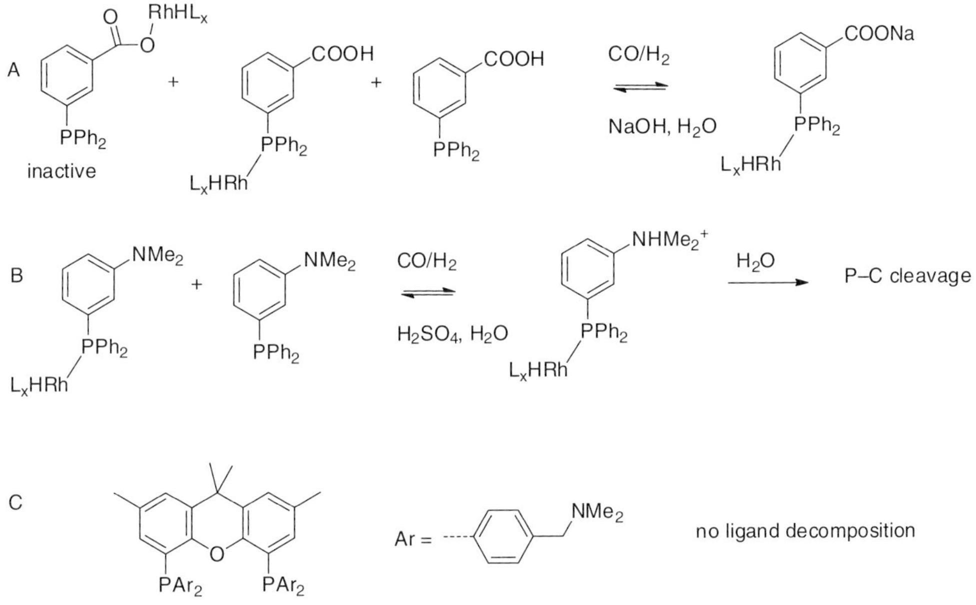

**Scheme 8.15** Disadvantageous reactions in one-phase catalysis, two-phase separation (A and B) and an example of a successful system (C).

A very large variety of immobilization and incapsulation techniques of hydroformylation catalysts has been reported and often "good stability and recyclability" was reported, but data on decomposition are virtually absent as catalytic runs on arylphosphine and aryl phosphites should be conducted for several days before conclusions can be drawn or decomposition products can be identified (for a recent report on incapsulation see Ref. [155] and references therein). Stability of an immobilized catalyst can depend on minor details, as for instance was shown by Jongsma and coworkers [136], who found that a polymeric phosphite grafted on silica gave a stable catalyst for 10 days in benzene as the solvent without loss of activity, but led to rapid leaching in toluene. This was assigned to the low "solubility" of the polymer chains in toluene, which led to incomplete complexation of rhodium and thus leaching. All immobilization techniques, introduced to solve the "problem of catalyst separation", seem to introduce new obstacles, which explains why very few of these techniques have found application.

In recent years room-temperature ionic liquids (IL) have proven to be attractive and alternative media for many homogeneously catalyzed hydroformylation reactions. The properties of ionic liquids can be tuned by adjustment of the cation–anion pair. For example, a plethora of imidazolium ionic liquids can be prepared by varying the anion and the hydrocarbyl groups on the imidazole ring. The combination of high density, high stability, immeasurable vapor pressure, and the possibilities for catalyst immobilization via ionic interactions allows easy separation of the catalyst by simple phase separation or distillation and recycling, as most products are not, or only slightly, soluble in the ionic phase [137]. The purity of the ILs is extremely important for their stability and Mehnert found that traces of acids reduce the stability considerably [138]. Using tppti* (tppti* = tri(m-sulfonyl) triphenyl phosphine 1,2-dimethyl-3-butyl-imidazolium salt) Mehnert found losses of rhodium in the semipolar organic phase, which could be reduced by choosing the best IL; [bmim][PF6] gave the best results due to its very low miscibility with polar substances.

The fastest and most selective catalyst in IL was reported by Bronger and coworkers (60 bar, 100 °C, TOF = 6200 $h^{-1}$, l/b = 44) [139]. A Xantphos bis-imidazolium salt of $PF_6^-$ was the catalyst of choice. No catalyst leaching (Rh-loss <0.07% of initial rhodium intake, P-loss <0.4% of the initial phosphorus intake) or losses in performance could be measured during 1-octene hydroformylation recycle experiments in 1-butyl-3-methylimidazolium hexafluorophosphate. At low catalyst loadings, activities and regioselectivities competitive with one-phase catalysis in conventional solvents were observed. At high catalyst loadings the system is extremely stable and has a long shelf-life as a result of the formation of stable, if inactive, rhodium dimers.

In the same way as aqueous catalysts have been used on a solid support (SAPC) as introduced by Arhancet and coworkers [140], ILs have also been brought onto solid supports to facilitate further the separation of catalyst and product and to enhance phase contact between the two phases by forming thin layers [141, 142] (SILP, supported ionic liquid phase catalysts). They were studied with regard to their stability in the continuous gas-phase hydroformylation of propene. Kinetic data were in good agreement with known results from biphasic hydroformylation in the liquid state, which confirmed previously published results on the homogeneous nature of the heterogenized Rh-SILP catalyst. Long-term stability exceeded 200 h time on stream with minor loss in selectivity (0.1% $h^{-1}$). The formation of high-boiling side-products dissolved in the ionic liquid layer was thought to be the cause of the slow deactivation. The small decrease in activity could be compensated by a vacuum procedure, regaining the initial activity, which showed that catalyst loss was not due to catalyst decomposition.

Biphasic catalysis in which one of the phases is a fluoro phase (FBC) was introduced independently by Vogt and by Horvath and Rabai [143]. The system is based on the immiscibility of fluorinated compounds with organic solvents, which is due to the polar C−F bond and low polarizability of fluorine making this a "non-interacting" solvent. Ligands functionalized with long fluorinated alkyl groups, the so-called "pony tails", are soluble in the fluorous phase, whereas the substrates generally dissolve in the organic phase. At low temperature the phases separate, while at higher temperatures one phase may form. Separation is usually not perfect and further extractions are needed. Other concerns are the cost of the fluorous liquids, catalyst leaching, and product contamination by fluorine compounds.

A detailed hydroformylation study was conducted by Horvath and coworkers using $P(C_2H_2C_6F_{13})_3$ as the ligand [144]. The hydroformylation of 1-decene was studied using rhodium (100 °C, 11 bar) in a 50/50 vol% toluene/$C_6F_{11}CF_3$ solvent mixture, which forms a homogeneous liquid phase above 100 °C. According to NMR spectroscopy at high pressure the solution structures of the rhodium complexes of this ligand are similar to those of triphenylphosphine **1** and **2**. The hydroformylation activity is slightly higher than that of the catalyst based on trioctylphosphine, but an order of magnitude below that of triphenylphosphine. Due to the electron-withdrawing character of the fluoro-tails the l/b ratio is similar to that of tpp.

Foster, Cole-Hamilton and coworkers reported on another successful fluorophase hydroformylation system [145]. The best ligand in their studies was $P(O-4-C_6H_4C_6F_{13})_3$, an aryl phosphite equipped with a fluoro-tail, while they used perfluoro-1,3-dimethylcyclohexane as the solvent. The catalysis was carried out at 70 °C, 20 bar, because at higher temperatures decomposition was observed, which was most likely phosphonate formation following low levels of aldol condensation (Scheme 8.11). It was found that 1-octene was completely miscible with the solvent under the conditions used for hydroformylation, but the product nonanal phase separated. Rhodium leaching was estimated to be at the level of 1 ppm, which is an impressive number, but still for a bulk process chemical that may not be good enough; tens of ppb may be the limit, depending on the, strongly fluctuating, rhodium price. Phosphine loss was about 3%, which is much lower than that of other processes. In their comparison of rates and selectivities with other processes this fluoro process comes out highly favorably, with a Rh/P ratio of only 3 (at 1 mM of Rh).

Mathivet, Monflier, and coworkers studied several fluoro-tail aryl phosphites as ligands in the rhodium-catalyzed hydroformylation of higher alkenes (1-decene) at 80 °C and 40 bar of syn gas [146]. The miscibility of 1-decene and its hydroformylation product with 1-H-perfluorooctane was studied at various temperatures and, interestingly, it was found that 1-decene concentrations are about 10 times higher than those of undecanal in $C_8HF_{17}$. The solubility of the latter at 75 °C was about 1%. Since the rhodium complex is also in the fluorophase the higher solubility of 1-decene is an advantage. The fluorophase needs to be recycled and unconverted alkene remaining in it is advantageous. The product layer will contain fluorooctane and 1-decene as well, and clearly either extraction or distillation is necessary. A key issue is the efficient recycling of the fluoro-tail phosphite and catalyst and this turned out to be the case as in the first four cycles the conversion was roughly the same (or even increased initially) but the regioselectivity dropped, which points most likely to some ligand decomposition.

Two typical ligands are shown in Figure 8.8. The first ligand tested, **14**, [147] showed rapid decomposition in water–thf mixtures (complete decomposition in 24 h),

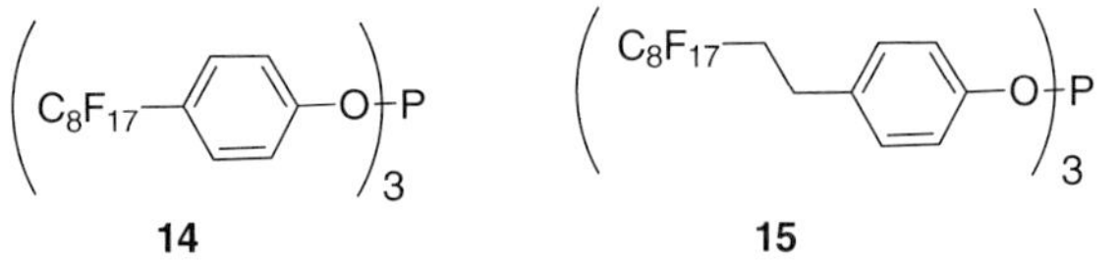

**Figure 8.8** Fluoro-tail phosphites.

while the second ligand, **15**, showed no hydrolysis at all after 48 h. The separation of the fluoro-tail from the phenol ring by two methylene groups sufficiently reduces the electron-withdrawing character of the perfluorooctyl group.

As initially no water arising from aldol condensation is present, decomposition of the ligand was proposed to take place via the formation of dioxophospholanes from phosphite and the aldehyde product (Scheme 8.12). Activities and selectivities are high, but more stable phosphites need to be developed.

### 8.3.8
### Hydroformylation by Nanoparticle Precursors

In view of the increasing interest in the use of metal nanoparticles (MNPs) in the last decade as catalysts or catalyst precursors a few reports related to hydroformylation will be mentioned here. There is no proof for a hydroformylation reaction taking place on the surface of an MNP. The selectivities and low rates obtained can be best explained by homogeneous catalysts formed by leaching, sometimes with the aid of an added ligand [148]. As a side-reaction, hydrogenation of the alkene substrate may occur, a reaction typical of heterogeneous Rh catalysts. Occasionally in solution the rhodium hydrido carbonyl complex formed with the added ligand was characterized in solution by NMR spectroscopy [149]. With regard to catalyst–product separation it suffers from the same drawbacks as immobilized catalysts, and probably the problems are more severe. Moreover, during the reaction only part of the expensive Rh catalyst is active, which is not very attractive either. MNPs will become interesting as a catalyst for Rh-catalyzed hydroformylation when active species on the metal surface are discovered.

### 8.4
### Palladium-Catalyzed Alkene–CO Reactions

### 8.4.1
### Introduction

Palladium-catalyzed alkene–CO reactions in protic solvents give rise to esters, oligomers (oligoketone esters), polyketone polymers, diesters, and amides, and many other compounds if other nucleophiles are present. Polyketones, perfectly alternating polymers of an alkene and CO, came into focus after the discovery of the effect of weakly coordinating ligands and bidentate phosphines in palladium catalysis by Sen [150] and Drent [151], respectively, in the early 1980s. Here we will focus on ester, oligomer, and polymer formation from alkene and CO. The early work was reviewed by Drent and Budzelaar in 1996 [152], but the mechanistic explanations are now outdated and even since the appearance of a more recent monograph [153] additional insights have been published [as reviewed in Ref. 154]. Polyketone is a desirable product as it is a high performance polymer that potentially can be made at a low price as the feedstocks are very cheap. It has material properties similar to nylon as a so-called engineering thermoplastic [155], but if it could take advantage of the large-scale of production it would be much cheaper than nylon. In particular it has a high chemical resistance, excellent barrier properties to hydrocarbons and small gases such as oxygen, it is less apt to deformation than polyamides, and the hydrolytic stability is particularly good for aliphatic polyketones, leading to performance advantages over many condensation polymers such as polyamides and polyesters.

Copolymerization of ethene and carbon monoxide has been known for a long time (1951), but catalyst activities were rather low until the 1980s. Coordination polymerization for ethene/CO was first reported by Reppe [156] using nickel cyanide catalysts. The molecular weights were very low and, in addition to the polymers, diethylketone and propanoic acid were produced. The first palladium catalyst (phosphine complexes of $PdCl_2$) producing an alternating polymer of CO and ethene (Figure 8.9) was reported in 1967 [157]. Molecular weights were high and its potential was recognized [158, 159], but the product contained large amounts of precipitated palladium.

The discoveries of Sen and Drent enabled the commercial synthesis of polyketone as the yields increased to values around tens of kg per gram of palladium. The copolymer of ethene and CO has a melting point of approximately 260 °C, which is too high for melt processing (e.g., extrusion) of the product without decomposition. A copolymer containing a few percent of propene as the third monomer led to melting points around 220 °C. The latter polymer was commercialized, many years later, by Shell in 1996 on a small scale in the UK (20 000 t a$^{-1}$), but the plant was closed in 2000. Many patents have expired by now, but entering the area of bulk production requires large investment, as usual in the bulk chemical business, not only from the producers but also from the users (e.g., car manufacturers). In addition, engineering thermoplastics based on polyesters, polyarylketones, or polyamides are available in many grades, from several producers, and for E/CO polyketone the market was (and still would be) limited.

Depending on the ligand used and the conditions, the same reaction of alkenes, CO and a protic component (water, alcohol) and a palladium catalyst leads to esters via a hydroxycarbonylation or methoxycarbonylation of alkenes. Intermediate cases will

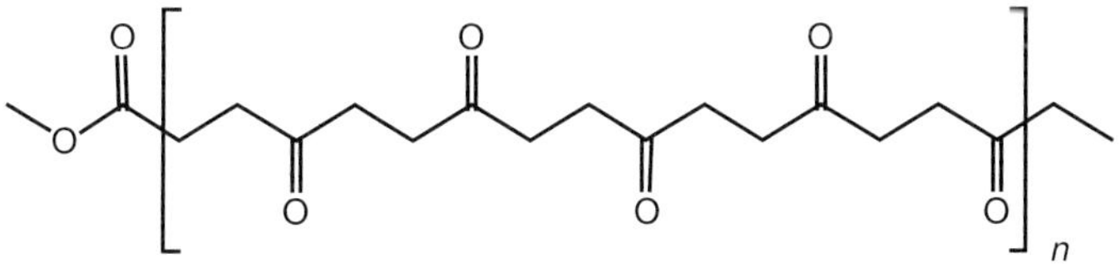

**Figure 8.9** Polyketone from ethene and CO.

lead to oligomers. The main products of Sen's tpp-based catalyst were actually methyl propanoate and oligomers. Extremely fast and selective catalysts for the formation of methyl propanoate (see Section 8.4.5) have been discovered and developed by Lucite International (formerly ICI) and this process was commercialized in Singapore in 2008 [160]. It is the starting material for the manufacture of methyl methacrylate (MMA), which alternatively is made from acetone via hydrocyanation. This route produces stoichiometric quantities of ammonium sulfate as a co-product, for which there is no use and it is dumped into the sea. The new catalytic route from ethene, methanol, CO, and formaldehyde (also from methanol and thus syn gas) is much more attractive, both from an environmental and a commercial point of view. Lucite is the world leader in acrylates; it is now fully owned by Mitsubishi Rayon Co. Ltd.

## 8.4.2
## Brief Mechanistic Overview

The catalytic cycle for the formation of esters, oligomers, or polymers can start by insertion of ethene into a palladium hydride bond, or with the formation of a carbomethoxy species. This was already recognized in the 1960s and 1970s when the first reports appeared [161]. Scheme 8.16 shows the two catalytic cycles for methoxycarbonylation of ethene (pre-coordination of substrates and equilibria have been omitted for most steps).

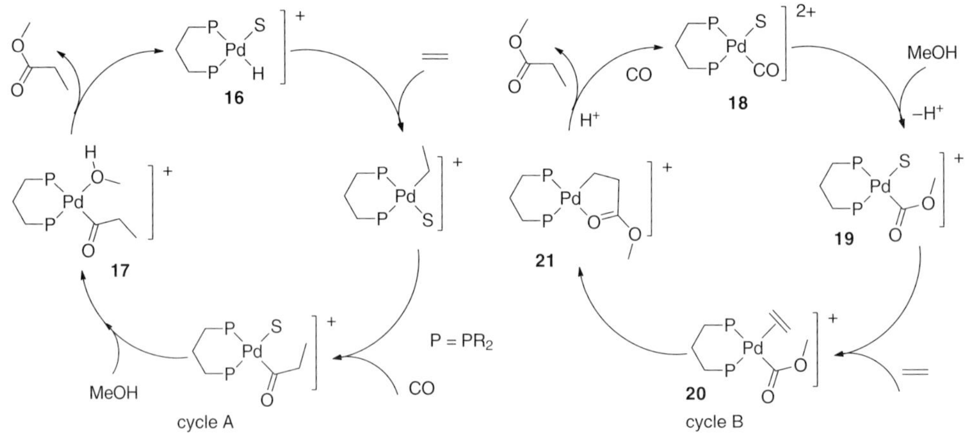

**Scheme 8.16** Two catalytic cycles for the methoxycarbonylation of ethene.

A cycle that starts with hydride **16**, will return to hydride **16**, and, unless another reaction occurs, palladium will remain in cycle A. Already in 1979 Toniolo and coworkers showed that in the alkoxycarbonylation of propene in butanol using $PdCl_2(PPh_3)_2$ as the catalyst it was advantageous to use a mixture of CO and $H_2$, and the latter was supposed to react with the dichloride producing a hydridochloropalladium species, thus initiating cycle A [162]. The two insertion reactions in cycle A are supposed to be fast. The reaction of the propionyl intermediate **17** to methanol to give product has been drawn with the methanol molecule coordinated to palladium, as a nucleophilic attack from outside the coordination sphere has not been observed so far [163]. Hydride **16** may undergo reductive elimination to give Pd(0) and HX, which is a reversible reaction for phosphine ligands. If the reaction is not reversed, palladium can be oxidized *in situ* to form a species such as **18** that can start a catalytic cycle B, and in this way reactivation takes place (see top of Scheme 8.17). Toniolo *et al.* [162] isolated a carbobutoxypalladium species (*trans*-BuO(O)CPdCl(PPh$_3$)$_2$ in the reaction mentioned above and, although the catalytic cycle involved cycle A, catalysis could also be initiated using this carbobutoxy species of type **19**.

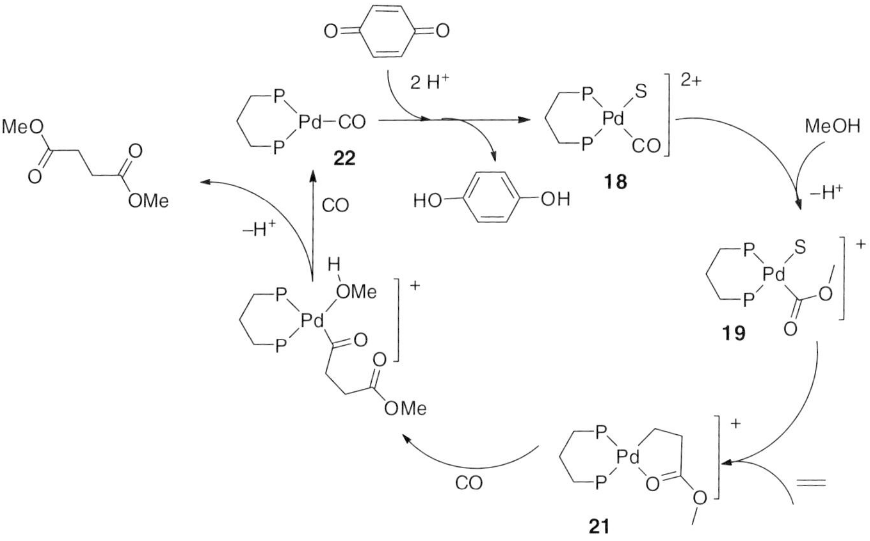

**Scheme 8.17**  Formation of succinic ester showing reoxidation of Pd(0) by quinone.

As regards cycle B, the conversion of **18** into **19** may involve coordination of methanol (methoxy) to palladium and the formation of carbomethoxy involves a migratory insertion; also here probably both mechanisms take place, depending on the system. Migratory insertion of ethene in **20** is a relatively slow step according to stoichiometric reactions [162, 164]. Intermediate **21**, may or may not be stabilized by an internal ketone coordination to palladium, often called "back-biting" [165]. In some cases the formal protonation **21** is not a direct reaction, but it was shown that first a β-hydride elimination occurs, followed by re-insertion giving a palladium enolate, which can be rapidly protonated to give the product [166]. Catalysis via cycle B can be interrupted if reduction to Pd(0) by CO occurs, for instance by hydride elimination from a carbohydroxy species formed from CO and water, and reductive elimination of HX. If this happens, and cycle B is the only productive mechanism, re-oxidation of Pd(0) to Pd(II) must be applied; a common oxidizing agent for this reaction is quinone together with HX. As quinone is often used to reactivate a palladium catalyst, an example of this sequence is shown in Scheme 8.17 for the formation of succinate.

When multiple insertions of CO and ethene occur, giving polymer or oligomer, the termination reaction, either the one from cycle A or cycle B, determines the initial state of the new cycle. Terminating a cycle with the elimination of an ester (cycle A) returns palladium to the palladium hydride species that will start the new cycle with alkene insertion to give an alkyl head group of the oligomer. Thus, if only one chain transfer mechanism is operative, all oligomers, in this instance, will have an alkyl head group and an ester end-group. Under mild conditions this was indeed found for catalyst systems based on dppp and Drent's conditions [167]. At higher temperatures oligomers having the same head and end groups and mixed ones were found in a statistical ratio. This means that two chain transfer mechanisms are operative, and since a growing chain "does not know" how it started it may have the same head and end group, but the total of alkyl and ester end groups is the same, unless the palladium initiators are oxidized or reduced, or otherwise transferred into their counterpart.

### 8.4.3
### Early Reports on Decomposition and Reactivation

A key problem of palladium catalysis in reactions consuming CO is the reduction of the active Pd(II) to metallic palladium by CO and water. This is very common in palladium catalysis (alkenes and water will do the same via a Wacker reaction) and it suffices here to quote a few reports that deal explicitly with this matter. Addition of extra $PPh_3$ in the alkoxycarbonylation of propene using $PdCl_2(PPh_3)_2$ gave less precipitation of palladium metal [168], but it did not change the regioselectivity (35% branched at 100 °C, 95 bar). Rates for the chloride-containing catalysts are modest. Addition of LiCl lowers the rate but leads to a higher preference for branched esters for aliphatic alkenes. Lower concentrations of alcohol gave lower rates but higher b/l selectivity. This was explained by assuming the presence of different catalytic species, of which the more sterically hindered ones would give the linear product. Alternatively, a slow catalyst may lead to equilibration of branched and linear alkanoyl species, leading to more branching, while fast catalysts may give the kinetic, linear product [169]. For ethene and CO, in addition to methyl propionate formation at 1 bar of ethene, oligomers and polymers were also obtained at higher pressures of ethene (40 bar) with an HCl-rich catalyst system [170] and $PPh_3$ as the ligand. The polymer contains an excess of ethylketone end-groups, for which the water-gas shift reaction was held responsible (**18** to **16**, Scheme 8.18).

Chloride plays an important role in maintaining a palladium(II) complex, because other anions give easy reduction of palladium, not only by water, and CO or alkene, but also via reduction by $PPh_3$ [171]. An excess of HCl can revert Pd(0) to the active PdHCl species via oxidative addition and many catalytic systems with high acid concentration have been reported [172]. Routinely, Drent used $Pd(OAc)_2$ as the palladium precursor and ptsa (*p*-toluene sulfonic acid) was added in excess to form a palladium salt with the ptsa anion as a weakly coordinating anion [152]. In methanol

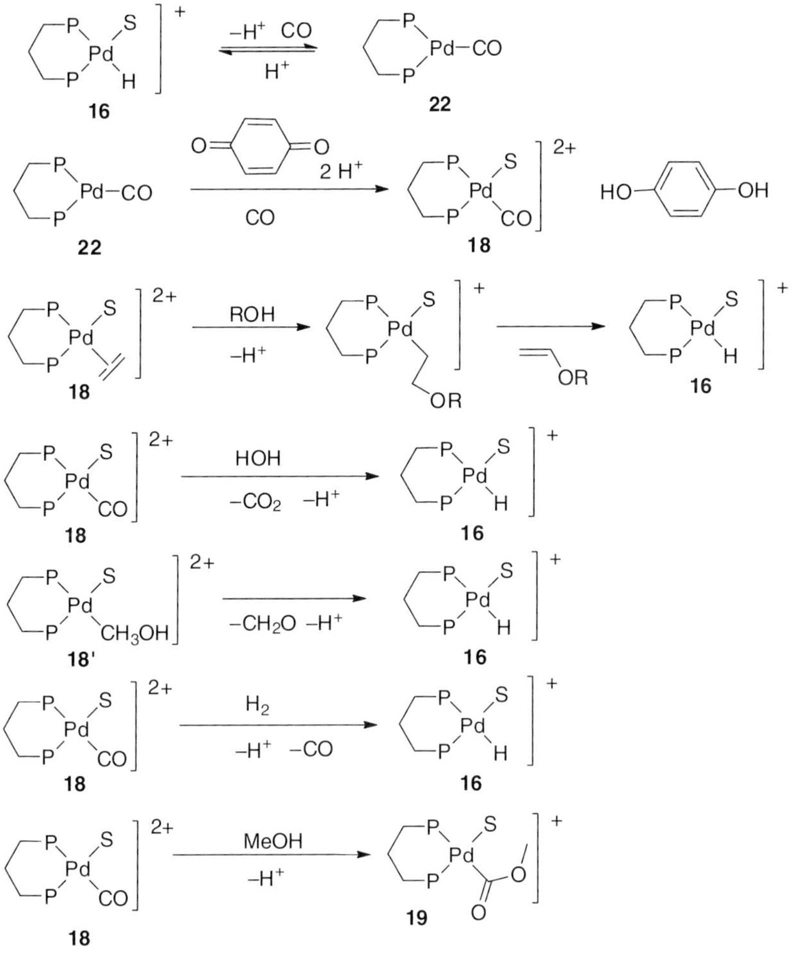

**Scheme 8.18** Conversion of palladium catalyst precursors.

the anion of ptsa is sufficiently weakly coordinating. For diphosphine ligands this is also a stable system, but for dinitrogen ligands this is not the case.

Interconversion of the potential precursors for cycles A and B plays an important role in palladium catalysis and methanol as a solvent and strong acids providing the anion were a fortuitous choice in the "Drent catalysis". Methanol in many cases aids the formation of the active species, be it a carbomethoxy species **19** or a hydride **16**, both starting from Pd(II). Methanol, water, and dihydrogen can also transform Pd(II) into hydride **16** via reaction with ethene or CO. The reactions are summarized in Scheme 8.18 and with these in hand one can understand the many findings reported on rate increases obtained with any of these additives. Which of the additives is effective as a reactivator reveals the dead end of the catalytic cycle, or which precursor is ineffective under certain conditions for certain substrates.

Palladium(II) species of type **18** are clearly more stable than palladium hydrides **16**, which may decompose to give Pd(0) species **22**, which are not active and have to be regenerated, either via the equilibrium **16** and **22** + HX, or via oxidation with quinones. In copolymerization reactions the most efficient additives are quinones, indicating that Pd(0) compounds present a dormant state. For catalysts based on phosphine ligands rate enhancements vary from 2–15 [173] showing that the ptsa present does not lead to sufficient stabilization of the hydride; most ligands used are arylphosphines or less strongly donating cycloalkylphosphines. The Pd(0) compounds formed are relatively stable and do not lead necessarily to metal deposition. In some instances they form dimers in a reaction with Pd(II) cations [174], one of which has been crystallographically characterized [175]. The dimers undergo oxidative addition with acid to give a hydride and a cationic Pd(II) complex and thus the formation of the dark orange dimers leads to a dormant state rather than to catalyst decomposition.

For dinitrogen ligands the situation is worse as neither hydrides nor Pd(0) complexes are stable and if the hydride intermediates do not rapidly undergo insertion of alkene and CO, they will decompose giving quaternary nitrogen salts and palladium metal. Therefore, for bipyridine and phenanthroline ligands rate enhancements of 200-fold were observed when quinones were added [176]. These rate enhancements are due to larger amounts of active catalysts and not to an increased rate of propagation, because the molecular weight of the polymers remains the same. For styrene copolymerization diimine ligands are favored and the polymerization always terminates via β-H elimination at high temperatures and as a result quinone must always be used to reoxidize Pd(0) to initiate the new chain via a carbomethoxy initiator. Thus, the end groups in these polymers are vinyl ketones and methyl esters. For polymerizations at room temperature or below different solutions have been reported not requiring quinones or methanol, see Section 8.4.4.

### 8.4.4
### Copolymerization

As we have seen above, methanol plays a crucial role in palladium carbonylation, both as an initiator and a chain transfer agent, and in methoxycarbonylation also as a reactant. Catalysts containing growing chains are more stable in the absence of methanol, not only because of its action as a chain transfer agent, but also because β-hydrogen elimination, in which methanol does not participate as a chain transfer agent, seems to be faster in methanol than in aprotic solvents. If one wants to do catalysis in aprotic solvents another mechanism of initiation is required. The most common technique is to start with alkylpalladium species, of which methylpalladium is preferred as it cannot decompose via β-hydrogen elimination. Monomethylation of a Pd(II) complex is not straightforward as most methylating agents will give dimethylpalladium species. It was found that $Me_4Sn$ provides selectively and quantitatively (dppp)Pd(CH$_3$)Cl and the like, which by reaction with Ag(I) salts of weakly coordinating anions can be converted to cationic catalyst precursors **23** [177] (see Scheme 8.19). A general precursor is (1,5-cod) Pd(CH$_3$)Cl, which can be reacted with the bidentate ligands of choice [178].

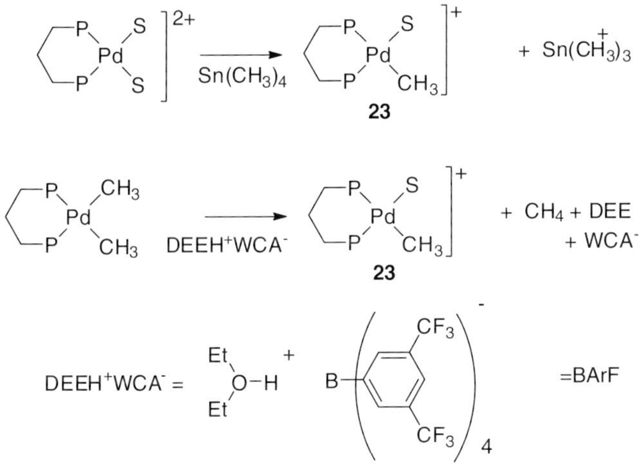

**Scheme 8.19** Formation of methylpalladium initiators.

Alternatively, dimethylpalladium complexes can be used as the starting complexes which are converted by acids with weakly coordinating anions (such as BArF, see Scheme 8.19) into active cationic monomethylpalladium complexes **23** [179]. The resulting species **23** containing a cis-coordinating bidentate is not very stable and it is usually generated *in situ* below room temperature. A bis-triphenylphosphine methylpalladium species is more stable, as now the phosphine donors occupy mutual trans-positions and stabilize the Pd–C bond cis to both P-donors (see Section 8.4.5). Methylpalladium complexes **23** were stable and useful precursors for kinetic studies of insertion reactions [179, 180].

As mentioned above, styrene is particularly prone to termination reactions in methanol and, at low temperatures, dichloromethane gives more stable systems and higher molecular weights. Particularly interesting is the catalyst development focusing on the use of 2,2,2-trifluoroethanol (TFE) as a solvent by Milani and coworkers [181]. TFE slows chain transfer reactions and palladium metal precipitation, but still aids in restarting the cycle by a carboalkoxypalladium species. Chain termination takes place as a β-hydrogen elimination and thus reoxidation of Pd(0) or PdH$^+$ is needed to achieve high production, which is effectively solved by the addition of quinones. In the absence of quinones alkyl-ketone and vinyl-ketone end-groups were observed. The use of an electron-withdrawing tetrafluoro-substituted phenanthroline ligand led to molecular weights as high as 1 000 000 for *p*-Me-styrene/CO (7000 repetitive units) [182].

The productivity of several diphosphine complexes was studied by Bianchini and coworkers [183]. They found a low productivity for dppe and palladium acetate and their *in situ* analysis showed that this was due to a disproportionation reaction leading to cationic complexes Pd(dppe)$_2^{2+}$ and palladium acetate. The latter quickly decomposes to palladium metal under the conditions and the former is inactive in the copolymerization. Also, in dichloromethane the dppe based catalyst was less active than the methyl-substituted 1,4-butane-diyl bridged diphosphines. The formal chain transfer via protonation of an alkyl chain end takes place via enolate formation (see next paragraph).

In Scheme 8.20 the polymer chain transfer mechanisms are summarized, essentially based on the two cycles shown in Scheme 8.16 for the methoxycarbonylation of ethene; chain transfer can take place via a formal protonation of the alkyl chain-end in **21p**, enforcing the start of the new chain via a carbomethoxy species, or via the alcoholysis of acylpalladium species **17p**, forming a hydride as the new initiator. A difference from Scheme 8.16 is that now a chain might start as **16**, but it can undergo the next chain transfer via **18**, thus forming two alkyl chain ends. As stated in the introduction, if one mechanism is much faster than the other, polymers with an

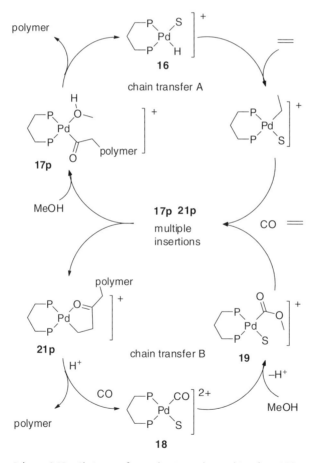

**Scheme 8.20** Chain transfer mechanisms observed in ethene/CO copolymerization.

ethylketone and a methyl ester end group will form exclusively, provided that the intermediate palladium species **16** and **18** do not undergo interconversion.

While dimethylpalladium compounds are readily protonated to give cationic methylpalladium complexes, the protonation of cationic methylpalladium species is extremely slow, and the same is true for complexes **21** and **21p**. Van Leeuwen found that the formal, simple protonation takes place at the enolate complex formed by β-hydride elimination and re-insertion of the enone in the palladium hydride bond [184] (see Scheme 8.21). The rate determining step is the β-hydride elimination reaction and thus the reaction was independent of the concentration of water, methanol, or acid, within the window studied.

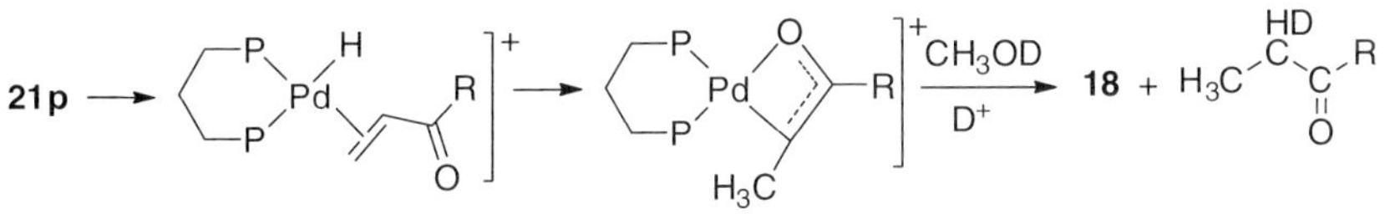

**Scheme 8.21**  Protonation of alkyl chain end via enolate.

The conversion of a palladium acyl species with methanol or water into the ester or acid has long been considered to be a second sphere nucleophilic attack of alcohol or water at the acyl carbon atom. In the next section on alkoxycarbonylation of alkenes, for which the reaction is more relevant perhaps, we will see that most likely the oxo nucleophile first coordinates to palladium before the ester is formed.

## 8.4.5
## Methoxy- and Hydroxy-carbonylation

In Section 8.4.2 we outlined mechanisms A ("hydride" mechanism) and B ("carbomethoxy" mechanism) proposed for alkoxycarbonylation of alkenes (see Scheme 8.16), named after the initiating species. For both mechanisms stoichiometric reactions of the individual steps were found that lend support to each one of them. From their studies on palladium chloride complexes containing triphenylphosphine Toniolo and coworkers suggested already in 1979 that catalysis based on the hydride mechanism A is much faster than that based on mechanism B, mainly because the stoichiometric insertion reaction of alkenes into carboalkoxypalladium chloride and two tpp ligands, was very slow, or did not take place [162]. Dekker and coworkers found that with the use of cis-coordinating bidentate phosphines, the preferred ligands for polyketone formation, insertion of alkenes took place even in the chloride acetylpalladium complexes and it was proposed that five-coordinate species may be involved in alkene [164] (and CO [180, 185]) insertion reactions. In cationic species obtained by exchanging chloride for a weakly coordinating anion (triflate being sufficient in methanol-containing media) the reaction was much faster. In the absence of CO the reaction was followed by rapid β-hydride elimination, unless the geometry disfavored this reaction, as for norbornene (for $PPh_3$ see [186]). For ligands with small bite angles, such as dppe, the fastest elimination was found.

The hydroesterification (methoxycarbonylation) of cyclohexene with $PPh_3$ complexes of palladium tosylate was studied in detail by Vavasori and coworkers [187]. The highest rate was obtained in the presence of small amounts of water and *p*-toluenesulfonic acid (TsOH), which clearly points to the hydride mechanism; water restores palladium hydrides from Pd(II) salts, and acid shifts the equilibrium $PdH^+$ versus Pd(0), via reductive elimination, towards the side of the hydride. The kinetics found were rather complex, with maximum rates passing through a maximum for concentrations of water, $PPh_3$, TsOH, and CO, and with an initial first order for methanol and cyclohexene. All features are in accord with the hydride mechanism, and with alcoholysis of a propionyl palladium complex as the rate-determining step (see also Milstein for alkoxycarbonylation of organic halides, as opposed to a reductive elimination of a hydrocarbyl group and a carbomethoxy species [188]). The cyclohexene concentration enters the rate equation because it determines the equilibrium concentration of the acyl intermediate. In line with this, the order of the rates for different alcohols was MeOH > EtOH > PrOH > 2-PrOH; this is the usual order for reactions involving nucleophilic attack of alcohols, because steric factors prevail. Also, the coordination behavior of alcohols towards Lewis acidic metals is dominated by steric factors and not by the electron density on oxygen. The highest TOF observed was almost $1000 \, mol \, mol^{-1} \, h^{-1}$ at $100 \, °C$. This work gives a good insight into how to improve the reaction rate, that is, how to make the maximum amount of palladium available in the active state, rather than in one of the dormant states. The detailed mechanism of the nucleophilic attack was not discussed. First, we will discuss a very fast methoxycarbonylation reaction and then we will return to the mechanism of alcoholysis.

For years the paradigm had been that monodentate phosphines lead primarily to ester formation, whereas diphosphines give polyketone [152], until Drent (Shell) reported that 1,3-bis(di-*t*-butylphosphino)propane as a ligand gave a fast reaction with ethene, CO, and MeOH to methyl propanoate [189]. The ligand contains the 1,3-propanediyl bridge, hitherto the ideal bridge of ligands producing polyketone with palladium catalysts. Subsequently, at ICI, Tooze and coworkers discovered a catalyst system that was even faster, based on 1,2-bis(di-*t*-butylphosphinomethyl)benzene (dtbpx) [190], a ligand reported before by Shaw as a very bulky ligand in platinum complexes [191], and later further developed by Spencer and coworkers [192]. As mentioned in the introduction (Section 8.4.1) this palladium catalyst system was developed into a commercial process by Lucite for the production of methyl propionate, the precursor for methyl methacrylate. The mechanism has been unraveled in a series of publications by Clegg, Tooze, Eastham, and coworkers [193].

As a ligand, dtbpx (Figure 8.10) is characterized by its steric bulk, a wide bite angle, and strong electron donors. The whole range of intermediate complexes of the hydride cycle was identified, in which the ligand acts as a cis-bidentate [193c]. Under oxidizing conditions with oxygen or quinone the dtbpx-based hydride **16x** was not oxidized, thus demonstrating the stability of the hydride. The strong electron donicity of dtbpx stabilizes the hydride species. All *in situ* measurements also pointed to the hydride cycle (cycle A, in Scheme 8.16).

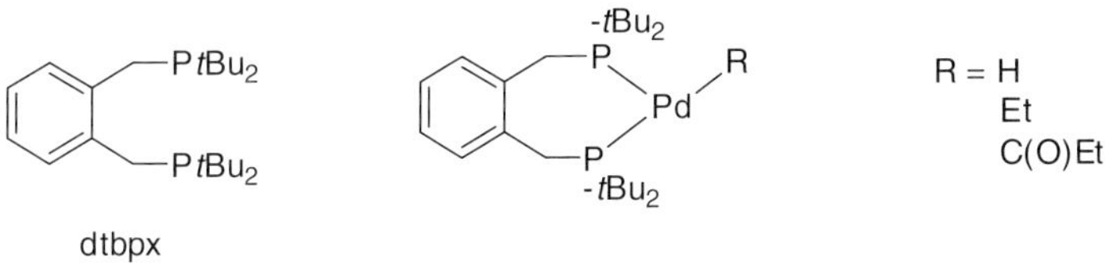

**Figure 8.10** Ligand and active complex of the Lucite methyl propanoate process.

Because of its steric bulk dtbpx was thought to facilitate a monodentate coordination mode or a trans-coordination mode in oligometallic complexes, as was observed in one acetate complex of 1,3-bis(di-*t*-butylphosphino)propane [193b]. These findings might explain the propensity for a fast alcoholysis and a higher barrier for alkene insertions. The small size of the hydride permits insertion of ethene to start the cycle, in spite of the steric bulk of the ligand (and ethene). Coordination of CO and migratory insertion are also fast reactions and propionyl-palladium forms. The propionyl group, however, is now sufficiently bulky to prohibit (or slow) coordination and migratory insertion of ethene and thus no polymerization takes place, and provided that alcoholysis can occur, only ester formation will take place. It was concluded that the steric properties were of paramount importance to obtain such high selectivity and rates [193a].

Hydroxycarbonylation (or hydrocarboxylation) in water was extensively studied by Verspui, Sheldon and coworkers with the use of water-soluble phosphine ligands: tppts as a monodentate and dppp-s (and derivatives thereof) as bidentates (see Figure 8.11). In water, as in methanol, monodentate phosphines led to the formation of acids (instead of methyl esters) and some oligomers [194], and bidentate ligands in water gave polyketone when ethene or propene was copolymerized with CO [195]. The use of weakly coordinating anions is not a necessity in water or water–acetic acid mixtures as the solvent, as the solvent will stabilize the dissociated anions via hydrogen bonds. Efficient polymerization catalysts were obtained in this way by Vavasori and coworkers [196].

In all cases the catalytic cycle is based on the hydride route, mechanism A. Most catalytic systems were stable up to 90 °C, depending strongly on the conditions. Acids of weakly coordinating anions work well, and the larger the excess often the better.

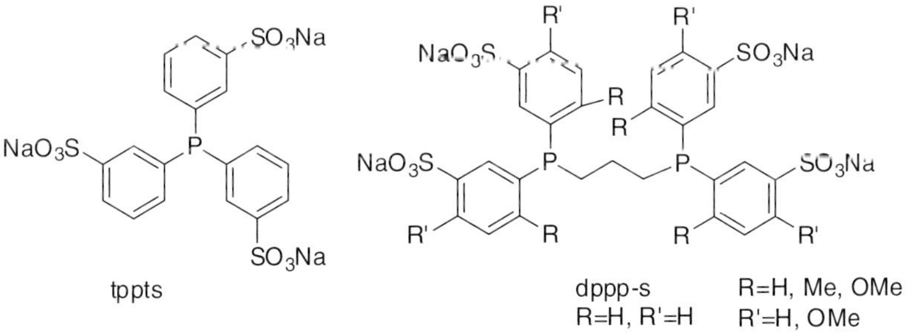

**Figure 8.11** Examples of water-soluble ligands used by Verspui *et al.*

Coordinating anions lead more quickly to reductive elimination and precipitation of metallic palladium. In basic media immediate precipitation of metal occurs. Higher alkenes react more slowly, because of their lower solubility, which affects the kinetics either by lowering the equilibrium concentration of the acyl resting state or by shifting the kinetics to a different rate-limiting step. For bidentate phosphines there is a strong dependence on ligand concentration (versus palladium concentration), as at a 2:1 ratio inert $Pd(diphosphine)^{2+}$ complexes form, whereas a ratio below 1:1 leads to palladium metal precipitation.

The polymers formed in water usually have higher contents of ketone end groups than expected statistically. There are two reasons for this. The first is general for all water-based systems, and that is that $Pd^{2+}$ is converted easily to $PdH^+$ via the well-known shift reaction –actually only half of it – converting CO into $CO_2$ and protons (from water) into hydrides (rather than to hydrogen as in the complete shift reaction). Under the acidic concentrations $PdH^+$ species are sufficiently stable to start the catalytic cycle with the insertion of ethene. Secondly, the enolate protonation reaction as the chain transfer reaction in polymerization (Scheme 8.21) is fast in water, the authors suggest [184, 195b] and thus termination also has a preference for the formation of ketone chain ends.

By NMR spectroscopy the formation of all species featuring in catalytic cycle A were identified for ethene as the substrate $(RPd(tppts)_3{}^+$, in which R = H, Et, C(O) Et), demonstrating the involvement of these water-soluble complexes as intermediates in the catalytic cycle of the Pd-catalyzed aqueous phase hydrocarboxylation of alkenes [197]. All intermediates were trans-phosphine complexes, as shown in Figure 8.12. Whereas ethene and CO are readily inserted at $-20\,°C$ in a Pd hydride and a Pd—ethyl bond, respectively, the hydrolysis of the Pd—acyl bond proceeds at elevated temperatures. Kinetic investigations showed that the conversion of a Pd—acyl to a Pd—hydride complex is a rate-determining, pseudo first-order reaction. Obviously, they concluded, this latter reaction is the rate-determining step in the catalytic cycle.

Up to 2003 it was thought that alcoholysis (or hydrolysis) of acylpalladium complexes was taking place in the trans-species depicted in Figure 8.12 as an attack of the alcohol nucleophile (or water) from outer sphere. Since insertion reactions are excluded in trans-species, this is the only reaction remaining for this species. The insertion of the first alkene and CO molecules should also take place in a cis-orientation of hydrogen/hydrocarbyl and inserting unsaturated molecule, but it was thought that this reaction could start via a 5-coordinate species. It should be noted that the preferred geometry of a bis-phosphine palladium complex **24** with R = H, Et,

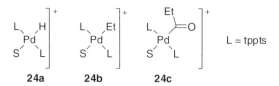

**Figure 8.12**  Complexes **24** identified by Verspui *et al.* for ttpts.

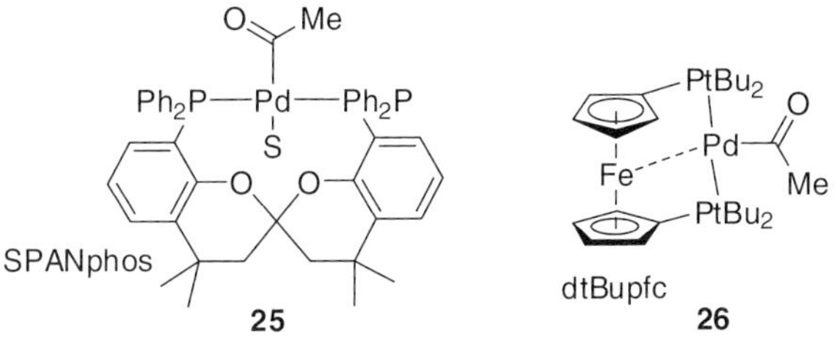

**Figure 8.13** Trans-diphosphine complexes **25** and **26** of acetylpalladium.

or C(O)Et is a trans-disposition of the phosphines because of the trans-influence. Indeed, tpp and tppts adopt this structure, as we have seen above, (if available, the fourth coordination site may be taken by a third phosphine trans to the anionic σ-donor ligand). Therefore, diphosphine ligands that favor a trans-coordination mode, will adopt the trans-conformation without exception.

Van Leeuwen and coworkers [163] studied the alcoholysis of a number of diphosphine acetylpalladium complexes, two of them having a preferred trans-phosphine disposition, either because of their backbone (SPANphos, **25**, Figure 8.13) or because of the steric bulk of their substituents, but having a backbone permitting the formation of trans-conformations (such as d-*t*-Bupfc, **26**).

Surprisingly, the ligand-enforced trans-cationic acetylpalladium complexes **25** and **26** did not react with methanol, not even after 24 h at room temperature, while all cis-complexes reacted with a few equivalents of methanol with a rate too fast to be measured. Trans-acetylpalladium complexes of flexible ligands showing a trans-configuration for the acetyl complexes, reacted with second-order kinetics with methanol. The order of nucleophilic attack of alcohols was the same as that reported by Toniolo [187], *viz.* MeOH > EtOH > *i*-PrOH > *t*-BuOH.

It was concluded that a trans-complex has to rearrange to a cis-intermediate, which then undergoes reductive elimination. In a trans-complex neither insertion reactions nor termination reactions take place! The sequence for a trans-complex is shown in Scheme 8.22. The group Z may be an ether oxygen atom in the ligand (or an iron atom if the bridge is ferrocene), or it may be a solvent molecule. The reductive elimination can be viewed as an intramolecular nucleophilic attack of the methoxy group on the

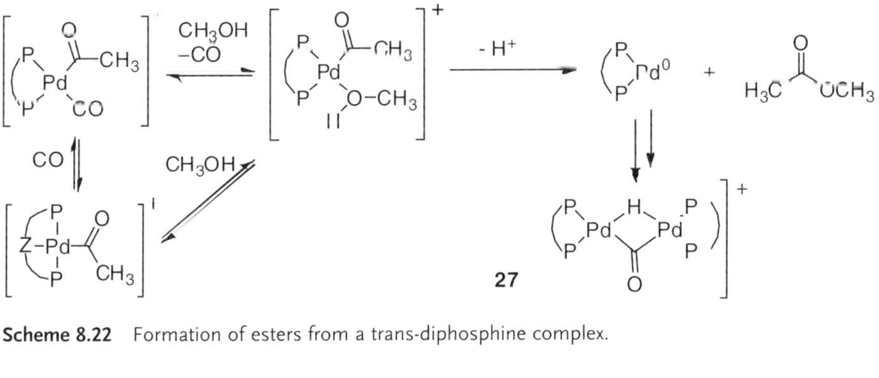

**Scheme 8.22** Formation of esters from a trans-diphosphine complex.

acyl carbon atom, or a "migratory reductive elimination". Palladium(0) is produced and this dimerizes with a proton and CO to give the orange dimer 27 (or a Pd(0) complex reacts with a PdH$^+$ species to give the same).

Thus complexes of flexible trans-ligands and monodentate phosphines will rearrange to the cis-conformation and then rapidly undergo alcoholysis. The rate of alcoholysis depends on the resulting steric bulk in the cis-complex, the more bulky the ligand set, the faster the alcoholysis. Under the conditions of catalysis, in the presence of ethene and CO, less bulky ligands will lead to insertion reactions and produce polymer. Cases in between give rise to oligomer formation. In Figure 8.14 is shown a series of examples of ligands and the molecular weights they give in this reaction. As might be expected, the resulting steric bulk for DPEphos and tpp is similar. The scheme also shows that with equal substituents at phosphorus (e.g., Ph) one can use the bite angle to correlate ligand and product distribution. Especially, ligands with small bite angles indicate that sterics dominate and not bite angles *per se*.

DFT calculations by Zuidema and coworkers support this mechanism and they also provide some ideas as to how MeOH as the attacking nucleophile might lose its proton via the participation of a number of methanol molecules [198]. This study also indicates that an attack of MeOH via a 5-coordinate species is highly unlikely,

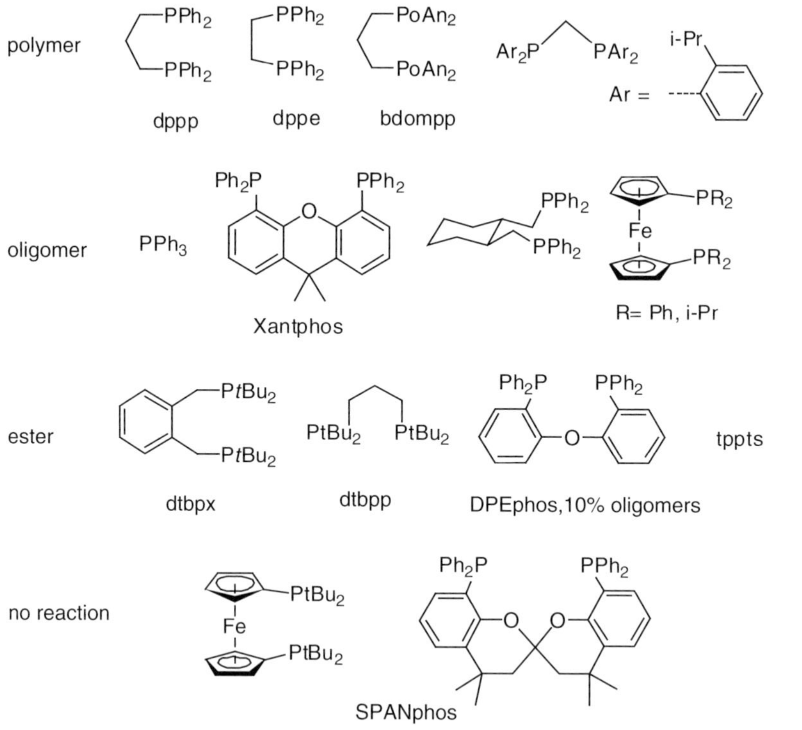

Figure 8.14 Products and ligand series for the reaction of Pd and CO, ethene, and MeOH.

and that an exchange in the square plane affords the intermediate containing a coordinated MeOH molecule. This study confirms that, for π-acceptor molecules such as CO and ethene, the participation of 5-coordinate species in the exchange process prior to migratory insertion is very likely.

A special case that deserves more attention is that of dtbpx, the ligand of the commercial Lucite process for methyl propanoate. While all studies were conducted with the triflate salts, the cationic triflate acetylpalladium complex of dtbpx was not sufficiently stable to allow the kinetic study of the alcoholysis reaction and instead the trifluoroacetate (TFA) derivative was used [199]. It was found that (dtbpx)Pd(TFA)C(=O)Me reacts extremely rapidly with methanol; at −90 °C the complex reacts with 10 equiv of MeOH to give complete conversion in less than 3 min (measured by NMR spectroscopy). Clegg and coworkers also reported that the reaction with propionylpalladium complexes of dtbpx was too fast to be measured [193a].

Thus, these results demonstrated that cis-acyl complexes can undergo an alcoholysis reaction millions of times faster than trans-complexes, and that a few trans-complexes do not react at all. The mechanism proposed for the alcoholysis is identical to that of the migratory reductive elimination in cross-coupling chemistry and steric factors play a dominant role in this (see references in [163]).

## 8.5
## Methanol Carbonylation

### 8.5.1
### Introduction

Acetic acid is a bulk chemical used industrially for a wide range of applications, as a solvent, and in the manufacture of cellulose acetate (for cigarette filters and photographic films) and vinyl acetate (paints, adhesives and textiles), as a bleach activator or in the manufacture of food, pharmaceuticals, pesticides, and so on [200]. The annual demand of 10 million tons includes 20% obtained by recycling (2008, BP). A large scale application as a solvent comprises the process for the oxidation of *p*-xylene to terephthalic acid with oxygen and a Co/Mn acetate/bromide catalyst. Terephthalic acid precipitates in the reactor which facilitates the separation of product and catalyst. As in hydroformylation catalysis the first catalytic process for the preparation of AcOH from MeOH and CO used cobalt catalysts at high pressure and temperature. In 1966 the rhodium-iodide process was discovered by Monsanto [201]. The first production plant based on this technology started operating in 1970 in Texas City. In 1986 the technology was transferred to BP Chemicals (British Petroleum). The major producers are BP and Celanese. Other metals are also active under similar conditions using the same iodide-based chemistry, such as nickel, palladium, and platinum [202]. In the mid 1990s BP introduced a new, more economic process using iridium as the catalyst, the CATIVA process (see Section 8.5.5). Since then many plants have been retrofitted for the use of this catalyst and new plants being built are based on the iridium technology. The market share of the CATIVA process amounts to 25% (2008, BP) [203].

8.5.2
**Mechanism and Side Reactions of the Monsanto Rhodium-Based Process**

The Monsanto process involves two interrelated catalytic cycles, an organometallic one based on rhodium ionic species, and an organic one in which iodide should be considered as the true catalyst (Scheme 8.23); it is a textbook example of mechanisms in homogeneous catalysis [204]. The two catalyst components are rhodium and iodide. Under the reaction conditions $RhI_3$ is reduced by $H_2O$ and CO to monovalent rhodium active species **28**, and methanol is converted to the iodo form, MeI.

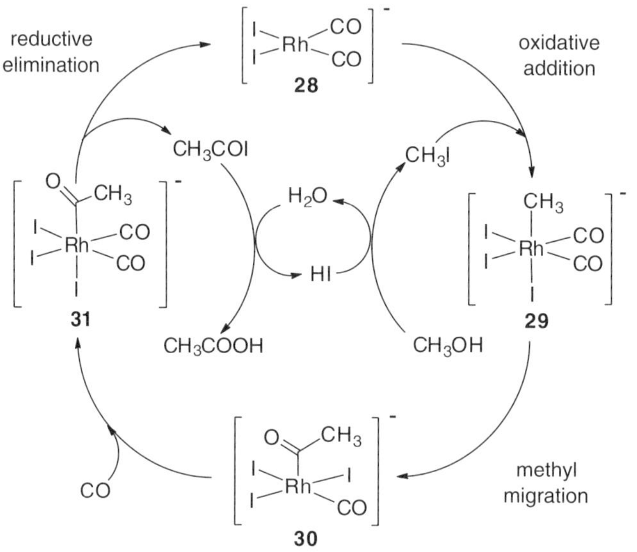

**Scheme 8.23** The mechanism of the Monsanto process.

The organometallic cycle comprises oxidative addition of MeI to $[RhI_2(CO)_2]^-$ **28** (generated *in situ*), which is considered to be the rate-limiting step of the Monsanto process. Ligand migration to generate the acetyl complex **30**, CO coordination and reductive elimination of acetyl iodide regenerates the rhodium active species **28**. Acetyl iodide enters the "organic" cycle where it is hydrolyzed to give acetic acid and HI which transforms MeOH into the more electrophilic MeI that enters the organometallic cycle.

It is important to note that reactions 8.1 and 8.2 involving iodide are quantitative, that is, the equilibrium lies completely on the right. As a result, all iodide in the system occurs as a methyl iodide, and, up to high conversion, the rate of AcOH production is independent of the methanol concentration [205].

$$CH_3OH + HI \rightleftharpoons H_2O + CH_3I \tag{8.1}$$

$$CH_3COI + H_2O \rightleftharpoons CH_3COOH + HI \tag{8.2}$$

One of the main drawbacks of the process is the loss of the expensive metal due to the formation of Rh(III) inactive species which, in addition, in areas of low CO pressure outside the reactor, precipitate as $RhI_3$. The inactive species $[RhI_4(CO)_2]^-$ can be produced by reaction of compounds **28** and **29** with HI according to reactions 8.3 and 8.4:

$$[Rh(CH_3)I_3(CO)_2]^- + HI \rightarrow CH_4 + [RhI_4(CO)_2]^- \tag{8.3}$$

$$[RhI_2(CO)_2]^- + 2HI \rightarrow H_2 + [RhI_4(CO)_2]^- \tag{8.4}$$

In the Monsanto process Rh precipitation is inhibited by keeping the water content relatively high, because water regenerates the Rh(I) active species from the labile $[RhI_4(CO)_2]^-$ via a partial water-gas shift reaction (8.5):

$$[RhI_4(CO)_2]^- + H_2O + CO \rightarrow [RhI_2(CO)_2]^- + 2HI + CO_2 \tag{8.5}$$

$$[RhI_4(CO)_2]^- + H_2 \rightarrow [RhI_2(CO)_2]^- + 2HI \tag{8.6}$$

$$CO + H_2O \rightarrow H_2 + CO_2 \tag{8.7}$$

In part, the reducing effect of water is counteracted by reactions 8.3 and 8.4 and this is reinforced by the high concentrations of HI resulting from high water concentrations. In the absence of water, added $H_2$ can also perform this role (reaction 8.6), because hydrogenolysis of $Rh-CH_3$ does not occur to a high extent as was reported in the early patents of Monsanto [206]. $H_2$ is also generated *in situ* in the reactor at high water concentrations through the water-gas shift reaction (8.7). Hydrogen builds up in the gas recycle and thus gases, for the major part being CO, have to be vented, which represents a loss.

The presence of water has contradictory effects; it is needed to maintain rhodium in the active Rh(I) form, but the reaction of $H_2O$ with CO via the water-gas shift reaction causes the loss of the CO feedstock and more HI leads to more "oxidation" (reactions 8.3 and 8.4). While the selectivity of the process in MeOH can be >97%, the selectivity in CO can be as low as 90%.

## 8.5.3
### The Mechanism of the Acetic Anhydride Process Using Rhodium as a Catalyst

A major part of the acetic acid produced is used as acetic anhydride and, therefore, immediately after the discovery of the Monsanto catalyst for acetic acid the search began for a process leading to acetic anhydride directly from methyl acetate. Indeed, the same rhodium catalyst converts methyl acetate into acetic anhydride but several problems had to be overcome before it could be applied industrially in 1983 by Halcon (later Eastman). The reaction uses also methyl iodide as the activator for the methyl group. The reaction scheme follows that of the Monsanto process except for the "organic" cycle, in which acetic acid replaces water, and methyl acetate replaces methanol (Equations 8.9 and 8.10):

$$CH_3COOCH_3 + HI \rightleftharpoons CH_3COOH + CH_3I \tag{8.9}$$

$$CH_3COI + CH_3COOH \rightleftharpoons CH_3COOOCCH_3 + HI \tag{8.10}$$

Reaction 8.9 generates methyl iodide for the oxidative addition, and reaction (8.10) converts the reductive elimination product acetyl iodide into the product and regenerates hydrogen iodide. There are, however, a few distinct differences [207] between the two processes:

1) The thermodynamics of the acetic anhydride formation are less favorable and the process is operated much closer to equilibrium.
2) The establishment of equilibria 8.9 and 8.10 is slow.
3) The equilibrium for reactions 8.9 and 8.10 does not lie on the right, as for reactions 8.1 and 8.2.
4) The reactions have an incubation time due to the slow reduction of Rh(III) salts.
5) Water cannot be present in the system to aid the reduction of Rh(III).

In the late 1980s, a major technological advance was achieved by Celanese, which circumvented the water issue [208]. First, salts (LiI or LiOAc) are added as promoters and in this way water levels can be kept low. Equations 8.9 and 8.10 are replaced by Equations 8.11 and 8.12 and read as follows:

$$CH_3COOCH_3 + LiI \rightleftharpoons CH_3COOLi + CH_3I \tag{8.11}$$

$$CH_3COI + CH_3COOLi \rightleftharpoons CH_3COOOCCH_3 + LiI \tag{8.12}$$

When using low water concentrations, the reduction of Rh(III) back to Rh(I) is slower (or does not take place at all), but so is the formation of $[RhI_4(CO)_2]^-$ due to the lower HI content available for reactions 8.3 and 8.4. The promotional effect of Li salts is attributed to the coordination of either acetate or iodide to **28**, forming a highly nucleophilic intermediate dianion, $[Rh(CO)_2I_2X]^{2-}$ (X = I or OAc), and also to the kinetic and thermodynamic parameters of this new organic cycle, in which water has been replaced by LiOAc [209, 210]. Secondly, dihydrogen (5% on CO) is added to reduce Rh(III) back to Rh(I), via the reverse of reaction (8.6). This can be done without increase of hydrogenation (hydrogenolysis) of the intermediate methyl complexes, which would lead to methane. The intermediate acetyl complexes do give some acetaldehyde which leads to ethylidene diacetate, $CH_3CH(CH_3O_2)_2$, as the by-product. This may decompose thermally to vinyl acetate and acetic acid, or it can be reduced to ethyl acetate, which in the recycle would lead eventually to propionic acid.

8.5.4
**Phosphine-Modified Rhodium Catalysts**

Since, for both processes, the oxidative addition is the rate-limiting step, it is clear that complexation of phosphine donors to a Rh(I) intermediate will strongly enhance this reaction. This has been well documented as indeed oxidative addition reactions to Rh (I) and Ir(I) complexes obey the textbook rules as no other reaction in organometallic chemistry. It generally follows a two-step $S_N2$ mechanism; nucleophilic attack by the metal on the methyl carbon to displace iodide, presumably with inversion of configuration at the carbon atom, and subsequent iodide coordination to the five-coordinate rhodium complex to give the methyl complex **29** [211, 212]. The product of this reaction has been fully characterized spectroscopically [213–215]. Observations that support this mechanism are the second-order rate-law observed (first-order in MeI and **28**) and the large negative activation entropies that indicate highly organized transition states. Complexes containing $PEt_3$ or dppe may show oxidative additions reactions 40–60 times faster at 25 °C than those of the unmodified complex **28** [216]. Very electron-rich rhodium complexes may undergo even faster oxidative addition reactions, but the result may be that reductive elimination from the phosphine analogue of compound **31** no longer occurs and thus the complex is not catalytically active [217]. Phosphine complexes usually afford active catalysts, and they are more active than **29**.

The $PEt_3$-rhodium modified system reported by Rankin and coworkers [218] (which undergoes the oxidative addition much faster than **28**) gave only an increase in activity of 1.8 under catalytic conditions (150 °C). At this temperature, loss of activity is observed after circa 10 min, due to degradation of the catalytically active system to $[Rh(CO)_2I_2]^-$. Decomposition proceeds via reaction of $[RhI_2(CO)(CH_3)(PEt_3)_2]$ with HI to generate $[RhI_3(CO)(PEt_3)_2]$, which reductively eliminates $[PEt_3I]^+$ leading to $OPEt_3$. A small amount of $Et_3PMeI$ was also observed. In the absence of water the catalytic reaction is slower and it was suggested that especially the reductive elimination seemed to slow down under these conditions, but phosphine decomposition was the cause of deactivation under all conditions studied.

Phosphonium salt formation is to be expected in these media and at least once the phosphonium salts were identified unambiguously [219]. Catalysts prepared from $1,2-(tBu_2PCH_2)_2C_6H_4$ and $[RhCl(CO)_2]_2$ catalyzed the carbonylation of methanol in the presence of MeI at rates faster than those in the absence of added phosphine, although they decompose. Methanol promotes the decomposition to $[RhI_2(CO)_2]^-$ and $[1,2-(MetBu_2PCH_2)_2C_6H_4)][I_3]_2$, which has been isolated and characterized. Eventually, phosphonium salts also give phosphine oxides. Although not extensively documented, this is the accepted fate of phosphine-modified catalysts and, therefore, industrial applications are not very likely.

Chelating symmetrical and unsymmetrical ligands were introduced to overcome the instability of the monophosphines under the harsh conditions of the process.

The use of symmetrical diphosphines was described in the patent literature [220, 221] and when using Xantphos derivatives as ligands for methanol carbonylation, they are stable systems that show a slightly higher activity than the Monsanto catalyst.

In this case a terdentate P—O—P coordination of the ligand was proposed. Carraz and coworkers [222] reported the use of asymmetrically substituted 1,2-ethanediyl diphosphines that are very stable under the reaction conditions, although they did not give an improvement in the catalytic activity compared to the Monsanto process. At the end of the reaction a mixture of diphosphine Rh(III) carbonyl complexes was isolated and characterized, and a second run was performed without loss of activity.

An alternative strategy to increase the activity and stability of the systems is the use of hemilabile phosphine ligands (P—X; X=P, O, S). Gonsalvi and coworkers found that oxidative addition of $CH_3I$ was enhanced 100-fold by using these ligands [223]. The ligands are supposed to stabilize the complex via the chelate effect and to increase the nucleophilicity of the rhodium by coordination of a hetero-donor atom (see Ref. 217).

The reactivity of dinuclear rhodium compounds toward MeI has been studied since the late 1970s [224, 225]. The oxidative addition/migration on the second metal center becomes more difficult after the first one has taken place, especially when the "open-book" conformation of the complexes is forced by the (bridging) ligands [226]. The possible involvement of dinuclear species in the Monsanto process was proposed in the earliest studies, where the dimerization equilibrium of the acetyl derivative $[RhI_3(COMe)(CO)]^-$ was considered [213]. The corresponding dinuclear species $[Rh_2I_6(CO)_2(COMe)_2]^{2-}$ was isolated in the form of its trimethylphenylammonium salt and characterized by X-ray diffraction. The possibility that neutral dinuclear methyl and acetyl Rh(III) species play an important role in the Monsanto process has very recently been reconsidered [227].

In an effort to develop new systems able to combine both stability and activity, Süss-Fink and coworkers reported several trans-spanning, but flexible diphosphines [228–230]. They concluded that trans-diphosphine complexes were better catalysts than the dimeric species isolated after the reaction. Surprisingly, dimeric complexes of structure **32** containing SPANphos (see Scheme 8.24) were found to be more active than monomeric complexes containing two donor phosphines in trans-positions [231]. Perhaps the bent structure of the dimeric complexes and the resulting interaction of the $d_{z2}$ orbitals can explain this. DFT calculations showed a slightly lower barrier for the dimeric complexes, but a clear orbital picture could not be derived from the calculations [232].

Currently the dimeric compounds based on SPANphos represent the fastest phosphine-based systems reported for methanol carbonylation. As mentioned above,

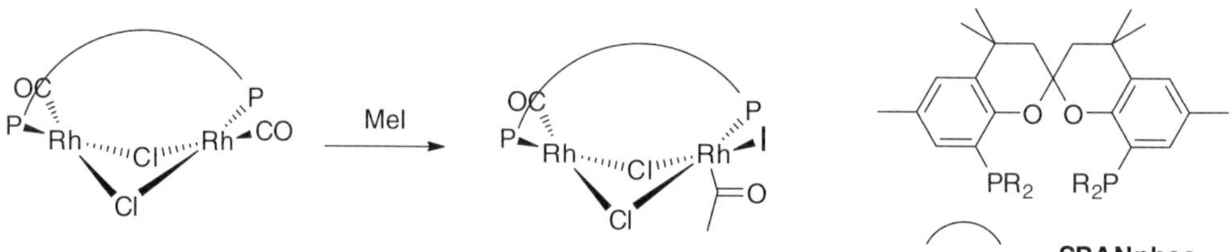

**Scheme 8.24** Dimeric catalysts for methanol carbonylation.

because of the expected quaternization and oxidation of phosphines the prospects for industrial application of these fast, phosphine-modified catalysts remain poor.

Clarke and coworkers studied the influence of $C_4$-bridged diphosphines on the rhodium-catalyzed carbonylation of methanol, with particular attention to the sensitivity of the process towards hydrogen [233]. At the time of its introduction in the 1960s the Monsanto process presented a major improvement in rate and selectivities compared to the BASF process based on cobalt catalysts. The amounts of by-products arising from hydrogenolysis reactions of the intermediate methyl and acetyl metal complexes, giving methane and acetaldehyde respectively, were considerably lower for rhodium (<1%) [234]. Cobalt produces much more methane and propionic acid. The latter is formed via carbonylation of ethanol, which is obtained via hydrogenation of acetaldehyde. In the BASF process the amount of propionic acid may be as much as 10 wt% of the acetic acid main product and once this was the major source of propionic acid. Although to a much lesser extent in the rhodium processes, propionic acid is still formed, it builds up in the catalyst recycle (distillation bottom) and it must be thoroughly removed from the main product. The formation of propionic acid increases when the hydrogen concentration increases. The CO feed is made from syn gas by separating it from hydrogen, a costly process, especially if a high purity is required. Thus, there is an incentive to search for modified rhodium catalysts that are less sensitive to hydrogen (hydrogen builds up in the gas recycle and large amounts are not desirable either). Clarke and coworkers addressed this goal and took as their starting point the outcome of a study by Moloy and Wegman, who found that certain phosphine modified rhodium catalysts, for example, those using dppp, were catalysts for the production of ethanal from syn gas and methanol (in the presence of ruthenium, ethanol was obtained), but that dppb was ineffective as a ligand in the methanol homologation reaction [235].

Their studies included the ligands dppb, BINAP and dppx (bisdiphenylphosphinoxylene). The complexes were studied as catalysts for methanol carbonylation using CO that contains hydrogen. By-product analysis of the organic carbonylation products revealed that the more rigid BINAP and dppx systems gave much lower proportions of acetaldehyde and methane side-products. These findings were supported by the study of the relative reactivity of the Rh(III) acetyls with hydrogen. The structure of the dppx complex suggests that the rigidity of the dppx backbone makes the diphenylphosphine group shield the acetyl from hydrogenolysis, which may be the ultimate origin of the greater hydrogen tolerance of the dppx carbonylation catalysts.

## 8.5.5
### Iridium Catalysts

An improved process using iridium (or a combination of iridium and another metal, usually ruthenium) came on stream in 1995, developed by BP [236]. The CATIVA and Monsanto processes are sufficiently similar that the same chemical plant may be used, which makes a retrofitting commercially highly attractive. The rhodium-based Monsanto process, after 25 years of successful operation, is gradually being replaced by the CATIVA-based process [236]. Iridium is now used in 25% of the processes, but other sources mention that 80% of production is now a CATIVA-based process [237]. In summary, the economic reasons for switching from rhodium to iridium are:

- The iridium catalyst works at low water concentrations, which means that one can reduce the number of drying columns. The low levels of liquid by-products formed are a significant improvement over those achieved with the high-water rhodium-based catalyst system and the quality of the product obtained under low water concentrations is exceptional [236].
- Less propionic acid by-product is produced which also reduces the work-up train.
- A higher concentration of catalyst can be used which increases the space–time yield. As a result investments are reduced by 30%.
- The catalyst system exhibits high stability allowing a wide range of process conditions and compositions to be accessed without catalyst precipitation.
- Iridium is several times cheaper than rhodium (on average by a factor of ten per kg over the last 5 years, which should be corrected for the atomic weight!); the prices of both show enormous fluctuation [238].
- The capacity of a Monsanto plant increases by 20–70% when the CATIVA catalyst is used instead.

Initial studies by Monsanto had shown iridium to be less active than rhodium for the carbonylation of methanol. Subsequent research, however, showed that the iridium catalyst can be promoted by added iodide salts, and this leads to a catalyst that is superior to the rhodium-based systems. The switch from rhodium to iridium also allows the use of less water in the reaction mixture. This change reduces the number of drying columns necessary, decreases by-product formation, and suppresses the water gas shift reaction. Compared with the Monsanto process, the Cativa process generates even less propionic acid by-product. Furthermore, the process allows a higher catalyst loading, while higher rhodium loadings, together with a high water content, lead to precipitation of rhodium salts, especially at lower CO pressures.

The iridium system shows high rates at low water concentrations. The catalyst system exhibits high stability, allowing a wide range of process conditions and compositions to be accessed without catalyst precipitation. In 2003 five plants were already in operation using this new catalyst.

In general, the oxidative addition to iridium [239] is much faster than that to the corresponding rhodium complexes (100–150). Also the equilibrium is on the side of the trivalent state. Thus, as expected reaction 8.2 is much faster for iridium. In itself this does not mean that the iridium catalyst is therefore faster than the rhodium catalyst, as the reductive elimination may be slower for iridium. Apparently, this situation has not yet been reached and it was reported that migration is now the slowest step [240], a common phenomenon for third row metals. In third row metals the metal-to-carbon σ-bonds are stronger, more localized, and more covalent than those in second-row metal complexes (a relativistic stabilization of the $Ir-CH_3$ bond). One can imagine that a more diffuse, electron-rich, σ-bonded hydrocarbyl migrates more easily.

In contrast to the rhodium process the most abundant iridium species in the BP process, the catalyst resting state, is not the low-valent Ir(I) iodide, but the product of the oxidative addition of MeI to this complex.

Two distinct classes of promoters have been identified for the reaction: simple iodide complexes of zinc, cadmium, mercury, indium, and gallium, and carbonyl complexes of tungsten, rhenium, ruthenium, especially $RuI_2(CO)_4$, osmium, and platinum [241]. Ionic salts such as LiI and $Bu_4NI$ are inhibitors. The promoters exhibit a unique synergy with iodide salts, such as lithium iodide, under low water conditions. Both main group and transition metal salts can influence the equilibria of the iodide species involved. A rate maximum exists at low water conditions, and optimization of the process parameters gives acetic acid with a selectivity in excess of 99% based upon methanol. IR-spectroscopic studies have shown that the salts added as activators abstract iodide from the ionic methyl-iridium species and that in the resulting neutral species the migration is 800 times faster (Scheme 8.25) [242]. Addition of transition metal complexes might also play a role in aiding the reduction of iridium(III) species, but no evidence for this has been reported. For cobalt catalysts activation by platinum has been reported for the reductive carbonylation of methanol to give acetaldehyde [243].

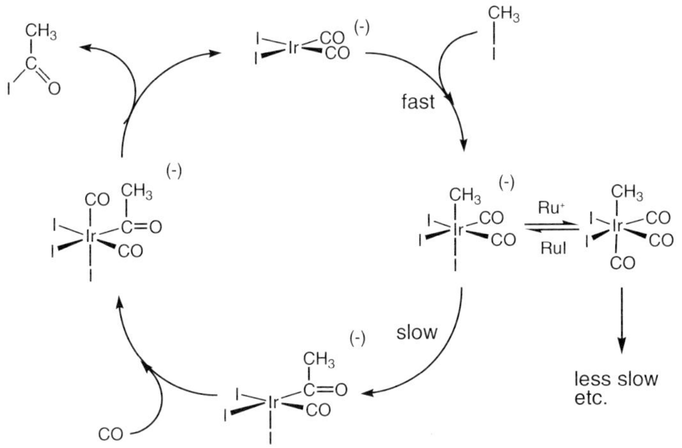

Scheme 8.25   BP's CATIVA process.

The CATIVA catalyst, in the way it is operated, is about 25% faster than the Monsanto rhodium catalyst. In addition, it was found that not only is the oxidative addition much faster than that of rhodium, but also it is no longer the rate-determining step. The migration of the methyl group to the co-ordinated carbon monoxide is now rate determining (Scheme 8.25).

Kalck and coworkers studied platinum salts as an additive and proved that for platinum iodide abstraction is also the mechanism of the rate enhancement [244]. In addition, a dinuclear IrPt complex was observed in the mass spectral analysis, a potential intermediate for this reaction. The metal additives function as "iodide" acceptors, and thus are Lewis acidic compounds. Higher valence states produce stronger Lewis acids, but these metal complexes should be stable with respect to the reducing CO environment, without consuming large amounts of reductant (CO, $H_2$) and oxidizing reagents (HI, $CH_3I$) in a non-productive way. Both $PtI_2(CO)_2$ and $RuI_2(CO)_4$ fulfill these requirements.

Studies on ligand-modified iridium complexes showed that, as is to be expected, phosphines accelerate oxidative addition, but for the CATIVA process this is not important. Interestingly, a P–S heterobidentate (dppms) leads to faster reaction than dppe, although the latter is the better donor ligand [223].

## 8.6
## Conclusions

In this chapter we have discussed four major reactions utilizing CO as a substrate. As in several other cases, the number of reports on these topics is extremely high, but the number of studies highlighting catalyst deactivation, catalyst recovery, and the measures to be taken in order to avoid decomposition is very small. For industrially applied reactions such as hydroformylation, alkene methoxycarbonylation, and carbonylation of methanol these issues are of high importance, and without doubt a lot of unpublished studies on deactivation mechanisms and their cures must be available in internal reports.

Hydroformylation catalyst decomposition is governed by two reactions: inhibition by unsaturates such as dienes or enones, and ligand decomposition. Inhibition by (hetero)dienes does not influence other equilibria and usually is observed as an incubation in batch reactions. The stable intermediates formed from (hetero)dienes present a dormant state of the catalyst, as the subsequent reactions (e.g., CO insertions, hydrogenation) are slow. Ligand decomposition may often take longer than the time for batch reactions and, therefore, often remains unnoticed. Dedicated studies on thermal ligand or catalyst decomposition in the absence of substrates proved to be more useful. Very minor changes in ligand structure have a large influence on the outcome. For phosphines and phosphites the decomposition pathways are very different.

For palladium carbonylation chemistry the elementary steps of activation and deactivation are well known and application of the principles derived from that knowledge has often led to more active catalysts. Often compromises must be reached between contradicting effects. Not much is known about phosphine decomposition, but from other reactions and unpublished results from industry it is known that P–C cleavage reactions play an important role in this.

Industrial applications of rhodium-catalyzed carbonylation of methanol use unmodified rhodium iodide carbonyl complexes and ligand decomposition is not an issue. Two key features are the formation of Rh(III) species that need to be reduced and precipitation of Rh salts in the low-pressure area of the plants. Solutions to these problems are available. Non-productive redox cycles give rise to $CO_2$ formation, which means a loss of CO feedstock.

The more recent CATIVA process based on iridium catalysts shows neither the latter drawback, nor the formation of insoluble salts, and as a result the system can be operated at higher catalyst concentration. Specific decomposition problems have not been reported so far.

## References

1 van Leeuwen, P.W.N.M. (2004) *Homogeneous Catalysis; Understanding the Art*, Chs. 6–8, 12, Kluwer Academic Publishers., Dordrecht, the Netherlands (now Springer).

2 Reppe, W. and Magin, A. (1951) U.S. Patent 2,577,208; (1952) *Chem. Abstr.*, **46**, 6143.

3 Roelen, O. (1948) *Angew. Chem.*, **A60**, 213.

4 Hebrard, F. and Kalck, P. (2009) *Chem. Rev.*, **109**, 4272–4282.

5 (a) Falbe, J. (ed.) (1980) *New Synthesis with Carbon Monoxide*, Springer-Verlag, Berlin; (b) Falbe, J. (ed.) (1970) *Synthesis with Carbon Monoxide*, Springer-Verlag, New York.

6 Toetsch, W., Arnoldi, D., Kaizik, A., and Trocha, M. (2003) PCT Int. Appl. WO 2003078365 (to Oxeno); (2003) *Chem. Abstr.*, **139**, 262474.

7 Nienburg, H.J., Kummer, R., Hohenschutz, H., Strohmeyer, M., and Tavs, P. (1973) DE 2206252, add. to Ger. Offen. 2,139,630 (to BASF); (1973) *Chem. Abstr.*, **79**, 136504.

8 van Driessche, E., van Vliet, A., Caers, R.F., Beckers, H., Garton, R., da Cruz, B., Lepagnol, M., and Kooke, E. (2008) PCT Int. Appl. WO 2008122526 (to ExxonMobil Chemical Patents Inc., USA); (2008) *Chem. Abstr.*, **149**, 452245.

9 Slaugh, L.H. and Mullineaux, R.D. (1966) U.S. Pat. 3,239,569 and 3,239,570 (to Shell); (1964) *Chem. Abstr.*, **64**, 15745 and 19420; (1968) *J. Organomet. Chem.*, **13**, 469.

10 Van Winkle, J.L., Lorenzo, S., Morris, R.C., and Mason, R.F. (1969) U.S. Patent 3,420,898. U.S. Appl. 1965-443703; (1967) *Chem. Abstr.*, **66**, 65101.

11 van Leeuwen, P.W.N.M. (2004) *Homogeneous Catalysis; Understanding the Art*, Ch. 7, Kluwer Academic Publishers.,

Dordrecht, the Netherlands (now Springer).

12 Crause, C., Bennie, L., Damoense, L., Dwyer, C.L., Grove, C., Grimmer, N., van Rensburg, W.J., Kirk, M.M., Mokheseng, K.M., Otto, S., and Steynberg, P.J. (2003) *Dalton Trans.*, 2036.

13 MacDougall, J.K., Simpson, M.C., Green, M.J., and Cole-Hamilton, D.J. (1996) *J. Chem. Soc. Dalton Trans.*, 1161–1172.

14 Young, J.F., Osborn, J.A., Jardine, F.A., and Wilkinson, G. (1965) *J. Chem. Soc. Chem., Commun.*, 131; Evans, D., Osborn, J.A., and Wilkinson, G. (1968) *J. Chem. Soc. (A)*, 3133; Evans, D., Yagupsky, G., and Wilkinson, G. (1968) *J. Chem. Soc. A*, 2660.

15 (a) Pruett, R.L. and Smith, J.A. (1969) *J. Org. Chem.*, **34**, 327; (b) Pruett, R.L., Smith, J.A., and African, S. (1968) Pat. 6804937 (to Union Carbide Corporation); (1969) *Chem. Abstr.*, **71**, 90819.

16 Konya, D., Almeida Leñero, K.Q., and Drent, E. (2006) *Organometallics*, **25**, 3166–3174.

17 (a) van Leeuwen, P.W.N.M. and Roobeek, C.F. (1983) *J. Organometal. Chem.*, **258**, 343; (b) Brit. Pat. 2,068,377 US Pat. 4,467,116 (to Shell Oil); (1984) *Chem. Abstr.*, **101**, 191142; (c) Jongsma, T., Challa, G., and van Leeuwen, P.W.N.M. (1991) *J. Organometal. Chem.*, **421**, 121; (d) van Rooy, A., Orij, E.N., Kamer, P.C.J., van den Aardweg, F., and van Leeuwen, P.W.N.M. (1991) *J. Chem. Soc., Chem. Commun.*, 1096–1097; (e) van Rooy, A., Orij, E.N., Kamer, P.C.J., and van Leeuwen, P.W.N.M. (1995) *Organometallics*, **14**, 34–43.

18 Billig, E., Abatjoglou, A.G., Bryant, D.R., Murray, R.E., and Maher, J.M. (1986) U.S.

Pat. 4,599,206 (to Union Carbide Corp.); (1989) *Chem. Abstr.*, **109**, 233177.

19 Billig, E., Abatjoglou, A.G., and Bryant, D.R.U.S. Pat. 4,769,498. U.S. Pat. 4,668,651; (1987) U.S. Pat. 4748261 (to Union Carbide Corp.); (1987) *Chem. Abstr.*, **107**, 7392r.

20 Yoshinura, N. and Tokito, Y. (1987) Eur. Pat. 223,103 (to Kuraray).

21 Diéguez, M., Pàmies, O., and Claver, C. (2004) *Chem. Rev.*, **104**, 3189–3216.

22 Sparta, M., Børve, K.J., and Jensen, V.R. (2007) *J. Am. Chem. Soc.*, **129**, 8487–8499.

23 (a) van Leeuwen, P.W.N.M., Casey, C.P., and Whiteker, G.T. (2000) *Rhodium Catalyzed Hydroformylation* (eds P.W.N.M. van Leeuwen and C. Claver), Kluwer Academic Publishers, Dordrecht, The Netherlands, Ch. 4; (b) van Leeuwen, P.W.N.M. (2004) *Homogeneous Catalysis: Understanding the Art*, Kluwer Academic Publishers, Dordrecht, The Netherlands, Ch. 8.

24 Evans, D., Yagupsky, G., and Wilkinson, G. (1968) *J. Chem. Soc. A*, 2660.

25 Castellanos-Páez, A., Castillón, S., Claver, C., van Leeuwen, P.W.N.M., and Lange, W.G.J. (1998) *Organometallics*, **17**, 2543.

26 Freixa, Z. and van Leeuwen, P.W.N.M. (2003) *Dalton Trans.*, 1890–1901.

27 van Leeuwen, P.W.N.M., Sandee, A.J., Reek, J.N.H., and Kamer, P.C.J. (2002) *J. Mol. Catal. A: Chem.*, **182–183**, 107–123.

28 van der Veen, L.A., Kamer, P.C.J., and van Leeuwen, P.W.N.M. (1999) *Organometallics*, **18**, 4765–4777.

29 Silva, S.M., Bronger, R.P.J., Freixa, Z., Dupont, J., and van Leeuwen, P.W.N.M. (2003) *New J. Chem.*, **27**, 1294–1296.

30 Ewalds, R., Eggeling, E.B., Hewat, A.C., Kamer, P.C.J., van Leeuwen, P.W.N.M., and Vogt, D. (2000) *Chem. Eur. J.*, **6**, 1496–1504.

31 (a) Castellanos-Paez, A., Castillon, S., Claver, C., van Leeuwen, P.W.N.M., and de Lange, W.G.J. (1998) *Organometallics*, **17**, 2543–2552; (b) Zuidema, E., Goudriaan, P.E., Swennenhuis, B.H.G., Kamer, P.C.J., van Leeuwen, P.W.N.M., Lutz, M., and Spek, A.L. (2010) *Organometallics*, **29**, 1210–1221.

32 Rafter, E., Gilheany, D., Reek, J.N.H., and van Leeuwen, P.W.N.M. (2010) *ChemSusChem*, **2**, 387–391.

33 (a) Heck, R.F. and Breslow, D.S. (1961) *J. Am. Chem. Soc.*, **83**, 4023; (b) Heck, R.F. (1969) *Acc. Chem. Res.*, **2**, 10–16.

34 (a) van der Veen, L.A., Boele, M.D.K., Bregman, F.R., Kamer, P.C.J., van Leeuwen, P.W.N.M., Goubitz, K., Fraanje, J., Schenk, H., and Bo, C. (1998) *J. Am. Chem. Soc.*, **120**, 11616–11626;(b) Zuidema, E., Escorihuela, L., Eichelsheim, T., Carbo, J.J., Bo, C., Kamer, P.C.J., and van Leeuwen, P.W.N.M. (2008) *Chem. Eur. J.*, **14**, 1843–1853.

35 van Rooy, A., Orij, E.N., Kamer, P.C.J., and van Leeuwen, P.W.N.M. (1995) *Organometallics*, **14**, 34–43.

36 (a) Coolen, H.K.A.C., Nolte, R.J.M., and van Leeuwen, P.W.N.M. (1995) *J. Organomet. Chem.*, **496**, 159–168; (b) Sielcken, O.E., Smits, H.A., Toth, I. (2002) Eur. Pat. Appl. 1249441 (to DSM N.V., Neth.); (2002) *Chem. Abstr.*, **137**, 303813.

37 Selent, D., Boerner, A., Wiese, K.-D., Hess, D., and Fridag, D. (2008) PCT Int. Appl. WO 2008141853 (to Evonik Oxeno G.m.b.H., Germany); (2008) *Chem. Abstr.*, **149**, 578125.

38 Hoegaerts, D. and Jacobs, P.A. (1999) *Tetrahedron: Asym.*, **10**, 3039–3043.

39 Coolen, H.K.A.C., Nolte, R.J.M., and van Leeuwen, P.W.N.M. (1996) *J. Org. Chem.*, **61**, 4739–4747.

40 Buisman, G.J.H., Martin, M.E., Vos, E.J., Klootwijk, A., Kamer, P.C.J., and van Leeuwen, P.W.N.M. (1995) *Tetrahedron: Asym.*, **6**, 719–738.

41 Kalck, P., Frances, J.M., Pfister, P.M., Southern, T.G., and Thorez, A. (1983) *J. Chem. Soc., Chem. Commun.*, 510.

42 (a) Orejon, A., Claver, C., Oro, L.A., Elduque, A., and Pinillos, M.T. (1998) *J. Mol. Catal. A: Chem.*, **136**, 279–284;(b) Vargas, R., Rivas, A.B., Suarez, J.D., Chaparros, I., Ortega, M.C., Pardey, A.J., Longo, C., Perez-Torrente, J.J., and Oro, L.A. (2009) *Catal. Lett.*, **130**, 470–475.

43 Davis, R., Epton, J.W., and Southern, T.G. (1992) *J. Mol. Catal.*, **77**, 159–163.

44 Diéguez, M., Claver, C., Masdeu-Bultó, A.M., Ruiz, A., van Leeuwen, P.W.N.M., and Schoemaker, G. (1999) *Organometallics*, **18**, 2107–2115.

45 Gao, H. and Angelici, R.J. (1998) *Organometallics*, **17**, 3063–3069.

46 Masdeu, A.M., Orejon, A., Ruiz, A., Castillon, S., and Claver, C. (1994) *J. Mol. Catal.*, **94**, 149–156.

47 Chiusoli, G.P., Venturello, C., and Merzoni, S. (1968) *Chem. Ind. (London, U. K.)*, 977.

48 Breuzard, J.A.J., Christ-Tommasino, M.L., and Lemaire, M. (2005) *Top. Organomet. Chem.*, **15**, 231–270.

49 Breuzard, J.A.J., Tommasino, M.L., Bonnet, M.C., and Lemaire, M. (2000) *C. R. Acad. Sci. Paris, Serie IIc, Chimie*, **3**, 557–561.

50 (a) Mieczynska, E., Trzeciak, A.M., Ziólkowski, J.J. (1993) *J. Mol.Catal.*, **80**, 189; (b) Buhling, A., Kamer, P.C.J., and van Leeuwen, P.W.N.M. (1995) *J. Mol. Catal. A: Chem.*, **98**, 69.

51 Sandee, A.J., Reek, J.N.H., Kamer, P.C.J., and van Leeuwen, P.W.N.M. (2001) *J. Am. Chem. Soc.*, **123**, 8468–8476.

52 Buhling, A., Kamer, P.C.J., van Leeuwen, P.W.N.M., Elgersma, J.W., Goubitz, K., and Fraanje, J. (1997) *Organometallics*, **16**, 3027–3037.

53 Andrieu, J., Camus, J.-M., Richard, P., Poli, R., Gonsalvi, L., Vizza, F., and Peruzzini, M. (2006) *Eur. J. Inorg. Chem.*, 51–61.

54 (a) For examples see: RajanBabu, T.V. and Ayers, T.A. (1994) *Tetrahedron Asym.*, **35**, 4295–4298; (b) Kwok, T.J. and Wink, D.J. (1993) *Organometallics*, **12**, 1954–1959; (c) Obora, Y., Liu, Y.K., Kubouchi, S., Tokunaga, M., and Tsuji, Y. (2006) *Eur. J. Inorg. Chem.*, 222–230; (d) Chen, J., Ajjou, A.N., Chanthateyanonth, R., and Alper, H. (1997) *Macromolecules*, **30**, 2897–2901; (e) Albers, J., Dinjus, E., Pitter, S., and Walter, O. (2004) *J. Mol. Catal. A: Chem.*, **219**, 41–46.

55 van den Beuken, E., de Lange, W.G.J., van Leeuwen, P.W.N.M., Veldman, N., Spek, A.L., and Feringa, B.L. (1996) *J. Chem. Soc., Dalton Trans.*, 3561.

56 van Leeuwen, P.W.N.M. and Roobeek, C.F. (1985) *J. Mol. Catal.*, **31**, 345.

57 van Rooy, A., de Bruijn, J.N.H., Roobeek, C.F., Kamer, P.C.J., and van Leeuwen, P.W.N.M. (1996) *J. Organomet. Chem.*, **507**, 69–73.

58 Fyhr, C. and Garland, M. (1993) *Organometallics*, **12**, 1753–1764.

59 (a) Billig, E., Abatjoglou, A.G., Bryant, D.R., Murray, R.E., and Maher, J.M. (1988) (to Union Carbide Corporation) U.S. Pat. 4,717,775; (1989) *Chem. Abstr.*, **109**, 233177; (b) Muilwijk, K.F., Kamer, P.C.J., and van Leeuwen, P.W.N.M. (1997) *J. Am. Oil Chem. Soc.*, **74**, 223–228.

60 Liu, G. and Garland, M. (2000) *J. Organomet. Chem.*, **608**, 76–85.

61 Barros, H.J.V., Guimaraes, C.C., dos Santos, E.N., and Gusevskaya, E.V. (2007) *Organometallics*, **26**, 2211–2218.

62 da Silva, J.G., Vieira, C.G., dos Santos, E.N., and Gusevskaya, E.V. (2009) *Appl. Catal. A: Gen.*, **365**, 231–236.

63 van Leeuwen, P.W.N.M. and Roobeek, C.F. (1985) *J. Mol. Catal.*, **31**, 345.

64 Deshpande, R.M., Divekar, S.S., Gholap, R.V., and Chaudhari, R.V. (1991) *J. Mol. Catal.*, **67**, 333–338.

65 van Leeuwen, P.W.N.M., Walczuk-Gusciora, E.B., Grimmer, N.E., Kamer, P.C.J. (2005) PCT Int. Appl. WO 2005049537; (2005) *Chem. Abstr.*, **143**, 9532.

66 Walczuk, E.B., Kamer, P.C.J., and van Leeuwen, P.W.N.M. (2003) *Angew. Chem. Int. Ed.*, **42**, 4665–4669.

67 Scheuermann né Taylor, C.J. and Jaekel, C. (2008) *Adv. Synth. Catal.*, **350**, 2708–2714.

68 Slough, G.A., Ashbaugh, J.R., and Zannoni, L.A. (1994) *Organometallics*, **13**, 3587–3593.

69 Liu, G. and Garland, M. (1999) *Organometallics*, **18**, 3457–3467.

70 Parshall, G.W., Knoth, W.H., and Schunn, R.A. (1969) *J. Am. Chem. Soc.*, **91**, 4990.

71 (a) O'Brien, C.J., Tellez, J.L., Nixon, Z.S., Kang, L.J., Carter, A.L., Kunkel, S.R., Przeworski, K.C., and Chass, G.A. (2009) *Angew. Chem. Int. Ed.*, **48**, 6836–6839; (b) Marsden, S.P. (2009) *Nature Chem.*, **1**, 685–687.

72 Aresta, M., Dibenedetto, A., Tommasi, I. (2001) *Eur. J. Inorg. Chem.*, 1801–1806.

73 Nicholas, K.M. (1980) *J. Organomet. Chem.*, **188**, C10–C12.

74 Larpent, C., Dabard, R., and Patin, H. (1987) *Inorg. Chem.*, **26**, 2922–2924.

75 Abatjoglou, A.G., Billig, E., and Bryant, D.R. (1984) *Organometallics*, **3**, 923–926.

76 Billig, E., Jamerson, J.D., Pruett, R.L. (1980) *J. Organomet. Chem.*, **192**, C49.

77 Miller, D.J., Bryant, D.R., Billig, E., and Shaw, B.L. (1990) U.S. Pat. 4,929,767 (to Union Carbide Chemicals and Plastics Co.); (1991) *Chem. Abstr.*, **113**, 85496.

78 Herrmann, W.A. and Kohlpaintner, C.W. (1993) *Angew. Chem. Int. Ed. Engl.*, **32**, 1524.

79 Kaneda, K., Sano, K., and Teranishi, S. (1979) *Chem. Lett.*, 821–822.

80 Chan, A.S.C., Caroll, W.E., and Willis, D.E. (1983) *J. Mol. Catal.*, **19**, 377.

81 (a) Matthews, R.C., Howell, D.K., Peng, W.-P., Train, S.G., Dale Treleaven, W., and Stanley, G.G. (1996) *Angew. Chem. Int. Ed. Engl.*, **35**, 2253; (b) Broussard, M.E., Juma, B., Train, S.G., Peng, W.-J., Laneman, S.A., and Stanley, G.G. (1993) *Science*, **260**, 1784.

82 Aubry, D.A., Monteil, A.R., Peng, W.-J., and Stanley, G.G. (2002) *C. R. Chim.*, **5**, 473–480.

83 Cheliatsidou, P., White, D.F.S., and Cole-Hamilton, D.J. (2004) *Dalton Trans.*, 3425–3427.

84 Reinius, H.K. and Krause, A.O.I. (2000) *J. Mol. Catal. A: Chem.*, **158**, 499–508.

85 Moser, W.R., Papile, C.J., and Weininger, S.J. (1987) *J. Mol. Catal.*, **41**, 293.

86 Reinius, H.K., Suomalainen, P., Riihimaki, H., Karvinen, E., Pursiainen, J., and Krause, A.O.I. (2001) *J. Catal.*, **199**, 302–308.

87 (a) Suomalainen, P., Laitinen, R., Jääskeläinen, S., Haukkaa, M., Pursiainen, J.T., and Pakkanen, T.A. (2002) *J. Mol. Catal. A: Chem.*, **179**, 93–100; (b) Suomalainen, P., Reinius, H.K., Riihimäki, H., Laitinen, R.H., Jääskeläinen, S., Haukka, M., Pursiainen, J.T., Pakkanen, T.A., and Krause, A.O.I. (2001) *J. Mol. Catal. A: Chem.*, **169**, 67–78.

88 Bergounhou, C., Neibecker, D., and Réau, R. (1988) *J. Chem. Soc., Chem. Commun.*, 1370–1371.

89 Bergounhou, C., Neibecker, D., and Mathieu, R. (2003) *Organometallics*, **22**, 782–786.

90 Buisman, G.J.H., van der Veen, L.A., Kamer, P.C.J., and van Leeuwen, P.W.N.M. (1997) *Organometallics*, **16**, 5681–5687.

91 Bergounhou, C., Neibecker, D., and Mathieu, R. (2004) *J. Mol. Catal. A: Chem.*, **220**, 167–182.

92 Niyomura, O., Iwasawa, T., Sawada, N., Tokunaga, M., Obora, Y., and Tsuji, Y (2005) *Organometallics*, **24**, 3468–3475.

93 Bianchini, C., Oberhauser, W., Orlandini, A., Giannelli, C., and Frediani, P. (2005) *Organometallics*, **24**, 3692–3702.

94 Rivas, A.B., Pérez-Torrentea, J.J., Pardey, A.J., Masdeu-Bultó, A.M., Diéguez, M., and Oro, L.A. (2009) *J. Mol. Catal. A: Chem.*, **300**, 121–131.

95 Borman, P.C. and Gelling, O.J. EP 96-203070 (to DSM N.V. Neth.); (1998) *Chem. Abstr.*, **128**, 323141.

96 Billig, E., Abatjoglou, A.G., Bryant, D.R., Murray, R.E., and Maher, J.M. (1988) (to Union Carbide Corporation) U.S. Pat. 4,717,775; (1989) *Chem. Abstr.*, **109**, 233177.

97 Ramirez, F., Bhatia, S.B., and Smith, C.P. (1967) *Tetrahedron*, **23**, 2067.

98 Billig, E., Abatjoglou, A.G., Bryant, D.R., Murray, R.E., and Maher, J.M. (1988) (to Union Carbide Corporation) U.S. Pat. 4,717,775; (1989) *Chem. Abstr.*, **109**, 233177.

99 Babin, J.E., Maher, J.M., and Billig, E. (1994) EP 590611 (to Union Carbide Chemicals and Plastics Technology Corp. USA); (1994) *Chem. Abstr.*, **121**, 208026.

100 Ueda, A., Fujita, Y., and Kawasaki, H. (2001) JP 2001342164 (to Mitsubishi Chemical Corp. Japan); (2001) *Chem. Abstr.*, **136**, 21214.

101 (a) Babin, J.E. and Whiteker, G.T. (1992) WO 93103830;(b) Buisman, G.J.H., Kamer, P.C.J., and van Leeuwen,

P.W.N.M. (1993) *Tetrahedron: Asym.*, **4**, 1625.

102 (a) Buisman, G.J.H., Vos, E.J., Kamer, P.C.J., and van Leeuwen, P.W.N.M. (1995) *J. Chem. Soc., Dalton Trans.*, 409–417; (b) Buisman, G.J.H., van der Veen, L.A., Klootwijk, A., de Lange, W.G.J., Kamer, P.C.J., van Leeuwen, P.W.N.M., and Vogt, D. (1997) *Organometallics*, **16**, 2929–2939.

103 Mathivet, T., Monflier, E., Castanet, Y., Mortreux, A., and Couturier, J.-L. (2002) *C. R. Chim.*, **5**, 417–424.

104 (a) van der Slot, S.C., Kamer, P.C.J., van Leeuwen, P.W.N.M., Fraanje, J., Goubitz, K., Lutz, M., and Spek, A.L. (2000) *Organometallics*, **19**, 2504–2515; (b) Baber, R.A., Clarke, M.L., Orpen, A.G., and Ratcliffe, D.A. (2003) *J. Organometal. Chem.*, **667**, 112–119; (c) van der Slot, S.C., Duran, J., Luten, J., Kamer, P.C.J., and van Leeuwen, P.W.N.M. (2002) *Organometallics*, **21**, 3873–3883; (d) Clarke, M.L., Ellis, D., Mason, K.L., Orpen, A.G., Pringle, P.G., Wingad, R.L., Zaher, D.A., and Baker, R.T. (2005) *Dalton Trans.*, 1294–1300;

105 (a) Sakai, N., Mano, S., Nozaki, K., and Takaya, H. (1993) *J. Am. Chem. Soc.*, **115**, 7033; (b) Horiuchi, T., Ohta, T., Shirakawa, E., Nozaki, K., and Takaya, H. (1997) *J. Org. Chem.*, **62**, 4285.

106 Breeden, S., Cole-Hamilton, D.J., Foster, D.F., Schwarz, G.J., and Wills, M. (2000) *Angew. Chem. Int. Ed. Engl.*, **39**, 4106.

107 (a) Clark, T.P., Landis, C.R., Freed, S.L., Klosin, J., and Abboud, K.A. (2005) *J. Am. Chem. Soc.*, **127**, 5040–5042; (b) Thomas, P.J., Axtell, A.T., Klosin, J., Peng, W., Rand, C.L., Clark, T.P., Landis, C.R., and Abboud, K.A. (2007) *Org. Lett.*, **9**, 2665–2668.

108 (a) Igau, A., Gruetzmacher, H., Baceiredo, A., and Bertrand, G. (1988) *J. Am. Chem. Soc.*, **110**, 6463; (b) Arduengo, A.J., Dias, H.V.R., Harlow, R.L., and Kline, M. (1992) *J. Am. Chem. Soc.*, **114**, 5530.

109 Viciu, M.S., Germaneau, R.F., and Nolan, S.P. (2002) *Org. Lett.*, **4**, 4053–4056.

110 Jackstell, R., Harkal, S., Jiao, H., Spannenberg, A., Borgmann, C., Roettger, D., Nierlich, F., Elliot, M., Niven, S., Kingsley, C., Navarro, O., Viciu, M.S., Nolan, S.P., and Beller, M. (2004) *Chem. Eur. J.*, **10**, 3891–3900.

111 (a) Cardin, D.J., Doyle, M.J., and Lappert, M.F. (1972) *J. Chem. Soc. Chem. Commun.*, 927; (b) Doyle, M.J. and Lappert, M.F. (1974) *J. Chem. Soc. Chem. Commun.*, 679.

112 Praetorius, J.M. and Crudden, C.M. (2008) *Dalton Trans.*, 4079–4094.

113 (a) Bortenschlager, M., Schutz, J., von Preysing, D., Nuyken, O., Herrmann, W.A., and Weberskirch, R. (2005) *J. Organomet. Chem.*, **690**, 6233; (b) Bortenschlager, M., Mayr, M., Nuyken, O., and Buchmeiser, M.R. (2005) *J. Mol. Catal. A: Chem.*, **233**, 67; (c) Zarka, M.T., Bortenschlager, M., Wurst, K., Nuyken, O., and Weberskirch, R. (2004) *Organometallics*, **23**, 4817.

114 Poyatos, M., Uriz, P., Mata, Y.A., Claver, C., Fernandez, E., and Peris, E. (2003) *Organometallics*, **22**, 440.

115 Douglas, S., Lowe, J.P., Mahon, M.F., Warren, J.E., and Whittlesey, M.K. (2005) *J. Organomet. Chem.*, **690**, 5027.

116 Praetorius, J.M., Kotyk, M.W., Webb, J.D., Wang, R.Y., and Crudden, C.M. (2007) *Organometallics*, **26**, 1057.

117 Laï, R., Daran, J.-C., Heumann, A., Zaragori-Benedetti, A., and Rafii, E. (2009) *Inorg. Chim. Acta*, **362**, 4849–4852.

118 Green, M., McMullin, C.L., Morton, G.J.P., Orpen, A.G., Wass, D.F., and Wingad, R.L. (2009) *Organometallics*, **28**, 1476–1479.

119 Dastgir, S., Coleman, K.S., Cowley, A.R., and Green, M.L.H. (2009) *Dalton Trans.*, 7203–7214.

120 Gil, W., Trzeciak, A.M., and Zio'lkowski, J.J. (2008) *Organometallics*, **27**, 4131–4138.

121 (a) van Leeuwen, P.W.N.M. and Roobeek, C.F. (1983) *J. Organomet. Chem.*, **258**, 343–350; (b) van Leeuwen, P.W.N.M. and Roobeek, C.F. (1980) Brit. Pat. 2 068 377 (to Shell); (1984) *Chem. Abstr.*, **101**, 191142.

122 Jeletic, M.S., Jan, M.T., Ghiviriga, I., Abboud, K.A., and Veige, A.S. (2009) *Dalton Trans.*, 64–76.

123 (a) McGuinness, D.S., Cavell, K.J., Skelton, B.W., and White, A.H. (1999) *Organometallics*, **18**, 1596; (b) McGuinness, D.S. and Cavell, K.J. (2000) *Organometallics*, **19**, 741; (c)McGuinness, D.S. and Cavell, K.J. (2000) *Organometallics*, **19**, 4918.

124 (a) van Rensburg, H., Tooze, R.P., Foster, D.F., and Slawin, A.M.Z. (2004) *Inorg. Chem.*, **43**, 2468; (b) van Rensburg, H., Tooze, R.P., Foster, D.F., and Otto, S. (2007) *Inorg. Chem.*, **46**, 1963.

125 Arnoldy, P. (1999) *Rhodium Catalyzed Hydroformylation* (eds P.W.N.M. van Leeuwen and C. Claver), Kluwer Academic Publishers, Dordrecht, Netherlands, Ch. 8, pp. 203–229.

126 Kuntz, E.G. (1977) Fr. Pat. 2314910 **C** (to Rhone-Poulenc); (1977) *Chem. Abstr.*, **87**, 101944.

127 Frohning, C.D. and Kohlpainter, C.W. (1996) *Applied Homogeneous Catalysis with Organometallic Compounds* (eds B. Cornils and W.A. Herrmann), VCH Verlag GmbH, Weinheim, pp. 29–104.

128 Herrmann, W.A. and Kohlpainter, C.W. (1993) *Angew. Chem.*, **105**, 1588.

129 Bexten, L., Cornils, B., and Kupies, D. (1986) Ger. Pat. 3431643 (to Ruhrchemie); (1986) *Chem. Abstr.*, **105**, 117009.

130 Albanese, G., Manetsberger, R., Herrmann, W.A., and Schwer, C. (1996) Eur. Pat. Appl. 704451 (to Hoechst); (1996) *Chem. Abstr.*, **125**, 11135.

131 Herwig, J. and Fischer, R. (1999) *Rhodium Catalyzed Hydroformylation* (eds P.W.N.M. van Leeuwen and C. Claver), Kluwer Academic Publishers, Dordrecht, Netherlands, Ch. 7, pp. 189–202.

132 Cornils, B., Konkol, W., Bach, H., Daembkes, G., Gick, W., Wiebus, E., and Bahrmann, H. (1985) Ger. Pat. 3415968 (to Ruhrchemie); (1986) *Chem. Abstr.*, **104**, 209147.

133 Schreuder Goedheijt, M., Hanson, B.E., Reek, J.N.H., Kamer, P.C.J., and van Leeuwen, P.W.N.M. (2000) *J. Am. Chem. Soc.*, **122**, 1650–1657.

134 Divekar, S.S., Bhanage, B.M., Deshpande, R.M., Gholap, R.V., and Chaudhari, R.V. (1994) *J. Mol. Catal.*, **91**, L1–L6.

135 (a) Sudheesh, N., Sharma, S.K., Shukla, R.S., and Jasra, R.V. (2008) *J. Mol. Catal. A: Chem.*, **296**, 61–70; (b) Artner, J., Bautz, H., Fan, F., Habicht, W., Walter, O., Döring, M., and Arnold, U. (2008) *J. Mol. Catal.*, **255**, 180–189.

136 Jongsma, T., van Aert, H., Fossen, M., Challa, G., and van Leeuwen, P.W.N.M. (1993) *J. Mol. Catal.*, **83**, 37–50.

137 Wasserscheid, P. and Keim, W. (2000) *Angew. Chem. Int. Ed.*, **39**, 3773–3789.

138 Mehnert, C.P., Cook, R.A., Dispenziere, N.C., and Mozeleski, E.J. (2004) *Polyhedron*, **23**, 2679–2688.

139 Bronger, R.P.J., Silva, S.M., Kamer, P.C.J., and van Leeuwen, P.W.N.M. (2004) *Dalton Trans.*, 1590–1596.

140 Arhancet, J.P., Davis, D.E., Merola, J.S., and Hanson, B.E. (1989) *Nature*, **339**, 454.

141 (a) Mehnert, C.P., Cook, R.A., Dispenziere, N.C., and Afeworki, M. (2002) *J. Am. Chem. Soc.*, **124**, 12932; (b) Riisager, A., Wasserscheid, P., van Hal, R., and Fehrmann, R. (2003) *J. Catal.*, **219**, 252.

142 Riisager, A., Fehrmann, R., Haumann, M., and Wasserscheid, P. (2006) *Eur. J. Inorg. Chem.*, 695–706.

143 (a) Vogt, M. (1991) Rheinisch-Westfälischen Technischen Hochschule, PhD. Thesis, Aachen, Germany; (b) Horvath, I.T. and Rabai, J. (1994) *Science*, **266**, 72.

144 Horvath, I.T., Kiss, G., Cook, R.A., Bond, J.E., Stevens, P.A., Rabai, J., and Mozeleski, E.J. (1998) *J. Am. Chem. Soc.*, **120**, 3133–3143.

145 Foster, D.F., Gudmunsen, D., Adams, D.J., Stuart, A.M., Hope, E.G., Cole-Hamilton, D.J., Schwarz, G.P., and Pogorzelec, P. (2002) *Tetrahedron*, **58**, 3901–3910.

146 Mathivet, T., Monflier, E., Castanet, Y., Mortreux, A., and Coutourier, J.-L. (2002) *Tetrahedron*, **58**, 3877–3888.

147 Mathivet, T., Monflier, E., Castanet, Y., Mortreux, A., and Coutourier, J.-L. (1999) *Tetrahedron Lett.*, **40**, 3885–3888; (1998) *Tetrahedron Lett.*, **39**, 9411–9414.

148 (a) Han, M. and Liu, H. (1996) *Macromol. Symp.*, **105**, 179–183; (b) Wen, F., Bönnemann, H., Jiang, J., Lu, D., Wang, Y., and Jin, Z. (2005) *Appl. Organometal. Chem.*, **19**, 81–89; (c) Tuchbreiter, L. and Mecking, S. (2007) *Macromol. Chem. Phys.*, **208**, 1688–1693;

(d) Bruss, A.J., Gelesky, M.A., Machado, G., and Dupont, J. (2006) *J. Mol. Catal. A*, **252**, 212–218.

149 Axet, M.R., Castillón, S., Claver, C., Philippot, K., Lecante, P., and Chaudret, B. (2008) *Eur. J. Inorg. Chem.*, 3460–3466.

150 (a) Sen, A. and Lai, T.W. (1982) *J. Am. Chem. Soc.*, **104**, 3520; (b) Lai, T.W. and Sen, A. (1984) *Organometallics*, **3**, 866.

151 (a) Drent, E. (1984) Eur. Pat. Appl. 121,965; (1985) *Chem. Abstr.*, **102**, 46423; (b) Drent, E., van Broekhoven, J.A.M., and Doyle, M.J. (1991) *J. Organomet. Chem.*, 417, 235.

152 Drent, E. and Budzelaar, P.H.M. (1996) *Chem. Rev.*, **96**, 663.

153 Sen, A. (2003) *Catalytic Synthesis of Alkene-Carbon Monoxide Copolymers and Cooligmers*, Catalysis by Metal Complexes, vol. 27 (eds B.R. James and P.W.N.M. van Leeuwen), Kluwer Academic Publishers, Dordrecht, Netherlands.

154 Cavinato, G., Toniolo, L., and Vavasori, A. (2006) *Top. Organomet. Chem.*, **18**, 125–164.

155 Ash, C.E. (1994) *J. Mater. Educ.*, **16**, 1–20 and (1995) *Int. J. Polym. Mater.*, **30**, 1–13.

156 Reppe, W. and Magin, A. (1951) U. S. Patent 2,577,208; (1952) *Chem. Abstr.*, **46**, 6143.

157 Gough, A. (1967) British Pat. 1,081,304; (1967) *Chem. Abstr.*, **67**, 100569.

158 Fenton, D.M. (1970) U.S. Pat. 3,530,109; (1970) *Chem. Abstr.*, **73**, 110466; (1978) U. S. Pat. 4,076,911; (1978) *Chem. Abstr.*, **88**, 153263.

159 Nozaki, K. (1972) U.S. Pat. 3,689,460; (1972) *Chem., Abstr.*, **77**, 152860; (1972) U.S. Pat. 3,694,412; (1972) *Chem. Abstr.*, 77, 165324; (1974) U.S. Pat. 3,835,123; (1975) *Chem. Abstr.*, **83**, 132273.

160 http://www.lucite.com/news.asp; 31-01-2010. Earlier announcements said Shanghai, 2005.

161 (a) Tsuji, J., Morikawa, M., and Kiji, J. (1963) *Tetrahedron Lett.*, 1437; (b) Fenton, D.M. (1973) *J. Org. Chem.*, **38**, 3192.

162 Bardi, R., Del Pra, A., Piazzesi, A.M., and Toniolo, L. (1979) *Inorg. Chim. Acta*, 35, L345–L346.

163 van Leeuwen, P.W.N.M., Zuideveld, M.A., Swennenhuis, B.H.G., Freixa, Z., Kamer, P.C.J., Goubitz, K., Fraanje, J., Lutz, M., and Spek, A.L. (2003) *J. Am. Chem. Soc.*, **125**, 5523–5540.

164 Dekker, G.P.C.M., Elsevier, C.J., Vrieze, K., van Leeuwen, P.W.N.M., and Roobeek, C.F. (1992) *J. Organomet. Chem.*, **430**, 357–372.

165 Brumbaugh, J., Whittle, R.R., Parvez, M., and Sen, A. (1990) *Organometallics*, **9**, 1735.

166 Zuideveld, M.A., Kamer, P.C.J., van Leeuwen, P.W.N.M., Klusener, P.A.A., Stil, H.A., and Roobeek, C.F. (1998) *J. Am. Chem. Soc.*, **120**, 7977–7978.

167 Drent, E., van Broekhoven, J.A.M., and Doyle, M.J. (1991) *J. Organomet. Chem.*, 417, 235–251.

168 Cavinato, G. and Toniolo, L. (1981) *J. Mol. Catal.*, 10, 161–170.

169 del Rio, I., Claver, C., and van Leeuwen, P.W.N.M. (2001) *Eur. J. Inorg. Chem.*, 2719–2738.

170 Cavinato, G., Vavasori, A., Amadio, E., and Toniolo, L. (2007) *J. Mol. Catal. A: Chem.*, **278**, 251–257.

171 Amatore, C., Jutand, A., and Medeiros, M.J. (1996) *New J. Chem.*, **20**, 1143–1148.

172 Guiu, E., Caporali, M., Munoz, B., Mueller, C., Lutz, M., Spek, A.L., Claver, C., and Van Leeuwen, P.W.N.M. (2006) *Organometallics*, **25**, 3102–3104.

173 Van Broekhoven, J.A.M. and Drent, E. (1987) Eur. Pat. Appl. 235,865 (to Shell); (1988) *Chem. Abstr.*, **108**, 76068.

174 Dekker, G.P.C.M., Elsevier, C.J., Vrieze, K., van Leeuwen, P.W.N.M., and Roobeek, C.F. (1992) *J. Organomet. Chem.*, **430**, 357.

175 Budzelaar, P.H.M., van Leeuwen, P.W.N.M., Roobeek, C.F., and Orpen, A.G. (1992) *Organometallics*, **11**, 23.

176 Drent, E. (1986) Eur. Pat. Appl. 229,408 (to Shell); (1988) *Chem. Abstr.*, **108**, 6617.

177 van Leeuwen, P.W.N.M. and Roobeek, C.F. (1990) Eur. Patent Appl. 380162 (to Shell Research); (1991) *Chem. Abstr.*, **114**, 62975.

178 Rülke, R.E., Han, I.M., Elsevier, C.J., Vrieze, K., van Leeuwen, P.W.N.M., Roobeek, C.F., Zoutberg, M.C., Wang, Y.F., and Stam, C.H. (1990) *Inorg. Chim. Acta*, **169**, 5.

179 Shultz, C.S., Ledfort, J., DeSimone, J.M., and Brookhart, M. (2000) *J. Am. Chem. Soc.*, **122**, 6351.

**180** Dekker, G.P.C.M., Elsevier, C.J., Vrieze, K., and van Leeuwen, P.W.N.M. (1992) *Organometallics*, **11**, 1598–1603.

**181** Scarel, A., Durand, J., Franchi, D., Zangrando, E., Mestroni, G., Milani, B., Gladiali, S., Carfagna, C., Binotti, B., Bronco, S., and Gragnoli, T. (2005) *J. Organomet. Chem.*, **690**, 2106–2120.

**182** Durand, J., Zangrando, E., Stener, M., Fronzoni, G., Carfagna, C., Binotti, B., Kamer, P.C.J., Müller, C., Caporali, M., van Leeuwen, P.W.N.M., Vogt, D., and Milani, B. (2006) *Chem. Eur. J.*, **12**, 7639–7651.

**183** Bianchini, C., Lee, H.M., Meli, A., Oberhauser, W., Peruzzini, M., and Vizza, F. (2002) *Organometallics*, **21**, 16–33.

**184** Zuideveld, M.A., Kamer, P.C.J., van Leeuwen, P.W.N.M., Klusener, P.A.A., Stil, H.A., and Roobeek, C.F. (1998) *J. Am. Chem. Soc.*, **120**, 7977.

**185** Markies, B.A., Wijkens, P., Boersma, J., Spek, A.L., and van Koten, G. (1991) *Recl. Trav. Chim. Pays–Bas*, **110**, 133.

**186** Sen, A. and Lai, T.-W. (1984) *J. Am. Chem. Soc.*, **106**, 866.

**187** Vavasori, A., Toniolo, L., and Cavinato, G. (2003) *J. Mol. Catal. A, Chem.*, **191**, 9–21.

**188** Milstein, D. (1986) *J. Chem. Soc., Chem. Commun.*, 817.

**189** Drent, E. and Kragtwijk, E. (1992) Eur. Pat. EP 495,548 (to Shell); (1992) *Chem. Abstr.*, **117**, 150569.

**190** Tooze, R.P., Eastham, G.R., Whiston, K., Wang, X.-L. (1996) PCT Int. Appl. WO 9619434 (to ICI); (1996) *Chem. Abstr.*, **125**, 145592.

**191** Moulton, S.J. and Shaw, B.L. (1976) *J. Chem. Soc., Chem. Commun.*, 365–366.

**192** Mole, L., Spencer, J.L., Carr, N., and Orpen, A.G. (1991) *Organometallics*, **10**, 49–52.

**193** (a) Clegg, W., Eastham, G.R., Elsegood, M.R.J., Heaton, B.T., Iggo, J.A., Tooze, R.P., Whyman, R., and Zacchini, S. (2002) *Organometallics*, **21**, 1832–1840; (b) Clegg, W., Eastham, G.R., Elsegood, M.R.J., Tooze, R.P., Wang, X.-L., and Whiston, K. (1999) *Chem. Commun.*, 1877–1878; (c) Eastham, G.R., Heaton, B.T., Iggo, J.A., Tooze, R.P., Whyman, R., and Zacchini, S. (2000) *Chem. Commun.*, 609–610.

**194** (a) Papadogianakis, G., Verspui, G., Maat, L., and Sheldon, R.A. (1997) *Catal. Lett.*, **47**, 43;(b) Chepaikin, E.G., Bezruchenko, A.P., Leshcheva, A.A., and Boiko, G.N. (1994) *Russ. Chem. Bull.*, **43**, 360.

**195** (a) Verspui, G., Schanssema, F., and Sheldon, R.A. (2000) *Angew. Chem. Int. Ed.*, **39**, 804–806; (b) Verspui, G., Schanssema, F., and Sheldon, R.A. (2000) *Appl. Catal. A*, **198**, 5–11.

**196** (a) Vavasori, A., Tonioli, L., Cavinato, G., and Visentin, F. (2003) *J. Mol. Catal. A Chem.*, **204**, 295; (b) Vavasori, A., Tonioli, L., and Cavinato, G. (2004) *Mol. Catal. A Chem.*, **215**, 63.

**197** Verspui, G., Moiseev, I.I., and Sheldon, R.A. (1999) *J. Organomet. Chem.*, **586**, 196.

**198** Zuidema, E., Bo, C., and van Leeuwen, P.W.N.M. (2007) *J. Am. Chem. Soc.*, **129**, 3989–4000.

**199** TFA was found to be an attractive anion for methoxycarbonylation of alkenes: Blanco, C., Ruiz, A., Godard, C., Fleury-Brégeot, N., Marinetti, A., and Claver, C. (2009) *Adv. Synth. Catal.*, **351**, 1813–1816.

**200** Haynes, A. (2001) *Educ. Chem.*, **38**, 99–101.

**201** (a) Forster, D. (1976) *J. Am. Chem. Soc.*, **98**, 846–848; (b) Forster, D. (1979) *Adv. Organomet. Chem.*, **17**, 255.

**202** (a) van Leeuwen, P.W.N.M. and Roobeek, C.F. (1985) (to Shell Internationale Research Maatschappij B. V., Neth.) Eur. Pat. Appl. EP 133331. (b) Yang, J., Haynes, A., and Maitlis, P.M. (1999) *Chem. Commun.*, 179–180; (c) Tonde, S.S., Kelkar, A.A., Bhadbhade, M.M., and Chaudhari, R.V. (2005) *J. Organomet. Chem.*, **690**, 1677–1681.

**203** http://www.bp.com/sectiongenericarticle.do?categoryId=9027101&contentId=7049636.

**204** van Leeuwen, P.W.N.M. (2004.) Chapter 6, in *Homogeneous Catalysis: Understanding the Art*, Springer, Dordrecht, pp. 109–124.

**205** Claver, C. and van Leeuwen, P.W.N.M. (2003) *Comprehensive Coordination*

*Chemistry II*, vol. 9, Elsevier, Amsterdam.

**206** Maitlis, P.M., Haynes, A., James, B.R., Catellani, M., and Chiusoli, G.P. (2004) *Dalton Trans.*, 3409–3419.

**207** Zoeller, J.R., Agreda, V.H., Cook, S.L., Lafferty, N.L., Polichnowski, S.W., and Pond, D.M. (1992) *Catal. Today*, **13**, 73–91.

**208** (a) Smith, B.L., Torrence, G.P., Murphy, M.A., and Aguiló, A. (1987) *J. Mol. Catal.*, **39**, 115–136; (b) Smith, B.L., Torrence, G.P., Aguiló, A., and Alder, J.S.(January 7, 1991) (Hoechst Celanese Corporation) US Patent 5,144,068.

**209** Murphy, M.A., Smith, B.L., Torrence, G.P., and Aguilo, A. (1986) *J. Organomet. Chem.*, **303**, 257–272.

**210** Kinnunen, T. and Laasonen, K. (2001) *J. Mol. Struct. (Theochem)*, **542**, 273–288.

**211** Griffin, T.R., Cook, D.B., Haynes, A., Pearson, J.M., Monti, D., and Morris, G. (1996) *J. Am. Chem. Soc.*, **118**, 3029–3030.

**212** Chauby, V., Daran, J-C., Berre, C.S.-B., Malbosc, F., Kalck, P., Gonzalez, O.D., Haslam, C.E., and Haynes, A. (2002) *Inorg. Chem.*, **41**, 3280–3290.

**213** Adamson, W., Daly, J.J., and Forster, D. (1974) *J. Organomet. Chem.*, **1**, C17–C19.

**214** Haynes, A., Mann, B.E., Gulliver, D.J., Morris, G.E., and Maitlis, P.M. (1991) *J. Am. Chem. Soc.*, **113**, 8567–8569.

**215** Haynes, A., Mann, B.E., Morris, G.E., and Maitlis, P.M. (1993) *J. Am. Chem. Soc.*, **115**, 4093–4100.

**216** van Leeuwen, P.W.N.M. and Freixa, Z. (2008) in *Modern Carbonylation Methods*, (ed. L. Kollár), Wiley, Weinheim, p. 1–25.

**217** McConnell, A.C., Pogorzelec, P.J., Slawin, A.M.Z., Williams, G.L., Elliott, P.I.P., Haynes, A., Marr, A.C., and Cole-Hamilton, D.J. (2006) *Dalton Trans.*, 91–107.

**218** (a) Rankin, J., Poole, A.D., Benyei, A.C., and Cole-Hamilton, D.J. (1997) *J. Chem. Commun.*, 1835–1836; (b) Rankin, J., Benyei, A.C., Poole, A.D., and Cole-Hamilton, D.J. (1999) *J. Chem. Soc., Dalton Trans.*, 3771–3782.

**219** Jiménez-Rodríguez, C., Pogorzelec, P.J., Eastham, G.R., Slawin, A.M.Z., and Cole-Hamilton, D.J. (2007) *Dalton Trans.*, 4160–4168.

**220** Bartish, C.M.(January 13, 1977) (Air Products & Chemicals, Inc.) U.S. Patent 4,102,920.

**221** (a) Gaemers, S. and Sunley, J.G. (2004) (BP Chemicals Limited) PCT Int. Appl. WO 2004/101487. (b) Gaemers, S. and Sunley, J.G. (2004) (BP Chemicals Limited) PCT Int. Appl. WO 2004/101488.

**222** (a) Carraz, C-A., Ditzel, E.J., Orpen, A.G., Ellis, D.D., Pringle, P.G., and Sunley, G.J. (2000) *Chem. Commun.*, 1277–1278; (b) Baker, M.J., Carraz, C-A., Ditzel, E.J., Pringle, P.G., and Sunley, G.J. (March 31, 1999) U.K. Pat. Appl. 2,336,154.

**223** Gonsalvi, L., Adams, H., Sunley, G.J., Ditzel, E., and Haynes, A. (2002) *J. Am. Chem. Soc.*, **124**, 13597–13612.

**224** Mayanza, A., Bonnet, J-J., Galy, J., Kalck, P., and Poilblanc, R. (1980) *J. Chem. Res. (S)*, 146.

**225** Doyle, M.J., Mayanza, A., Bonnet, J-J., Kalck, P., and Poilblanc, R. (1978) *J. Organomet. Chem.*, **146**, 293–310.

**226** Jiménez, M.V., Sola, E., Egea, M.A., Huet, A., Francisco, A.C., Lahoz, F.J., and Oro, L.A. (2000) *Inorg. Chem.*, **39**, 4868–4878.

**227** Haynes, A., Maitlis, P.M., Stanbridge, I.A., Haak, S., Pearson, J.M., Adams, H., and Bailey, N.A. (2004) *Inorg. Chim. Acta*, **357**, 3027–3037.

**228** Burger, S., Therrien, B., and Süss-Fink, G. (2005) *Helv. Chim. Acta*, **88**, 478–486.

**229** Thomas, C.M. and Süss-Fink, G. (2003) *Coord. Chem. Rev.*, **243**, 125–142.

**230** Thomas, C.M., Mafia, R., Therrien, B., Rusanov, E., Stúckli-Evans, H., and Süss-Fink, G. (2002) *Chem. Eur. J.*, **8**, 3343–3352.

**231** Freixa, Z., Kamer, P.C.J., Lutz, M., Spek, A.L., and van Leeuwen, P.W.N.M. (2005) *Angew. Chem., Int. Ed.*, **44**, 4385–4388.

**232** Feliz, M., Freixa, Z., van Leeuwen, P.W.N.M., and Bo, C. (2005) *Organometallics*, **24**, 5718–5723.

**233** Lamb, G., Clarke, M., Slawin, A.M.Z., Williams, B., and Key, L. (2007) *Dalton Trans.*, 5582–5589.

**234** Eby, R.T. and Singleton, T.C. (1983) *Applied Industrial Catalysis*, vol. 1, Academic Press, London, p. 275.

**235** Moloy, K.G. and Wegman, R.W. (1989) *Organometallics*, **8**, 2883–2892.

**236** (a) (1996) *Chem. Br.*, **32**, 7; (b) (March 3, 1997) C&EN; (c) Jones, J.H. (2000) *Platinum Met. Rev.*, **44**, 94–105.

**237** Alperowicz, N.(January 28, 2008) *Chem. Week*, 30.

**238** See www.engelhard.com/eibprices.

**239** (a) Ellis, P.R., Pearson, J.M., Haynes, A., Adams, H., Bailey, N.A., and Maitlis, P.M. (1994) *Organometallics*, **13**, 3215; (b) Griffin, T.R., Cook, D.B., Haynes, A., Pearson, J.M., Monti, D., and Morris, G.E. (1996) *J. Am. Chem. Soc.*, **118**, 3029.

**240** (a) Sunley, G.J. and Watson, D.J. (2000) *Catal. Today*, **58**, 293; (b) Ghaffar, T., Charmant, J.P.H., Sunley, G.J., Morris, G.E., Haynes, A., and Maitlis, P.M. (2000) *Inorg. Chem. Commun.*, **3**, 11.

**241** Haynes, A., Maitlis, P.M., Morris, G.E., Sunley, G.J., Adams, H., Badger, P.W., Bowers, C.M., Cook, D.B., Elliott, P.I.P., Ghaffar, T., Green, H., Griffin, T.R., Payne, M., Pearson, J.M., Taylor, M.J., Vickers, P.W., and Watt, R.J. (2004) *J. Am. Chem. Soc.*, **126**, 2847–2861.

**242** Wright, A.P. (2001) Abstracts of Papers 222nd ACS National Meeting, Chicago, IL, U.S, CATL-044; (2001) *Chem. Abstr.*, AN 637430.

**243** Steinmetz, G.R. (1984) *J. Mol. Catal.*, **26**, 145–148.

**244** Gautron, S., Lassauque, N., Le Berre, C., Azam, L., Giordano, R., Serp, P., Laurenczy, G., Daran, J.C., Duhayon, C., Thiébaut, D., and Kalck, P. (2006) *Organometallics*, **25**, 5894–5905.

# 9
# Metal-Catalyzed Cross-Coupling Reactions

## 9.1
## Introduction; A Few Historic Notes

Metal-catalyzed cross-coupling reactions have grown enormously in popularity since the introduction of the Suzuki–Miyaura reaction for C–C bond formation and the Buchwald–Hartwig protocols for C–N, C–O, and C–S bond formation, both in the laboratory and in industry [1, 2]. More recently other carbon–heteroatom bond formations were added, such as C–P and C–B reactions. The catalytic formation of C–C bonds involves the reaction of a hydrocarbyl metal reagent with an organic halide or pseudohalide, most often containing an $sp^2$ hybridized carbon atom. The organometallic reagent can be based on a wide range of metals and often the reaction is named after its inventor (neglecting the catalytic metal): Mg (Kumada–Corriu), Cu (non-catalytic Castro–Stephens; catalytic Sonogashira), Zn, Al, Zr (Negishi), Sn (Stille), B (Suzuki–Miyaura), and Si (Hiyama). The elegance of the Buchwald–Hartwig protocol is that the conjugate acids of the heteroatom reagents are used and that the acid formed from this proton and the halide is neutralized by an added base. The coupling of aromatic halides with alkenes, the Heck–Mizoroki reaction, is considered part of the cross-coupling concept. The importance of these findings was underscored in 2010 by the Nobel prize for chemistry, which was granted to Heck, Negishi, and Suzuki.

The formation of new C–C bonds with the use of Grignard reagents was discovered by Grignard more than 100 years ago and concerned the addition reaction of RMgX reagents to carbonyl compounds [3]. Nucleophilic substitution of reactive organic halides such as benzylic halides also proceeds with good yields, but coupling of Grignard reagents with alkyl or aryl halides to form new C–C bonds gives low selectivities, reactions are very slow, or do not proceed at all. Homo-coupling of phenylmagnesium bromide was observed almost a century ago by Bennett and Turner (1914) when they added $CrCl_3$ to a solution of the Grignard reagent [4]. The formation of biphenyl is stoichiometric and not catalytic. It would seem that the reaction takes place at Cr and that a radical pathway (leading to benzene) is only a minor route. In the absence of Cr, diphenyl was also observed due

*Homogeneous Catalysts: Activity – Stability – Deactivation*, First Edition. Piet W.N.M. van Leeuwen and John C. Chadwick.

to impurities in Mg. Attempts to cross couple two different Grignard reagents gave the two homo-coupling products (hexane and diphenyl). Addition of $FeCl_3$ to PhMgBr gave diphenyl in a low yield. A few years later (1919) Krizewsky and Turner obtained diphenyl in 85% yield together with CuI from PhMgI by adding $CuSO_4$, thus showing that Cu is a much better coupling agent [5].

In general, RMgX reagents and alkyl halides give C−C bond formation in low yield and elimination also occurs. Alternatively, alkane/alkene products can arise from radical disproportionation after single electron transfer; indeed, reactions of similar alkyllithium compounds with alkyl halides show CIDNP (chemically induced dynamic nuclear polarization), proving the formation of radical pairs to some extent [6]. The low reactivity between RX and RMgX is also key to the synthesis of Grignard reagents during the reaction of RX with metallic Mg; if this reaction were fast, the synthesis of Grignard reagents would not be possible. Aryl halides and Grignard reagents give halide/magnesium exchange, often used for making Grignard derivatives, and do not undergo coupling to new C−C bonds. The exchange reaction also leads to homo-coupled products. For more ionic hetero-atom centered halides, such as Si and P, nucleophilic substitution by Grignard reagents can be very efficient.

The influence of transition metal salts in catalytic amounts on reactions of Grignard reagents was already reported by Kharasch and coworkers in 1941 [7]. They noticed that the course of the coupling reaction could be changed by the addition of catalytic amounts of first row transition element chlorides. Magnesium had to be purified thoroughly by sublimation in order to conduct these experiments. The first reaction reported is the addition of $i$BuMgBr to benzophenone in the presence of $MnCl_2$ or $CrCl_3$. In the absence of transition metal, the yield of the expected diphenylmethanol, which is the product of a two-electron transfer, was >90%, but addition of as little as 1% of $MnCl_2$ gave 90% yield of the pinacol, the product of a one-electron transfer. $CrCl_3$ was half as active and the reaction was less selective, this also occurred at 2% of transition metal [7a]. MeMgBr gave pinacol in the presence of $CoCl_2$, but otherwise it gave the C−C bond forming product 1,1-diphenylethanol in high yield [7c].

For isophorone (3,5,5-trimethyl-2-cyclohexen-1-one) and MeMgBr in the absence of transition metals the products obtained were the 1,2 methyl addition product (addition to C=O) together with the dehydration product of the latter [7b]. In the presence of $FeCl_3$ the main reaction was probably double bond isomerization to 3,5,5-trimethylcyclohex-3-enone, the less stable isomer. Addition of 1% of $CoCl_2$ or $NiCl_2$ gave the pinacol product, the result of a one-electron transfer. Addition of 1% of CuCl gave the 1,4 addition product 3,3,5,5-tetramethylcyclohexanone in >80% yield, a reaction that later has become an important synthetic tool.

The reaction of PhMgBr with alkyl or aryl bromides and $CoCl_2$ as the "catalyst" gave mainly the homo-coupling product diphenyl, presumably via radical reactions, while in the absence of transition metals hardly any reaction took place [7d]. Most other metals gave lower yields of diphenyl, except $NiCl_2$, but unfortunately this reaction was only performed with bromobenzene as the halide component and we cannot distinguish between homo- and hetero-coupling.

In this context earlier work by Gilman and Lichtenwalter [8] on stoichiometric reactions that represent cross-coupling reactions should be mentioned. In stoichiometric reactions of PhMgI and metal halides of the Groups 8–10, high yields of the reductive elimination product diphenyl were obtained for Fe, Co, Ni, Ru, Rh, and Pd (all >97%). The oldest report (1914) on this chemistry is probably from Bennett and Turner [9], who found that PhMgBr reacted with $CrCl_3$ at 35 °C (3 h) to give diphenyl quantitatively; the reaction was assumed to proceed via $Ph_xCrCl_{3-x}$ species. During the synthesis of PhMgBr substantial quantities of diphenyl were formed, showing that the Mg used was less pure than that used nowadays. Several aromatic Grignard reagents reacted this way with $CrCl_3$, but aliphatic ones did not. An attempt to achieve cross-coupling between two distinct aromatic Grignard reagents gave only homo-coupling products.

Publications on the coupling reaction of Grignard reagents with organohalides by Kochi (Ag, Cu, Fe) [10–12], Kumada (Ni) [13], and Corriu (Ni) [14] set the pace for modern cross-coupling catalysis. Kumada and coworkers found that addition of 0.1–1% of a nickel phosphine chloride catalyst gave 98% yield and selectivity in the reaction of EtMgBr and chlorobenzene. They found that complexes of bidentate phosphines afforded much better catalysts than monophosphines and that dppp gave a better catalyst than dppe. Only $sp^2$-C organic halides could be used, but, remarkably, chloride worked very well. Since the elementary steps involved were already known at the time as stoichiometric reactions in organometallic chemistry, the mechanism they proposed is still valid today.

The Sonogashira reaction involves the coupling of terminal alkynes with aryl or vinyl halides with a Pd/Cu catalyst [15]. This reaction was first reported by Sonogashira and Hagihara in 1975, and without Cu by Cassar and Heck [16]. Stoichiometric coupling of alkynylcopper with iodobenzene was established by Castro and Stephens [17]. With a palladium catalyst added, $PdCl_2(PPh_3)_2$, the most popular catalyst in the 1970s, the reaction was not only much faster, but also the alkyne could be used in the presence of a base to eliminate the hydrogen atom. For instance, 1,4-diiodobenzene was coupled with phenylethyne, 1% of Pd catalyst, 2% of CuI, $Et_2NH$ as the base, at room temperature for 3 h, in 98% yield. Nowadays versions without Cu are also known [18].

The first metal-catalyzed Negishi cross-coupling reaction by Baba and Negishi was reported in 1976 and concerned the stereospecific reaction of alkenylalanes with alkenyl halides using nickel and palladium catalysts. The Ni catalyst was prepared in situ from $Ni(acac)_2$ and DIBAL in the presence of 4 mol $PPh_3$, while the Pd catalyst was prepared from $PdCl_2(PPh_3)_2$ and DIBAL [19]. The zinc variant followed a year later and showed the coupling of iodo and bromo aromatic compounds with phenyl and benzyl zinc organometallic reagents. With p-bromobenzonitrile high yields were reported, but even p-iodonitrobenzene gave 90% yield of the coupling product [20].

The first examples of the Stille reaction were published in 1977 [21] by Migita, and in 1978 by Milstein and Stille. One of the hydrocarbyl groups of $R_4Sn$ (or later $RSnMe_3$, $RSnBu_3$) was coupled to acid chlorides yielding ketones in high yield when a Pd catalyst was used, $PhCH_2PdCl(PPh_3)_2$ [22]. $Pd(PPh_3)_4$ was less effective as the saturated species slows the oxidative addition. The use of the alkylating agent is not efficient and organotin compounds are highly toxic, but nevertheless the reaction was very popular.

The Miyaura–Suzuki reaction was published in 1979 and its initial version was the reaction of alkenylboranes with alkenyl halides giving dienes [23]. The limitations of the use of the reactive metal alkyls used by Negishi were realized and it was thought that boranes would be an attractive alternative in organic synthesis, being compatible with a large number of functional groups. Boranes themselves are not such strongly alkylating agents and therefore an anionic base such as NaOEt was added to make the boronate esters *in situ*. The latter are more electron-rich and they were indeed capable of alkylating the palladium halide intermediate formed after oxidative addition (see Section 9.2). The catalyst used was $PdCl_2(PPh_3)_2$, which was sometimes reduced with $NaBH_4$. Suzuki started this work entering from the hydroboration area and thus the alkenylboranes were initially obtained via hydroboration of alkynes [24]. It was followed by the coupling of alkenylboranes with aromatic halides. The reaction of arylboronates with aryl halides, nowadays best known, was published the year after [25]. An enormous amount of variations have been published [26], including many useful applications in synthesis, and the reaction is used industrially. A group of examples are the sartan drugs –for treatment of hypertension – which are all based on a diphenyl backbone equipped with various nitrogen functions [27].

Like boranes, silanes as providers of the hydrocarbyl group in cross-coupling reactions also suffer from the low alkylating power. In the Hiyama cross-coupling reaction this was solved by using a fluoride donor to enhance the release of the hydrocarbyl group of a silane. Thus, vinyl, allyl, and ethynyl silanes were coupled with aryl, vinyl, and allyl halides in high yields with TASF (trisdimethylaminosulfonium difluorotrimethylsilicate) as the fluoride donor, and $(\eta^3\text{-allylPdCl})_2$ as the catalyst [28]. The more common $F^-$ donor in silyl ether chemistry, TBAF (tetrabutylammonium fluoride), was found to be less effective.

The first examples of the Buchwald–Hartwig $C-N$ coupling reaction appeared simultaneously in 1995 from the two inventors [29]. Both publications followed earlier work in which tin amides were used as the source of the nitrogen nucleophile [30], inspired by the work of Migita, ten years before [31].

The Mizoroki–Heck reaction, first reported by Mizoroki [32, 33], is usually also considered under cross-coupling reactions, although there is no nucleophile involved in this reaction. Instead, after oxidative addition of an aryl halide, insertion of an alkene occurs, and the catalyst returns to zero-valent metal via β-elimination and elimination of HX (see Section 9.2).

The palladium-catalyzed allylic alkylation, an important tool for organic chemists, can also be regarded as a cross-coupling reaction. It has been developed largely by Trost [34a]. The allylic substitution in simple allylpalladium chloride was published by Tsuji [34b]. The method has been applied in the synthesis of many complex organic molecules [35]. During the reaction a new $C-C$ or $C-N$ bond is formed but, unlike the cross-coupling reactions above, the nucleophile exerts an outersphere attack and does not coordinate to Pd.

Among other catalytic reactions, cross-coupling catalysis occupies an exceptional position in deactivation studies, because the number of studies devoted to deactivation as such represents only a very small fraction of the total number of publications in this area, which may well be of the order of five to ten thousand. There are several reasons for this, one being that the majority of the publications concerns applications of more or less known catalytic systems on organic synthetic problems, including optimization but without focusing on deactivation. Researchers in the area of coordination and organometallic chemistry focusing on the design and synthesis of new Pd, Ni, and Cu complexes (especially Pd) often add a screening of the Miyaura–Suzuki reaction as an application of their chemistry to their own work; these studies are not of importance here.

In the early years of exploration there was, as usual, no attention yet given to practical, economic catalysts giving a high turnover number (TON). Carbon–heteroatom bond forming was more challenging than C–C bond formation. In more recent years, a few research groups, such as the Buchwald group, have addressed this target, and, evidently, if high TONs are a target, catalyst stability becomes an issue. Mechanistic studies, such as those from the Hartwig and Buchwald groups, have contributed enormously to our understanding of the course of the reaction and this is used to arrive at more effective catalysts. Kinetic studies have played an important role as well. It was recognized that catalyst initiation is the key issue in many cross-coupling reactions, much more so than in polymerization and carbonylation catalysis, for which a standard recipe works well for a few metals and a wide range of ligands. Therefore, ample attention will be paid to initiation, also because many solutions are now at hand. There are a few results on ligand decomposition that will be discussed. Precipitation of palladium metal is a general phenomenon that will be mentioned sporadically. Metal purity and metal nanoparticles (MNPs) as precursors will also be dealt with.

## 9.2
## On the Mechanism of Initiation and Precursors

In the following sections an attempt has been made to separate the ways in which the reactions are started, how the next steps take place, and how the reductive elimination is influenced by the catalyst obtained. This turned out to lead to arbitrary choices. For example, an oxidative addition reaction of an aryl halide to a Pd(0) precursor (Section 9.2.1) may lead to one of the precursors treated in Section 9.2.2 and the topic may occur in either part.

## 9.2.1
## Initiation via Oxidative Addition to Pd(0)

There is general consensus about the overall mechanism for C–C bond formation, which in summary comprises the following steps (Scheme 9.1) [36]: (i) conversion of the precursor into one of the intermediates participating in the

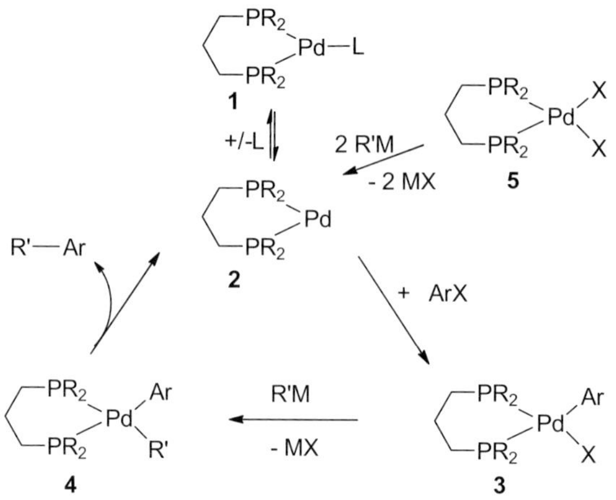

**Scheme 9.1** Reaction sequence for C–C bond formation.

catalytic cycle, (ii) the oxidative addition of the aryl (vinyl) halide to a Pd(0) intermediate, (iii) the replacement of the halide (or pseudohalide) by the hydrocarbyl group or transmetallation, and (iv) the product-forming reductive elimination. The details omitted such as metals, ligand, (pseudo)halide, bases, solvents, and so on are extremely important, as they have a strong influence on rates, turnovers, and selectivities. In Scheme 9.1 a cis bidentate ligand has been drawn to reduce the number of complexes that may form to cis ones only.

As mentioned in Section 9.1 the common precursors in the first decade were PdCl$_2$(PPh$_3$)$_2$ and Pd(PPh$_3$)$_4$. Meanwhile, we have learnt that the nature of the precursor is of the utmost importance and many results could have been improved if the proper precursor complex had been applied. The use of a stable zerovalent Pd species requires dissociation of ligands, which may be slow, or the equilibrium between **1** and **2** (Scheme 9.1) may be well on the side of **1**, thus reducing the efficiency of the catalyst. When Pd(II) precursors **5** are used, reduction is needed to form the starting Pd(0) compound. For the reactions using metal alkyls this is not a problem, as the first reduction will take place easily and only a small amount of homo-coupled R'–R' will form as a side-product. The less strong alkylating agents based on B and Si require the use of an activating base, which will be the same one that is used in the coupling reaction itself. It should be noted that the by-products formed during the activation, although present only in the quantities of the catalyst, may still influence the catalysis either positively or negatively. For example the presence of halides or acetates may influence the oxidative addition and may even have a beneficial effect, as was studied by Amatore [37]. He found that a Br$^-$ ion coordinated to Pd(0) complexes of PPh$_3$ accelerates oxidative addition.

Catalysts with ligands other than PPh$_3$ can be made *in situ* from metal salts such as Pd(OAc)$_2$, PdCl$_2$(RCN)$_2$, Ni(acac)$_2$, and so on. Aged samples of Pd(OAc)$_2$ may lead to inferior results, unless used with strong acids; it is a trimeric compound that is soluble in organic solvents, and if not completely soluble, it should be recrystallized.

As mentioned in Section 9.1 metal salts can also be reduced in advance in the presence of the ligand to give the precursors containing the desired ligand, for example, with DIBAL [19]. Often the reduction was left to chance and phosphine was sacrificed as the reducing agent (in the presence of water this works well) or the reduction reaction was of an unknown nature. The reduction of $Pd(OAc)_2$ with $PPh_3$ and water was studied by Amatore and coworkers [38].

Water was added deliberately by Buchwald and coworkers [39] to obtain Pd(0) from $Pd(OAc)_2$ in the presence of 2.2 mol of XPhos for the coupling of chloroanisole and aniline.

As a Pd(0) precursor $Pd_2(dba)_3$ is often used, but the displacement of dba by the desired ligand may be slow and lead to an incubation time. The high temperatures regularly applied in cross-coupling catalysis may have been necessary in several instances only to enforce the initial reduction or the displacement of ligands, as needed before oxidative addition can take place. Furthermore, the presence of dba afterwards may still block part of the catalyst [40].

Hartwig and coworkers found that in amination reactions with primary alkyl and aryl amines, secondary cyclic alkylamines, and secondary arylalkylamines with bromoarenes using $Pd(BINAP)_2$ and its dppf analogue as the catalysts, the reaction was zero order in all substrates and that the rate-limiting step was the loss of one ligand from the precursor [41]. At higher conversions substrate inhibition was observed for dppf as the ligand and aniline as the substrate, which originated from the reverse oxidative addition of the product.

Recently, Hartwig, Blackmond, Buchwald and coworkers published a revised mechanism for the amination of aryl halides catalyzed by Pd/BINAP complexes [42]. The new feature in this mechanism compared to previous findings is that they placed $Pd(BINAP)_2$ outside the catalytic cycle, which explains the induction period that occurs when $Pd(BINAP)_2$ is used as the precursor, thus underpinning this phenomenon in Scheme 9.1 ($1 \rightarrow 2$). Clearly, these problems might have been avoided if a complex such as ArPdBr(BINAP) **3** had been used as the catalyst. For Xantphos an inhibiting influence of an excess ligand was also reported by Buchwald and coworkers [43].

Monophosphine/1,6-diene palladium complexes of *o*-phenylenephosphine ligands are convenient precursors for the oxidative addition, as reported by Beller and coworkers [44], complexes of simple diallyl ether being the most effective ones in Suzuki reactions. Apparently diallyl ether falls off easily, because the complexes are more active than the catalysts formed *in situ* from $Pd(OAc)_2$ or $Pd_2(dba)_3$.

Oxidative addition of $p\text{-}tBuC_6H_4Br$ to $Pd(oTol_3P)_2$ turned out to be much faster than addition to $Pd(PPh_3)_4$, one of the common Pd(0) precursors at the time [45]. The reaction was shown to take place after dissociation of phosphine from the two-coordinate Pd(0) complex, forming a highly unsaturated, reactive species. At the concentrations studied, the reaction was first order in arene bromide and minus first order in the free phosphine concentration. Free phosphine is produced during the reaction, because the product is the dimer $[ArPdBr(oTol_3P)]_2$.

The oxidative addition of all four aryl halides (of practical importance) to Pd(0) complexes was studied by Barrios-Landeros and Hartwig [46] for one of the modern,

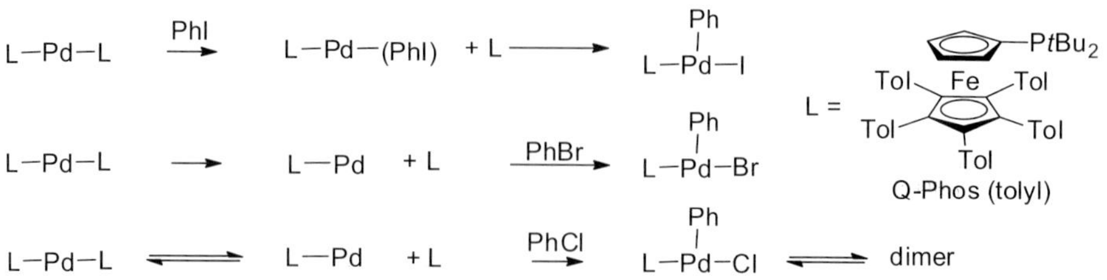

**Scheme 9.2** Kinetics of the oxidative addition of PhX (I, Br, Cl) to Pd(Q-Phos)$_2$.

bulky, electron-rich monophosphines, Q-Phos (Ph$_5$FcP$t$Bu$_2$, see also Section 9.5). Q-Phos gave highly active catalysts for a wide range of cross-coupling reactions and its bisphosphine complex can be used as the precursor. The oxidative addition of PhI, PhBr, and PhCl to Pd(Q-Phos)$_2$ at 30–65 °C in neat PhX produced [Pd(Q-phos)(Ph)(X)] quantitatively with one molecule of free ligand (tolyl instead of phenyl derivatives of Q-Phos were used for solubility reasons). The complexes are monomeric, except the Cl derivative, which is a dimer in apolar solvents and in the solid state. The kinetic results are summarized in Scheme 9.2.

Surprisingly, different kinetics were found for the three substrates. Addition of iodobenzene is first order in [PhI] and the reaction is independent of ligand concentration (the single arrows denote irreversible reactions within the concentration limits studied). Addition of bromobenzene is independent of the concentrations of both ligand and bromobenzene; after dissociation of a ligand the reactive species is captured by PhBr and reactions for other aryl bromides (anisyl, tolyl) showed exactly the same rate. Furthermore, the rate was unaffected by large excesses of L. Chlorobenzene is less reactive and ligand dissociation becomes a reversible process; the rate constant for addition of PhCl depended positively on [PhCl] and inversely on [L].

The study was extended to the ligands Pd(P$t$Bu$_3$)$_2$, Pd(1-AdP$t$Bu$_2$)$_2$, Pd (CyP$t$Bu$_2$)$_2$, and Pd(PCy$_3$)$_2$ to include steric effects [47]. The conclusions remained basically the same; the rate constants depend more strongly on the identity of the halide than on the steric bulk of the ligand and less reactive halides need a lower degree of phosphine coordination. Only for aryl chlorides did the concentration of L matter. In general, reaction rates may differ enormously from one substrate, ligand, and set of conditions to the other, but as a trend these kinetic findings are very useful.

Indeed, in a study by Shekhar and Hartwig on the effects of bases and halides on the amination of chloroarenes catalyzed by Pd(P$t$Bu$_3$)$_2$, the kinetic behavior was rather complicated [48]. Not unexpectedly, for aryl chlorides the oxidative addition is the rate-limiting step, electron-poor aryl chlorides showing the slowest reaction. The extent to which the rates depended on the concentration and identity of the bases used, was determined by the electronic properties of the aryl chlorides. The rates of reactions of electron-rich and electron-neutral chloroarenes were independent of the concentration of the bulky alkoxide base OCEt$_3$$^-$, but they were dependent on the concentrations of the less hindered O$t$Bu$^-$ base and the softer 2,4,6-tri-*tert*-butyl-phenoxide base. The recommendation would be to use the latter. It was suggested that probably more species were involved in the initiation of the catalytic cycle.

In the absence of strong base the oxidative addition of PhBr to Pd(P$t$Bu$_3$)$_2$ was catalyzed by Pd(P$t$Bu$_3$)$_2$HBr within the frame of conditions studied [49]. This result means that, curiously, addition or metathesis of PhBr with the hydride complex is faster than oxidative addition to Pd(P$t$Bu$_3$)$_2$, which proceeds via rate limiting dissociation of the ligand. Oxidative addition of Br(HP$t$Bu$_3$) to Pd(P$t$Bu$_3$)$_2$ must be a fast reaction as well, see Scheme 9.3.

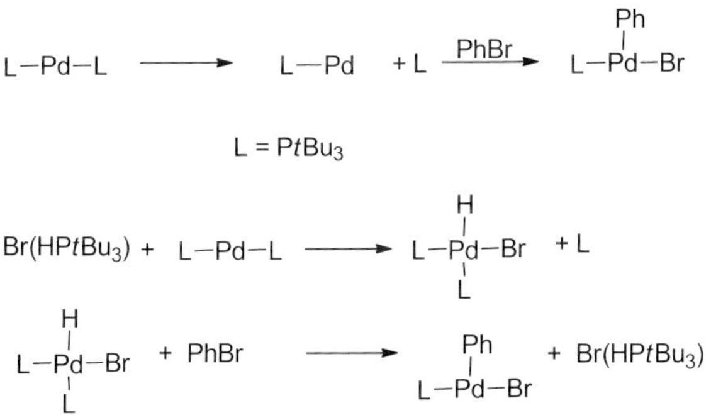

**Scheme 9.3**  Pd(P$t$Bu$_3$)$_2$HBr as a catalyst in the oxidative addition of PhBr to Pd(P$t$Bu$_3$)$_2$.

The mechanism for the Buchwald–Hartwig reaction, and in general C–Y (Y = N, O, S, or P) coupling reactions, is basically the same as that shown in Scheme 9.1 for C–C bond forming reactions, but there are a few distinct differences that will be outlined. In Scheme 9.4 the mechanism of C–N bond formation is summarized for complexes containing diphosphines. Since X is used throughout for halide or pseudohalide, Y will be used for N, O, S, or P, the atoms indicating primary or secondary amines, alcohols, thiols, and HPR$_2$ compounds, respectively. The reactions shown for **3**, **4**, and **6** are particularly valid for Y = N.

The mechanistic aspects of C–Y bond formation have been studied in depth by Hartwig and coworkers [50]. Since C–Y bond formation takes place in the absence of an alkylating agent, reduction of Pd(II) must take a different course. Louie and Hartwig studied the formation of palladium(0) in amination reactions using tin amides, and found that β-hydride elimination at the amido group followed by base-aided reductive elimination of HX led to reduction of palladium [51]. See Scheme 9.5 for an example; the conversion of **3** to **4** may be a multi-step reaction, including equilibria as well. The β-hydride elimination as the mechanism for arene formation was proven by deuteration studies with the use of fully deuterated dimethylamidotin derivatives. In addition to ArD, ArH was also formed, not stemming from the solvent. Other groups, perhaps from the ligand, may therefore participate in side reactions giving arenes, ArH.

In 1995 Buchwald and Guram [30b] already disclosed the formation of arenes via β-elimination of hydrogen from the amine and reductive elimination of ArH, which

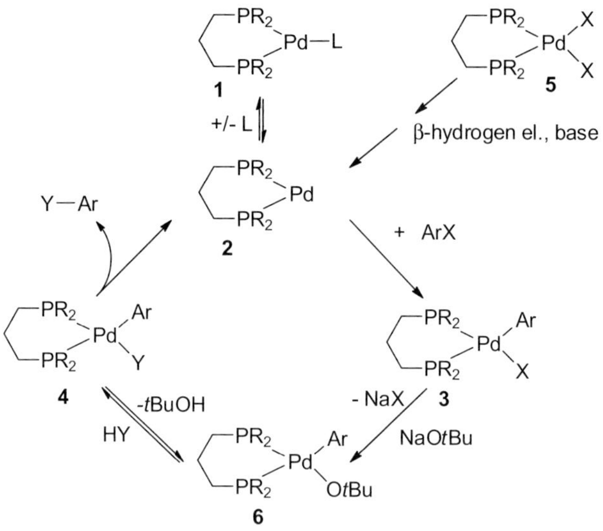

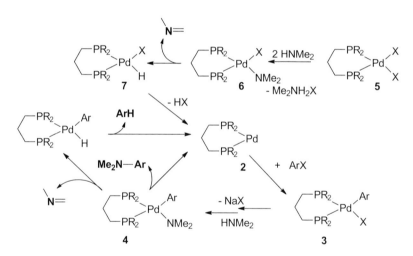

**Scheme 9.4**  Simplified mechanism for C−Y bond formation (Y = N).

**Scheme 9.5**  β-Hydrogen elimination forming Pd(0) and causing ArH formation, highlighting the products.

was the dominating pathway when PPh$_3$ was used as the ligand, while P(*o*-tolyl)$_3$ gave cleanly the cross-coupled amine. For iridium Hartwig also noted that more bulky ligands relatively suppressed β-hydride elimination and enhanced reductive elimination [52].

For the coupling reaction of *p*-bromotoluene and piperazine with Pd/BINAP as the catalyst, it was found that arene formation was higher in polar, aprotic solvents such as NMP than in apolar solvents; *m*-xylene was the best solvent in spite of the limited solubility of the base NaO*t*Bu [53].

An excess of piperazine also reduced arene formation. Beletskaya proposed a transfer of a β-hydrogen from a coordinated amine to Pd and the aryl fragment, because stronger bases, giving a faster deprotonation, gave less arene formation [54]. Notably Et₃N, which cannot form amides, can also reduce Pd(OAc)₂ to Pd(0) [38].

Perhaps the solvent effect is related to the inhibition shown by NaI, formed in coupling reactions of aryl halides and amines, on these reactions, as found by Buchwald and coworkers [55]. By switching to a solvent in which the iodide salt was insoluble, no inhibition was observed.

*In situ* reduction was achieved when a bulky Josiphos derivative (1-Cy₂P,2-*t*Bu₂P- α-ethylferrocene) was used in the very effective N−C coupling of aryl chlorides and primary amines [56]. Pd(OAc)₂ and the free ligand were used as the initiators. Although the ligand is stable towards oxygen, even in solution, an even more stable one-component catalyst was reported, namely the corresponding dichloride palladium adduct. Apparently, this is reduced rapidly by primary amines (aniline included) [57].

The use of amines as an *in situ* reducing agent is not always satisfactory and other mild reducing agents may be needed. In an optimization study on a 4 kg scale, workers at Pfizer deliberately added 2 mole of phenylboronic acid to a palladium acetate/Davephos catalyst (as used before by Buchwald), which was found to be the best out of many other catalyst activation protocols [58]. The catalyst was prepared outside the reactor, leaving the reagents together for 15 m at room temperature (thus more strongly reducing agents could have been used). The reagents of the coupling reaction are shown in Scheme 9.6; the β-aminoacid precursor is chiral.

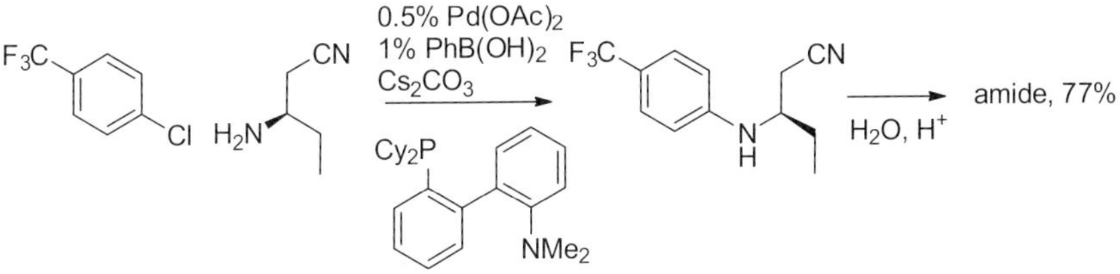

**Scheme 9.6** Reduction of Pd(II) by phenylboronic acid.

The order of the ease of oxidative addition as regards the halides is well established: I > Br > Cl, with triflate similar to Br. The order Br/OTf varies and depends on the catalyst system and even the nucleophile used. Kamikawa and Hayashi did competition experiments for *p*-bromophenyl triflate and phenylmagnesium bromide [59] in the Kumada coupling using PdCl₂ phosphine complexes. Bidentate phosphines such as dppp and dppb gave, selectively, replacement of the triflate group, while monodentate phosphines gave mainly cleavage of the C−Br bond over the C−OTf bond. As was often found, the addition of LiBr had a favorable effect on the rate of reaction involving the triflate cleavage. Brown and coworkers discovered that the course of the cleavage in aryl bromides and aryl triflates depended on the nature of the nucleophile [60], that is, whether the metal was Mg or Zn on the one hand or a boric acid on the other. The reaction of *m*-bromophenyl triflate in a Kumada and Negishi coupling with dppp as the ligand occurred by substitution of triflate, as found by Hayashi, while Suzuki reactions led to displacement of bromide irrespective of whether dppp or monodentate P*t*Bu$_3$ was used. Stille and Mizoroki–Heck reactions also led to displacement of triflate, thus leaving the boron reagents as the exception. This is rather curious as oxidative addition of ArBr or ArB(OH)$_2$ to PdL or PdL$_2$ (L is phosphine) should not be influenced by the next step, X/R exchange, because a reversible oxidative addition is rather unlikely. Thus, the question was how does a boronic acid or boronate influence the oxidative addition such that the selectivity entirely changes. The most likely explanation given is that actually the oxidative addition takes place onto a Pd species ligated to an anion, as reported by Jutand and Amatore [37] (note also the influence of LiBr added to systems void of Br). Brown proposed that the boronate anion present modifies the Pd center by forming a complex, such that subsequently the C−Br bond will be cleaved preferentially. This bond between Pd−boronate, if not broken after the oxidative addition, facilitates the transfer of the aryl group from B to Pd, as proposed earlier for Rh by Hartwig.

### 9.2.2
### Hydrocarbyl Pd Halide Initiators

Convenient starting materials are compounds of type **3**, which can be formed by oxidative addition of an aryl or vinyl halide to Pd(0). Usually they are stable compounds, monometallic with bidentate ligands and bridged bimetallic species for monodentate ligands, or, for very bulky monodentates, they are also monometallic. They can be made by reacting Pd$_2$(dba)$_3$ with R'X in the presence of the phosphorus or nitrogen ligand to be studied. This reaction may be slow, but for the on-purpose complex synthesis this is no setback. Displacement of a weak ligand in a complex **3** or a dimer of the general formula (ArLPdBr)$_2$, in which L is a monodentate ligand, by the desired ligand is often a convenient route (see Figure 9.1). Except for compounds **9** and **11**, numerous examples can be envisioned.

A good precursor is ($\eta^3$-allylPdCl)$_2$ (or its ligated derivative **10**), as used by Hiyama [28], or better still perhaps ($\eta^3$-2-methallylPdCl)$_2$, which has a longer shelf-life. It is almost always used as the catalyst precursor in allylic alkylation, amination of dienes [61], and often in polymerization chemistry, but its usefulness [62] as a ligand-free, hydrocarbyl-containing Pd(II) compound is underestimated in cross-coupling catalysis.

Also for C−Y formation starting from Pd(II) precursors, phosphines have been suggested as the reducing agent for Pd(II). For instance, in the presence of hydroxide or acetate, BINAP will be oxidized by Pd(II) to its monoxide yielding Pd(0) [63]. Of the two most effective representatives for diphosphines and monophosphines, however, neither oxidation of Xantphos nor that of the diarylphosphine ligands of Buchwald is a rapid process. On the contrary, the uncatalyzed oxidation by dioxygen of substituted *o*-phenylenephenyl-dicylohexylphosphine (as in **11**) is strongly hampered by the

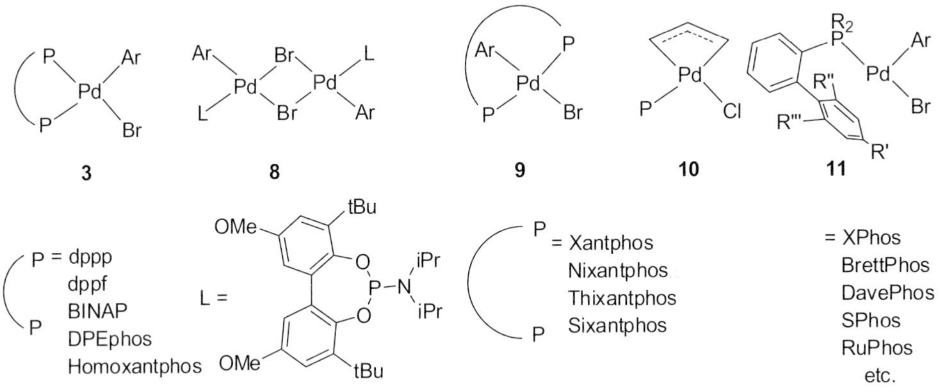

Figure 9.1 Examples of hydrocarbyl Pd halide precursors; P represents a phosphine.

steric bulk of the ligand [64], in spite of the presence of strongly donating Cy groups, but data for Pd-catalyzed oxidation by base are limited. Thus, leaving reduction of palladium to chance, even when using the optimal ligands, is not an ideal proposition.

The use of intermediates **3** as catalyst is very useful for C−Y formation as well; the issues concerning the conversion of the precursors into one of the intermediates occurring in the catalytic cycle ("the catalyst") can be avoided by using arylpalladium complexes **3**. Several authors showed for the Buchwald–Hartwig coupling reaction that precursors **8–11**, as shown in Scheme 9.5 give rise to immediate initiation of the catalytic reaction without incubation, and likewise for the Heck reaction [65, 66]. For example, C−N coupling can be carried out at room temperature with a turnover frequency as high as 140 mol mol$^{-1}$h$^{-1}$ [67–69], while many investigators would use temperatures as high as 120 °C for such a reaction. Likewise, for XPhos and related ligands, the preferred precursor is **11**. Dimers of monovalent Pd(I) are even faster catalyst precursors, as they disproportionate rapidly into Pd(0) and Pd(II) with a Pd/P ratio of 1 : 1 [70]. In this way half of the Pd may remain unused, unless reduction of PdBr$_2$ takes place, as in Suzuki, Kumada, and Negishi reactions for instance (Scheme 9.7).

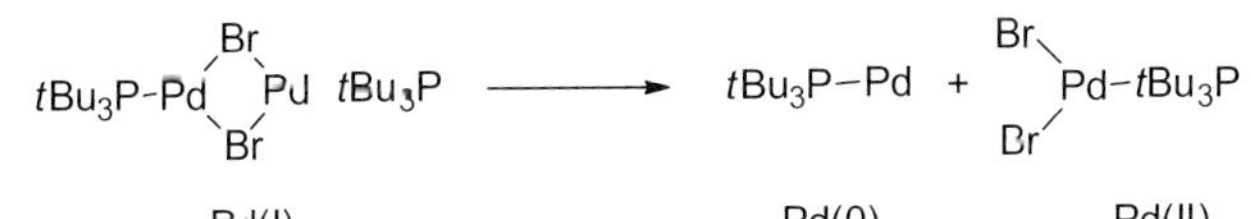

Scheme 9.7 Disproportionation of a Pd(I) dimer.

Complex **8** of the successful ligand P-$o$-tol$_3$ can be conveniently made from Pd (P-$o$-tol$_3$)$_2$ by addition of the desired aryl bromide [71]. The Pd(0) starting material can be made from Pd$_3$(dba)2/Pd(dab)$_3$ by adding P-$o$-tol$_3$, which works well because the product has a low solubility in benzene/ethyl ether. The complex is a dimer both in the solid and in solution. While direct synthesis of **3**, **8**, and **9** would seem to be an easy entry into catalysts not showing incubation, the direct synthesis from $Pd_2(dba)_3$, the desired ligand, and aryl bromide often leads to side reactions and low yields. For instance dba may undergo addition reactions and β-hydride elimination, and "PdArX" may disproportionate into $PdX_2$ (containing the ligand under study) and Pd metal. A replacement reaction of P-*o*-tol$_3$ in **8** by the desired monodentate or bidentate phosphine is often a better alternative [72, 73].

When NHC ligands were used, even $Pd_2(dba)_3$ was found to be an effective precursor and cross-coupling of aryl chlorides and secondary amines was achieved at room temperature [74]. This reactivity was tentatively assigned to the combination of strongly electron donating ability and severe steric demands of the NHC used. NHCs gave good results with the use of NHC complexes made from $(\eta^3$-allylPdCl)$_2$ as the precursor for Buchwald–Hartwig reactions at room temperature, although at slightly higher temperatures the reaction proceeded much faster, which was later ascribed to a somewhat slow initiation [75]. Therefore, a faster initiator was sought and 1-phenyl substituted allyl (cinnamyl) palladium chloride appeared an excellent precursor at room temperature for Suzuki–Miyaura and Buchwald–Hartwig reactions [68]. The reaction between morpholine and bromomesitylene was conducted at room temperature with 0.1 mol% catalyst, reaching completion in a few hours. Moreover, with 10 ppm catalyst the reaction reached completion at 80 °C after 30 h, a TON of 100 000.

Starting with an arylpalladium halide ligand complex, however, does not always lead to the fast catalysis expected. Alvaro and Hartwig [76] showed in a study of C−S bond formation that each elementary step occurred rapidly, but that the overall catalytic process was slower. The ligand used is a Josiphos-type ligand, CyPF-*t*Bu, that was shown to be a very successful ligand in many cross-coupling reactions by the Hartwig group. The ligand is very bulky, which probably contributes to a fast reductive elimination, and it is strongly electron-donating to palladium, facilitating aryl chloride additions and avoiding metal–ligand dissociation. First, it was shown that the use of $Pd(OAc)_2$ and the ligand gave a low activity because only a small amount of Pd(II) was converted to Pd(0) to start the catalytic cycle. When $Pd_3(dba)_2$ was used as the precursor the strong back donation to dba led to a slow initiation of the reaction. The corresponding aryl halide intermediate **3** was obtained from Pd(*o*tolP)$_2$, CyPF-*t*Bu and aryl bromide, and was rapidly converted with thiolate and base into the desired product. Yet, the overall catalytic reaction was slower than all individual steps studied, the reason being that in the catalytic system palladium was tied up in unreactive species outside the catalytic cycle (see Scheme 9.8).

The Cy-*t*-BuPF-based catalysts nevertheless function very well and low catalyst loading could be used; it was concluded that compared to other ligands they resisted decomposition reactions with thiols and thiolates rather well [77].

Colacot and coworkers reported a versatile synthesis of $PdL_2$ (L = phosphine) species starting from (1,5-cod)Pd(II) salts in methanol and base [78]. An intermediate metallated hydrocarbyl Pd halide is formed (next section), which undergoes elimination to form the desired Pd(0) catalyst (Scheme 9.9).

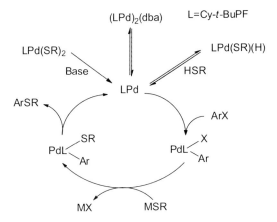

**Scheme 9.8** The reaction scheme for C—S bond formation and the unreactive species formed.

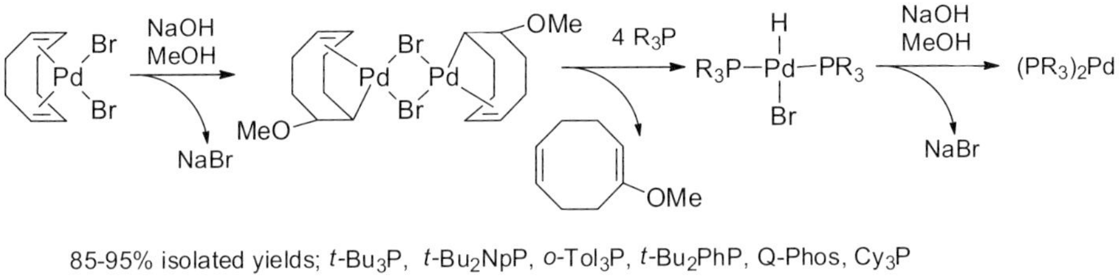

85-95% isolated yields; $t$-Bu$_3$P, $t$-Bu$_2$NpP, $o$-Tol$_3$P, $t$-Bu$_2$PhP, Q-Phos, Cy$_3$P

**Scheme 9.9** Convenient synthesis of PdL$_2$ complexes.

## 9.2.3
### Metallated Hydrocarbyl Pd Halide Initiators

Metallated hydrocarbyl Pd halide complexes are effective catalysts or catalyst precursors and the area was critically reviewed by Beletskaya and Cheprakov [79]. They concluded that several promises as regards recyclability and enantiomeric control have not materialized, but that the publications on well-defined, metallated catalysts initiated an explosive interest in this field of catalysis. Especially, the recovery of the unchanged catalyst after the reaction created great expectations.

In 1995, Herrmann, Beller, and coworkers published the use of palladacycle **12**, which afforded unprecedented TONs. The new catalyst was developed because of the interest of Hoechst to extend the Heck reaction to the cheaper and more readily available aryl chlorides with a Pd catalyst that could be reused or discarded. Especially for bromides, high TONs were achieved. For example, in the Heck coupling of $p$-bromobenzaldehyde with acrylates, TONs of $2 \times 10^5$ were achieved [80]. At such a low level of Pd the catalyst need not be recovered for economical reasons, although for other reasons one might want to remove most of it (toxicity, or as a decomposition catalyst). The catalyst is very robust and incubation takes place at low temperature. A solution of Pd(OAc)$_2$ in the presence of *o*-tolylphosphine in toluene gives metallacycle **12** in 3 min at 50 °C [81] (see Scheme 9.10). Initially the mechanism of the coupling catalysis by the palladacycle was proposed, both by Herrmann and others, to involve a switch between Pd(II) and Pd(IV) [82]. There are many examples of Pd(II–IV) cycles, albeit with different ligand systems [83], but in cross-coupling or Heck reactions no examples presenting a clear proof have been reported, and rather the contrary has found experimental support.

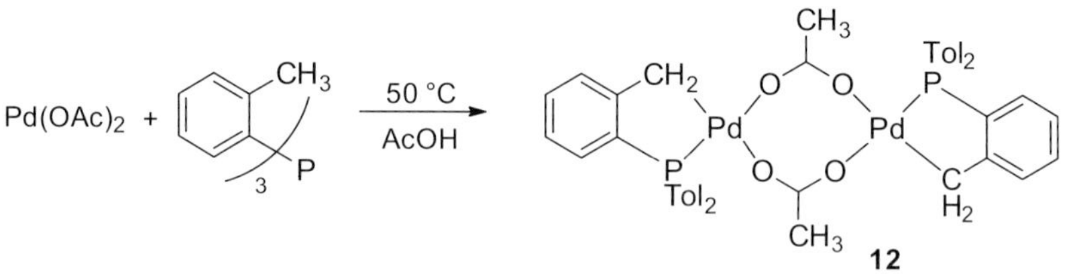

**Scheme 9.10**  Formation of cyclometallated precursor **12**.

A year later Louie and Hartwig [84] showed that catalytically active Pd(0) complexes can be formed from complex **12** by two different routes: β-hydrogen elimination of a Pd amide or C−C bond-forming reductive elimination involving a Pd–Ph substituent (see Scheme 9.11). Addition of diethylamine to **12** gives bridge splitting and formation of the hydrogen-bonded monomeric **13**. Treatment of **13** with NaO*t*Bu gave Pd(P-*o*-Tol$_3$)$_2$ **14**, together with Pd if no excess of ligand was present, and full conversion to **14** in the presence of extra ligand. The process first leads to deprotonation of diethylamine giving the amide palladium complex, which via β-H elimination and reductive elimination produces *o*-tolylphosphine.

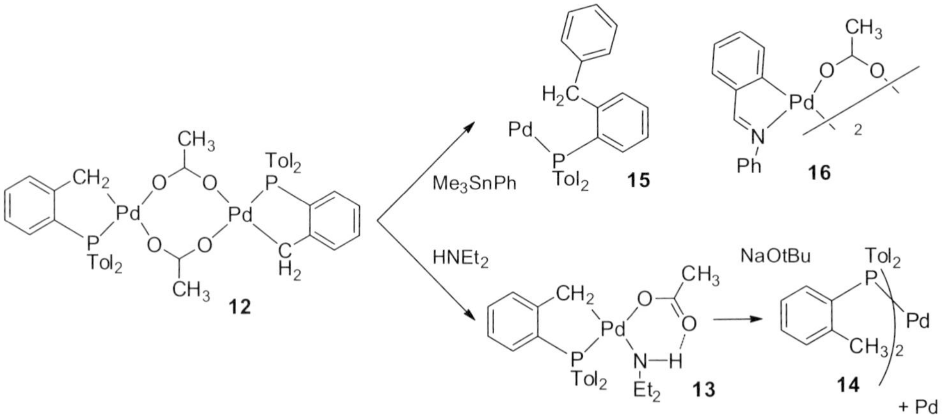

**Scheme 9.11**  Generation of Pd(0) **13**, **14**, and **15** from cyclometallated precursor **12**, and imine metallacycle **16**.

Compound **12** did not give cross-coupling of aryl bromides with diphenylamine, which lacks β-hydrogen atoms, under conditions that **14** does. In a Suzuki reaction using phenylboronic acid formation of **15** was observed, showing a pathway for this reaction that connects metallocycle **12** with the systems known thus far. These results were not considered conclusive at the time, because it was not proven that these reactions actually occur in the catalytic system. Nowadays, there is consensus that the mechanism passes through the "normal" Pd(0)/Pd(II) species.

The advantages of o-tolylphosphine especially in the Heck and Suzuki–Miyaura reaction remain: low catalyst loading and high stability. For aryl chlorides slightly higher catalyst loadings are needed. Compared to PPh₃-based catalysts ligand loading is also extremely low, probably because in the metallacyclic structure less P–C cleavage occurs (Herrmann). In spite of the low concentrations one can imagine that at the reaction temperature the unstable Pd(0) undergoes ligand metallation, when the halide substrate does not rapidly capture the unstable intermediate Pd(0). In this way a stable catalyst precursor is formed that might enter catalysis again via one of the cross-coupling steps (for the Heck reaction it remains less clear how this may happen). As in the ligand-free system (Section 9.7.1) the metallacyclic catalyst is also sensitive to the presence of halide, which may be added deliberately as tetraalkylammonium halide or which may form during the reaction from the halide of the substrate [81].

The favorable performance of o-tolylphosphine was already described by Heck in the 1970s, but he ascribed the high activity to the steric properties of the ligand as metallation was not investigated [85]. Spencer reported high TONs of o-tolylphosphine in the Heck–Mizoroki reaction of p-nitrobromobenzene and ethyl acrylate at 130 °C in DMF (134 000), but no mechanistic details were presented [86]. A few years later metallation was explicitly refuted by Heck [87], but the work of Herrmann, Beller, and coworkers has clearly shown the outstanding properties of the metallacyclic precursors.

Blackmond, Pfaltz, and coworkers studied in detail the kinetics of metallacyclic precursors in the Heck reaction of reactive substrates by calorimetry [88]. They found an incubation process, which was assigned to the removal of the metallacycle from the Pd precursor (see also [89]). Incubation was shorter for imine heterocycle **16** than for **12**. The presence of water shortened the incubation time [88, 90]. It was not known which ligand remained coordinated to Pd after the initiation; for **12** this could be tri-o-tolylphosphine. After incubation the rate of reaction appeared independent of aryl halide concentration, first order in acrylate and the square root of the palladium concentration. The same kinetics, without incubation time, were reported before by Van Leeuwen and coworkers, using a dimer of type **8** (containing tri-o-tolylphosphine, or bulky phosphoramidites or phosphites) for the same reaction [65] (for the scope of bulky phosphites see [91]). In this case simple kinetic measurements at low conversions were sufficient to obtain the kinetic equation. The kinetics proved that either insertion or complexation of the alkene was rate determining and that not the electron-rich phosphines but rather the electron-poor phosphites and amidites provided the fastest catalysts. Aryl chlorides do not show activity, neither when bulky amidites are used nor when aryl-substituted Xantphos ligands are used in cross-coupling reactions, unless the chlorides are highly activated.

Also for dppp/Pd/iodide complexes **3** the rate-limiting step in the Heck reaction of iodobenzene and methyl acrylate or styrene is the complexation/insertion of the alkene [92].

Milstein and coworkers used a strongly donating PCP-pincer ligand in a very stable complex 2,6-($i$Pr$_2$PCH$_2$)$_2$C$_6$H$_3$Pd(TFA) as catalyst (precursor) for the Heck reaction of iodobenzene and bromobenzene with butyl acrylate (140 °C) and obtained TONs of 100 000–500 000 [93]. The control experiments they did strongly suggested that a Pd(0) species was not involved in the catalytic cycle of this (pre)catalyst. They noted that the recovered catalyst contained iodide instead of triflate, the pincer complex otherwise being intact. Perhaps traces of Pd were extracted from the strongly binding ligand to give free Pd atoms (solvated and complexed to iodide ions), which are the actual catalysts (*cf.* PdNPs see Section 9.7.2).

Beller and Riermeier postulated that metallacycles **12** rapidly converted into dimers of structure **8**, which were intermediates in the catalytic cycle of a Heck reaction producing three-substituted alkenes [94]; the nature of the reducing agent leading to the intermediate Pd(0) compound was not clear. Since the behavior of metallacyclic complexes and L$_2$Pd(0) was slightly different in the Heck reaction, Herrmann proposed that after reduction of Pd(II) to Pd(0) the metallacycle remains coordinated to Pd(0) and this anionic species is highly reactive towards oxidative addition [95]. In the presence of the reducing agents in the cross-coupling systems there was no reason to assume other species than regular Pd(0) species formed by the reductant, as described above.

Replacement of the acetate in **12** by an acetylacetonate anion gives a monomeric catalyst (precursor) which is also highly active in the Heck olefination of aryl halides with styrene [96].

Another extremely reactive metallated precursor was reported by Bedford, who obtained millions of turnovers with metallated bulky phosphite or phosphinite Pd complexes in Heck and Suzuki–Miyaura reactions [97] with reactive substrates such as *p*-bromoacetophenone. For less reactive substrates, TONs were lower, but still with catalyst loadings well below 1%, which seems a popular starting point in catalysis research. Phosphite-based ligands are often somewhat slower because oxidative addition becomes rate limiting and phosphinite ligands in the work of Bedford seem to afford the right balance between oxidative and reductive properties. Figure 9.2 shows three representative catalysts (**17–19**) displaying TONs of several millions [98]! Metallated palladium complexes had been known for many years, but their use as catalysts or precursors was limited [99]; as Bedford mentions "the study of metallacyclic catalysts has almost certainly been hampered by the misconception that such species are likely to be deactivation products and that the metallation renders them catalytically inactive". Lewis found that the orthometallated complex [PdCl{$\kappa^2$-*P,C*–P(OC$_6$H$_4$)(OPh)$_2$}{P(OPh)$_3$}] could be used as an active precatalyst in the hydrogenation of alkynes and alkenes [100]. Metallated rhodium complexes of *o-t*Bu-phenyl phosphite are convenient precursors for Rh-catalyzed hydrogenation and hydroformylation catalysis (see Chapter 8 [101]). Herrmann's, Beller's, and Bedford's

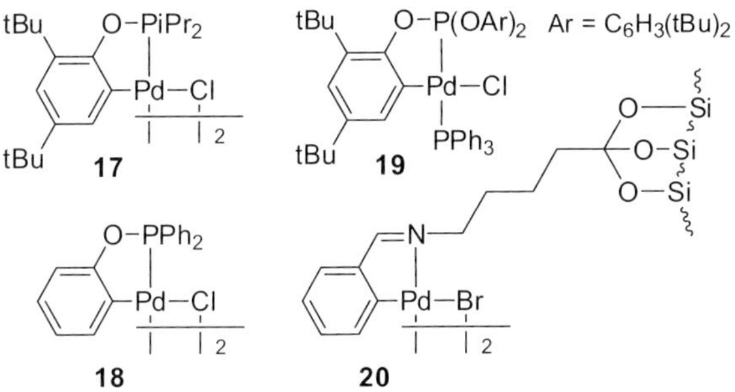

**Figure 9.2** Three of Bedford's precursors (**17–19**) giving catalysts with very high TONs in Suzuki–Miyaura reactions and an imine-based immobilized catalyst **20** (Nowotny).

publications stimulated the application of a broad range of metallated Pd species containing amine, imine, oxazoline [102], thioether, and so on functionalitites, many of them showing very high TONs in catalysis (for a review see [103]).

All attempts to obtain enantiomeric excesses in cross-coupling or Heck reactions with the use of chiral palladacycles failed [103]. Several of these precursors were immobilized on polymers (Bergbreiter [104]) or solids (Nowotny [105]) and all these results showed that the recyclability of the catalysts was low and TONs decreased rapidly with each cycle, the remaining liquid phase showing activity without incubation time. All these features support the idea that metallated species in cross-coupling chemistry are precursors rather than catalysts, although the mechanisms remain to be unraveled in several instances.

For the Suzuki–Miyaura reaction the formation of the active Pd center was as to be expected, the same as reported for the "Herrmann" catalyst by Hartwig [84]. Both the arylated phosphinite and phenylphenol were identified in a reaction of the precursor and phenylboronic acid [98] (Scheme 9.12). Aryl chlorides do not show particularly high activities but this was solved by adding $Cy_3P$ to the catalyst precursor in stoichiometric amounts; the isolated bridge-cleaved adducts were also used.

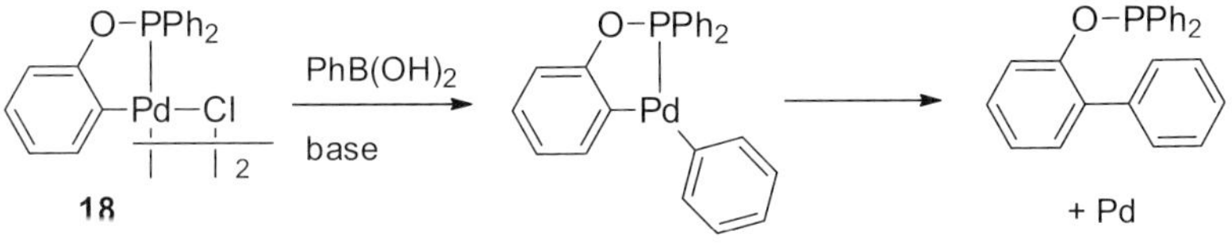

**Scheme 9.12** Conversion of metallated phosphinites in active catalysts.

Bedford, Brown and coworkers studied the effect of ligand hydrolysis on the performance of the Bedford catalysts [106]. Hydrolysis did play a role, but the effect depended also on the catalyst precursor and the ligand. One of the metallated precursors remained by far the best catalyst, better than the catalysts derived from transformed ligands. Thus, a detailed explanation of why with the use of these catalysts millions of turnovers can be achieved in the Suzuki–Miyaura reaction is still missing.

Above we mentioned a range of P-based ligands that give very high TONs. It should be mentioned that NHC ligands can give very high TONs in Heck–Mizoroki and Suzuki–Miyaura reactions. For instance pyridine- or ester-functionalized NHC ligands developed by McGuinness and Cavell gave TONs as high as 1 700 000 in the Heck reaction of *p*-bromoacetophenone and butyl acrylate [107]. Catalysis may well be due to ligand-free palladium, but this was not investigated. An advantage of the use of metallacyclic precursors is the exact 1/1 ratio, when metallated phosphines are used with Pd and P, as an excess for many ligands slows the rate of reaction [108].

Buchwald and coworkers reported a metallated, fast precatalyst **21** that did not show any incubation [109] and which was very stable. The system is four-coordinate and it contains one of the successful monodentates (SPhos, RuPhos, XPhos), a halide, and a metallated phenetyl amine ligand that falls off completely when a base is added (Scheme 9.13). The catalyst is active for various unreactive aromatic chlorides, with TONs over 1000 in less than 10 min at 100 °C, and is still active at −10 °C. Thus, these catalysts are very useful for compounds containing sensitive functional groups that need to be handled at low temperature.

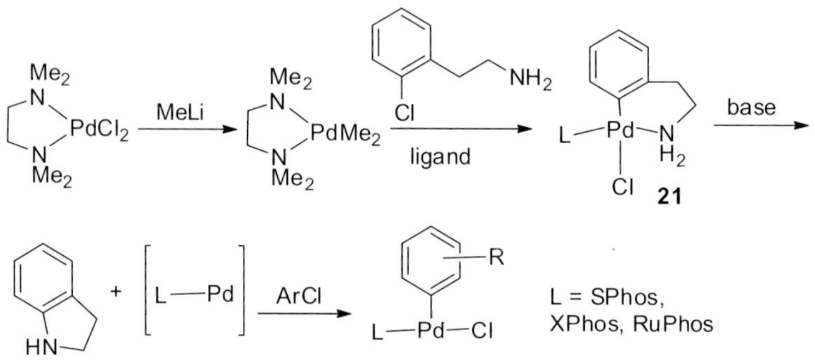

**Scheme 9.13** Stable, metallated precursor.

Few kinetic studies have been carried out on C−C or C−N coupling reactions, but also the ones on C−N bond formation show that the oxidative addition is not rate determining when aryl bromides are used. The high rate at room temperature for the formation of 2-methoxy-*N*-(*p*-tolyl)aniline indicates that oxidative addition must also be fast and thus the higher temperatures required in other reactions must have a different origin [69]. The few reports on successful use of aryl chlorides in diphosphine-based catalysts suggest that for aryl chlorides oxidative addition remains a slow, if not inaccessible, route. Milstein and coworkers reported on the successful use of dippp and dippb (1,3-bis-*i*-Pr$_2$P-propane and the butane analog) in the Heck–Mizoroki reaction with aryl chlorides as the substrates [110]. For aryl chlorides the best systems available at present are those based on bulky, biaryl monophosphines and the initially reported necessity of very electron-rich ligands (e.g., *t*-Bu$_3$P [111]) seems no longer valid. While most experiments have been done with the alkyl analogues, it was shown by Buchwald and coworkers that the diphenyl analogue of SPhos is also highly active in the Suzuki–Miyaura reaction at room temperature, thus showing that steric bulk is more important than electron richness [112]. Since oxidative addition of aryl chlorides in the present reactions is the turnover limiting step, one might expect that Heck, Suzuki and amination reactions for one substrate would have the same rates. This is not the case as the conditions (temperature, catalyst loading) are often completely different. This is caused by the effect of ligands and bases present which may or may not activate the palladium zero complex, as was studied in detail by Hartwig for systems containing $t$Bu$_3$P as the ligand [113] and for PPh$_3$ by Jutand and Amatore [37], and by the formation of "dormant" Pd species (Scheme 9.8).

Weck and Jones critically analyzed a range of metallated precursors by adding a polymeric Pd trap to the catalytic system and they showed that catalytic activity ceased when the trap was added for a large number of systems [114].

## 9.3
## Transmetallation

This reaction involves the exchange of the Pd bound (pseudo)halide ion with the nucleophile, which in the case of a C-based nucleophile is called transmetallation. This reaction was also presented in Section 9.2.1, because it may well be used as a means to reduce Pd(II) to Pd(0).

Replacement of the halide coordinated to palladium by the C, N, O, or S-based nucleophile is a reaction scarcely amenable to general conclusions, as the reactions depend strongly on the character of the nucleophile, its cation (if it is anionic), and the solvent. Clearly, when the nucleophile is a hydrocarbyl anion (metal = Sn, B, Mg, Zn, Li, and so on) both kinetics and thermodynamics are favorable as a result of the formation of the inorganic metal halide, the driving force of the reaction. When the nucleophile is nitrogen-based, energetics and kinetics are less favorable, and even less so when the nucleophile is an alkoxide. The order for elimination of the heteroatom-based fragments is $P > S > N > O$ [50c]. Kinetic measurements of catalytic reactions show that for C−N bond formation the kinetics may change during the reaction because of the changes in the medium [69, 115]. Most likely, in those cases the rate-limiting step remains the reductive elimination, while the kinetic equation includes a pre-equilibrium of the halide–nucleophile exchange.

As concerns a possible ligand effect on such substitution reactions, no data are available, but it seems likely that exchange processes involving five-coordinate complexes are more readily accessible when wide bite angle diphosphines or monodentates are used, be they bis- or mono-phosphine complexes. Halide exchange was found to be an associative process for allylpalladium halide complexes [116]. For halide exchange between two transition metals of simple complexes dimeric intermediates were proposed [117]. In the majority of exchange processes on Pd an associative mechanism was found or suggested, although other reactions may involve the creation of a vacant site, such as β-hydride elimination. There are indications though, that even insertion and β-elimination reactions in Pd and Pt may involve five-coordinate species instead of a species with a vacant site.

DFT calculations on Pd complexes support the involvement of five-coordinate species when π-bonding ligands are involved, while for replacement by hard ligands such as methanol dissociative pathways were preferred [118] (and references therein for experimental evidence).

Farina proposed 14-electron species for alkyl-halide exchange reactions [119], although this was rejected by Espinet and coworkers [120]. Farina and Krishnan reported on the Stille reaction with phenyl iodide and vinyltributyltin as the reagents and $Pd_3(dba)_2$ as the catalyst precursor. An enhancement of the rate of reaction by two to three orders of magnitude was achieved by replacing the classic ligand $PPh_3$ by $AsPh_3$ or $P(2\text{-furanyl})_3$. Triphenyl phosphite also provided a fast catalyst, and oxidative addition was not the rate-limiting step in these systems with PhI as the substrate, as for most of the 20 ligands reported the rate equation only contained a first order in vinyltin concentration. Of the group of bidentate ligands, dppp provided a much faster catalyst than dppe and dppf. The excellent performance of triphenyl-arsine has not found much follow-up in cross-coupling chemistry and probably when used with other aryl halides its performance is not as good. Today the proposal of 14-electron intermediates formed via ligand dissociation would not be so surprising, although most systems showing T-shaped intermediates contain much more volu-minous ligands than $P(2\text{-furanyl})_3$. Besides, vinyltributyltin is ideally suited for an associative exchange reaction via vinyl coordination. It would be interesting to study this reaction (with aryl bromides) using precursors pictured in Figure 9.1 to have exact Pd/L ratios of 1 or 2. Amatore, Jutand, and coworkers also found that transmetallation is the rate-limiting step in the Stille reaction of PhI and vinyltin compounds [121].

Associative exchange of halide and alkyl groups was demonstrated by Espinet and coworkers in their studies on a model Stille reaction in complexes containing electron-poor aryl groups ($C_6Cl_2F_3$), $AsPh_3$, and halides or triflates as the anions [122]. It was concluded that for halides as the anions a cyclic intermediate is formed via a $2 + 2$ addition of PdX and SnR, leading to exchange.

Reversible transfer of a furyl group from Sn to Pd was observed by Cotter and coworkers [123]. The tin compound is $Bu_3Sn(furyl)$ and the palladium compound is a cationic PCP pincer complex. An intermediate was observed in which Pd coordinates first to the furyl group, which then is transferred to Pd under formation of $Bu_3Sn(OTf)$. Involvement of such a π-interaction may be relevant to phenyl or vinyl transfer in the transmetallation step in other reactions. A cationic Pd species is not a likely entity under cross-coupling conditions, but in relative terms Pd undergoing arylation or vinylation is the most electrophilic and π-acidic of the two metals. In trans-*alkyl*-ation such an intermediate cannot form.

Transfer of a phenyl group from phenylboronic acid to Pd does not occur spontaneously and an important characteristic of the Suzuki–Miyaura reaction is the addition of base to aid the halide/phenyl exchange between Pd and B. The base may be anhydrous such as NaO*t*Bu or even aqueous $K_2CO_3$. The base probably replaces the halide ion on palladium and another base anion adds to phenylboronic acid to form a boronate, for example, $PhB(OH)_3^-$ in the simplest case. Subsequently the more electron-rich boronate exchanges the phenyl group with an OH group in a bimolecular reaction. A more detailed picture of how this happens is not known. In academic research an excess of boronic compound is used (up to 50%), because considerable homo-coupling may take place; this is caused by the reaction of two boronic compounds with Pd(II) species followed by reductive elimination. For industrial applications excessive homo-coupling needs to be avoided.

There is a lot of scattered information on the suppression of homo-coupling in the Suzuki–Miyaura reaction, which is hard to retrieve or summarize. Here we will give just a few typical examples. The stoichiometry in the case of homo-coupling requires the presence of an oxidizing agent and indeed there are examples in which the use of oxidizing agents leads to complete homo-coupling [124]. Thus, one measure to avoid homo-coupling may be the exclusion of oxygen [125] or other oxidizing agents, although there are examples where even under exclusion of oxygen homo-coupling still occurs. In polar solvents more homo-coupling was observed by Blum and coworkers [126]. In line with the reports of oxidizing agents Miller and coworkers at Eli Lilly were able to reduce the formation of homo-coupling products by the addition of a weakly reducing agent, potassium formate, and the use of a facile nitrogen subsurface sparge prior to introduction of the catalyst [127]. Bryce and coworkers also observed a reduction of the homo-coupled product when small amounts of potassium formate were added [128], although in some cases competing protodeboronation was observed. Homocoupling of arylboronic acids was successfully carried out by Pd/C in water/2-propanol (9:1 volume ratio) under air to obtain symmetric biaryls in good yield [129]. The base-free, ligand-free system and the heterogeneous nature of the catalysts allow a practical and environmentally friendly operation, although the actual catalysis takes place in solution.

High yields of homo-coupled products were obtained with the use of NHC Pd complexes [130], but the complete reaction equation was not given by the authors, as the reaction might involve diborane formation or an oxidizing agent.

Kedia and Mitchell from GlaxoSmithKline [131] applied the kinetic analysis provided by Blackmond [132] to reduce the amount of homo-coupled products, which in their case were PCBs. PCB concentrations should be kept below 50 ppm according to US law, and thus the incentive to reduce the level was even greater than usual. The analysis showed that homo-coupling occurred especially towards the end of a batch reaction, when the resting state/rate-limiting step changed from reductive elimination to oxidative addition. Usually, in this situation, oxidation of Pd(0) takes place leading to homo-coupling [133], but in their case this was excluded and oxidative addition of chorophenylboronic acid to Pd(0) might explain the results.

Purposely high yields of homo-coupling of arylboronic acids cannot only be achieved by adding oxidants, but also by using Cu [134], and perhaps in this case diboranes are formed. Aryl halides or triflates can be effectively homo-coupled with the use of diboranes (bis(pinacolato)diboron) as the reducing agent [135].

Organotrifluoroborates present an alternative to hydrocarbylboronic acids in Suzuki–Miyaura coupling reactions, avoiding the activation by base of the neutral boranes. They were first used by Genêt and coworkers in the palladium-catalyzed cross-coupling of a variety of organotrifluoroborates with aryldiazonium salts [136, 137]. It took some time before they were also applied in cross-coupling reactions with organohalides or reactive esters [138]. Organotrifluoroborates are easily accessible by a variety of one-pot synthetic routes from readily available, inexpensive starting materials [139, 140]. The process with the use of aryltrifluoroborates turned out to be extremely general and the scope in substrates and ligands was very broad. A most impressive demonstration of biaryl synthesis was disclosed in the preparation of trityrosine. The analogous pinacol boronate gave none of the double coupling product, while the aryltrifluoroborate afforded the desired product in 74% overall yield (Scheme 9.14).

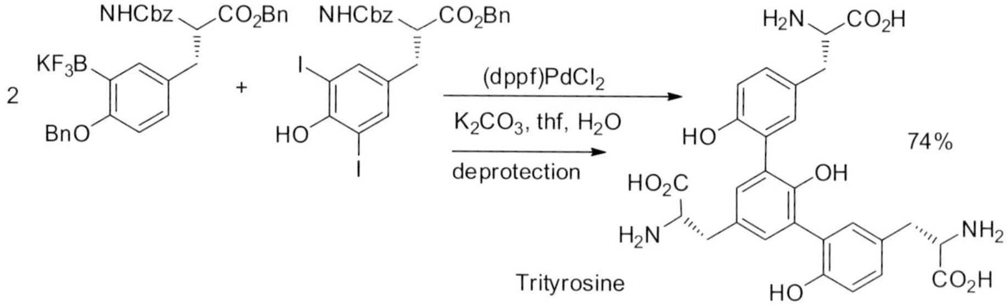

**Scheme 9.14** An example of the use of organotrifluoroborate in a coupling reaction.

As is the case for aryl boronic acids, specialized ligands may be employed to promote coupling in specific systems.

Of particular note is the use of Buchwald's SPhos ligand for the efficient coupling of a variety of aromatic and heteroaromatic chlorides, including electron-rich and sterically hindered substrates (Scheme 9.15).

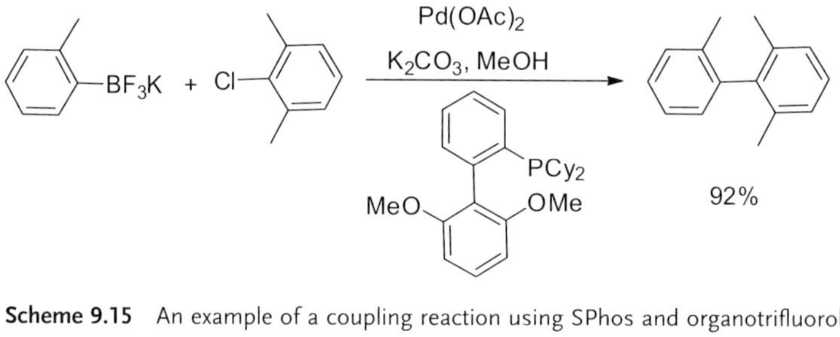

**Scheme 9.15** An example of a coupling reaction using SPhos and organotrifluoroborate.

Billingsley and Buchwald applied borate anions (lithium triisopropyl 2-pyridylborate) as alkylating agents for 2-pyridyl groups [141] in the Miyaura–Suzuki reaction. Other borylated 2-pyridine precursors gave low yields, because electron-deficient heteroaryl boron derivatives undergo transmetallation at a relatively low rate, and these reagents rapidly decompose by a proton transfer reaction yielding pyridine. In many cases very high yields were obtained, and, interestingly, SPOs (secondary phosphine oxides) were used as the ligands [142], as they gave better yields than the biaryl phosphines. SPOs usually form hydrogen-bonded bridged dimeric monoanions. They may also react with borates (giving pyridine in this case), but the catalysts were not identified.

## 9.4
## Reductive Elimination

### 9.4.1
### Monodentate vs Bidentate Phosphines and Reductive Elimination

Basically, there are two mechanisms available to accelerate reductive elimination reactions, one is via three-coordinate species and the other one is via wide bite angle bidentate ligands. Initial studies involved $sp^3$-hybridized hydrocarbyl groups, for which reductive elimination is relatively slow [143–145], while for $sp^2$-hybridized hydrocarbyl groups very fast reductive eliminations are observed [146]. Reductive elimination of ethane from dimethylpalladium complexes required dissociation of one of the phosphines, as was reported by Stille and Yamamoto [147]. There is now ample evidence for such three-coordinate T-shaped complexes that undergo rapid reductive elimination, as many bulky phosphine ligands and the o-phenylene type ligands (also named biarylphosphines) show such structures and ligand dissociation is no longer a prerequisite [50a,148, for a review on monoligated Pd complexes see 149]. It is interesting to note how relevant and important the basic studies by Stille on reductive elimination have become in the last decade, as he already distinguished the two basic mechanisms. The monodentate systems and the bidentate systems have in common the asymmetric nature of the $C-X$ elimination (X not being in this case another aryl group); this will be outlined in the later paragraphs dealing with bidentates.

While, initially, wide bite angle ligands such as BINAP, dppf, DPEphos, and Xantphos were the best performing ligands for the Kumada–Corriu and the Buchwald–Hartwig reaction, nowadays the biarylphosphines as developed mainly by the Buchwald group are the ligands of choice, especially because they can deal with aryl chlorides as the substrate. Surely, there are still niches for bidentate ligands [150], including bulky bidentate phosphines [151], and also NHC ligands [152]. The biarylphosphines are readily prepared in a range of substitution patterns and Buchwald adapted the ligands to obtain optimal yields for a large variety of cross-coupling reactions. For diphosphines these modifications are more difficult to achieve and far less variations have been developed. Ligands based on diphenyl ether are easily accessible and a few variations have been reported; lithiation of commercially available ditolyl or diphenyl ether, followed by reaction with $R_2PCl$ leads to these ligands [153]. In Scheme 9.16 we have depicted the elegant synthesis of the biaryl phosphine ligands, which can be carried out as a one-pot reaction [154]. A 1,2-dihalobenzene is treated with Mg or alternatively with 1 mole of Grignard reagent (or BuLi [155]) that leads to the benzyne derivative (in this first step a simple Grignard reagent may be used, instead of the more expensive one that may be needed in the second step). A (second) molecule of Grignard adds to the benzyne derivative to

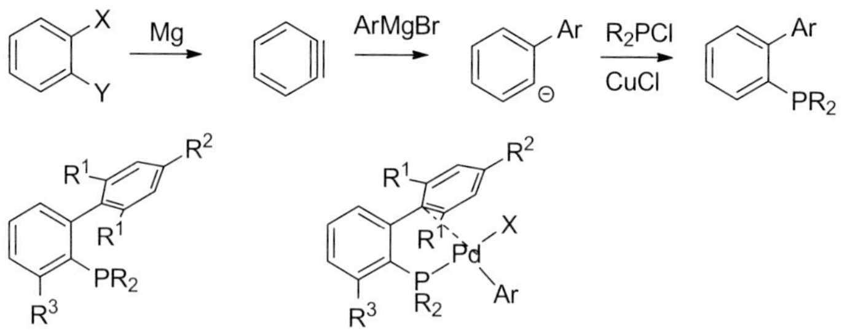

**Scheme 9.16** Diarylphosphine ligand synthesis and a monoligated Pd complex.

give the diaryl anion, which will react with ClPR$_2$ to give the diarylphosphine. A copper catalyst increases the yield [156]. Other routes have been reported [157], but the benzyne route remains the most attractive one. Synthesis on a 10 kg scale was demonstrated for MePhos and XPhos via this route, while DavePhos required some modifications [158]. A key point in the large-scale synthesis is the exothermicity of the Grignard formation, the benzyne formation, and the reaction with R$_2$PCl. One can already note a heat effect on a g scale synthesis of Grignard reagents and thus heat exchange for a reactor of several m$^3$ will be critical (the surface to volume ratio from a Schlenk vessel to such a reactor decreases by a factor of 100). Availability of reagents on a scale of hundreds of kg can also be a limitation for large-scale production. Mauger and Mignani at Rhodia worked out the exothermicity of the synthesis of one of these ligands and also that of a cross-coupling reaction for the production of hydrazones [159]. Indeed, all steps are highly exothermic and one has to carry out the reactions in a well controlled way under slow addition of the reagents, making sure that the reaction proceeds continuously.

A few variations were reported utilizing "shielding" phenyl groups in the ligand's structure that fall beyond the basic structure of Buchwald's ligands (for example see [160]).

The history was not planned this way perhaps, but in hindsight one can summarize the key points of the biaryl ligands in terms of the substituent effects in the different positions of the basic structure [161]. The R groups on phosphorus are usually strongly electron donating and they accelerate or enable (for aryl chlorides) the oxidative addition of the aryl halide substrate. Their steric bulk contributes to a faster reductive elimination. Groups R$^1$ (one may be H) prevent metallation of the aryl group by Pd and thus more catalyst is retained in an active state [162]. Large R$^1$ groups contribute (as does the aryl group) to the preferential formation of monophosphine palladium complexes. Several structures were found in which the aryl ring coordinates weakly to Pd with its *ipso* C-atom, which enhances the formation of mono-ligand complexes. The shielding of the phosphorus atom by the aryl group diminishes the reactivity of the Buchwald ligands towards oxidation by O$_2$ [163]. R$^3$ locks the PR$_2$ group, as in BrettPhos (Figure 9.3), such that Pd is fixed in close proximity to the aryl ring, which strengthens the aforementioned effects [164]. The obtained T-shaped intermediate strongly accelerates reductive elimination.

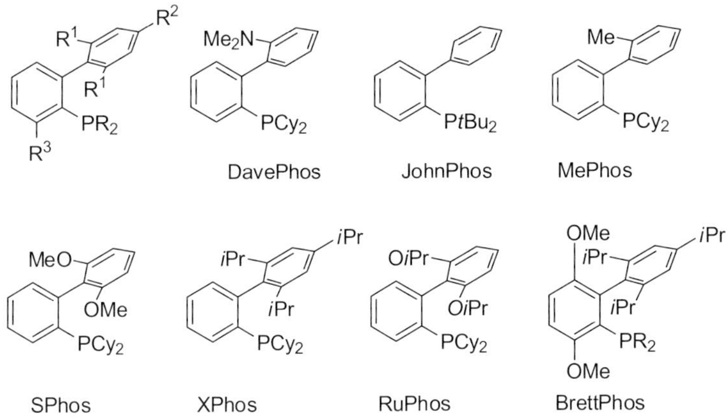

**Figure 9.3** Several of the Buchwald ligands.

Initially, it was thought that a heteroatom stabilizes the intermediate palladium complexes such as the amine group in DavePhos or the ether groups in SPhos. The activities of the catalysts based on XPhos and MePhos clearly showed that this was not the case, although there still may be examples in which such a weak stabilization does contribute to the catalyst's performance. Likewise, ligand rigidity may not be advantageous for all steps as we know from other catalytic processes; for instance we know little about the transmetallation step and since this is an associative process, more flexibility may be required in this step. For reductive elimination and oxidative addition more space and flexibility may be needed to lower the energies [165] close to the arene–Pd interaction, which may also be hampered by too rigid systems. All catalysts (ligands) differ slightly in their behavior and it seems that for each substrate a new set of reaction parameters, base, solvent, and an optimized ligand is needed. Thus, in spite of the high level of understanding, optimization remains empirical. Martin and Buchwald attributed the longevity of catalysts based on SPhos to two main factors. First, it stabilizes Pd(0) intermediates by favorable interactions of the aromatic $\pi$ system with the Pd center, as supported by X-ray crystallography of the SPhosPd(0)dba complex. This complex possesses a Pd(0) $\eta^1$-arene interaction with the *ipso* carbon [166]. They also postulated that the high activity of catalysts based on SPhos is due to the ability of this ligand to stabilize and maximize the concentration of the mono-palladium intermediates with a relatively small ligand. As mentioned above, these intermediates would be expected to be particularly reactive in oxidative addition and transmetallation processes.

Reductive elimination in stoichiometric reactions can be enhanced by the addition of electron-withdrawing alkenes [167]. Although this concept has often been quoted, its practical use in catalysis is a relatively recent development [168]. An example of such an alkene is dba, often present as part of the precursor (slowing oxidative addition!) or *p*-fluorostyrene added on purpose. The area was reviewed by Fairlamb [169]. Lei and coworkers introduced a new concept to enhance reductive elimination in a Negishi reaction by incorporating an electron-poor alkene in an ortho-position of one of the phenyl rings of triphenylphosphine [170] and very effective catalysis was accomplished. The kinetics were studied in a collaboration with Marder and coworkers and the reaction studied is presented in Scheme 9.17 [171].

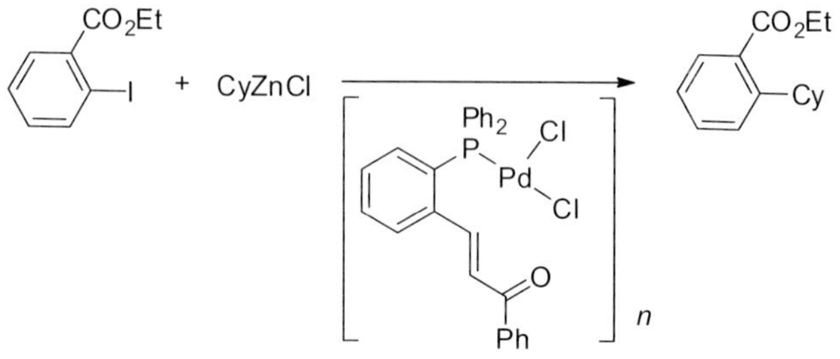

CO$_2$Et + CyZnCl →  [ Pd complex with Ph$_2$P, Cl, Cl ]$_n$ → CO$_2$Et—Cy

**Scheme 9.17** Phosphino-alkene ligand enhancing reductive elimination.

The substrates used are relatively reactive ones, in both the oxidative addition step and the nucleophile substitution step, *viz.* ethyl 2-iodobenzoates and cyclohexylzinc chloride, but the reductive elimination of a C$_{sp3}$–C$_{sp2}$ coupling is always very slow in conventional systems. The kinetics was complicated by incubation times, which was not understood since reduction of the palladium dichloride phosphine complex by the alkylzinc reagents should be fast. Dimers or PdNPs were not involved because the reaction showed a first order dependence in Pd concentration. TONs of 100 000 were reported and at the steepest part of the s-curve of the reaction the TOF was as high as 1000 mol mol$^{-1}$ s$^{-1}$, while the rate of reaction was independent of the concentrations of the substrates in this area. Thus, the adjacent alkene increased the rate of reaction dramatically. After decades in which the oxidative addition was found or supposed to be the rate-limiting step, there are now many examples in which nucleophilic substitution, alkene insertion (Heck–Mizoroki reaction), or reductive elimination are rate limiting, all dependent on the exact nature of the catalytic system.

The subtle differences in ligands, substrates, and conditions have to do with deactivation processes or the occupation of dormant states, but it is unlikely that we will learn all details about all these systems; perhaps a couple of large applications will be singled out and for these one might obtain more details, as is generally the case in catalysis.

Reductive elimination in square planar nickel and palladium(II) complexes is strongly enhanced by cis wide bite angle bidentate phosphines [143,144a,172–177]. Simple MO pictures [178] explain that wide bite angles stabilize the zero-valent state while bite angles close to 90° will stabilize divalent, square planar complexes. Also, electron-donating ancillary ligands will stabilize palladium(II) species and slow down reductive elimination (and the reverse for electron-withdrawing ligands). In the vast amount of organometallic chemistry this seems one of the rules best conserved throughout [179]. The fast eliminations reported in the last decade displayed by *t*-butyl-substituted phosphines pose some questions (*vide infra*).

Examples of cross-coupling reactions enhanced by wider bite angles were already known before bite angle effects received attention [174]. The first explicit mention of a bite angle effect on reductive elimination including experimental justification is probably from Brown and Guiry [177]. They reported that replacing iron by ruthenium in dppf to give dppr, which at the time represented one of the ligands with the widest bite angle, accelerated the reductive elimination.

Electronic effects for both ligands and substrates on reductive elimination of C–heteroatom moieties were studied extensively by Hartwig and coworkers and have been reviewed by Hartwig [50c]. The reductive elimination of C–heteroatom moieties is an asymmetric phenomenon, not symmetric as is the elimination of ethane from a dimethyl complex, and its character has often been described as a migratory reductive elimination in which the heteroatom-based nucleophile migrates to the aryl group. Electronic effects of the substituents on the aryl group support this description; electron acceptors on the aryl group accelerate reductive elimination (Scheme 9.18) [51, 180]. Asymmetrically substituted bidentate ligands might therefore favor elimination, but Hartwig showed that the unfavored isomer may prevail [181]; the same happens in migratory insertion reactions when asymmetric ligands such as phosphine-amines are used [182] (due to the trans influence the nucleophile prefers the position cis to the strongest donor group and thus does not gain extra nucleophilicity; on the contrary the unsaturated "receptor" becomes more electron-rich). Reductive elimination is faster for stronger electron-donor eliminating groups [51b]; thus reductive elimination involving a $CF_3$ group was, until recently, very rare (see Section 9.4.2).

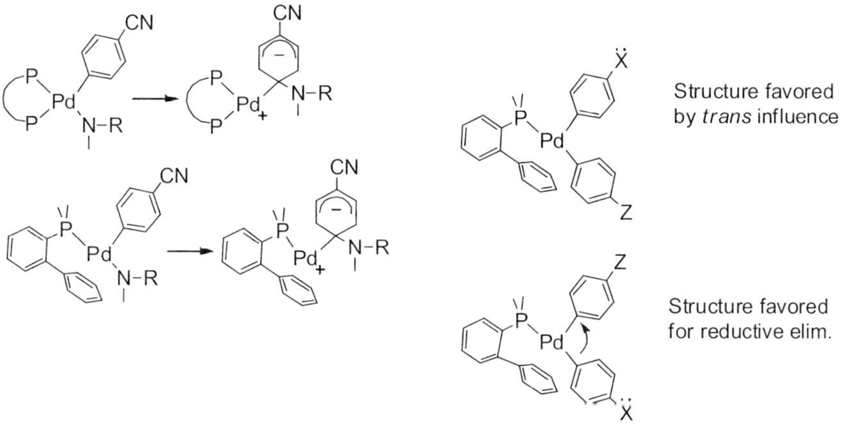

Scheme 9.18 Mechanism of aryl-N reductive elimination from a Pd-diphosphine complex and a T-shaped intermediate.

Thus, as concerns N-based nucleophiles, also more electron-rich amido groups show faster elimination [51b]. As mentioned above, when the aryl group contains electron-withdrawing groups the elimination is often faster, which seems in contrast to the general rule that the eliminating groups should be electron rich. For the "migratory reductive elimination" process a Meisenheimer-type stabilization of the intermediate may accelerate the reaction (Scheme 9.21) [180]. This is the reverse of the mechanism proposed many years ago by Fitton and Rick for the oxidative addition of aryl halides to palladium zero [183]. By the same token, heteroaromatics can be aminated at a rate three orders of magnitude higher when a Lewis acid such as BEt₃ coordinates to the nitrogen atom in the 4-pyridyl group in a stoichiometric reaction using DPPBz as the ligand [184]. This behavior is very similar to the enhanced reductive elimination by Lewis acids in the hydrocyanation process [185]. Likewise, dppf complexes of 2- and 3-furyl and 2- and 3-thiophenyl heteroaromatics underwent faster reductive elimination (amination) for the isomers having the less electron-rich carbon attached to palladium, demonstrating again that the carbon atom undergoes nucleophilic attack [186].

Another example that exemplifies the lower reactivity of weaker nucleophiles is the amidation of aryl halides [176c], amidates being the weaker nucleophile. In addition, Pd-amidates tend to form $\kappa^2$-complexes, and for this coordination mode an even lower reactivity towards reductive elimination was expected. Thus Xantphos worked especially well for this coupling reaction (using aryl bromides), while monodentates did not. Later, Buchwald and coworkers discovered a monodentate biarylphosphine that is highly active for amide–aryl chloride coupling (using aryl chlorides), the essence being that this ligand (also) prohibits the bidentate coordination of the amidate [187]. This is an example of the subtlety of this reaction, as one single methyl group ortho to the phosphine moiety is capable of affording a stable, active catalyst (see Scheme 9.19).

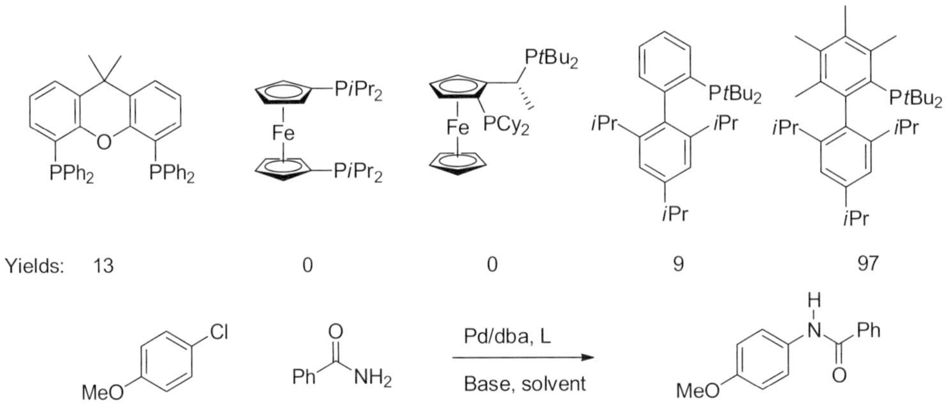

Yields:   13                    0                    0                    9                    97

**Scheme 9.19**  Ligand effect on yields of amidation of aryl chlorides.

A similar effect as in amidates may have occurred in enolate intermediates, which are intermediates in the α-arylation of ketones reported by Buchwald and coworkers [150a]. In this case as well, bidentates such as BINAP and Xantphos were the preferred ligands.

Reductive elimination from arylpalladium complexes containing functionalized alkyl groups, such as cyanoalkyl and enolates, as studied by Culkin and Hartwig [188], is enhanced by electron-withdrawing, bulky, wide bite angle bidentate phosphines, as well as electron-withdrawing groups on the aryl moiety. In this series the nature of the substituted alkyl group had the largest influence on the rate; an electron-withdrawing group on the alkyl, which stabilizes the anion and decreases its nucleophilicity, for example the cyanomethyl group, gave much slower reductive elimination, as was supported by theoretical studies [173].

Rates of reductive eliminations of alkyl cyanides and methyl esters (from methoxy acyl complexes) span many orders of magnitude upon bite angle variation [172, 189]. Computational studies confirm the trends of the experimental results [118, 173]; even small $PH_2$ groups showed this trend and this was interpreted as a genuine electronic bite angle effect. The larger $PMe_2$ and $PPh_2$ groups treated by QM/MM methods give similar energy differences, but the large magnitudes of the differences found for the stoichiometric, experimental models [173] or carbonylation catalysis [118] cannot be fully reproduced. Unfortunately, neither larger groups, for example, t-butyl groups, nor sufficiently wide bite angles were considered.

In addition to the electronic effect, ligands having wider bite angles also exert more steric interaction with the substrates, as is the important factor in rhodium hydroformylation [190]. Bulky substituents at the phosphorus atom also lead to more steric repulsion between the two phosphines, thus widening the bite angle. In recent years a number of bulky alkyl diphosphines with less wide bite angles have been used successfully, such as Josiphos and dppf type ligands [144c, 191]. Smooth trends of steric or bite angle effects are rare for palladium, because small changes may lead to population of other intermediates or inactive states and as a result changes in kinetics are more drastic. In hydrocyanation, for instance, slightly smaller bite angles than those of Xantphos or Sixantphos give nickel dicyanides as a side reaction and the catalyst is inactive. Replacement of phenyl groups in dppf by o-tolyl groups in the arylation of ketones is an example in which more steric bulk (either directly or via a wider bite angle) doubles the rate; well considered this is a small change for such an increase in cone angle [192].

There are several indications that t-butyl groups at both phosphorus ligands will widen the bite angle due to mutual repulsion. If the bridge consists of five (Xantphos, DPEphos) or six atoms (BISBI) this may lead to trans complexes and both the widening and the steric bulk result in an adverse effect, because trans complexes do not undergo reductive elimination (see [189] and references therein). Thus, a three-atom bridged bidentate ligand (Josiphos) [193], ferrocene ligands [194, 195], or four-atom bridged ligand such as dtbpx (Figure 9.4) [196] may be more effective, because they do not form dormant trans complexes. Complexes of dtbpf are inactive in methoxycarbonylation, as their palladium alkyl and acyl complexes have trans geometry [189], but in cross-coupling they are active and apparently the amide–aryl complexes have cis geometry, as do the alkoxide complexes [50b]. Xantphos forms trans-aryl halide palladium complexes of structure **9** [176e, 73, 198], but the active cis isomers remain accessible and trans bromide complexes **9** are more active in cross-coupling reactions than the cis bromide complexes **3**.

It has been suggested that oxygen coordination of Xantphos might aid reductive elimination [198]; this seems unlikely since DPEphos should be more prone to this

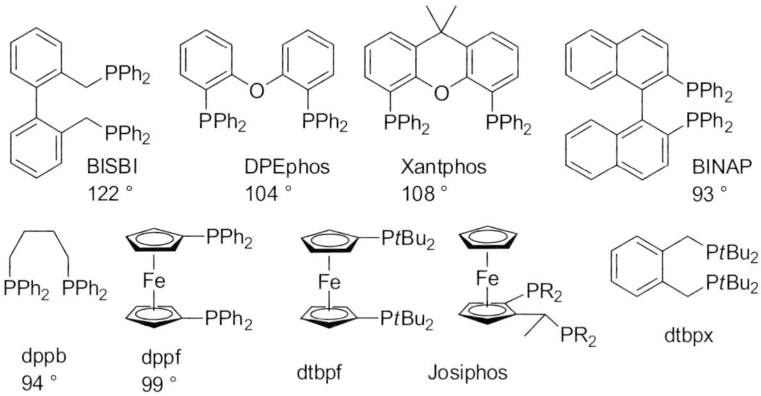

**Figure 9.4** Ligands mentioned in this section.

and Xantphos usually gives better results in C−N coupling reactions than DPEphos [150b]. Also, ether coordination to palladium zero is not expected to be of importance energetically.

On several occasions, especially in methoxycarbonylation [196], the possibility of monodentate coordination for bulky bidentate ligands has been proposed, which would connect the two mechanisms of T-shaped intermediates and wide bite angle, bulky intermediates. For cross-coupling reactions there are no data available, but indications are against it, as experimental data for alkyl cyanide elimination and CO/ethene/MeOH catalysis do not support the arm-off mechanism. The data of Marcone and Moloy [172] include the study of the Thorpe–Ingold effect (gem dimethyl effect) and substitution at the 2-carbon of the 1,3-propanediyl bridge shows that the reductive elimination reaction is not retarded by the substituents and thus dissociation of one phosphine ligand can be excluded. Interestingly, and supporting this conclusion, Goldberg and Moloy and coworkers [199] did observe a 100-fold retardation of the reductive elimination of ethane in platinum(IV) complexes for gem dialkyl-substituted propane bridged ligands, which proves nicely that a Thorpe–Ingold effect exists in reactions of this type. It was known that reductive elimination from Pt(IV) complexes containing monophosphines involves phosphine dissociation.

In CO/ethene polymerization chain transfer involves, in part, a reductive elimination reaction. Thus, a gem dimethyl effect should lead to higher molecular weights if dissociation of one phosphine took place in such a process. The opposite is observed [200], as the rate of polymerization increases and the molecular weight slightly decreases when gem dimethyls were applied in the ligand. The lower MW is in accord with a more bulky ligand that causes a faster reductive elimination. In cross-coupling chemistry not many detailed studies of this type exist, but if one assumes that dissociation of one phosphine ("arm-off" mechanism) should be easier for DPEphos than for Xantphos, cross-coupling results do not support this mechanism. Hamann and Hartwig suggested a dissociation mechanism to explain the formation of side-products requiring β-elimination reactions using DPEphos and Xantphos and their *o*-tolyl derivatives – dppf was the most effective ligand in this instance [201]. Kranenburg suggested that wider bite angles in the reactions discussed here might give five-coordinate species that lead to side product formation, for example via β-hydride elimination [175]. In a more recent publication Hartwig proposed that the Josiphos ligands derive their effectiveness from the tight binding of Pd to the rigid bidentate ligand and thus there is little support for an arm-off mechanism in reductive elimination reactions of Pd.

In diphosphine complexes reductive elimination takes place from the cis complex and yet trans complexes were found to be effective catalysts. As indicated above, a range of PdArBr complexes containing bidentate ligands afforded both trans complexes and cis complexes, depending on the bite angle of the diphosphine, but for this series Van Leeuwen's group found that the trans complexes were the fastest catalysts in N−C bond formation [69, 73]. The explanation given is that the X-ray studies concern the halide complexes and not the amide–aryl complexes that undergo reductive elimination; upon substitution of halide for amide a cis complex forms and the wider the bite angle, the faster the reductive elimination.

Gelman and coworkers described the use of a trans-chelated palladium complex derived from 1,8-bis-(4-(diphenylphosphino)phenyl)anthracene as a catalyst in carbonylative Suzuki coupling and methoxycarbonylation of aryl iodides and bromides. The selectivity was attributed to the unique structural features of the trans-chelating ligands [202]. Yields are excellent and side reactions and catalyst decomposition do not play a role as 0.01% catalyst could be used (an example is shown in Scheme 9.20). The high temperature required would suggest that, as in Xantphos, a transient cis complex should form before reductive elimination can take place.

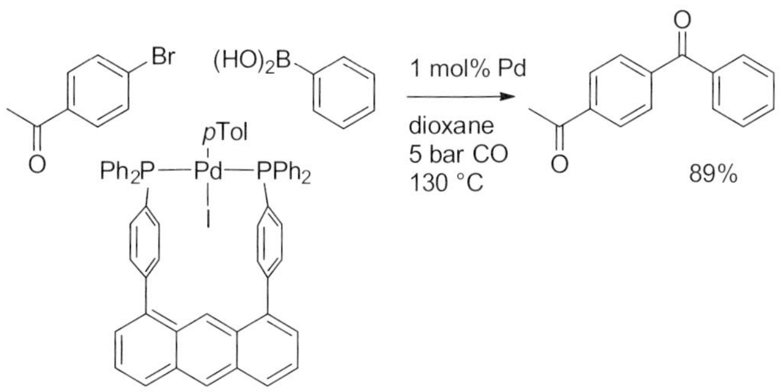

**Scheme 9.20**  A trans-precursor for carbonylative Suzuki–Miyaura coupling.

Terphenyl-based diphosphines (Figure 9.5) also form trans Pd complexes which were active as catalysts in the Suzuki–Miyaura and Heck–Mizoroki reactions of reactive aryl bromides and iodides with phenylboronic acid and styrene, respectively [203]. The bulky alkyl-substituted diphosphines were not active in coupling reactions of chlorobenzene at 100–120 °C. While the Suzuki reaction worked well for all three ligands, the Heck reaction worked best for the diphenylphosphino-substituted ligand. Since the oxidative addition did take place in the Suzuki reaction

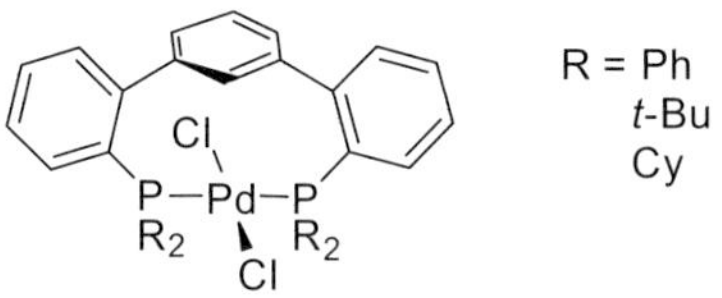

R = Ph
t-Bu
Cy

**Figure 9.5** Terphenyl-derived diphosphines and their trans-complexes.

for all three ligands, the insertion of styrene must be the slow step (for *t*Bu) or hampered step (for Cy) in the Heck reaction. It was thought that the ligand is sufficiently flexible to form cis complexes that undergo reductive elimination in the cross-coupling reaction. As cited above (Section 9.1), alkene arylation proceeds better with electron-poor ligands [65].

A large difference between cyclohexyl and *t*-butyl substituents in the Heck reaction was also noted by Hills and Fu, who compared $Cy_3P$ and $tBu_3P$ in the coupling of chlorobenzene and methyl acrylate [204]. They found that the nature of the base used was important, as for the slower $Cy_3P$-based catalysts $L_2PdHCl$ hydride species turned out to be the resting state. Thus, in electron-rich Pd complexes HCl elimination may be rate limiting. Interestingly, in complexes of $tBu_3P$ reductive elimination can be accomplished with weaker bases than in complexes of $Cy_3P$. Crystal structures showed that the complex of the former is highly distorted from a pure trans configuration due to the difference in diameter of H and Cl anions; this distortion enhances reductive elimination.

We return to the bidentate systems. There is a balance between bite angle and steric bulk of the ligand [205]; while for phenylphosphine analogues Xantphos often comes out as best (but not for aryl chlorides, as oxidative addition does not take place), Josiphos derivative CyPF-*t*-Bu (Figure 9.6) performs much better than Xantphos with the same substituents at phosphorus in C−N bond forming reactions (93% vs. <5% in one reaction involving 3-chloropyridine). The efficiency of palladium catalysts containing this ligand for the amination process was believed to result from, first, an unusually rigid backbone, which causes the ligand to bind tightly to palladium. Secondly, strong electron donation promotes the oxidative addition of less reactive chloroarenes, and thirdly it was thought that the steric bulk that disfavors diarylation, facilitates the generation of the Pd(0) intermediate, and enhances reductive elimination from the arylpalladium amido complexes. Thus, these factors strongly resemble those mentioned for the diaryl monophosphine systems. For several reactions very high TONs have been obtained with the use of the Josiphos derivatives.

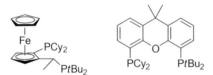

**Figure 9.6** Josiphos (CyPF-*t*-Bu) and Xantphos homologues.

CyPF-*t*-Bu, the Josiphos analogue of Figure 9.6, which forms active catalysts for ammonia [206], alkylamine, and arylamine amination of aryl halides, showed a very slow reaction when ammonia was used as the nucleophile. Klinkenberg and Hartwig studied the stoichiometric reductive elimination of aniline from the phenylpalladium amide (NH$_2$) intermediate [207]. Although the parent amide (NH$_2$) is a better nucleophile than ArNH, its reductive elimination was very slow. The reason brought forward is that the steric bulk of the arylamide (and alkylamide) destabilizes the intermediate to such an extent that a much faster reductive elimination takes place.

## 9.4.2
### Reductive Elimination of C—F Bonds

The formation of aryl–fluorine bonds presents a special case from a scientific point of view, but also from the point of view of health care and the pharmaceutical industry as the number of ArF-containing drugs is substantial. The elementary steps of an envisaged process were studied in detail by Grushin, who very early on described the concepts and the attractiveness of catalytic aromatic fluorination [208]. After many years of catalytic attempts that did not lead to results, his attention was directed towards the fundamental steps and the synthesis of the presumed intermediates. Overall thermodynamics allow the metathesis of ArBr and MF to obtain ArF; for example the nucleophilic displacement in electron-poor aromatics is an industrial process (F replacing Cl, the Halex reaction). Oxidative addition of ArBr to Pd proceeds without difficulty, but the exchange of Br by the nucleophile F remained to be proven, as was the reductive elimination of ArF. The first complex containing a Pd—F bond and a phenyl group was reported in 1997 by Grushin and coworkers [209], which opened the way to the study of reductive elimination. We do not consider here the reductive elimination from Pd(IV) [210], either as starting material or as an intermediate in "F$^+$" initiated reactions [211], as these systems are hard to imagine in a catalytic scheme (after the reductive elimination of ArF a PdX$_2$ species remains that has to be reduced first).

Thus, while thermodynamically feasible, kinetically the elimination of ArF from Pd(II) is still a rare process. The F anion is very basic in the phosphine-stabilized Pd complex, as is evident from several hydrogen-bonded complexes [212]. However, migratory reductive elimination does not take place as the Pd—F bond is very strong. Heating (PPh$_3$)$_2$PdPhF did not give reductive elimination of PhF, but led to fluoride transfer to phosphorus and formation of diphenyl. Most likely the mechanism involves formation of a metallophosphorane, without changing the valence state of Pd (Grushin, V. V. Personal communication). The final results might be compared with an aryl/alkoxide interchange on metal and phosphorus, as was found for Pt [213], although at the time no metallophosphorane intermediacy was proposed as the mechanism. Both reactions are shown in Scheme 9.21. The basic character of F should also aid the attack at the ipso C-atom leading to reductive elimination, but it does not as the reaction at the P-atom is faster. Several examples are known for Ar/F exchange on noble-metal/phosphorus-ligand complexes [212].

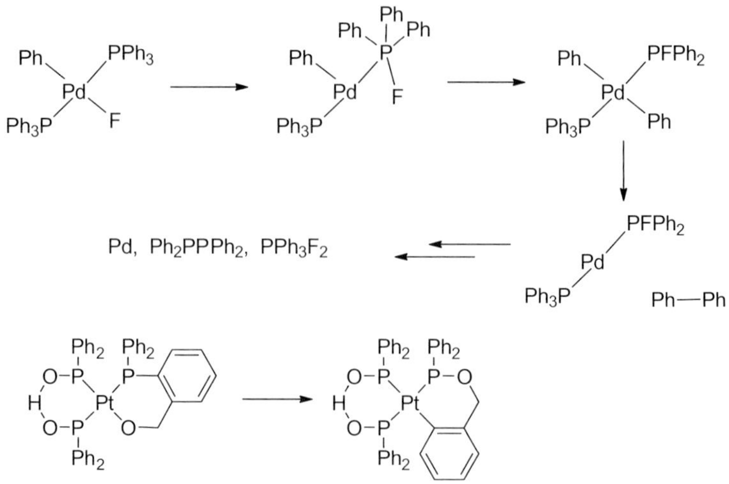

**Scheme 9.21**   C—F/O interchange between P and M (Pd, Pt).

Macgregor, Grushin, and coworkers studied Ph/F exchange on the fluoride homologue of Wilkinson's complex [214] and established that the phosphorane mechanism is the most likely one, which means that not the Rh-atom but the P-atom undergoes a valence state change during this reaction. The other important conclusion was that unlike in an oxidative addition reaction of Ph–PPh$_2$ to a metal, no vacancy on the metal is needed for this reaction [215]. This was also established experimentally; addition of extra phosphine did not slow P–F bond formation. DFT calculations confirmed the metallophosphorane mechanism [216].

Monophosphine complexes were also considered as they may more easily lead to reductive elimination (see Section 9.2.5) and they were found to form dimeric compounds [LPd(Ar)F]$_2$, including bulky ligands such as $p$-tolyl$_3$P and P$t$Bu$_3$ [217]. Upon heating no ArF formation was found. Taking into account the faster reductive elimination of $p$-nitro-phenyl moieties, Yandulov and Tran tested a range of ligands based on $p$-tolyl$_3$P that were added to the dimer. For one of the Buchwald biaryl phosphines ($t$Bu-XPhos) they found that upon heating of this mixture some 10% of $p$-nitrofluorobenzene was formed.

Grushin and Marshall reported that the same compounds containing Ph, $p$-tolyl, or $p$-anisyl in such a reaction with excess $t$Bu-XPhos did not yield any ArF formation [218]. A wide variety of monophosphines (including rigid bicyclic ones, BINAP monoxide, and so on) and diphosphines were tested (see the review by Grushin [212]), but none of them gave Ph—F formation, not even (Xantphos)Pd(Ph)F, which did give C$_6$H$_5$CF$_3$ smoothly when the CF$_3$ congener was used (*vide infra*). NHCs in complexes of the structure (NHC)(PPh$_3$)Pd(Ph)Cl gave reductive elimination of Ph-NHC chloride at room temperature and thus NHCs were discarded [219]. While reactions with Pd were not successful, CuF$_2$ did stoichiometrically convert PhI into PhF at high temperature [220].

Recently Buchwald and coworkers scored an important success in the area of palladium-catalyzed aromatic fluorination [221], which is rather surprising after so many attempts at stoichiometric fluorination reactions had failed. The reaction follows the concept outlined by Grushin and resulted from one of the biaryl phosphine ligands, BrettPhos (Scheme 9.22). First, it was shown that (BrettPhos) Pd(Ar)F, indeed giving monomeric complexes, underwent reductive elimination to give 15–25% of ArF, which could be increased by the addition of ArBr to 45–55%. It was suggested that the lower yield in the absence of added aryl bromide was due to a reaction of the (Brettphos)Pd(0) complex formed with the remaining (Brettphos) PdAr(F) complex. Hartwig and coworkers also observed increased product yields in the reductive elimination of C–S bonds from Pd(II) centers in stoichiometric reactions when Pd(0) scavengers (in this instance PPh$_3$ as a ligand) were added to the reaction [144a].

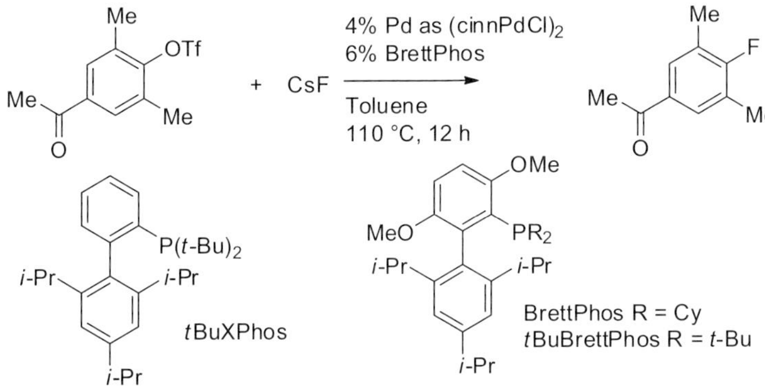

**Scheme 9.22** Fluorination of aryl triflates.

The Pd-catalyzed fluorination worked well with the use of Brettphos and *t*BuBrettPhos for aryl triflates. An example is shown in Scheme 9.22. The reaction proceeds with 1–4% of Pd (administered as the cinnamyl choride complex dimer) and 50% excess of BrettPhos (note that the excess needed will depend on catalyst concentrations), temperature 80–130 °C, and CsF as the fluoride donor. Apolar solvents (cyclohexene) worked best, as polar solvents gave more reduction to ArH and formation of regional isomers of ArF. No explanation was given, but the formation of isomers did not follow the pattern of that of benzyne as intermediates. Meta- and ortho-substituted aromatics with Me or MeO donor groups gave high selectivities to the desired product, but para-substituted tolyl and anisyl triflates gave large amounts of meta-isomers (in toluene, 36% and 70% respectively), This hints to an attack of F⁻ to the more electrophilic meta-position of the aryl group bonded to Pd instead of to the ipso carbon atom, followed by a rate-determining transfer of hydrogen from the ortho-position to the ipso carbon, as the overall reaction rate to by-products showed a normal H/D isotope effect.

As mentioned above, reductive elimination of ArCF$_3$ was not an easy process either. Although the bonding properties of F and CF$_3$ are very different, the Taft constants and inductive Hammett constants of F and CF$_3$ substituents are rather similar. Therefore, Grushin turned to the CF$_3$ elimination reaction to learn more about the backgrounds of nucleophilic fluorine substitution. This reaction also turned out to be rather difficult and it was first observed for the complex containing the wide bite angle ligand Xantphos [222]. Recently, a catalytic route for the introduction of $CF_3$ groups was reported by Buchwald and coworkers [223], an important breakthrough since $CF_3$ is also a frequent substituent in pharmaceuticals. With the use of 6% Pd and 9% of BrettPhos, many aryl chlorides were coupled in high yield at 130 °C with $Et_3SiCF_3$ as the source for $CF_3$. Still, the conditions and low TONs demonstrate that, even for this system, electron-poor groups, such as $CF_3$, do not undergo elimination easily.

## 9.5
## Phosphine Decomposition

### 9.5.1
### Phosphine Oxidation

Before addressing P−C cleavage reactions a few paragraphs will be devoted to other mechanisms of phosphine decomposition such as oxidation. In the solid state aryl phosphines (unless abundantly substituted with MeO donor groups) are not very sensitive to aerial oxidation. In solution though, they are more sensitive. Alkyl phosphines are very prone to oxidation by air and they are the ligands of choice when aryl chlorides are used as the substrate, as palladium complexes of arylphosphines usually do not undergo oxidative addition of aryl chlorides. Aryl chlorides are cheaper and more abundantly available than aryl bromides and since the penalties associated with waste disposal go by weight rather than by mole, the coproduction of NaCl is much more attractive than the coproduction of KBr (molecular weights 58 and 119, respectively). Thus, many attempts in the last decade have been directed towards the use of aryl chlorides in cross-coupling chemistry, with ample success.

In view of the oxidation sensitivity, the protocol of Fu for the use of $tBu_3P$ as the ligand is useful [224]. The ligand is added to the reaction mixture as the phosphonium salt (e.g., with $BF_4^-$ as the counterion), which can be handled in air. The proton is removed by a base, liberating the free phosphine, once the system is brought under an inert gas. Various applications have been demonstrated. In a recent modification, 3 equiv of $KF \cdot 2H_2O$ were added to $Pd_3(dba)_2$ and $[HP(t\text{-}Bu)_3]BF_4$ [225].

An accidental, highly favorable property of the Buchwald ligands is their resistance towards oxidation by molecular oxygen, in spite of the presence of two strongly donating Cy or $t$-Bu groups. Barder and Buchwald studied this phenomenon in detail [226], and they found that in the preferred conformation of the ligand the lone pair points to the 2,6-substituted second aryl ring, which inhibits the approach of a second phosphine to a $R_3P \cdots O−O\bullet$ phosphine-dioxygen intermediate. Thus, the 2,6-substitution plays an important role in this. Unsubstituted biphenylphosphines are oxidized at least ten times faster than the isopropyl-substituted XPhos.

Several other oxygen-containing reagents can oxidize phosphines, aided by Pd complexes as the catalyst. Even water can oxidize a phosphine, producing phosphine oxide and hydrogen, which is thermodynamically feasible (even for PPh₃ and thus certainly for alkylphosphines). Several hard bases such as acetate and hydroxy groups can oxidize phosphines, such as BINAP, in the presence of Pd [227].

### 9.5.2
### P–C Cleavage of Ligands

Reductive cleavage of P–C bonds was already used as an established step in phosphine synthesis in the 1960s and phosphine decomposition via reductive cleavage on a transition metal (Rh, Co) was reported in the early 1970s. Replacement of Ph-groups in PPh₃ by MeO can be carried out catalytically with high TOFs with Rh as the catalyst [228]. Aryl exchange from phosphorus to Pd, and from there into the product, was observed in the seventies by several groups [229] (for the mechanisms and references thereof see Sections 1.4.3–1.4.6). For instance in an attempted Heck–Mizoroki reaction of 2-chlorophenol and ethyl acrylate with Pd(PPh₃)₄ as the catalyst at 150 °C only ethyl cinnamate was obtained as the product in low yield and chlorophenol was completely recovered. An excess of PPh₃ prevented the reaction. The mechanism proposed was a reversible oxidative addition of arylphosphines [230], which explains the retarding effect of the excess ligand (Scheme 9.23). An example of a complex **3**, PdI(Ar′)(PPh₃)₂, showed exchange of aryl and phenyl groups in relatively mild conditions [231]. Aryl exchange on phosphorus was also used as a synthetic tool for phosphine synthesis, but the yields rarely exceed 60% [232]. For the preparation of phosphonium salts from triphenylphosphine and aryl halide the yields can be much higher (95%, 1% Pd/dba, 145 °C, 1–24 h) [233] (*cf.* Scheme 9.25).

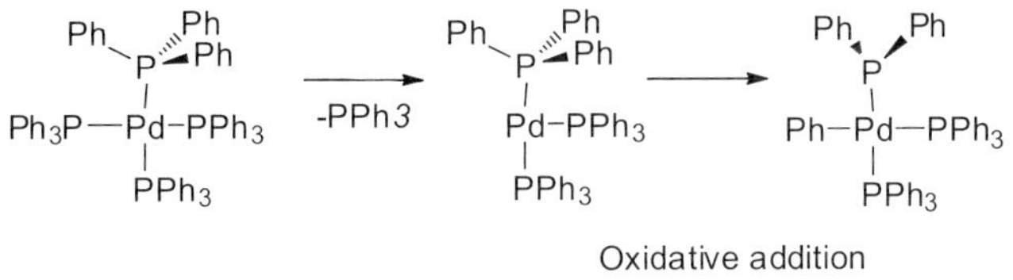

Scheme 9.23 Oxidative addition of arylphosphines after creating a vacancy.

As early as 1972, Matsuda and coworkers used PPh₃ as the source of the aryl group, in what one might call a Heck–Mizoroki reaction, the arylation of alkenes (styrene, acrylates, 1-octene, cyclohexene) with the use of stoichiometric amounts of Pd(OAc)₂ and PPh₃ [231a,b] in AcOH as the solvent at 50 °C. Reaction of the alkenes with *p*-tolylphosphine complex gave the corresponding *p*-tolyl derivatives. As the mechanism, they proposed a nucleophilic attack by, for example, acetate at the coordinated P-atom, with simultaneous migration of phenyl to Pd. The phosphorus by-products were PhP(O)(OH)₂ and Ph₂P(O)OH, but Ph₃PO was also observed, resulting from a non-productive oxidation [234] (Scheme 9.24).

Hartwig and coworkers studied the enantioselective α-arylation of ketones with aryl triflates catalyzed by Difluorphos complexes of Pd and Ni (and related atropisomeric ligands) [235]. Triflates gave much higher yields and ees than the halide

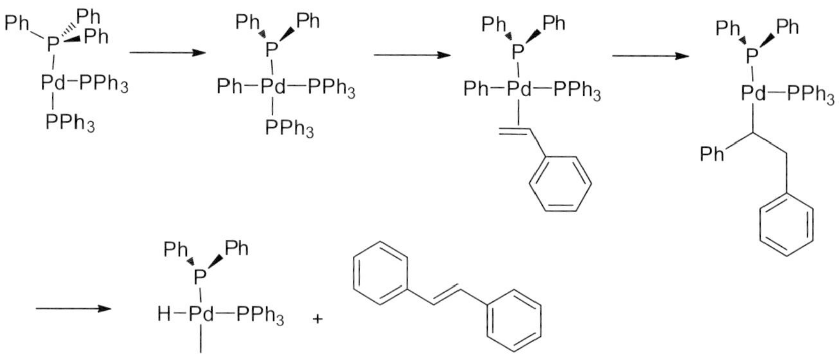

**Scheme 9.24** Heck reaction with PPh$_3$ as the aryl source.

precursors and iodides especially gave poor results. The stability of the ligand toward P—C bond cleavage, which occurs more readily with iodides, affected the reaction yield and the stability of the catalyst toward formation of achiral or less selective chiral catalysts perhaps also led to lower ees. P—C cleavage of BINAP was observed previously during amination of aryl bromides [41].

In catalytic reactions concerning low molecular weight products a small amount of aryl exchange between phosphine and substrate may not be important, but in polymer synthesis, via Suzuki–Miyaura cross-coupling for instance, it may mean that each polymer molecule contains an aryl group stemming from the catalyst, or each chain may be terminated by a phosphine (or phosphonium) endcap, as was discovered by Novak and coworkers [236]. The aryl–aryl interchange reaction of ArPdL$_2$I with L = P(4-FC$_6$H$_4$)$_3$ and Ar = 4-MeOC$_6$H$_4$ was found to follow pseudo-first-order kinetics. A marked inhibition in the presence of excess phosphine and/or excess iodide was observed, suggesting that a dissociative pathway was involved. An excess of phosphine to prevent the dissociative pathway could not be applied either, because another mechanism took over for the formation of phosphonium salts (see Scheme 9.25).

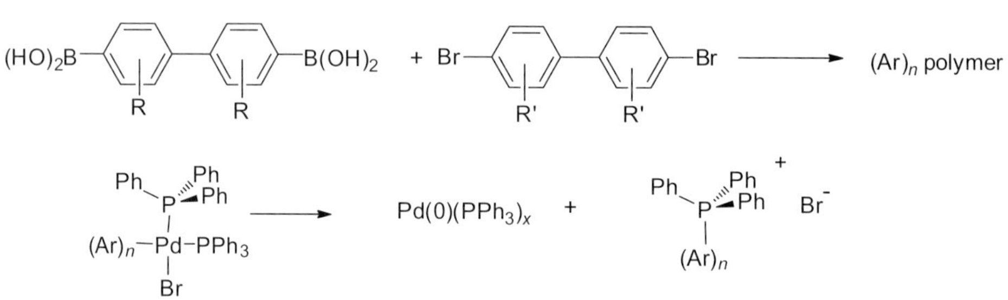

**Scheme 9.25** Polymer endcapping by phosphonium salts.

The interchange reaction proceeded via reductive elimination to form the phosphonium salt and oxidative addition again of a different P—C bond, suggesting that excess phosphine was acting as a trap for intermediate palladium(0) species, preventing the generation of the interchanged palladium(II) complex. Polar solvents enhanced the exchange reaction. Substituent effect studies of the interchange reaction indicated that electron-withdrawing groups on both the phosphine and palladium-bound aryl groups inhibited the exchange and so increased the steric bulk on both the phosphine and palladium-bound aryl groups. All three factors may be related to the ease of phosphonium ion formation.

Grushin studied the thermal stability and reactivity toward the Pd—Ph/P—Ph exchange reactions in all halide complexes of the type $(Ph_3P)_2Pd(Ph)X$ [237]. Iodides are by far the most labile. Kinetic studies of the aryl–aryl exchange reactions of $(Ph_3P)_2Pd(C_6D_5)X$ in benzene-$d6$ demonstrated that the rate of exchange decreases in the order $I > Br > Cl$, (100 : 4 : 1). The exchange was facilitated by a decrease in the concentration of the complex, polar media, and Lewis acids. Unlike the findings above by Novak, the concentration effect indicated that a dissociative process was involved, most likely dissociation of a phosphine in apolar media and an anion in polar media. Thus the use of apolar media may be advantageous in catalytic applications. The higher reactivity of iodides in the oxidative addition reaction may be offset by the higher propensity of iodide to side-reactions. Grushin also suggested running the reactions in the presence of extra phosphine or halide, and at the highest possible catalyst concentration, since the aryl–aryl exchange is initiated by a dissociation reaction.

In a palladium-catalyzed P—C bond formation using benzyl bromide and H-phosphonate diesters, considerable phosphonium salt formation was observed for $PPh_3$ as the ligand. In catalytic reactions $PPh_3$ gave poor results, while wide-bite angle bidentates such as Xantphos showed excellent results, indicating that reductive elimination is rate limiting (Scheme 9.26). Oxidative addition of benzyl bromide was indeed fast. For Xantphos phosphonium salt formation was much lower, because both oxidative addition to Pd(0) and reductive elimination were faster in this case, concluded the authors from stoichiometric reactions (in both instances carried out with a Pd/P ratio of 1 : 2) [238].

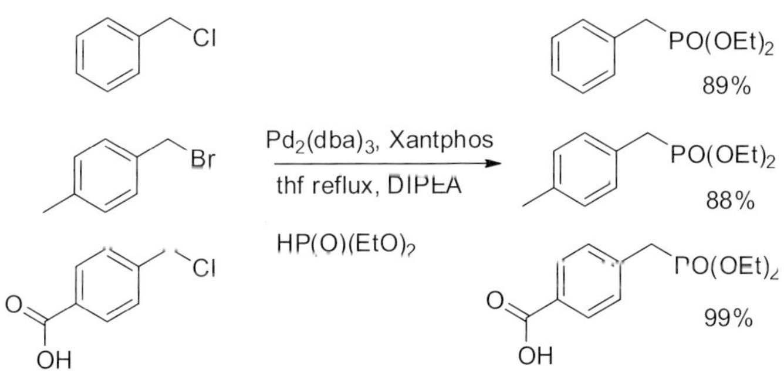

**Scheme 9.26**  Benzyl phosphonation with Pd/Xantphos.

Apart from the effects of ligands and solvents to prevent P—C exchange, Wallow and Novak also presented a phosphine-free Suzuki–Miyaura coupling reaction [239].

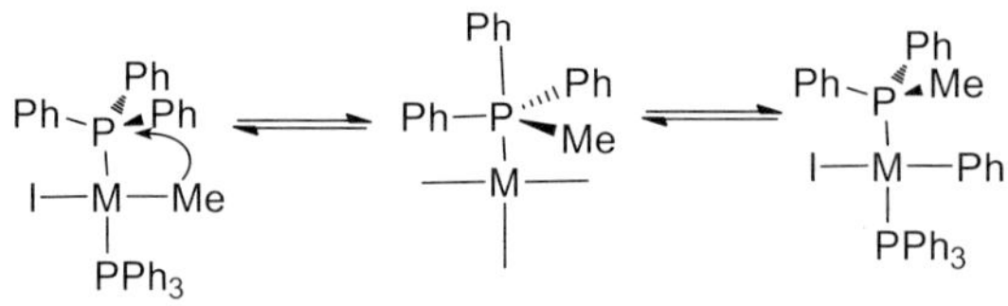

**Scheme 9.27** Alkyl/aryl exchange at Pd/P under mild conditions.

They found that a system using $[(\eta^3\text{-}C_3H_5)PdCl]_2$ as the phosphine-free precursor was two orders of magnitude more active than phosphine-supported catalytic systems such as $Pd(PPh_3)_4$ for aryl iodide substrates.

Formation of phosphonium salts during a Heck–Mizoroki reaction of bromobenzene and styrene was already observed in 1978 by Ziegler and Heck [240] without aryl exchange or P–C bond cleavage. Reductive elimination of $ArPh_3PBr$ from $ArPd$ $(PPh_3)_2Br$ was proposed as the mechanism. They also reported that this side-reaction did not take place when $P(o\text{-tolyl})_3$ was used as the ligand. For aryl iodides $Pd(OAc)_2$ was used successfully without the need for a phosphine ligand.

Alkyl/aryl exchange between Pd and P does not involve phosphonium salts, as was proven by Norton and coworkers [241]. They found that the methyl ligand of trans-$CH_3Pd(PPh_3)_2I$ exchanged with a phenyl group of $PPh_3$ to give $PhPd(PPh_3)(PMePh_2)I$. The $PMePh_2$ formed exchanged with the $PPh_3$ of the starting material. The rearrangement is irreversible, does not involve a free phosphonium cation, and does not require phosphine dissociation (Scheme 9.27). Such a rearrangement may involve phosphorane formation, that is, phosphorus changes its valence rather than Pd, as we have seen in Sections 1.4.3–1.4.6. At the time of the report, this mechanism was not yet known.

Herrmann and coworkers reported on the aryl exchange in complexes of the structure $L_2PdArCl$ formed by oxidative addition of aryl chloride to the corresponding Pd(0) phosphine complexes [242]. Chlorobenzene and electron-poor aryl chlorides $ClC_6H_4X$ (X = 4-$NO_2$, 4-CHO, 4-CN, 4-H) were added oxidatively to $Pd(PPh_3)_4$ or $Pd(PPh_3)_2(dba)$ at 100–140 °C to give complexes of the type trans-Pd $(PPh_3)_2(C_6H_4\text{-}p\text{-}X)Cl$. Oxidative addition of electron-rich aryl chlorides $ClC_6H_4Y$ (Y = 4-$CH_3$, 4-$CH_3O$) to $Pd(PPh_3)_4$ was expected to lead to the complexes trans-Pd $(PPh_3)_2(C_6H_4Y)Cl$, but instead the reaction gave, in almost quantitative total yield, mixtures consisting of 90% trans-$Pd(PPh_3)_2(Ph)Cl$ and 10% of the expected product trans-$Pd(PPh_3)\{PPh_2(C_6H_4\text{-}p\text{-}Y)\}(Ph)Cl$. The stability of the primary oxidative addition products was examined and it was shown that the complexes with electron-rich Ar groups, synthesized independently by halogen-exchange from the iodo-derivatives, underwent a facile aryl–aryl exchange between the palladium center and the coordinated phosphine ligands. Most likely the mechanism involves the migration of the electron-rich aryl groups to the electrophilic phosphorus atom. Subsequent intermolecular phosphine scrambling led to further isomerization, resulting in the formation of the observed reaction products. Complexes with chelating phosphine ligands (P∩P) of the type cis-$Pd(P∩P)(Ph)Cl$ (P∩P = dppe, dppp) were obtained by oxidative addition of chlorobenzene to Pd(P∩P)(dba) or by simple phosphine exchange with PPh₃. For the bidentate ligands no aryl exchange was reported, which may be due to the rigidity of the structures containing a bridging ligand.

Aryl exchange followed by catalyst decomposition that limited the TON of Suzuki–Miyaura and Heck–Mizoroki reactions led to the development of the *o*-tolylphosphine cyclometallated catalysts by Herrmann and coworkers (as discussed before in Section 9.2.3) [80]. For example, it was reported that P–C bond cleavage played an important role in the deactivation of the arylation of *n*-butyl acrylate [243, 244]. The temperatures needed were rather high, >120 °C, for both chlorides and bromides. When PPh₃ was used as the ligand and electron-rich 4-bromoanisole as the substrate, considerable amounts of butyl cinnamate were found (i.e., the aromatic group stems from the ligand used). For *o*-tolylphosphine this was not the case.

Marcuccio and coworkers [245] found incorporation of phenyl groups from PPh₃ in the coupling product of halobenzenes and arylboronic acids in quantities up to 1/3 of the desired product. In another experiment they showed that as much as 76% of the phenyl groups of PPh₃ could be converted to diphenyl (Scheme 9.28). In a search for triarylphosphines that would give less phosphine aryl incorporation they found that tris(2-MeO-phenyl)phosphine performed best, with only 3% of anisole-containing product. Addition of extra phosphine reduced by-product formation, as reported by Kong and Cheng [246] and several others cited in this chapter, but it was not practical as the rate of reaction was reduced considerably.

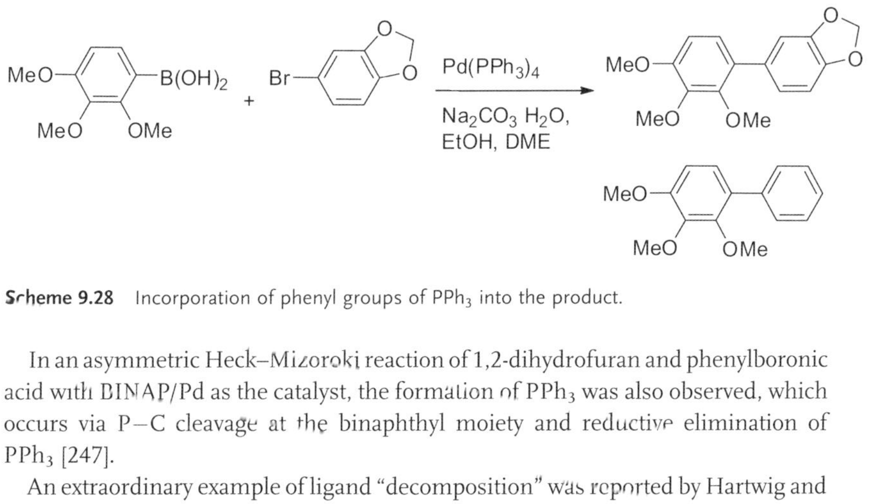

**Scheme 9.28** Incorporation of phenyl groups of PPh₃ into the product.

In an asymmetric Heck–Mizoroki reaction of 1,2-dihydrofuran and phenylboronic acid with BINAP/Pd as the catalyst, the formation of PPh₃ was also observed, which occurs via P–C cleavage at the binaphthyl moiety and reductive elimination of PPh₃ [247].

An extraordinary example of ligand "decomposition" was reported by Hartwig and coworkers for the monodentate ligand *t*Bu₂PFc (Fc = ferrocene) used in Pd-catalyzed cross-coupling of aryl halides and aryloxides [248]. During the reaction arylation of the unsubstituted Cp-ring took place and, surprisingly, the resulting arylated ligand appeared to be a more active catalyst. Independent synthesis afforded Ph₅FcP*t*Bu₂ containing a pentaphenylated cyclopentadienyl ring (Scheme 9.29), which indeed gave a fast catalyst for C–O bond formation.

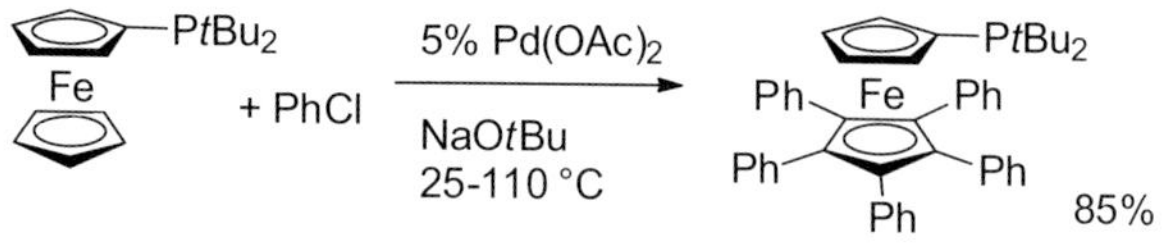

**Scheme 9.29** Formation of $Ph_5FcPtBu_2$ (Q-Phos) via phenylation with PhCl.

Grushin, MacGregor, and coworkers studied the mechanisms of catalyst poisoning in Pd-catalyzed cyanation of haloarenes [249]. Especially important was that each step of the standard catalytic cycle can be disrupted by excess cyanide, leading to the facile formation of inactive $[(CN)_4Pd]^{2-}$, $[(CN)_3PdR]^{2-}$, and $[(CN)_3PdH]^{2-}$. Water was particularly harmful to the catalysis because of ready protonolysis of $CN^-$ to HCN, which is highly reactive toward Pd(0). Depending on the conditions, the reaction of $[(Ph_3P)_4Pd]$ with HCN in the presence of extra $CN^-$ can give rise to $[(CN)_4Pd]^{2-}$ or the remarkably stable new hydride $[(CN)_3PdH]^{2-}$. When $Bu_4N^+$ was used as cation for $CN^-$ in this system an N–C cleavage was observed, as was apparent from the formation of $[(CN)_3PdBu]^{2-}$. Mechanistically the C–N cleavage resembles P–C cleavage in phosphonium salts. Both Hofmann degradation (by strong base resulting from water and $CN^-$, forming $NBu_3$ and 1-butene) and Pd-catalyzed C–N cleavage occurred. The often-used NaCN and KCN sources of $CN^-$ are very hygroscopic and therefore their use was not recommended. Zinc cyanide and $K_4[Fe(CN)_6]$ were considered less capable of generating $CN^-$ and they have been used with success in catalytic cyanation reactions [250, 251]. Protic solvents should be avoided. Any excess of $CN^-$ leads to the formation of $[(CN)_3PdAr]^{2-}$ via displacement of $PPh_3$ by $CN^-$ [252], which does not undergo reductive elimination. Thus, the paper explains nicely why the cyanation reaction, at first sight a simple modification of a cross-coupling reaction, may easily derail.

## 9.6
## Metal Impurities

Metal impurities affect the synthesis of the Grignard reagent and the cross-coupling reaction itself, as we have seen in the introduction (Section 9.1) [7]. Transition metals present during the Grignard synthesis lead to homocoupling of the organic fragment through radical formation and also in the coupling reaction with ketones they lead to one-electron transfer processes, as studied by Kharasch and coworkers. Ahsby and coworkers studied, in a series of publications, the reaction of MeMgBr with ketones and nitriles and found that the kinetics were complicated by the formation of a number of complexes and that the selectivity depended on both ketone and Grignard concentrations [253]. Side-product formation (e.g., pinacol derivatives) was completely avoided when ultrapure magnesium was used. When a large excess of ketone was used no by-products were detected either, independent of the purity of the magnesium employed, but this is not a practical solution.

The early results from Kochi and coworkers on the Grignard coupling reaction with alkyl halides in the presence of salts of Ag, Cu, and Fe showed a clear effect of the transition metal catalysts, but selectivities were poor [10, 11, 12]. In conclusion, the presence of metal impurities is worth studying. Note that in C−C coupling reactions the boronate, silane, or tin-based metal organic components are also synthesized from a Grignard reagent and, therefore, an effective synthesis of the latter is important. The practical solution, both in the laboratory and in industry, is usually to test several grades of magnesium from different suppliers.

One industrial application was analyzed in detail by Giordano [254], Scheme 9.30. It concerns the synthesis of Diflunisal, a non-steroidal anti-inflammatory drug. Sales price is around €300/kg and a few hundred tons are made worldwide per annum.

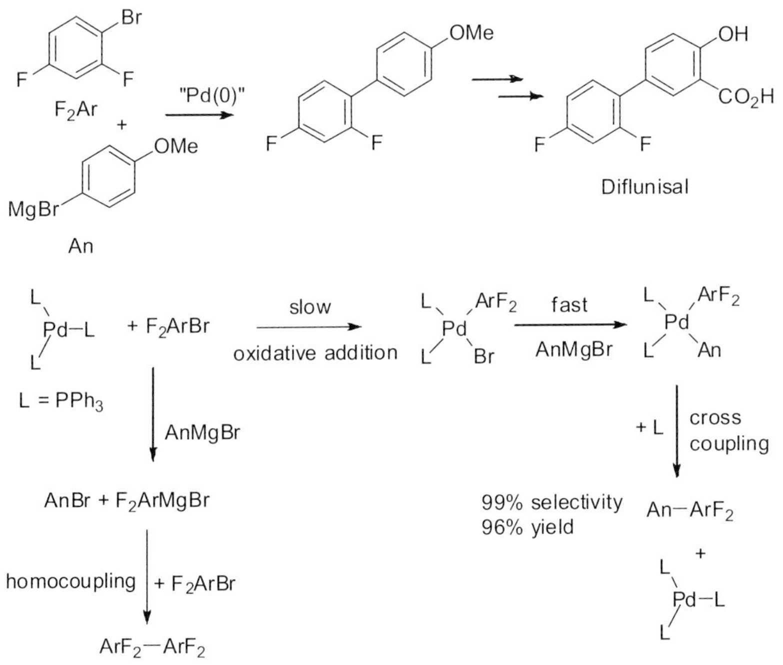

**Scheme 9.30** Synthesis of Diflunisal via a Kumada–Corriu cross-coupling.

A very pure product is required. The palladium content may not exceed 10 ppm and the impurities should be below 0.1%. Classical syntheses such as the Ullmann reaction cannot be applied because they produce a lot of by-products. Two steps are critical: the cross-coupling itself, and making the Grignard of 4-bromoanisole. The Grignard is made from 4-bromoanisole, Mg (3% excess), $I_2$ catalyst, in tetrahydro furan (3 M), at 70 °C. The major problem is the formation of the "homo"-coupled product during the Grignard synthesis. Up to 6% of 4,4′-dimethoxybiphenyl may be formed. The amount depended on the type of magnesium that was used. The impurities found were Cu (10–100 ppm), Fe (30–300 ppm), Ni (5–10 ppm), and Mn (35–400 ppm). It was established that high yields of the desired Grignard could only be obtained if the metal impurities for all these metals were at the lowest levels of the ranges indicated, otherwise they would catalyze a cross-coupling reaction!

The next step is the coupling of the Grignard reagent with 1,3-difluoro-4-bromo-benzene using Pd(PPh$_3$)$_4$ (0.1%) as the catalyst. The catalyst is generated *in situ* from palladium acetate and triphenyl phosphine. The starting material for difluorobro-mobenzene is difluoroaniline that is converted to the bromide via diazotation. The Grignard (3 M) is added to a solution of the bromide. Substantial amounts of homocoupled 4,4′-dimethoxybiphenyl are formed when the Grignard is added too quickly. If oxidative addition is the slowest step in the reaction sequence, the concentration of the aryl bromide should be maximized. Oxidative addition is the rate-limiting step in this instance due to the use of Pd(PPh$_3$)$_3$ as the *in situ* formed precursor. The metallation reaction at palladium, replacing the bromide ion by the anisyl group, is relatively fast and thus the concentration of the Grignard may be kept low. Side-reactions are all due to reactions of the Grignard reagent, anisyl magnesium bromide, which is added over a period of 4 h. Indeed, the reaction is operated in such a way that there is hardly any Grignard reagent in solution. It turned out that higher temperatures gave the best results (85 °C). The total turnover is more than 3000 at high selectivity and yield. The reaction is run in 12 m$^3$ stainless steel, glass-lined reactors. The conversion to Diflusinal is completed by removing the methoxy group by refluxing in HBr/AcOH and the carboxylic acid group is introduced via a base-catalyzed reaction with CO$_2$.

Studies with iron catalysts related to those of Kochi (see next paragraph) were carried out by Kim and coworkers [255]. They studied the effect of transition metal ions (iron in this case) on the reaction of benzyl bromide with MeMgI to give either ethylbenzene as the desired cross-coupling product or dibenzyl, the homo-coupling product. When MeMgI prepared with pure magnesium was used, the ratio of ethylbenzene to dibenzyl was 22 to78, and with reagent grade magnesium, the ratio became 33 to 67. This indicated that metallic impurities in magnesium affect the reaction mechanism leading to less homo-coupling (the impurities were not iden-tified). Surprisingly, when ferric chloride was added to the reaction mixture in catalytic amounts the ratio of ethylbenzene to dibenzyl reached 80 to 20, thus much less product was formed via a radical pathway. The authors speculated that the reaction in the presence of ferric ion seemed to follow mainly an ionic mechanism involving an iron-benzyl bromide π-complex. This would imply an oxidative addition to a low-valent Fe species. The complex formation was expected to enhance ionic attack of MeMgI on the benzyl carbon to give more ethylbenzene.

In the last decade the use of iron complexes as catalysts in cross-coupling reactions has undergone a sudden, spectacular growth [256], building on the early work of Kochi [12]. Occasionally, addition of iron salts enhanced homo-coupling via radical formation, as revealed by CIDNP [6], but on other occasions selective cross-coupling was obtained (this is not in contradiction, as observation of CIDNP requires a radical pathway only for part of the reagents). Tamura and Kochi found that a rapid coupling reaction took place at room temperature between alkyl Grignard reagents and alkenyl bromides with retention of the *E*- or *Z*-nature of the alkenyl compound (*Z*-1-bromopropene giving *Z*-2-butene) with FeCl$_3$ as the catalyst. On the other hand, the reaction of MeMgBr with alkyl halides gave disproportionation to alkene/alkane, clearly following a radical pathway, similar to silver catalysts. In recent years, a plethora of selective cross-coupling reactions with Fe has been published, which were highly selective if carried out under proper control of the details (source of Fe [257], solvent [258], additives [259, 260]) [256, 261]. The groups of Fürstner [262] and Hayashi [263] published several examples of, especially, triflate and chloride coupling reactions, in which the iron-based catalysts outperformed by far the "classic" Ni and Pd catalysts in selectivity. This takes away the doubts that traces of Group 10 metals might be responsible for the activity of the Fe catalysts (the common contaminant in Fe is Mn, but this is not active in cross-coupling reactions).

Nakamura and coworkers reported variants of the Suzuki–Miyaura and Negishi reactions using 1–5% Fe(dppbz)Cl$_2$ as the catalyst [264, 265] (dppbz = 1,2-bis (diphenylphosphino)benzene) at 0–40 °C. Nagashima, Nakamura, and coworkers found a profound effect of TMEDA on iron-catalyzed coupling reactions of ArMgX with alkyl halides [266]. TMEDA coordinated to Fe in this reaction rather than to Li or Mg, which adds to the complexity of the mechanism. The mechanism most likely varies from one system to the other and, as yet, we must refrain from explanations. Bedford and coworkers reported a variant of the Negishi reaction catalyzed by Fe using Zn alkyl reagents [267]. They used FeCl$_2$(dppbz)$_2$ (or dppp) as the catalyst. In this case also the role of the alkylating agent is more complex than just acting as an alkylating agent.

Excluding the participation of traces of Ni or Pd in the Fe-catalyzed coupling reactions is not an easy task. Simply replacing Fe by Ni changes the circumstances for Ni considerably compared to the ppm levels at which they might be present in the Fe catalyst. The way to go would be to use ultrapure Fe and Mg (or Zn, Al, or Li) and add Ni or Pd at the ppm level to this system and study the effect of these additions. Particularly in view of the high TONs (millions!) obtained by Herrmann and Bedford with their catalyst systems, it seems important to establish the role of trace metals. The attractiveness of the use of Fe instead of Pd is obvious and, therefore, the first results on a Suzuki–Miyaura reaction with Fe as the catalyst were regarded as an important breakthrough. Franzén and coworkers reported that FeCl$_2$(Py)$_4$ is an effective catalyst in this reaction for the coupling of aryl bromides with phenylboronic acid [268]. Bedford, Nakamura, and coworkers could not reproduce these results and showed in addition that with a substrate ratio to Pd of 1 000 000 complete coupling was obtained [269]. With p-bromoacetophenone, activities with Pd can be achieved at the ppb level [98]. Thus, caution must be exercised before concluding that new metals catalyze the classic cross-coupling reactions.

Substituted diaryliron complexes react in a stoichiometric fashion with alkyl iodides, as was found by Knochel and Wunderlich [270]. The organometallic iron starting materials were conveniently prepared by ferration of the substituted arene with tmp$_2$Fe·2MgCl$_2$·4 LiCl (see Scheme 9.31). When FeCl$_3$ was used in the ferration reaction the same Fe(II) complex was obtained. With the use of highly pure Fe the yield of the coupling with octyl iodide was only 25%, but addition of 0.5% of NiCl$_2$ raised the yield to 94%, the same as that obtained with Fe of 98% purity only. Metal chlorides such as divalent Co, Mn, or Cu did not improve the reaction, nor did FeCl$_3$. In addition 10 mol% of p-fluorostyrene was used to accelerate reductive elimination.

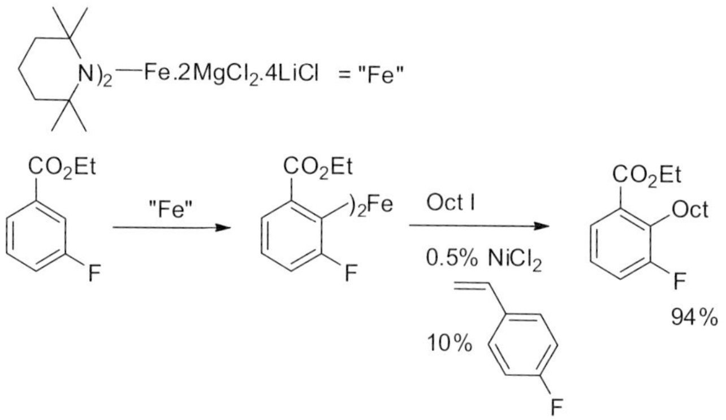

**Scheme 9.31** Ni-catalyzed coupling of aryliron compounds with alkyl iodides.

Fluorostyrene was known to enhance reductive eliminations in Ni-catalyzed cross-coupling reactions [168a] (Section 9.4.1).

Leadbeater and coworkers showed that 50 ppb of Pd present in $Na_2CO_3$ used as the base were able to catalyze a Suzuki–Miyaura reaction between phenylboronic acid and 4-bromoacetophenone to yield the coupling product at 98% yield in a few minutes in a microwave oven at 150 °C [271]. The reaction took place in the absence of phosphine ligand. Initially, it was thought that the coupling reaction could be established in the absence of Pd, because the Pd level was below the level of detection. In this context we should note that De Vries and coworkers [272] showed that the Heck reaction can be run with the addition of what Beletskaya termed "homeopathic" quantities of palladium catalysts [196d], also in ligand-free systems. We shall return to this when discussing the use of PdNPs as catalysts (precursors), Section 9.7.

"Palladium-free" Sonogashira reactions were considered by Plenio [273] and he concluded that care should be taken with such claims, as several less active metals are used in 10% quantities while Pd may show TONs exceeding 10 000. Espinet, Echavarren, and coworkers investigated a potential gold-catalyzed Sonogashira reaction and they arrived at the conclusion that the actual catalyst was Pd [274]. Mechanistic studies revealed that several reactions known for Pd(0)–Pd(II) couples did not take place for the Au(I)–Au(III) couple. It was concluded that the Au-catalyzed Sonogashira reactions reported [275] might well proceed thanks to Pd contamination of any of the compounds involved in the reaction.

Correa and Bolm reported on iron-catalyzed cross-couplings leading to arylated amides, phenols, thiols, and alkynes [276], using relatively high $FeCl_3$ loadings (10%) in combination with 20 mol% of a ligand (a diamine or a diketone) in a solvent such as toluene at 135 °C. Both the Bolm group and the Buchwald group noted that the success of the reaction depended heavily on the source of iron [277]. By running the reactions with $FeCl_3$ from different sources and different purities, and by adding 5–100 ppm of $Cu_2O$, they proved that the actual catalytic metal was Cu.

So far we have considered that only one metal is responsible for the catalytic coupling of the organometal fragment and the hydrocarbyl halide compound.

This turned out to be rather difficult and the actual situation may be more complicated. Transition metal complexes may catalyze the formation of Grignard reagents, or catalyze their decomposition during the formation. Likewise, more metals may be involved in the cross-coupling process, for example, a second metal may aid the oxidative addition, or the hydrocarbyl/halide exchange on the "catalytic" metal. It will be very time-consuming to solve all questions being raised concerning the ever increasing amount of new cross-coupling catalysts [278].

## 9.7
## Metal Nanoparticles and Supported Metal Catalysts

### 9.7.1
### Supported Metal Catalysts

Many examples are known of cross-coupling reactions that use metallic Pd on a support, especially in Heck–Mizoroki and Suzuki–Miyaura reactions at higher temperatures [279] (supported MNPs will be mentioned in Section 9.7.2). Heterogeneous catalysts offer several advantages, such as high stability of the catalyst, easy removal of the catalyst from the reaction mixture by filtration, and reusability of the catalyst for several times. Disadvantages are that Pd catalysts on solid supports often require higher reaction temperatures and lack the possibility of introducing stereospecific control. Pd catalysts supported by inorganic materials are very robust with respect to water and oxygen. The most popular catalysts comprise Pd on carbon.

We will assume here that the actual catalysis takes place in solution rather than at the surface of the metal (for more details see below). For instance Pd on carbon was utilized as the source for Pd and after the reaction no apparent loss of Pd was noted [280], which means that either the amount that goes into solution is very small, or most of it precipitates again after the reaction is over (i.e., all organyl halide has been consumed). The average TOF or TON are not very high; for Pd/C often the TON is below 100, and Zecca's review gives a TON of about 500 for Pd on oxidic supports [280], which is rather low compared to the millions obtained with several other systems. Many aspects were discussed in Zecca's review. During the reaction one can do a filtration test and ICP analysis to see whether there is Pd in solution [281] or whether the solution shows activity [282]. The activity of the effluent of a flow system containing the solid catalyst often came out positive.

Arai and coworkers investigated the leaching process thoroughly and we will summarize a few important findings [283]. They observed that catalysts such as 10% Pd/C and a 1% Pd/SiO$_2$ underwent extensive leaching during the Heck reaction of iodobenzene and methyl acrylate at 75 °C, in the presence of NEt$_3$ in N-ethylpyrrolidinone. Maximum activity in solution coincided with the highest concentrations in solution. They noted that Pd precipitated on the support at lower concentrations of PhI, towards the end of the reaction. Re-precipitation was more effective on C than on oxidic supports and best results were obtained by heating to 160 °C. Arai noted that deactivation occurred for solid catalysts, even if re-precipitation was complete.

Potential causes for this may be sintering of the Pd particles, or covering of Pd by carbonaceous products and by the salts co-produced in the reaction. As in heterogeneous catalysis, this fouling may be reversible; the salts can be removed by washing and polymeric organic material can be removed by burning, although this will be deleterious for the size of the Pd particles.

Conlon and coworkers at Merck reported an example of a Suzuki–Miyaura reaction in which Pd was reprecipitated on C leaving less than 4 ppm of Pd in solution [284]. Köhler also reported excellent recovery of Pd/C in a Heck–Mizoroki reaction with TONs as high as 18 000 in 2 h [285]. Improved TONs in the Heck reaction using Pd metal supported catalysts were reported by the same author and coworkers [286]. In this instance they used Pd on $TiO_2$, $Al_2O_3$, and zeolite NaY as the catalyst precursor for the reaction of aryl halides and styrene at 140 °C in NMP. For bromobenzene a TON of 100 000 was achieved in 6 h and chlorobenzene was about 10 times slower. They monitored the rate of reaction and the concentration in solution and found a clear correlation between the two. In this case the use of Pd is much more efficient than usual.

Ligand-free systems are closely related to systems using metallic Pd or Pd nanoparticles (PdNPs) in the absence of ligands. This theme also received ample attention in the last two decades for Heck, Suzuki, and Sonogashira reactions. Ligand-free systems were studied as early as 1983 by Spencer from Ciba–Geigy showing the industrial interest in cheap, ligand-free catalysts for the Heck–Mizoroki reaction [86]. TONs up to 10 000 were obtained for p-cyanobromobenzene and various alkenes, but as mentioned above in the presence of o-tolylphosphine their TONs were much higher. Spencer and Blaser showed that substituted benzoyl chloride could be used as the substrate instead of aryl halides and that under loss of CO a Heck–Mizoroki reaction could be accomplished under somewhat milder conditions than the previous example of ligand-free arylation of vinyl substrates [287] to give cinnamic acid esters and stilbenes. Oxidative addition of aroyl chlorides occurred more readily than that of aryl bromides (Scheme 9.32).

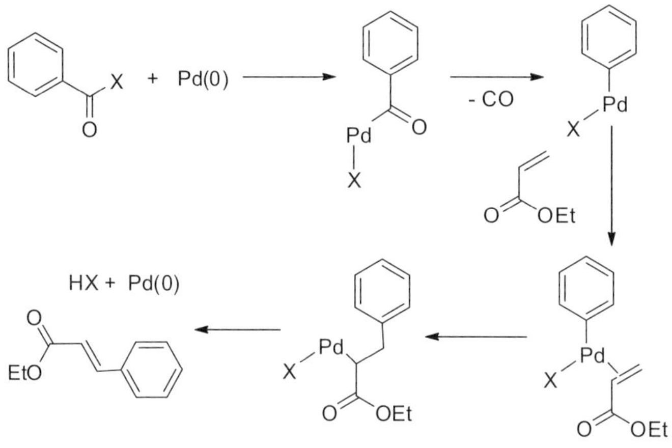

**Scheme 9.32** Decarbonylative Heck–Mizoroki reaction (X=$PhCO_2$, Cl, OH).

It is worth mentioning in this context that De Vries and coworkers used carboxylic anhydrides as a source of the aryl group in the Heck–Mizoroki reaction (Scheme 9.32) in which in an "aroyl carboxylate" adds oxidatively to Pd(0). Decarbonylation of the aroylpalladium fragment affords the arylpalladium used in the coupling reaction [288]. Co-produced are CO and carboxylic acid, which can be recycled. As arylcarboxylic acids are made industrially by oxidation of methyl aromatics (or aldehydes), this route presents a halide-free and salt-free alternative in cross-coupling chemistry. Su and coworkers used the carboxylic acid directly under loss of CO with benzoquinone as the oxidant, without converting the acid to the anhydride [289].

Why are ligand-free catalysts derived from supported Pd catalysts such good catalysts one might wonder, after so much evidence for the need for very specific ligands has been generated in the area of cross-coupling chemistry? Not always, but in many cases the temperature is very high for the ligand-free or Pd metal-based systems, which may limit their use in fine chemical synthesis. As we have seen in Section 9.2.1 oxidative addition need not be the rate-limiting step, and if it is, it is because there is an excess of ligand present such as $PPh_3$ or dba to stabilize Pd(0). In ligand-free systems this is not the case, although in the absence of P-donor ligands, the neutral Pd atom is less electron-rich and less prone to undergo oxidative addition. Because of its highly unsaturated state, ligand-free Pd may be more reactive towards oxidative addition than a ligated Pd complex. In a cross-coupling reaction reductive elimination may then be rate limiting (for all halides except chlorides), but reductive elimination should be faster from the ligand-free Pd intermediates than from Pd-phosphine complexes, as the net effect of phosphines is electron donation.

For a ligand-free Heck–Mizoroki reaction, reductive elimination of HX can be controlled by the base added, as mentioned in Section 9.4.1, and one of the other steps may be rate limiting. Carrow and Hartwig studied the nature of the ligand-free Pd species in this reaction directly synthesized from a homogeneous precursor, avoiding PdNPs as the source [290]. They identified dimers of anionic $ArPdBr_2^-$ as the species undergoing alkene insertion, most likely after dissociation, forming the arylated alkene and a Pd precipitate at room temperature. The reaction is faster than that of a $PPh_3$-ligated Pd intermediate, which was ascribed to steric effects. The anionic species were also active as catalysts for the reaction of iodobenzene and methyl acrylate at $30\,°C$; for less reactive substrates they could be used at $130\,°C$ as well. The neutral complexes of $PtBu_3$ reacted more slowly than the anionic ligand-free catalysts with alkenes. Calculations showed that Pd in anionic complexes is more electron-rich than Pd in the neutral $PtBu_3$ complex and, therefore, also in this case, the higher reactivity was assigned to less steric hindrance in the anionic complex.

Thus, an anionic, phosphine-free complex was identified, which may well be the complex generated from Pd/support (or PdNPs) by the aryl halide and other salts present. If a sufficient amount of Pd goes "into solution", ligand-free catalysts may indeed be relatively active and several systems showing high TOFs and TONs are cited above. Many reports showed low TONs calculated over the total of Pd in the system when solid metal precursors were used. In conclusion, there might be a role for nanoparticles, which have a high surface to bulk ratio and thus contain a high proportion of the metal exposed to the oxidizing aryl halide and halide ions in solution.

## 9.7.2
### Metal Nanoparticles as Catalysts

A very wide range of MNPs has been synthesized in the last decades for various purposes and their use as catalysts is one of them. There are many ways to obtain MNPs and stabilize them [291], for which we refer to review articles and books [292–294]. Initially MNPs and colloids were thought to present an intermediate phase between homogeneous and heterogeneous catalysts [295], and perhaps a source for new catalytic reactions. The latter has not been the case yet, it would seem. Reactions typical of heterogeneous catalysts, such as the hydrogenation of aromatics and CO, may take place on the surface of the MNPs, and most likely the MNPs remain intact in those reactions [296]. Traditionally, in heterogeneous catalysis many efforts were devoted to the synthesis of supported metal particles of smaller size in order to increase the efficiency of the metal usage because of the higher surface to volume ratio (together with a high surface area of the support). For reactions that are run at high temperatures, very small particles may not be stable and restructuring of the particles may occur. This process can be enhanced by substrates such as CO that can provide a way to transport the metal atoms as carbonyls. Very small particles may become less active per surface atom than larger particles because the reaction takes place at structural features such as kinks and steps that are not present in very small particles. For example, the intrinsic activity of Co per weight as a Fischer–Tropsch catalyst increases drastically with the decreasing size of the CoNPs, but below 5 nm the activity drops and the selectivity for methane increases. More specifically, normalized for the number of surface atoms the activity is constant above 6 nm [297]. Actually, one might say that this size effect presents a case of a "new" reaction or selectivity change for MNPs as compared to "bulk" metal.

Reactions with PdNP catalysts that strongly resemble homogeneous catalytic processes, such as cross-coupling reactions, the Heck–Mizoroki reaction, and allylic alkylation have been the subject of much discussion, whether the PdNP serves as the catalyst or as a sink/precursor for monometallic complexes; extensive critical reviews have been published [298]. Ligand-free palladium "atoms" (solvated, though) are probably very active catalysts in C–C coupling reactions and this may explain why nanoparticles can lead to active catalysts, and even to "efficient" recycling, as only a very small amount of the catalyst precursor is consumed in each cycle. Asymmetric PdNPs catalysts have been reported, and examples include Pd-catalyzed hydrosilylation of styrene [299], and Pd-catalyzed allylic alkylation of racemic substrates [300]. Modification of surfaces with chiral molecules has been known for several decades to give rise to enantioselective catalysis [301], but the similarity of ligands used in homogeneous catalysis and the recent MNP-based enantioselective catalysis seems suspect. The bulkiness of the ligands and their narrow bite angles make it even less likely that the enantioselective reactions take place at the surface. Evidence is growing that the latter reactions are catalyzed by homogeneous complexes [298].

A key issue for homogeneous catalysis is that MNPs can form in a reversible manner, as long as the molecular species are not removed from the reactor, as happens in a flow system, while the formation of larger metal particles is often irreversible, both thermodynamically and kinetically.

MNPs may be involved in three ways in cross-coupling chemistry, either as the fate of a homogeneous catalyst during or after its use, or as catalyst precursors, or catalysts themselves. Precipitation of the catalyst at the end of a reaction may be done intentionally as a means to accomplish separation of the catalyst or it may happen prematurely, before the catalysis is complete. Often palladium precipitation at the end of a reaction has not been reported. The opinion that MNPs are a reservoir for homogeneous catalysts in a similar way as was found for the Pd metal catalysts mentioned in Section 9.7.1 seems to prevail in recent years [302].

The first reports concerning the Heck reaction and PdNPs were published by the groups of Reetz and Beller (Hoechst) in 1996 [303, 304]. Reetz and coworkers prepared colloidal Pd or PdNPs electrochemically in the presence of tetraalkyl ammonium salts ($Oct_4NBr$) and Pd/Ni NPs stabilized by $Bu_4NBr$, a method used before for Pt by Grätzel [305], more specifically the Jeffery protocol in this instance [306]. The catalysts were used for the Suzuki–Miyaura reaction of relatively reactive substrates (aryl bromides, p-nitrochlorobenzene) and phenylboronic acid at 120 °C with modest TONs, while chlorobenzene did not react. Polar solvents such as dimethylacetamide worked best. The Heck–Mizoroki reaction was carried out with butyl acrylate and iodobenzene. Beller and coworkers studied the reaction of p-bromoacetophenone and butyl acrylate in the presence of in situ made PdNPs at 140 °C in dimethylacetamide. The best results were obtained by adding PdNPs prepared by the method of Bönneman [307] (reduction of $PdCl_2$ by $Oct_4N.BHEt_3$) to the reagents at the reaction temperature. TOFs as high as 24 000 h$^{-1}$ were observed. Less reactive chlorides did not react at this temperature. The TONs were not as high as those for the metallacyclic catalysts published at about the same time [80] (Section 9.2.3).

In a subsequent publication Reetz and coworkers described [308] the use of N,N-dimethylglycine to stabilize PdNPs used in a phosphine-free Heck reaction, for which they achieved TONs of 100 000 and TOFs of over 1000 under the conditions applied. It was one of the most active systems to date. At that time catalysis was supposed to take place on the PdNPs. As can be seen in Scheme 9.33 the most interesting feature of this system is that rates and TONs increase with lower concentrations of Pd. Dimethylglycine has a stabilizing role, although at lower concentrations the effect fades. One explanation for the increasing TOF could be that at lower concentrations smaller particles are formed and that a higher surface to bulk ratio is obtained, the reaction taking place at the surface, as was supposed at the time. Another explanation is that at low concentrations a larger proportion of Pd (if not all) is present in the solution as monometallic or oligomeric palladium complexes, that undergo oxidative addition more rapidly than the surface atoms.

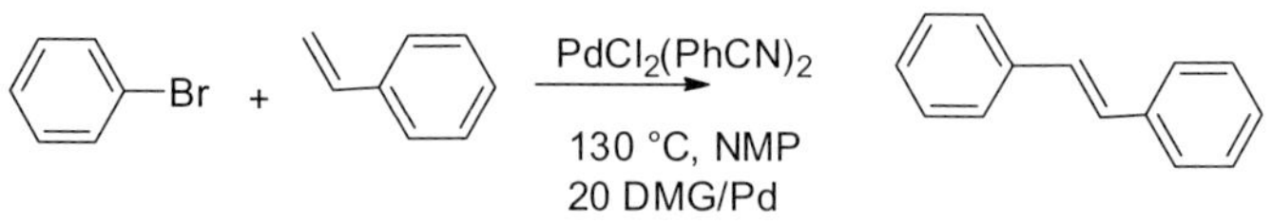

Influence of N,N-dimethylglycine (DMG) on phosphine-free Heck reaction

| | Pd(OAc)$_2$ | | | PdCl$_2$(PhCN)$_2$ / 20 DMG | | |
|---|---|---|---|---|---|---|
| [Pd] | conv.% | TON | TOF | conv.% | TON | TOF |
| 1.5 | 51 | 33 | 1 | 98 | 65 | 6.5 |
| 0.1 | | | | 96 | 960 | 96 |
| 0.01 | 77 | 8600 | 360 | 98 | 9800 | 408 |
| 0.0009 | 85 | 94000 | 980 | 96 | 106700 | 1100 |

**Scheme 9.33** Early reports on PdNPs showing higher activity at lower concentrations.

Dissolution of PdNPs via oxidative addition would also benefit from a higher surface to bulk ratio, while particle growth is second order or higher in Pd concentration. As a result, much smaller particles will be present. Aryl bromides containing electron-withdrawing substituents (*p*-nitro, *p*-cyano) gave much higher conversions than electron-releasing groups (*p*-dimethylamino), showing that the oxidative step in those systems with high Pd concentrations is indeed rate limiting, which with homogeneous catalysts is not always the case [65].

The concentration/particle size phenomenon is not unique and probably has been observed more often in systems in which Pd metal formation occurs. An example of Pd in methanol and CO comes to mind, for which also higher TOFs were found at lower concentrations for the carbonylation of methanol to acetic acid [309].

Reetz and Westermann later showed that indeed in such ligand-free Pd Heck catalysts PdNPs were present [310], but they did not decide yet on their role, that is, whether they were true catalysts or serving as a reservoir of Pd. Useful precursors are Pd(OAc)$_2$, PdCl$_2$, and palladacycles not containing phosphines. The so-called Jeffery procedure using Pd(OAc)$_2$ in the presence of NBu$_4$Cl as a phase transfer agent is very effective [306].

The authors pointed out that three differently *in situ*-generated nanosized Pd colloids showed similar behavior in catalysis. They generalized that other phosphine-free catalysts were also based on PdNPs, including Pd-catalyzed Ullmann reactions, as suggested earlier [311]. They concluded that although the catalysts contained "solutions" of colloids, the catalyst character was not that of a traditional homogeneous catalyst and that catalysis more likely occurred at defect sites, steps, and kinks on the surface of the nanoparticles. The latter processes belong rather to the realm of heterogeneous catalysis. As mentioned before, however, the occurrence of kinks and steps decreases with particle size.

In today's view, indeed the first step may take place at thermodynamically less stable Pd atoms at the surface. Vinyl or aryl halides dissolve the metal oxidatively, reminiscent of the formation of the Grignard reagents. One can imagine that a "corrosive" medium will aid the process of oxidative dissolution of Pd and anions can play a similar role as in homogeneous, phosphine-based systems.

The Heck–Mizoroki reaction was reviewed by Beletskaya and Cheprakov for all systems known to date, but of particular interest is their contribution on ligand-free and PdNPs-based systems [312]. The enormous flexibility of the reaction was exemplified by the wide variety of catalyst precursors and substrates that can be used. We quote: "The catalyst is often anything containing palladium, even in *homeopathic* doses, and other metals can perform the task in the 'absence' of palladium." Also for Kumada–Corriu reactions such catalysts using low doses of Pd were developed [313].

Diéguez and coworkers demonstrated that PdNPs in the presence of chiral oxazolinyl-phosphite ligands derived their activity from complexes in solution in asymmetric allylic alkylation and the Heck–Mizoroki reaction [314]. The PdNMPs had an average diameter of 2.5 nm and contained a relatively large amount of ligand (one bidentate ligand per eight Pd atoms). The reactions were carried out in a continuous-flow membrane reactor with a cut-off MW of 700 Da and the effluent had the same activity as that found in the reactor, thus proving that the reaction took place in solution.

Baiker and coworkers studied the homogenous/heterogeneous nature of the catalyst $Pd/Al_2O_3$-BINAP in the AAA reaction of 1,3-diphenylallyl acetate and sodium dimethyl malonate [315]. They speculated how a heterogeneous catalyst might lead to ees of 60%, but their conclusion was that most likely homogeneously dissolved Pd species led to the catalytic activity observed. Oxidized surface Pd was reduced again by the ligand or the THF solvent, but no proof was found for activity of PdNP itself.

Köhler and coworkers studied the behavior of PdNPs on $Al_2O_3$ as a catalyst in the Suzuki–Miyaura reaction at 65 °C [316]. They also asserted that the reaction took place in solution with a homogeneous catalyst. In this system under these conditions the particle size was constant at ~2 nm; this was considered accidental due to an equal dissolution and re-precipitation of Pd on the PdNPs. In a Sonogashira reaction using PdNPs Rothenberg and coworkers found that the size was reduced during the reaction [317], in a system that also showed leaching.

While there seems to be agreement that the PdNPs studied in detail so far function as a source for Pd species in solution responsible for the activity in cross-coupling reactions, involvement of PdNPs should not be excluded. There still may be phosphine-free systems that catalyze cross-coupling reactions at the surface of Pd; Pd atoms in edges or steps may have two coordination sites, which is sufficient to undergo oxidative addition, nucleophilic replacement, and reductive elimination. A mechanism requiring two or more coordination sites should not be excluded either. For example, the hydrogenation of aryl halides to arenes on $Pd/SiO_2$ and related supports might take place on the surface of Pd [318]. Thus, MNPs still hold a promise for new reactions to be discovered. As reservoirs for molecular catalysts, they have shown advantages, but the control of their size during catalysis seems an intrinsic problem that cannot be solved easily.

## 9.7.3
## Metal Precipitation

Metal precipitation in homogeneous catalysis is nowadays a less recurrent problem than it was decades ago, as many ligands are now available that form stable complexes and the high activities often allow one to work at low temperatures. The formation of palladium metal is a common phenomenon in palladium catalysis, and it is usually referred to as palladium black formation. Cross-coupling is very prone to palladium precipitation, but also in this area it is less dominant nowadays than before. Towards the end of the reaction Pd metal still might precipitate, which one may use as a means of separation of metal and organic product [319], as mentioned above. If precipitation is not desired, one might prevent it by working with an excess of aryl halide. The presence of CO as one of the reactants to accomplish the synthesis of amides or esters is notorious for Pd precipitation. Buchwald's interest in the synthesis of Weinreb amides (N-methoxy-N-methyl amides) directed him towards their synthesis from aryl bromides, amines, and CO [320, 321]. In this instance, bidentate phosphines were found to be the best, rather than biaryl monophosphines, probably because the former stabilize Pd complexes more efficiently against metal precipitation. Importantly, the reactions could be performed at only 1 bar when Xantphos was used. BINAP or Josiphos [322] ligands afford active catalysts only at elevated pressure. The flexibility of Xantphos, forming cis- and trans-complexes, was held responsible for this activity, and perhaps one may add that it also allows the formation of five-coordinate complexes with CO entering the coordination sphere in an associative manner [118, 173]. Apparently a flexible bite angle ligand leads to a more stable catalyst.

Formation of metal agglomerates starts with dimer and trimer formation and occasionally this has been observed via mass spectroscopy [319] or EXAFS. Van Strijdonck and coworkers studied the decomposition of Pd allylic alkylation catalysts with time-resolved EXAFS on a time-scale with intervals of seconds and they found that dimethylallyl-palladium(Xantphos)(TfO) formed dimeric and trimeric species before clusters were observed [323]. Related organometallic compounds are known and contain Pd(I), for instance a dimer containing two Pd-Xantphos moieties bound to one allyl fragment. The decomposition reactions were also monitored by UV–vis spectroscopy and in the visible region two species were observed while the solution turned red, which disappeared when a black precipitate formed, in accordance with the EXAFS studies.

## 9.8
## Conclusions

In the last two decades cross-coupling catalysis has become undoubtedly the largest application of homogeneous catalysts, probably surpassing the areas of carbonylation and metathesis, and approaching perhaps polymerization. Handbooks and reviews need continuous updating and it seems impossible to stay aware of all material published on even one single cross-coupling reaction. Catalyst deactivation is seldom addressed explicitly and the many tables containing results for catalyst systems, metal, ligands, the two key substrates, and conditions have to be analyzed in order to obtain useful data on what may be related to catalyst activity or decomposition. Most often a useful analysis is not feasible without further experimentation and there was no solution to that while writing this chapter. One may take a different approach, as Farina did, who reviewed high-yielding Pd catalysts [324]. What is good for one system may not solve the next problem, as many in this area have accepted that most likely each couple of substrates has its own solution. In this chapter we have concentrated on those contributions that go beyond the mere reports on yields and TONs, although also many useful recipes have been cited that do not offer much explanation. We hope that by the examples selected and described a substantial part of the concepts that should be borne in mind in the efforts to find the best catalyst have been covered.

To learn more in detail about dormant states, unexpectedly slow steps or side-reactions, for the thousands of examples described is an impossible task. Only a few important reactions can be subjected to further study and one reason to do so is their industrial application. A couple of industrial investigations (also by academic researchers) were mentioned above and several recent accounts stress that this is currently happening [158, 325].

Little attention has been paid above to catalyst recovery or removal of metal and ligand traces from the product, except for mentioning conditions that might enhance precipitation of Pd as the metal on a support after the reaction. To scavenge precious metals from waste streams or the effluent of a reactor containing a supported catalyst, workers at Johnson Matthey developed scavenger beds that can reduce the metal content down to a few ppm [326].

Cross-coupling chemistry has attained a key position in organic chemistry; it affords valuable shortcuts avoiding protection and deprotection steps, new conversions or conversions under much milder conditions, much more selective reactions thus reducing waste production, and so on. Ligand variation leads to control of reaction pathways, for example, in ring closure reactions as reported by Tsvelikhovsky and Buchwald [327]. The industrial importance, the crucial role in organic synthesis in the laboratory, and the many intriguing scientific issues will ensure that the many workers in cross-coupling chemistry will continue to surprise us with new inventions and more sophisticated explanations.

## References

1 de Meijere, A. and Diederich, F. (eds) (2004) *Metal-Catalyzed Cross-Coupling Reactions*, 2nd edn, Wiley-VCH Verlag GmbH, Weinheim.

2 Hartwig, J.F. (2002) *Handbook of Organopalladium Chemistry for Organic Synthesis*, vol. 1 (ed. E.I. Negishi), Wiley-Interscience, New York, pp. 1051 and 1097.

3 (a) Grignard, V. (1900) *C. R. Hebd. Seances Acad. Sci.*, **130**, 1322; (b) Grignard, V. (1901) *Ann. Chim.*, **24**, 433; cited in (c) Shinokubo, H. and Oshima, K. (2004) *Eur. J. Org. Chem*, 2081–2091.

4 Bennett, G.M. and Turner, E.E. (1914) *J. Chem. Soc.*, **105**, 1057–1062.

5 Krizewsky, J. and Turner, E.E. (1919) *J. Chem. Soc.*, **110**, 559–561.

6 (a) Ward, H.R. and Lawler, R.G. (1967) *J. Am. Chem. Soc.*, **89**, 5518; (b) Lawler, R.G. (1967) *J. Am. Chem. Soc.*, **89**, 6519.

7 (a) Kharasch, M.S., Kleiger, R., Martin, J.A., and Mayo, F.R. (1941) *J. Am. Chem. Soc.*, **63**, 2305–2307; (b) Kharasch, M.S. and Tawney, P.O. (1941) *J. Am. Chem. Soc.*, **63**, 2308–2315; (c) Kharasch, M.S. and Lambert, F.L. (1941) *J. Am. Chem. Soc.*, **63**, 2315–2316; (d) Kharasch, M.S. and Fields, E.K. (1941) *J. Am. Chem. Soc.*, **63**, 2316–2320.

8 Gilman, H. and Lichtenwalter, M. (1939) *J. Am. Chem. Soc.*, **61**, 2316–2320.

9 Bennett, G.M. and Turner, E.E. (1914) *J. Chem. Soc.Trans.*, 1057–1062.

10 Kochi, J.K. and Tamura, M. (1971) *J. Am. Chem. Soc.*, **93**, 1483–1485.

11 Kochi, J.K. and Tamura, M. (1971) *J. Am. Chem. Soc.*, **93**, 1485–1487.

12 Tamura, M. and Kochi, J.K. (1971) *J. Am. Chem. Soc.*, **93**, 1487–1489.

13 Tamao, K., Sumitani, K., and Kumada, M. (1972) *J. Am. Chem. Soc.*, **94**, 4374–4376.

14 Corriu, R.J.P. and Masse, J.P. (1972) *J. Chem. Soc., Chem. Commun.*, 144.

15 Chinchilla, R. and Nájera, C. (2007) *Chem. Rev.*, **107**, 874–922.

16 (a) Sonogashira, K., Tohda, Y., and Hagihara, N. (1975) *Tetrahedron Lett.*, **16**, 4467–4470; (b) Cassar, L. (1975) *J. Organomet. Chem.*, **93**, 253; (c) Dieck, H.A. and Heck, R.F. (1975) *J. Organomet. Chem.*, **93**, 259.

17 Stephens, R.D. and Castro, C.E. (1963) *J. Org. Chem.*, **28**, 3313–3315.

18 (a) Tianrui, R., Ye, Z., Weiwen, Z., and Jiaju, Z. (2007) *Synth. Commun.*, **37**, 3279–3290; (b) Fukuyama, T., Shinmen, M., Nishitani, S., Sato, M., and Ryu, I. (2002) *Org. Lett.*, **10**, 1691–1694; (c) Roya, S. and Plenioa, H. (2010) *Adv. Synth. Catal.*, **352**, 1014–1022.

19 Baba, S. and Negishi, E. (1976) *J. Am. Chem. Soc.*, **98**, 6729–6731.

20 Negishi, E., King, A.O., and Okukado, N. (1977) *J. Org. Chem.*, **42**, 1821–1823.

21 Kosugi, M., Shimizu, Y., and Migita, T. (1977) *Chem. Lett.*, 1423.

22 Milstein, D. and Stille, J.K. (1978) *J. Am. Chem. Soc.*, **100**, 3636–3638.

23 Miyaura, N. and Suzuki, A. (1979) *J. Chem. Soc., Chem. Commun.*, 866–867.

24 Suzuki, A. (1982) *Acc. Chem. Res.*, **15**, 178–184.

25 Yanagi, T., Miyaura, N., and Suzuki, A. (1981) *Synth. Commun.*, **11**, 513.

26 Miyaura, N. and Suzuki, A. (1995) *Chem. Rev.*, **95**, 2457–2483.

27 Smith, G.B., Dezeny, G.C., Hughes, D.L., King, A.O., and Verhoeven, T.R. (1994) *J. Org. Chem.*, **59**, 8151–8156.

28 Hatanaka, Y. and Hiyama, T. (1988) *J. Org. Chem.*, **53**, 918–920.

29 (a) Louie, J. and Hartwig, J.F. (1995) *Tetrahedron Lett.*, **36**, 3609; (b) Guram, A.S., Rennels, R.A., and Buchwald, S.L. (1995) *Angew. Chem., Int. Ed. Engl.*, **34**, 1348; (c) Wolfe, J.P., Wagaw, S., and Buchwald, S.L. (1996) *J. Am. Chem. Soc.*, **118**, 7215–7216; (d) Driver, M.S. and Hartwig, J.F. (1996) *J. Am. Chem. Soc.*, **118**, 7217–7218.

30 (a) Paul, F., Patt, J., and Hartwig, J.F. (1994) *J. Am.Chem. Soc.*, **116**, 5969–5970; (b) Guram, A.S. and Buchwald, S.L. (1994) *J. Am. Chem. Soc.*, **116**, 7901–7902.

31 Kosugi, M., Kameyama., M., and Migita, T. (1983) *Chem. Lett.*, 927–928.

32 Heck, R.F. and Nolley, J.P. (1972) *J. Org. Chem.*, **37**, 2320.

33 Mizoroki, T., Mori, K., and Ozaki, A. (1971) *Bull. Chem. Soc. Jpn.*, **44**, 581.

34 (a) Trost, B.M. and Verhoeven, T.R. (1982) *Comprehensive Organometallic Chemistry*, vol. 8 (eds G. Wilkinson, E.W. Able, and F.G.A. Stone), p. 799; (b) Tsuji, J., Takahashi, H., and Morikawa, M. (1965) *Tetrahedron Lett.*, 4387.

35 Trost, B.M. and Crawley, M.L. (2003) *Chem. Rev.*, **103**, 2921.

36 Echavarren, A.M. and Cárdenas, D.J. (2004) *Metal-Catalyzed Cross-Coupling Reactions*, 2nd edn (eds A. de Meijere and F. Diederich), Wiley-VCH Verlag GmbH, Weinheim, pp. 1–40.

37 Amatore, C. and Jutand, A. (2000) *Acc. Chem. Res.*, **33**, 314.

38 Amatore, C., Carre, E., Jutand, A., and M'Barki, M.A. (1995) *Organometallics*, **14**, 1818.

39 Fors, B.P., Krattiger, P., Strieter, E., and Buchwald, S.L. (2008) *Org. Lett.*, **10**, 3505–3508.

40 Mace, Y., Kapdi, A.R., Fairlamb, I.J.S., and Jutand, A. (2006) *Organometallics*, **25**, 1795.

41 Alcazar-Roman, L.M., Hartwig, J.F., Rheingold, A.L., Liable-Sands, L.M., and Guzei, A. (2000) *J. Am. Chem. Soc.*, **122**, 4618–4630.

42 Shekhar, S., Ryberg, P., Hartwig, J.F., Mathew, J.S., Blackmond, D.G., Strieter, E.R., and Buchwald, S.L. (2006) *J. Am. Chem. Soc.*, **128**, 3584.

43 Klingensmith, L.M., Strieter, E.R., Barder, T.E., and Buchwald, S.L. (2006) *Organometallics*, **25**, 82–91.

44 Gómez Andreu, M., Zapf, A., and Beller, M. (2000) *Chem. Commun.*, 2475–2476.

45 Hartwig, J.F. and Paul, F. (1995) *J. Am. Chem. Soc.*, **117**, 5373–5374.

46 Barrios-Landeros, F. and Hartwig, J.F. (2005) *J. Am. Chem. Soc.*, **127**, 6944–6945.

47 Barrios-Landeros, F., Carrow, B.P., and Hartwig, J.F. (2009) *J. Am. Chem. Soc.*, **131**, 8141–8154.

48 Shekhar, S. and Hartwig, J.F. (2007) *Organometallics*, **26**, 340–351.

49 Barrios-Landeros, F., Carrow, B.P., and Hartwig, J.F. (2008) *J. Am. Chem. Soc.*, **130**, 5842–5843.

50 (a) Yamashita, M. and Hartwig, J.F. (2004) *J. Am. Chem. Soc.*, **126**, 5344; (b) Mann, G., Shelby, Q., Roy, A.H., and Hartwig, J.F. (2003) *Organometallics*, **22**, 2775; (c) Hartwig, J.F. (2007) *Inorg. Chem.*, **46**, 1936.

51 (a) Louie, J. and Hartwig, J.F. (1996) *Angew. Chem., Int. Ed.*, **35**, 2359–2361; (b) Hartwig, J.F., Richards, S., Barañano, D., and Paul, F. (1996) *J. Am. Chem. Soc.*, **118**, 3626.

52 Hartwig, J.F. (1996) *J. Am. Chem. Soc.*, **118**, 7010–7011.

53 Christensen, H., Kiil, S., Dam-Johansen, K., Nielsen, O., and Sommer, M.B. (2006) *Org. Process Res. Dev.*, **10**, 762–769.

54 Beletskaya, I.P., Bessmertnykh, A.G., and Guilard, R. (1999) *Tetrahedron Lett.*, **40**, 6393–6397.

55 Fors, B.P., Davis, N.R., and Buchwald, S.L. (2009) *J. Am. Chem. Soc.*, **131**, 5766–5768.

56 Shen, Q., Shekhar, S., Stambuli, J.P., and Hartwig, J.F. (2005) *Angew. Chem. Int. Ed.*, **44**, 1371–1375.

57 Shen, Q. and Hartwig, J.F. (2008) *Org. Lett.*, **10**, 4109–4112.

58 (a) Damon, D.B., Dugger, R.W., Hubbs, S.E., Scott, J.M., and Scott, R.W. (2006) *Org. Process Res. Dev.*, **10**, 472; (b) Huang, X., Anderson, K.W., Zim, D., Jiang, L., Klapars, A., and Buchwald, S.L. (2003) *J. Am. Chem. Soc.*, **125**, 6653–6655.

59 Kamikawa, T. and Hayashi, T. (1997) *Tetrahedron Lett.*, **38**, 7087–7090.

60 Espino, G., Kurbangalieva, A., and Brown, J.M. (2007) *Chem. Commun.*, 1742–1744.

61 Johns, A.M., Utsunomiya, M., Incarvito, C.D., and Hartwig, J.F. (2006) *J. Am. Chem. Soc.*, **128**, 1828–1839.

62 Vo, G.D. and Hartwig, J.F. (2008) *Angew. Chem. Int. Ed.*, **47**, 2127–2130.

63 Ozawa, F., Kubo, A., and Hayashi, T. (1992) *Chem. Lett.*, 2177.

64 Barder, T.E. and Buchwald, S.L. (2007) *J. Am. Chem. Soc.*, **129**, 5096–5101.

65 Van Strijdonck, G.P.F., Boele, M.D.K., Kamer, P.C.J., de Vries, J.G., and van Leeuwen, P.W.N.M. (1999) *Eur. J. Inorg. Chem.*, 1073.

66 Batsanov, A.S., Knowles, J.P., and Whiting, A. (2007) *J. Org. Chem.*, **72**, 2525–2532.

67 Ogata, T. and Hartwig, J.F. (2008) *J. Am. Chem. Soc.*, **130**, 13848–13849.

68 Marion, N., Navarro, O., Mei, J., Stevens, E.D., Scott, N.M., and Nolan, S.P. (2006) *J. Am. Chem. Soc.*, **128**, 4101.

69 Guari, Y., van Strijdonck, G.P.F., Boele, M.D.K., Reek, J.N.H., Kamer, P.C.J., and van Leeuwen, P.W.N.M. (2001) *Chem. Eur. J.*, **7**, 475.

70 Stambuli, J.P., Kuwano, R., and Hartwig, J.F. (2002) *Angew. Chem. Int. Ed.*, **41**, 4746–4748.

71 Paul, F., Patt, J., and Hartwig, J.F. (1995) *Organometallics*, **14**, 3030–3039.

72 Widenhoefer, R.A., Zhong, H.A., and Buchwald, S.L. (1997) *J. Am. Chem.Soc.*, **119**, 6787–6795.

73 Zuideveld, M.A., Swennenhuis, B.H.G., Boele, M.D.K., Guari, Y., van Strijdonck, G.P.F., Reek, J.N.H., Kamer, P.C.J., Goubitz, K., Fraanje, J., Lutz, M., Spek, A.L., and van Leeuwen, P.W.N.M. (2002) *J. Chem. Soc., Dalton Trans.*, 2308–2318.

74 Stauffer, S.R., Lee, S., Stambuli, J.P., Hauck, S.I., and Hartwig, J.F. (2000) *Org. Lett.*, **2**, 1423–1426.

75 Viciu, M.S., Germaneau, R.F., Navarro-Fernandez, O., Stevens, E.D., and Nolan, S.P. (2002) *Organometallics*, **21**, 5470–5472.

76 Alvaro, E. and Hartwig, J.F. (2009) *J. Am. Chem. Soc.*, **131**, 7858–7868.

77 Fernández-Rodríguez, M.A. and Hartwig, J.F. (2009) *J. Org. Chem.*, **74**, 1663–1672.

78 Li, H., Grasa, G.A., and Colacot, T.J. (2010) *Org. Lett.*, **12**, 3332–3335.

79 Beletskaya, I.P. and Cheprakov, A.V. (2004) *J. Organomet. Chem.*, **689**, 4055–4082.

80 Herrmann, W.A., Brossmer, C., Oefele, K., Reisinger, C.-P., Priermeier, T., Beller, M., and Fischer, H. (1995) *Angew. Chem. Int. Ed. Engl.*, **34**, 1844–1848.

81 Herrmann, W.A., Brossmer, C., Reisinger, C.-P., Riermeier, T.H., Oefele, K., and Beller, M. (1997) *Chem.Eur. J.*, **3**, 1357–1364.

82 Shaw, B.L., Perera, S.D., and Staley, E.A. (1998) *Chem. Commun.*, 1362–1363.

83 Canty, A.J. (1992) *Acc. Chem. Res.*, **25**, 83.

84 Louie, J. and Hartwig, J.F. (1996) *Angew. Chem. Int. Ed.*, **35**, 2359–2361.

85 Heck, R.F. (1979) *Acc. Chem. Res.*, **12**, 146.

86 Spencer, A. (1983) *J. Organomet. Chem.*, **258**, 101–108.

87 Mitsudo, T., Fischetti, W., and Heck, R.F. (1984) *J. Org. Chem.*, **49**, 1640.

88 Rosner, T., Le Bars, J., Pfaltz, A., and Blackmond, D.G. (2001) *J. Am. Chem. Soc.*, **123**, 1848–1855.

89 Nadri, S., Joshaghani, M., and Rafiee, E. (2009) *Organometallics*, **28**, 6281–6287.

90 Rosner, T., Pfaltz, A., and Blackmond, D.G. (2001) *J. Am. Chem. Soc.*, **123**, 4621–4622.

91 Jung, E., Park, K., Kim, J., Jung, H.-T., Oh, I.-K., and Lee, S. (2010) *Inorg. Chem. Commun.*, **13**, 1329–1331.

92 Amatore, C., Godin, B., Jutand, A., and Lemaitre, F. (2007) *Chem. Eur. J.*, **13**, 2002–2011.

93 Ohff, M., Ohff, A., van der Boom, M.E., and Milstein, D. (1997) *J. Am. Chem. Soc.*, **119**, 11687–11688.

94 Beller, M. and Riermeier, T.H. (1998) *Eur. J. Inorg. Chem.*, 29–35.

95 Herrmann, W.A., Oefele, K., von Preysing, D., and Schneider, S.K. (2003) *J. Organomet. Chem.*, **687**, 229–248.

96 Frey, G.D., Reisinger, C.-P., Herdtweck, E., and Herrmann, W.A. (2005) *J. Organomet. Chem.*, **690**, 3193–3201.

97 Albisson, D.A., Bedford, R.B., Lawrence, S.E., and Scully, P.N. (1998) *Chem. Commun*, 2095.

98 Bedford, R.B., Hazelwood (née Welch), S.L., Horton, P.N., and Hursthouse, M.B. (2003) *Dalton Trans.*, 4164–4174.

99 Bruce, M.L., Goodall, B.L., and Stone, F.G.A. (1973) *J. Chem. Soc., Chem. Commun.*, 558–559.

100 Lewis, L.N. (1986) *J. Am. Chem. Soc.*, **108**, 743–749.

101 Coolen, H.K.A.C., Nolte, R.J.M., and van Leeuwen, P.W.N.M. (1995) *J. Organomet. Chem.*, **496**, 159–168.

102 Ohff, M., Ohff, A., and Milstein, D. (1999) *Chem. Commun.*, 357–358.

103 Bedford, R.B. (2003) *Chem. Commun.*, 1787–1796.

104 Bergbreiter, D.E., Osburn, P.L., Wilson, A., and Sink, E.M. (2000) *J. Am. Chem.Soc.*, **122**, 9058.

105 Nowotny, M., Hanefeld, U., van Koningsveld, H., and Maschmeyer, T. (2000) *Chem. Commun.*, 1877.

106 Bedford, R.B., Hazelwood, S.L., Limmert, M.E., Brown, J.M., Ramdeehul, S., Cowley, A.R., Coles, S.J., and Hursthouse, M.B. (2003) *Organometallics*, **22**, 1364–1371.

107 McGuinness, D.S. and Cavell, K.J. (2000) *Organometallics*, **19** (5), 741–748.

108  Zapf, A. and Beller, M. (2001) *Chem. Eur. J.*, **7**, 2908–2915.

109  Biscoe, M.R., Fors, B.P., and Buchwald, S.L. (2008) *J. Am. Chem. Soc.*, **130**, 6686–6687.

110  Portnoy, M., Ben-David, Y., and Milstein, D. (1993) *Organometallics*, **12**, 4734.

111  (a) Reddy, N.P. and Tanaka, M. (1997) *Tetrahedron Lett.*, **38**, 4807; (b) Nishiyama, M., Yamamoto, T., and Koie, Y. (1998) *Tetrahedron Lett.*, **39**, 617; (c) Yamamoto, T., Nishiyama, M., and Koie, Y. (1998) *Tetrahedron Lett.*, **39**, 2367; (d) Old, D.W., Wolfe, J.P., and Buchwald, S.L. (1998) *J. Am. Chem. Soc.*, **120**, 9722; (e) Littke, A.F. and Fu, G.C. (1998) *Angew. Chem. Int. Ed.*, **37**, 3387.

112  Barder, T.E., Walker, S.D., Martinelli, J.R., and Buchwald, S.L. (2005) *J. Am. Chem. Soc.*, **127**, 4685.

113  Alcazar-Roman, L.M. and Hartwig, J.F. (2001) *J. Am. Chem. Soc.*, **123**, 12905.

114  Weck, M. and Jones, C.W. (2007) *Inorg. Chem.*, **46**, 1865–1875.

115  Guari, Y., van Es, D.S., Reek, J.N.H., Kamer, P.C.J., and van Leeuwen, P.W.N.M. (1999) *Tetrahedron Lett.*, **40**, 3789.

116  Vrieze, K., Volger, H.C., and van Leeuwen, P.W.N.M. (1969) *Inorg. Chim. Acta, Rev.*, **3**, 109–128.

117  Masters, C. and Visser, J.P. (1974) *J. Chem. Soc., Chem. Commun.*, 932–933.

118  Zuidema, E., van Leeuwen, P.W.N.M., and Bo, C. (2007) *J. Am. Chem. Soc.*, **129**, 3989–4000.

119  Farina, V. and Krishnan, B. (1991) *J. Am. Chem. Soc.*, **113**, 9585–9595.

120  Casares, J.A., Espinet, P., and Salas, G. (2002) *Chem. Eur. J.*, **8**, 4844–4853.

121  Amatore, C., Bahsoun, A.A., Jutand, A., Meyer, G., Ndedi Ntepe, A., and Ricard, L. (2003) *J. Am. Chem. Soc.*, **125**, 4212–4222.

122  Casado, A.L., Espinet, P., Gallego, A.M., and Martínez-Ilarduya, J.M. (2000) *J. Am. Chem. Soc.*, **122**, 11771–11782.

123  Cotter, W.D., Barbour, L., McNamara, K.L., Hechter, R., and Lachicotte, R.J. (1998) *J. Am. Chem. Soc.*, **120**, 11016–11017.

124  Adamo, C., Amatore, C., Ciofini, I., Jutand, A., and Lakmini, H. (2006) *J. Am. Chem. Soc.*, **128**, 6829–6836.

125  Rodríguez, N., Cuenca, A., Ramírez de Arellano, C., Medio-Simon, M., and Asensio, G. (2003) *Org. Lett.*, **10**, 1705–1708.

126  Talhami, A., Penn, L., Jaber, N., Hamza, K., and Blum, J. (2006) *Appl. Catal. A: Gen.*, **312**, 115–119.

127  Miller, W.D., Fray, A.H., Quatroche, J.T., and Sturgill, C.D. (2007) *Org. Proc. Res. Dev.*, **11**, 359–364.

128  Clapham, K.M., Batsanov, A.S., Bryce, M.R., and Tarbit, B. (2009) *Org. Biomol. Chem.*, **7**, 2155–2161.

129  Cravotto, G., Palmisano, G., Tollari, S., Nano, G.M., and Penoni, A. (2005) *Ultrasonics Sonochem.*, **12**, 91–94.

130  Jin, Z., Guo, S.-X., Gu, X.-P., Qiu, L.-L., Song, H.-B., and Fanga, J.-X. (2009) *Adv. Synth. Catal.*, **351**, 1575–1585.

131  Kedia, S.B. and Mitchell, M.B. (2009) *Org. Proc. Res. Dev.*, **13**, 420–428.

132  Blackmond, D. (2006) *J. Org. Chem.*, **71**, 4711.

133  Coifini, I. (2006) *J. Am. Chem. Soc.*, **128**, 6829.

134  Demir, A.S., Reis, O., and Emrullahoglu, M. (2004) Abstracts of Papers, 228th ACS National Meeting, Philadelphia, PA, United States, August 22–26, 2004 ORGN-064.

135  Brimble, M.A. and Lai, M.Y.H. (2003) *Org. Biomol. Chem.*, **1**, 2084–2095.

136  Genêt, J.-P., Darses, S., Brayer, J.-L., and Demoute, J.-P. (1997) *Tetrahedron Lett.*, **25**, 4393–4396.

137  Darses, S. and Genet, J.-P. (2008) *Chem. Rev.*, **108**, 288–325.

138  Batey, R.A. and Thadani, A.N. (2002) *Org. Lett.*, **4**, 3827–3830.

139  Vedejs, E., Chapman, R.W., Fields, S.C., Lin, S., and Schrimpf, M.R. (1995) *J. Org. Chem.*, **60**, 3020–3027.

140  Molander, G.A. and Ellis, N. (2007) *Acc. Chem. Res.*, **40**, 275–286.

141  Billingsley, K.L. and Buchwald, S.L. (2008) *Angew. Chem. Int. Ed.*, **47**, 4695–4698.

142  (a) Li, G.Y. and Marshall, W.J. (2002) *Organometallics*, **21**, 590–591; (b) Li, G.Y.

(2001) *Angew. Chem. Int. Ed.*, **40**, 1513–1516.

143 Calhorda, M.J., Brown, J.M., and Cooley, N.A. (1991) *Organometallics*, **10**, 1431.

144 (a) Mann, G., Baranano, D., Hartwig, J.F., Rheingold, A.L., and Guzei, I.A. (1998) *J. Am. Chem. Soc.*, **120**, 9205; (b) Kondo, T. and Mitsudo, T.-a. (2000) *Chem. Rev.*, **100**, 3205; (c) Fernandez-Rodriguez, M.A., Shen, Q., and Hartwig, J.F. (2006) *Chem. Eur. J.*, 7782.

145 Kantchev, E.A.B., O'Brien, C.J., and Organ, M.G. (2007) *Angew. Chem. Int. Ed.*, **46**, 2768.

146 Jin, L., Zhang, H., Li, P., Sowa, J.R. Jr., and Lei, A. (2009) *J. Am. Chem. Soc.*, **131**, 9892–9893.

147 (a) Gillie, A. and Stille, J.K. (1980) *J. Am. Chem. Soc.*, **102**, 4933; (b) Moravskiy, A. and Stille, J.K. (1981) *J. Am. Chem. Soc.*, **103**, 4147; (c) Ozawa, F., Ito, T., and Yamamoto, A. (1980) *J. Am. Chem. Soc.*, **102**, 6457; (d) Tatsumi, K., Hoffmann, R., Yamamoto, A., and Stille, J.K. (1981) *Bull. Chem. Soc. Jpn.*, **54**, 1857.

148 (a) Stambuli, J.P., Incarvito, C.D., Buhl, M., and Hartwig, J.F. (2004) *J. Am. Chem. Soc.*, **126**, 1184; (b) Yamashita, M., Takamiya, I., Jin, K., and Nozaki, K. (2006) *J. Organomet. Chem.*, **691**, 3189.

149 Christmann, U. and Vilar, R. (2005) *Angew. Chem. Int. Ed.*, **44**, 366–374.

150 (a) Fox, J.M., Huang, X., Chieffi, A., and Buchwald, S.L. (2000) *J. Am. Chem. Soc.*, **122**, 1360–1370; (b) Birkholz Gensow, M.-N., Freixa, Z., and van Leeuwen, P.W.N.M. (2009) *Chem. Soc. Rev.*, **38**, 1099–1118.

151 Hartwig, J.F. (2008) *Acc. Chem. Res.*, **41**, 1534–1544.

152 Díez-González, S. and Nolan, S.P. (2007) *Top. Organomet. Chem.*, **21**, 47–82.

153 Caporali, M., Mueller, C., Staal, B.B.P., Tooke, D.M., Spek, A.L., and van Leeuwen, P.W.N.M. (2005) *Chem. Commun.*, 3478–3480.

154 Wolfe, J.P., Singer, R.A., Yang, B.H., and Buchwald, S.L. (1999) *J. Am. Chem. Soc.*, **121**, 9550–9561.

155 Tomori, H., Fox, J.M., and Buchwald, S.L. (2000) *J. Org. Chem.*, **65**, 5334–5341.

156 Kaye, S., Fox, J.M., Hicks, F.A., and Buchwald, S.L. (2001) *Adv. Synth. Catal.*, **343**, 789–794.

157 (a) Nishida, G., Noguchi, K., Hirano, M., and Tanaka, K. (2007) *Angew. Chem.*, **119**, 4025–4028; (b) Kondoh, A., Yorimitsu, H., and Oshima, K. (2007) *J. Am. Chem. Soc.*, **129**, 6996–6997; (c) Ashburn, B.O., Carter, R.G., and Zakharov, L.N. (2007) *J. Am. Chem. Soc.*, **129**, 9109–9116; (d) Ashburn, B.O. and Carter, R.G. (2006) *Angew. Chem.*, **118**, 6889–6893.

158 Buchwald, S.L., Mauger, C., Mignani, G., and Scholz, U. (2006) *Adv. Synth. Catal.*, **348**, 23–39.

159 Mauger, C.C. and Mignani, G.A. (2004) *Org. Proc. Res. Dev.*, **8**, 1065–1071.

160 (a) Singer, R.A., Dore, M., Sieser, J.E., and Berliner, M.A. (2006) *Tetrahedron Lett.*, **47**, 3727–3731; (b) Rataboul, F., Zapf, A., Jackstell, R., Harkal, S., Riermeier, T., Monsees, A., Dingerdissen, U., and Beller, M. (2004) *Chem. Eur. J.*, **10**, 2983–2990; (c) So, C.M., Lau, C.P., and Kwong, F.Y. (2007) *Org. Lett.*, **9**, 2795–2798; (d) Harkal, S., Rataboul, F., Zapf, A., Fuhrmann, C., Riermeier, T., Monsees, A., and Beller, M. (2004) *Adv. Synth. Catal.*, **346**, 1742–1748; (d) Littke, A.F., Dai, C., and Fu, G.C. (2000) *J. Am. Chem. Soc.*, **122**, 4020–4028.

161 Martin, R., Anderson, K.W., Tundel, R.E., Ikawa, T., Altman, R.A., and Buchwald, S.L. (2006) *Angew. Chem. Int. Ed.*, **45**, 6523–6527.

162 Strieter, E.R. and Buchwald, S.L. (2006) *Angew. Chem. Int. Ed.*, **45**, 925–928.

163 Barder, T.E., and Buchwald, S.L. (2007) *J. Am. Chem. Soc.*, **129**, 5096–5101.

164 Fors, B.P., Watson, D.A., Biscoe, M.R., and Buchwald, S.L. (2008) *J. Am. Chem. Soc.*, **130**, 13552–13554.

165 Barder, T.E. and Buchwald, S.L. (2007) *J. Am. Chem. Soc.*, **129**, 12003–12010.

166 Barder, T.E., Walker, S.D., Martinelli, J.R., and Buchwald, S.L. (2005) *J. Am. Chem. Soc.*, **127**, 4685–4696.

167 (a) Yamamoto, T., Yamamoto, A., and Ikeda, S. (1971) *J. Am. Chem. Soc.*, **93**,

3350–3359; (b) Jensen, A.E. and Knochel, P. (2002) *J. Org. Chem.*, **67**, 79–85.

168 (a) Giovannini, R., Stüdemann, T., Dussin, G., and Knochel, P. (1998) *Angew. Chem. Int. Ed.*, **37**, 2387–2390; (b) Grundl, M.A., Kennedy-Smith, J.J., and Trauner, D. (2005) *Organometallics*, **24**, 2831–2833.

169 Fairlamb, I.J.S. (2008) *Org. Biomol. Chem.*, **6**, 3645–3656.

170 Luo, X., Zhang, H., Duan, H., Liu, Q., Zhu, L., Zhang, T., and Lei, A. (2007) *Org. Lett.*, **9**, 4571–4574.

171 Zhang, H., Luo, X., Wongkhan, K., Duan, H., Li, Q., Zhu, L., Wang, J., Batsanov, A.S., Howard, J.A.K., Marder, T.B., and Lei, A. (2009) *Chem. Eur. J.*, **15**, 3823–3829.

172 Marcone, J.E. and Moloy, K.G. (1998) *J. Am. Chem. Soc.*, **120**, 8527.

173 Zuidema, E., van Leeuwen, P.W.N.M., and Bo, C. (2005) *Organometallics*, **24**, 3703.

174 (a) Hayashi, T., Konishi, M., Kobori, Y., Kumada, M., Higushi, T., and Hirotsu, K. (1984) *J. Am. Chem. Soc.*, **106**, 158; (b) Ogasawara, M., Yoshida, K., and Hayashi, T. (2000) *Organometallics*, **19**, 1567.

175 Kranenburg, M., Kamer, P.C.J., and van Leeuwen, P.W.N.M. (1998) *Eur. J. Inorg. Chem.*, 155.

176 (a) Reductive elimination: Brown, J.M. and Cooley, N.A. (1988) *Chem. Rev.*, **88**, 1031; (b) Hartwig, J.F. (1998) *Acc. Chem. Res.*, **31**, 852; (c) Fujita, K.-i., Yamashita, M., Puschmann, F., Alvarez-Falcon, M.M., Incarvito, C.D., and Hartwig, J.F. (2006) *J. Am. Chem. Soc.*, **128**, 9044–9045.

177 Bite angle effect on reductive elimination: Brown, J.M. and Guiry, P.J. (1994) *Inorg. Chim. Acta*, **220**, 249.

178 Otsuka, S. (1980) *J. Organomet.Chem.*, **200**, 191.

179 For platinum see: Abis, L., Santi, R., and Halpern, J. (1981) *J. Organomet. Chem.*, **215**, 263.

180 (a) Barañano, D. and Hartwig, J.F. (1995) *J. Am. Chem. Soc.*, **117**, 2937; (b) Widenhoefer, R.A., Zhong, H.A., and Buchwald, S.L. (1997) *J. Am. Chem. Soc.*,

119, 6787; (c) Driver, M.S. and Hartwig, J.F. (1997) *J. Am. Chem. Soc.*, **119**, 8232.

181 Yamashita, M., Cuevas Vicario, J.V., and Hartwig, J.F. (2003) *J. Am. Chem. Soc.*, **125**, 16347.

182 (a) Dekker, G.P.C.M., Buijs, A., Elsevier, C.J., Vrieze, K., van Leeuwen, P.W.N.M., Smeets, W.J.J., Spek, A.L., Wang, Y.F., and Stam, C.H. (1992) *Organometallics*, **11**, 1937; (b) van Leeuwen, P.W.N.M., Roobeek, C.F., and van der Heijden, H. (1994) *J. Am. Chem. Soc.*, **116**, 12117; (c) van Leeuwen, P.W.N.M. and Roobeek, C.F. (1995) *Rec. Trav. Chim. Pays-Bas*, **114**, 73.

183 Fitton, P. and Rick, E.A. (1971) *J. Organomet. Chem.*, **28**, 287.

184 Shen, Q. and Hartwig, J.F. (2007) *J. Am. Chem. Soc.*, **129**, 7734.

185 Tolman, C.A., Seidel, W.C., Druliner, J.D., and Domaille, P.J. (1984) *Organometallics*, **3**, 33.

186 Hooper, M.W. and Hartwig, J.F. (2003) *Organometallics*, **22**, 3394.

187 Ikawa, T., Barder, T.E., Biscoe, M.R., and Buchwald, S.L. (2007) *J. Am. Chem. Soc.*, **129**, 13001–13007.

188 Culkin, D.A. and Hartwig, J.F. (2004) *Organometallics*, **23**, 3398.

189 van Leeuwen, P.W.N.M., Zuideveld, M.A., Swennenhuis, B.H.G., Freixa, Z., Kamer, P.C.J., Goubitz, K., Fraanje, J., Lutz, M., and Spek, A.L. (2003) *J. Am. Chem. Soc.*, **125**, 5523.

190 (a) van der Veen, L.A., Boele, M.D.K., Bregman, F.R., Kamer, P.C.J., van Leeuwen, P.W.N.M., Goubitz, K., Fraanje, J., Schenk, H., and Bo, C. (1998) *J. Am. Chem. Soc.*, **120**, 11616; (b) Carbo, J.J., Maseras, F., Bo, C., and van Leeuwen, P.W.N.M. (2001) *J. Am. Chem. Soc.*, **123**, 7630; (c) Zuidema, E., Escorihuela, L., Fichelsheim, T., Carbó, J.J., Bo, C., Kamer, P.C.J., and van Leeuwen, P.W.N.M. (2008) *Chem. Eur. J.*, **14**, 1843; (d) Zuidema, E., Daura-Oller, E., Carbo, J.J., Bo, C., and van Leeuwen, P.W.N.M. (2007) *Organometallics*, **26**, 2234.

191 Murata, M. and Buchwald, S.L. (2004) *Tetrahedron*, **60**, 7397.

192 Hamann, B.C. and Hartwig, J.F. (1997) *J. Am. Chem. Soc.*, **119**, 12382.

193 Roy, A.H. and Hartwig, J.F. (2004) *Organometallics*, **23**, 194.

194 Fihri, A., Meunier, P., and Hierso, J.-C. (2007) *Coord. Chem. Rev.*, **251**, 2017.

195 (a) Hartwig, J.F. (1999) *Pure Appl. Chem.*, **8**, 1417; (b) Prim, D., Campagne, J.-M., Joseph, D., and Andrioletti, B. (2002) *Tetrahedron*, **58**, 2041; (c) Ley, S.V. and Thomas, A.W. (2003) *Angew. Chem. Int. Ed.*, **42**, 5400; (d) Beletskaya, I.P. and Cheprakov, A.V. (2004) *Chem. Rev.*, **248**, 2337; (e) Beletskaya, I.P. (2005) *Pure Appl. Chem.*, **77**, 2021.

196 (a) Eastham, G.R., Heaton, B.T., Iggo, J.A., Tooze, R.P., Whyman, R., and Zacchini, S. (2000) *Chem. Commun.*, 609; (b) Clegg, W., Eastham, G.R., Elsegood, M.R.J., Heaton, B.T., Iggo, J.A., Tooze, R.P., Whyman, R., and Zacchini, S. (2002) *Organometallics*, **21**, 1832.

197 Yin, J. and Buchwald, S.L. (2002) *J. Am. Chem. Soc.*, **124**, 6043.

198 Zheng, N., McWilliams, J.C., Fleitz, F.J., Armstrong, J.D. III, and Volante, R.P. (1998) *J. Org. Chem.*, **63**, 9606.

199 Arthur, K.L., Wang, Q.L., Bregel, D.M., Smythe, N.A., O'Neill, B.A., Goldberg, K.I., and Moloy, K.G. (2005) *Organometallics*, **24**, 4624.

200 Mul, M.P., van der Made, A.W., Smaardijk, A.B., and Drent, E. (2003) *Catalytic Synthesis of Alkene-Carbon Monoxide Copolymers and Cooligomers* (ed. A. Sen), Kluwer Academic Publishers, Dordrecht, pp. 87–140.

201 Hamann, B.C. and Hartwig, J.F. (1998) *J. Am. Chem. Soc.*, **120**, 3694.

202 Kaganovsky, L., Gelman, D., and Rueck-Braun, K. (2010) *J. Organometal. Chem.*, **695**, 260–266.

203 Smith, R.C., Bodner, C.R., Earl, M.J., Sears, N.C., Hill, N.E., Bishop, L.M., Sizemore, N., Hehemann, D.T., Bohn, J.J., and Protasiewicz, J.D. (2005) *J. Organometal. Chem.*, **690**, 477–481.

204 Hills, I.D. and Fu, G.C. (2004) *J. Am. Chem. Soc.*, **126**, 13178–13179.

205 Shen, Q., Ogata, T., and Hartwig, J.F. (2008) *J. Am. Chem. Soc.*, **130**, 6586–6596.

206 Vo, G.D. and Hartwig, J.F. (2009) *J. Am. Chem. Soc.*, **131**, 11049–11061.

207 Klinkenberg, J.L. and Hartwig, J.F. (2010) *J. Am. Chem. Soc.*, **132**, 11830–11833.

208 Grushin, V.V. (2002) *Chem. Eur. J.*, **8**, 1006–1014.

209 Fraser, S.L., Antipin, M.Yu., Khroustalyov, V.N., and Grushin, V.V. (1997) *J. Am. Chem. Soc.*, **119**, 4769–4770.

210 Furuya, T. and Ritter, T. (2008) *J. Am. Chem. Soc.*, **130**, 10060–10061.

211 Ball, N.D. and Sanford, M.S. (2009) *J. Am. Chem. Soc.*, **131**, 3796–3797.

212 Grushin, V.V. (2010) *Acc. Chem. Res.*, **43**, 160–171.

213 van Leeuwen, P.W.N.M., Roobeek, C.F., and Orpen, A.G. (1990) *Organometallics*, **9**, 2179.

214 Macgregor, S.A., Roe, D.C., Marshall, W.J., Bloch, K.M., Bakhmutov, V.I., and Grushin, V.V. (2005) *J. Am. Chem. Soc.*, **127**, 15304–15321.

215 Macgregor, S.A. (2007) *Chem. Soc. Rev.*, **36**, 67–76.

216 Macgregor, S.A. and Wondimagegn, T. (2007) *Organometallics*, **26**, 1143–1149.

217 Yandulov, D.V. and Tran, N.T. (2007) *J. Am. Chem. Soc.*, **129**, 1342–1358.

218 Grushin, V.V. and Marshall, W.J. (2007) *Organometallics*, **26**, 4997–5002.

219 Marshall, W.J. and Grushin, V.V. (2003) *Organometallics*, **22**, 1591–1593.

220 Grushin, V. (2007) U.S. Patent 7, 202,388 (to DuPont); (2006) *Chem. Abstr.*, **144**, 317007.

221 Watson, D.A., Su, M., Teverovskiy, G., Zhang, Y., García-Fortanet, J., Kinzel, T., and Buchwald, S.L. (2009) *Science*, **325**, 1661–1664.

222 Grushin, V.V. and Marshall, W.J. (2006) *J. Am. Chem. Soc.*, **128**, 12644.

223 Cho, E.J., Senecal, T.D., Kinzel, T., Zhang, Y., Watson, D.A., and Buchwald, S.L. (2010) *Science*, **328**, 1679–1681.

224 Netherton, M.R. and Fu, G.C. (2001) *Org. Lett.*, **3**, 4295–4298.

**225** Lou, S. and Fu, G.C. (2010) *Adv. Synth. Catal.*, **352**, 2081–2084.

**226** Barder, T.E. and Buchwald, S.L. (2007) *J. Am. Chem. Soc.*, **129**, 5096–5601.

**227** Ozawa, F., Kubo, A., and Hayashi, T. (1992) *Chem. Lett.*, 2177.

**228** Kaneda, K., Sano, K., and Teranishi, S. (1979) *Chem. Lett.*, 821–822.

**229** (a) Kikukawa, K., Yamane, T., Takagi, M., and Matsuda, T. (1972) *J. Chem. Soc., Chem. Commun.*, 695–696; (b) Yamane, T., Kikukawa, K., Takagi, M., and Matsuda, T. (1973) *Tetrahedron*, **29**, 955; (c) Asano, R., Moritani, I., Fujiwara, Y., and Teranishi, S. (1973) *Bull. Chem. Soc. Jpn.*, **46**, 2910.

**230** Fahey, D.R. and Mahan, J.E. (1976) *J. Am. Chem. Soc.*, **98**, 4499–4503.

**231** Kong, K.-C. and Cheng, C.-H. (1991) *J. Am. Chem. Soc.*, **113**, 6313–6315.

**232** Wang, Y., Lai, C.W., Kwong, F.Y., Jia, W., and Chan, K.S. (2004) *Tetrahedron*, **60**, 9433–9439.

**233** Marcoux, D. and Charette, A.B. (2008) *J. Org. Chem.*, **73**, 590–593.

**234** Kikukawa, K., Takagi, M., and Matsuda, T. (1979) *Bull. Chem. Soc. Jpn.*, **52**, 1493–1497.

**235** Liao, X., Weng, Z., and Hartwig, J.F. (2008) *J. Am. Chem. Soc.*, **130**, 195–200.

**236** (a) Wallow, T.I., Seery, T.A.P., Goodson, F.E., and Novak, B.M. (1994) *Polym. Prepr. Am. Chem. Soc. Div.Polym. Chem.*, **35**, 710; (b) Novak, B.M., Wallow, T.I., Goodson, F.E., and Loos, K. (1995) *Polym. Prepr. Am. Chem. Soc. Div. Polym. Chem.*, **36**, 693; (c) Goodson, F.E., Wallow, T.I., and Novak, B.M. (1997) *J. Am. Chem. Soc.*, **119**, 12441–12453.

**237** Grushin, V.V. (2000) *Organometallics*, **19**, 1888–1900.

**238** Laven, G., Kalek, M., Jezowska, M., and Stawinski, J. (2010) *New J. Chem.*, **34**, 967–975.

**239** Wallow, T.I. and Novak, B.M. (1994) *J. Org. Chem.*, **59**, 5034–5037.

**240** Ziegler, C.B. Jr. and Heck, R.F. (1978) *J. Org. Chem.*, **43**, 2941–2946.

**241** Morita, D.K., Stille, J.K., and Norton, J.R. (1995) *J. Am. Chem. Soc.*, **117**, 8576–8581.

**242** Herrmann, W.A., Brossmer, C., Priermeier, T., and Oefele, K. (1994) *J. Organomet. Chem.*, **481**, 97–108.

**243** Herrmann, W.A., Brossmer, C., Oefele, K., Beller, M., and Fischer, H. (1995) *J. Organomet. Chem.*, **491**, C1–C4.

**244** Herrmann, W.A., Brossmer, C., Oefele, K., Beller, M., and Fischer, H. (1995) *J. Mol. Catal. A: Chem.*, **103**, 133–146.

**245** O'Keefe, D.F., Dannock, M.C., and Marcuccio, S.M. (1992) *Tetrahedron Lett.*, **33**, 6679–6680.

**246** Kong, K.-C. and Cheng, C.-H. (1991) *J. Am. Chem. Soc.*, **113**, 6313–6315.

**247** Penn, L., Shpruhman, A., and Gelman, D. (2007) *J. Org. Chem.*, **72**, 3875–3879.

**248** Shelby, Q., Kataoka, N., Mann, G., and Hartwig, J.F. (2000) *J. Am. Chem. Soc.*, **122**, 10718–10719.

**249** Erhardt, S., Grushin, V.V., Kilpatrick, A.H., Macgregor, S.A., Marshall, W.J., and Roe, D.C. (2008) *J. Am. Chem. Soc.*, **130**, 4828–4845.

**250** Tschaen, D.M., Desmond, R., King, A.O., Fortin, M.C., Pipik, B., King, S., and Verhoeven, T.R. (1994) *Synth. Commun.*, **24**, 887–890.

**251** Schareina, T., Zapf, A., and Beller, M. (2004) *Chem. Commun.*, 1388–1389.

**252** Dobbs, K.D., Marshall, W.J., and Grushin, V.V. (2007) *J. Am. Chem. Soc.*, **129**, 30–31.

**253** Ashby, E.C., Neumann, H.M., Walker, F.W., Laemmle, J., and Chao, L.-C. (1973) *J. Am. Chem. Soc.*, **95**, 3330–3337.

**254** Giordano, C. (1995) Peñiscola Meeting Cataluña Network on Homogeneous Catalysis; Giordano, C., Coppi, L., and Minisci, F. (1992) (to Zambon Group S p A,), Eur. Pat. Appl. EP 494419.(1992) *Chem. Abstr.*, **117**, 633603.

**255** Kim, J.C., Koh, Y.S., Yoon, U.C., and Kim, M.S. (1993) *J, Korean Chem. Soc.*, **37**, 228–236 (Journal written in Korean); *Chem Abstr* (1993) **119**, 95575.

**256** (a) Sherry, B.D. and Fürstner, A. (2008) *Acc. Chem. Res.*, **41**, 1500–1511; (b) Bolm, C., Legros, J., Le Paih, J., and Zani, L. (2004) *Chem. Rev.*, **104**,

6217–6254; (c) Fürstner, A. and Martin, R. (2005) *Chem. Lett.*, **34**, 624–629.

257 Neumann, S.M. and Kochi, J.K. (1975) *J. Org. Chem.*, **40**, 599–606.

258 Molander, G.A., Rahn, B.J., Shubert, D.C., and Bonde, S.E. (1983) *Tetrahedron Lett.*, **24**, 5449–5452.

259 Cahiez, G. and Avedissian, H. (1998) *Synthesis*, 1199–1205.

260 Nakamura, M., Matsuo, K., Ito, S., and Nakamura, E. (2004) *J. Am. Chem. Soc.*, **126**, 3686–3687.

261 Fürstner, A., Martin, R., Krause, H., Seidel, G., Goddard, R., and Lehmann, C.W. (2008) *J. Am. Chem. Soc.*, **130**, 8773–8787.

262 (a) Fürstner, A., Leitner, A., Méndez, M., and Krause, H. (2002) *J. Am. Chem. Soc.*, **124**, 13856–13863; (b) Fürstner, A. and Leitner, A. (2002) *Angew. Chem., Int. Ed.*, **41**, 609–612.

263 Berthon-Gelloz, G. and Hayashi, T. (2006) *J. Org. Chem.*, **71**, 8957–8960.

264 Hatakeyama, T., Hashimoto, T., Kondo, Y., Fujiwara, Y., Seike, H., Takaya, H., Tamada, Y., Ono, T., and Nakamura, M. (2010) *J. Am. Chem. Soc.*, **132**, 10674–10676.

265 Kawamura, S., Ishizuka, K., Takaya, H., and Nakamura, M. (2010) *Chem. Commun.*, **46**, 6054–6056.

266 Noda, D., Sunada, Y., Hatakeyama, T., Nakamura, M., and Nagashima, H. (2009) *J. Am. Chem. Soc.*, **131**, 6078–6079.

267 Bedford, R.B., Huwe, M., and Wilkinson, M.C. (2009) *Chem. Commun.*, 600–602.

268 Kylmälä, T., Valkonen, A., Rissanen, K., Xu, Y., and Franzén, R. (2008) *Tetrahedron Lett.*, **49**, 6679. Meanwhile the paper has been retracted.

269 Bedford, R.B., Nakamura, M., Gower, N.J., Haddow, M.F., Hall, M.A., Huwea, M., Hashimoto, T., and Okopie, R.A. (2009) *Tetrahedron Lett.*, **50**, 6110–6111.

270 Wunderlich, S.H. and Knochel, P. (2009) *Angew. Chem. Int. Ed.*, **48**, 9717–9720.

271 Arvela, R.K., Leadbeater, N.E., Sangi, M.S., Williams, V.A., Granados, P., and Singer, R.D. (2005) *J. Org. Chem.*, **70**, 161–168.

272 de Vries, A.H.M., Mulders, J.M.C.A., Mommers, J.H.M., Henderickx, H.J.W., and de Vries, J.G. (2003) *Org. Lett.*, **5**, 3285–3288.

273 Plenio, H. (2008) *Angew. Chem. Int. Ed.*, **47**, 6954–6956.

274 Lauterbach, T., Livendahl, M., Rosellon, A., Espinet, P., and Echavarren, A.M. (2010) *Org. Lett.*, **12**, 3006–3009.

275 Gonzalez-Arellano, C., Abad, A., Corma, A., Garcia, H., Iglesias, M., and Sanchez, F. (2007) *Angew. Chem. Int. Ed.*, **46**, 1536–1538.

276 Correa, A. and Bolm, C. (2007) *Angew. Chem. Int. Ed.*, **46**, 8862.

277 Buchwald, S.L. and Bolm, C. (2009) *Angew. Chem. Int. Ed.*, **48**, 5586–5587.

278 Nakamura, M. (2009) *Kagaku to Kogyo*, **62**, 994–995; (2009) *Chem. Abstr.*, **151**, 538052.

279 Yin, L. and Liebscher, J. (2007) *Chem. Rev.*, **107**, 133–173.

280 Biffis, A., Zecca, M., and Basato, M. (2001) *J. Mol. Catal. A: Chem.*, **173**, 249–274.

281 Hamlin, J.A., Hirai, K., Millan, A., and Maitlis, P.M. (1980) *J. Mol. Catal.*, **7**, 543.

282 Ohff, M., Ohff, A., and Milstein, D. (1999) *Chem. Commun.*, 357.

283 Zhao, F., Bhanage, B.M., Shirai, M., and Arai, M. (2000) *Chem. Eur. J.*, **6**, 843–848.

284 Conlon, D.A., Pipik, B., Ferdinand, S., LeBlond, C.R., Sowa, J.R., Izzo, B., Collins, P., Ho, G.J., Williams, J.M., Shi, Y.J., and Sun, Y.K. (2003) *Adv. Synth. Catal.*, **345**, 931–935.

285 Köhler, K., Heidenreich, R.G., Krauter, J.G.E., and Pietsch, M. (2002) *Chem. Eur. J.*, **8**, 622–631.

286 Pröckl, S.S., Kleist, W., Gruber, M.A., and Köhler, K. (2004) *Angew. Chem. Int. Ed.*, **43**, 1881–1882.

287 Blaser, H.-U. and Spencer, A. (1982) *J. Organomet. Chem.*, **233**, 267–274.

288 Stephan, M.S., Teunissen, A.J.J.M., Verzijl, G.K.M., and de Vries, J.G. (1998) *Angew. Chem. Int. Ed.*, **37**, 662–664.

289 Hu, P., Kan, J., Su, W., and Hong, M. (2009) *Org. Lett.*, **11**, 2341–2344.

290 Carrow, B.P. and Hartwig, J.F. (2010) *J. Am. Chem. Soc.*, **132**, 79–81.

291 Starkey Ott, L. and Finke, R.G. (2007)
*Coord. Chem. Rev.*, **251**, 1075–1100.

292 (a) Astruc, D. (ed.) (2008) *Nanoparticles
and Catalysis*, Wiley-VCH Verlag GmbH,
Weinheim; (b) Schmid, G. (2004)
*Nanoparticles, from Theory to Application*,
Wiley-VCH, Weinheim.

293 Somorjai, G.A. and Park, J.Y. (2008)
*Angew. Chem. Int. Ed.*, **47**, 9212–9228.

294 Durand, J., Teuma, E., and Gómez, M.
(2008) *Eur. J. Inorg. Chem.*, 3577–3586.

295 Astruc, D., Fu, J., and Aranzaes, J.R.
(2005) *Angew. Chem. Int. Ed.*, **44**,
7852–7872.

296 Sablong, R., Schlotterbeck, U., Vogt, D.,
and Mecking, S. (2003) *Adv. Synth. Catal.*,
**345**, 333.

297 Bezemer, G.L., Bitter, J.H.,
Kuipers, H.P.C.E., Oosterbeek, H.,
Holewijn, J.E., Xu, X., Kapteijn, F.,
van Dillen, A.J., and de Jong, K.P.
(2006)
*J. Am. Chem. Soc.*, **128**, 3956–3964.

298 (a) de Vries, J.G. (2006) *Dalton Trans.*, 421;
(b) Phan, N.T.S., van der Sluys, M., and
Jones, C.W. (2006) *Adv. Synth. Catal.*,
**348**, 609–679; (c) Durand, J., Teuma, E.,
and Gomez, M. (2008) *Eur. J.
Inorg. Chem.*, 3577–3586; (d) Duran
Pachon, L. and Rothenberg, G. (2008)
*Applied Organomet. Chem.*, **22**,
288–299; (c) Trzeciak, A.M. and
Ziólkowski, J.J. (2007) *Coord. Chem.
Rev.*, **251**, 1281–1293; (d) Djakovitch, L.,
Koehler, K., and de Vries, J.G. (2008)
*Nanoparticles and Catalysis* (ed. D.
Astruc), Wiley-VCH Verlag GmbH,
Weinheim, pp. 303–348; (e) Moreno-
Mañas, M. and Pleixats, R. (2003)
*Acc. Chem. Res.*, **36**, 638–643.

299 Tamura, M. and Fujihara, H. (2003)
*J. Am. Chem. Soc.*, **125**, 15742.

300 (a) Jansat, S., Gómez, M., Phillipot, K.,
Muller, G., Guiu, E., Claver, C.,
Castillón, S., and Chaudret, B. (2004)
*J. Am. Chem. Soc.*, **126**, 1592;
(b) Favier, I., Gómez, M., Muller, G.,
Axet, M.A., Castillón, S., Claver, C.,
Jansat, S., Chaudret, B., and Philippot, K.
(2007) *Adv. Synth. Cat.*, **349**, 2459.

301 Klabunovskii, E., Smith, G.V., and
Zsigmond, A. (2006) Heterogeneous
enantioselective hydrogenation, theory

and practice, in *Catalysis by Metal
Complexes*, vol. 31 (eds B.R. James and
P.W.N.M. van Leeuwen), Springer,
Dordrecht, the Netherlands.

302 (a) Rocaboy, C. and Gladysz, J.A. (2003)
*New J. Chem.*, **27**, 39–49; (b) Nowotny, M.,
Hanefeld, U., van Koningsveld, H., and
Maschmeyer, T. (2000) *Chem. Commun*,
1877–1878; (c) Beletskaya, I.P.,
Kashin, A.N., Karlstedt, N.B.,
Mitin, A.V., Cheprakov, A.V., and
Kazankov, G.M. (2001) *J.
Organomet. Chem.*, **622**, 89–96;
(d) Astruc, D. (2007) *Inorg. Chem.*, **46**,
1884–1894.

303 Reetz, M.T., Breinbauer, R., and
Wanninger, K. (1996) *Tetrahedron Lett.*,
**37**, 4499–4502.

304 Beller, M., Fischer, H., Kühlein, K.,
Reisinger, C.-P., and Herrmann, W.A.
(1996) *J. Organomet. Chem.*, **520**,
257–259.

305 Kiwi, J. and Grätzel, M. (1979) *J. Am.
Chem. Soc.*, **101**, 7214.

306 (a) Jeffery, T. (1984) *J. Chem. Soc.
Chem.Commun.*, 1287–1289;
(b) Jeffery, T. and David, M. (1998)
*Tetrahedron Lett.*, **39**, 5751–5754.

307 Bönnemann, H., Brijoux, W.,
Brinkmann, R., Dinjus, E., Fretzen, R.,
Joussen, T., and Korall, B. (1991)
*Angew. Chem. Int. Ed. Engl.*, **30**, 1312.

308 Reetz, M.T., Westermann, E., Lohmer, R.,
and Lohmer, G. (1998) *Tetrahedron
Lett.*, **39**, 8449–8452.

309 van Leeuwen, P.W.N.M. (1983) Eur. Pat.
Appl. 90443 (to Shell); (1984) *Chem.
Abstr.*, **100**, 191388.

310 Reetz, M.T. and Westermann, E. (2000)
*Angew. Chem. Int. Ed.*, **39**, 165–168.

311 Dyker, G. and Kellner, A. (1998) *J.
Organomet. Chem.*, **555**,
141–144.

312 Beletskaya, I.P. and Cheprakov, A.V.
(2000) *Chem. Rev.*, **100**, 3009–3066.

313 Alimardanov, A., Schmieder-van de
Vondervoort, L., de Vries, A.H.M., and
de Vries, J.G. (2004) *Adv. Synth. Catal.*,
**346**, 1812–1817.

314 Diéguez, M., Pàmies, O., Mata, Y.,
Teuma, E., Gómez, M., Ribaudo, F.,
and van Leeuwen, P.W.N.M. (2008) *Adv.
Synth. Catal.*, **350**, 2583–2598.

**315** Reimann, S., Grunwaldt, J.-D., Mallat, T., and Baiker, A. (2010) *Chem. Eur. J.*, **16**, 9658–9668.

**316** Soomro, S.S., Ansari, F.L., Chatziapostolou, K., and Köhler, K. (2010) *J. Catal.*, **273**, 138–146.

**317** Thathagar, M.B., Kooyman, P.J., Boerleider, R., Jansen, E., Elsevier, C.J., and Rothenberg, G. (2005) *Adv. Synth. Catal.*, **347**, 1965–1968.

**318** Aramendía, M.A., Borau, V., García, I.M., Jiménez, C., Marinas, A., Marinas, J.M., and Urbano, F.J. (2000) *C. R. Acad. Sci. Chem.*, **3**, 465–470.

**319** de Vries, A.H.M., Parlevliet, F.J., Schmeider-van de Vondervoort, L., Hommers, J.H.M., Henderickx, H.J.W., Walet, M.A.M., and de Vries, J.G. (2002) *Adv. Synth. Catal.*, **344**, 996–1002.

**320** Martinelli, J.R., Freckmann, D.M.M., and Buchwald, S.L. (2006) *Org. Lett.*, **8**, 4843–4846.

**321** Martinelli, J.R., Watson, D.A., Freckmann, D.M.M., Barder, T.E., and Buchwald, S.L. (2008) *J. Org. Chem.*, **73**, 7102–7107.

**322** Cai, C., Rivera, N.R., Balsells, J., Sidler, R.R., McWilliams, J.C., Shultz, C.S., and Sun, Y. (2006) *Org. Lett.*, **8**, 5161–5164.

**323** Tromp, M., Sietsma, J.R.A., van Bokhoven, J.A., van Strijdonck, G.P.F., van Haaren, R.J., van der Eerden, A.M.J., van Leeuwen, P.W.N.M., and Koningsberger, D.C. (2003) *Chem. Commun.*, 128–129.

**324** Farina, V. (2004) *Adv. Synth. Catal.*, **346**, 1553–1582.

**325** (a) Slagt, V.F., de Vries, A.H.M., de Vries, J.G., and Kellogg, R.M. (2010) *Org. Process Res. Dev.*, **14**, 30–47; (b) Beller, M., Zapf, A., and Mägerlein, W. (2001) *Chem. Eng. Technol.*, **24**, 575–582.

**326** Frankham, J. and Kauppinen, P. (2010) *Platinum Metals Rev.*, **54**, 200–202.

**327** Tsvelikhovsky, D. and Buchwald, S.L. (2010) *J. Am. Chem. Soc.*, **132**, 14048–14051.

# 10
# Alkene Metathesis

## 10.1
## Introduction

Alkene metathesis was first reported by Eleuterio [1] and the reaction concerned metathesis of light olefins over $MoO_3$ on alumina at high temperature, $160\,°C$. Heterogeneous applications involve the conversion of propene into ethene and 2-butene, or the reverse reaction, depending on demand and availability of the feedstocks. The ethene–propene–butene process was known as the Phillips Triolefin process and it was on stream for a number of years during the 1960s for the conversion of propene. In a BASF–FINA plant in Texas, which came on stream in 2001, the reverse process is carried out in order to increase the amount of propene coming from the cracker. It is named the OCT (Olefin Conversion Technology) process and more plants have been built since then. Another large scale industrial application of heterogeneous metathesis catalysis is part of the SHOP process, namely the step in which all undesirable α-olefins (1-alkenes) are converted to detergent-range internal alkenes via a combination of isomerization and metathesis [2]. Other heterogeneous catalysts involved oxides of tungsten and rhenium as the active metals.

The first homogeneous catalysts were also discovered in the 1960s in the search for new Ziegler–Natta catalysts and they were early transition metal (ETM) halides, especially $WCl_6$ or $WOCl_4$ treated with alkylating agents such as $AlEt_3$ and $AlEt_2Cl$ (DEAC) [3]. In the attempted polymerization of cyclopentene it was found that the double bonds were retained in the polymer and a new type of polymerization was found, ring-opening metathesis polymerization (ROMP). The polymerization of cyclooctene was commercialized by Hüls (Germany) and they brought trans-poly(1-octylene) onto the market in 1982 as Vestenamer 8012. Tungsten catalysts were used formed from $WCl_6$ and substituted phenols with DEAC as the activator. Alkylation of tungsten halides leads to dialkyl species which, by α-elimination give, metal-alkylidene initiators.

The activity of ruthenium for ring-opening metathesis polymerization has been known for a long time. Natta reported in 1965 that cyclobutene and 3-Me-cyclobutene can be polymerized by ruthenium chloride in protic media in a ROMP mechanism [4]. Reports on ROMP of norbornene and ruthenium in protic media appeared

*Homogeneous Catalysts: Activity – Stability – Deactivation*, First Edition. Piet W.N.M. van Leeuwen and John C. Chadwick.
© 2011 Wiley-VCH Verlag GmbH & Co. KGaA. Published 2011 by Wiley-VCH Verlag GmbH & Co. KGaA.

in the same year [5]. The resistance of ruthenium metathesis catalysts towards polar substrates and even protic solvents remained relatively unnoticed until, in 1988, Novak and Grubbs re-examined the ruthenium-catalyzed ROMP in protic media with 7-oxanorbornene as the substrate [6].

In the 1960s and1970s the mechanism of the metathesis reaction received a great deal of attention and was subject of much debate, as clearly the reaction should involve elementary steps hitherto unknown in organometallic chemistry. The "carbene" mechanism was first published by Hérisson and Chauvin [7]. Their proposal was based on the observation that, initially, in the ring-opening polymerization of cyclopentene in the presence of 2-pentene a mixture of compounds was obtained containing two ethylidenes, or two propylidenes, or one of each, rather than the latter as the single product for pair-wise reaction of cyclopentene and 2-pentene. As few people were aware of this work in the early1970s the discussion about the mechanism continued for another 5 years in the literature before definite proof had been discovered! Early reports about involvement of metal alkylidene complexes in a chain growth polymerization reaction and how metal alkylidenes may form via elimination reactions include suggestions by Dolgoplosk and coworkers [8]. In a subsequent paper they initiated the ring-opening metathesis reaction of cyclopentene or cyclooctadiene by the addition of (diazomethyl)benzene to tungsten hexachloride [9]. Other methods to generate the metal alkylidene species involve alkylidene transfer from phosphoranes [10] or ring-opening of cyclopropenes [11].

While very early on several industrial applications were introduced, since then metathesis applications developed slowly if steadily for 20 years, until around 1990 when new boosts accelerated the developments enormously. During those years Schrock initiated and developed the area of alkylidene and alkylidyne complexes of early transition metals, a breakthrough in organometallic chemistry and the *in situ* generation of ETM metathesis catalysts was replaced by the use of well-defined, highly active metal complexes. Initially his research focused on tantalum complexes of the type $CpTaCl_2R_2$, which after α-elimination (Scheme 10.1) led to alkylidene complexes $CpCl_2Ta-CHR'$ [12].

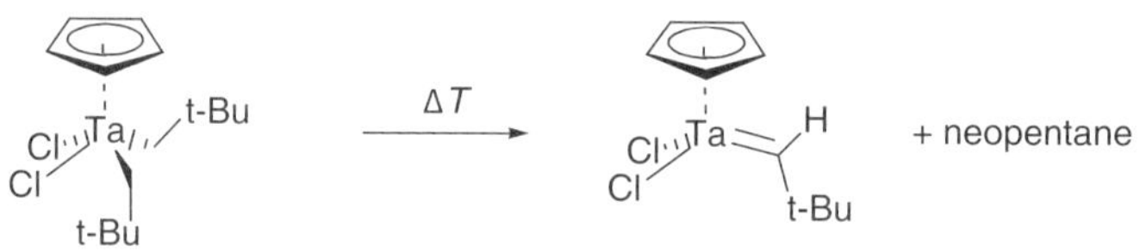

**Scheme 10.1** Formation of well-defined alkylidene complexes.

The putative intermediate for the metathesis reaction of a metal alkylidene complex and an alkene is a metallacyclobutane complex. Grubbs studied titanium complexes and he found that biscyclopentadienyl-titanium complexes are moderately active as metathesis catalysts; the stable resting state of the catalyst is a titanacyclobutane, rather than a titanium alkylidene complex [13] (Scheme 10.2). The alkylidene complex stabilized with $AlMe_2Cl$ is called Tebbe's reagent [14].

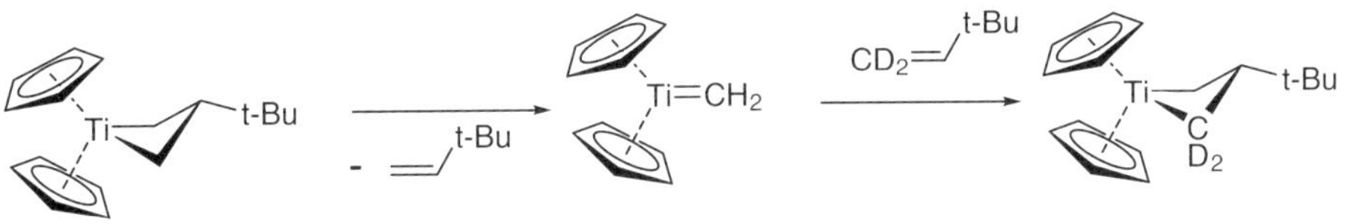

**Scheme 10.2**  Grubbs's titanacyclobutane catalyst for metathesis.

In the 1990s Grubbs introduced well-characterized ruthenium catalysts, the start of a new era in metathesis. New impulses for metathesis came from the work of both Schrock and Grubbs who both presented well-defined catalysts that showed activity for functionalized alkenes. Enantioselective metathesis was introduced by Grubbs, and also Schrock and Hoveyda, which led to even more applications in organic synthesis. In recent years metathesis has become an indispensable technique in the synthetic organic toolbox, together with several other organometallic catalytic reactions. Since mostly nowadays well-characterized catalysts are used we will focus on the decomposition reactions of these catalysts and leave the *in situ* prepared catalysts aside.

An important development in the last decade has been the immobilization of well-defined organometallic complexes and metathesis catalysts on oxide surfaces (silica) by Copéret and Basset. With the aid of sophisticated NMR techniques the surface species can be studied and characterized in detail (SOMC, surface organometallic chemistry, started by Basset more than 20 years ago). This has led to the most active heterogeneous catalysts known today and containing a percentage of up to 70 active species of the total amount of metal present, much higher than the best inorganic catalysts, which contain at most 2% active species (often much less). In Table 10.1 we give an example of the best catalysts for each metal taken from a recent review [15].

## 10.2
## Molybdenum and Tungsten Catalysts

### 10.2.1
### Decomposition Routes of Alkene Metathesis Catalysts

The oxophilic character of the initial ETM catalyst systems prohibited the use of alkenes containing functional groups, even very simple ones such as carboxylic esters, amides or ethers, as they will coordinate to the electrophilic metal, or react with the alkylating agents and the metal alkylidene catalyst. The search for functional-group resistant catalysts has always been a key issue in metathesis research. Less practical procedures were introduced, such as the addition of stoichiometric amounts of Lewis acids to block the polar donor group of the substrate. In the absence of proof, it was accepted that the metal alkylidene reactive group would react with C=O double bonds forming alkenes and inactive M=O species, as in Schrock alkylidenes the

**Table 10.1** Highlights of SOMC propene metathesis catalysts.

**Catalyst**

| | 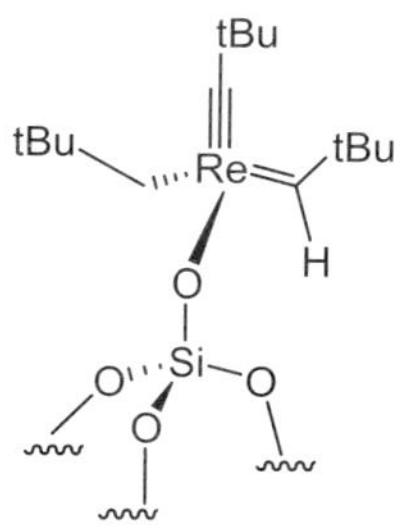 | 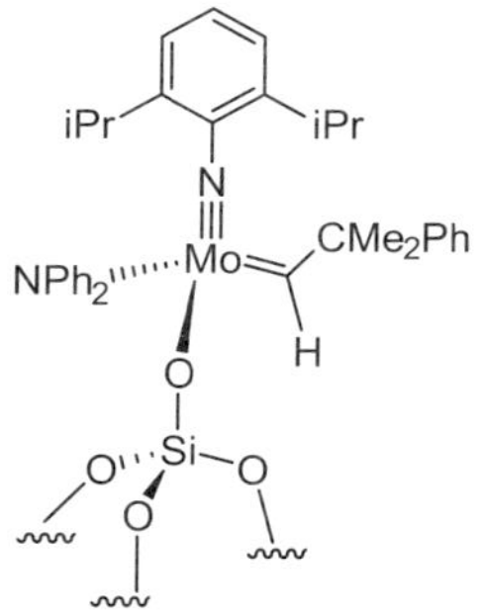 | 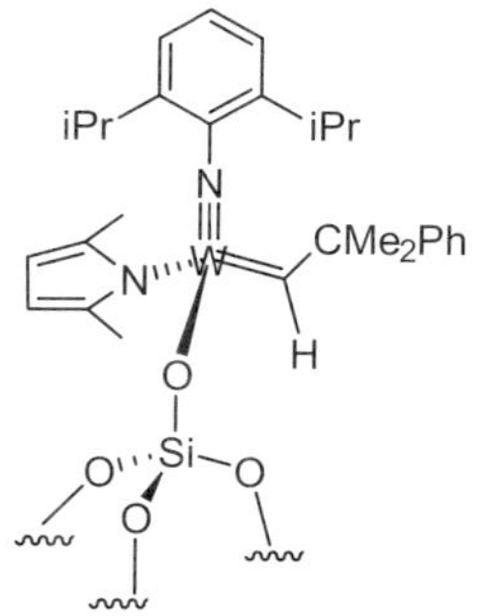 |
|---|---|---|---|
| Initial rate (h$^{-1}$) | 7,200 | 22,500 | 1400 |
| TON | 6,000 | 138,000 | 25,000 |
| Selectivity (%) | 96 | 99.9 | 99.9 |

(References: Re [16], Mo [17], W [18]).

metal is a high-valent, electrophilic metal atom and the alkylidene carbon atom carries a negative charge.

A first breakthrough, showing that it was not intrinsically impossible to accomplish the metathesis of, for example, carboxylic esters, even with early transition metals, was the metathesis of methyl oleic esters with the use of tungsten halide catalysts activated with tetramethyltin by Boelhouwer and coworkers [19]. Boelhouwer, an oleochemist, recognized the potential importance of converting oleochemicals into other chemical feedstocks via metathesis, which could be either self-metathesis or metathesis with other alkenes. Tetramethyltin reacts with tungsten hexachloride to form the initial carbene-tungsten species, but it is not sufficiently active as a metal alkyl species to react with methyl esters. Deuterium labeling studies by Grubbs demonstrated that the initial methylidene indeed stems from the methyl tin alkylating agent [20]. The tungsten complexes were deactivated towards nucleophiles by partial replacement of the halide anions by phenols. Turnover numbers were only of the order of 100–500, which was low compared to the cyclooctene and cyclopentene polymerization catalysts based on W and Mo, but many of today's catalysts applied in organic syntheses do not show higher turnover numbers.

Early transition metal based catalysts react with a variety of polar substrates and impurities, except the molybdenum ones substituted electronically and sterically in such a way that they become less oxophilic. In general, early transition metal alkylidenes will, for instance, react with aldehydes to form a metal oxo species and an alkene. A clean example of this reaction was published in 1990 by Schrock and coworkers (**1**, Scheme 10.3) [21]. This type of heteroatom metathesis reaction is very common. Even the highly stabilized catalyst shown in Scheme 10.3 still reacts cleanly with benzaldehyde within 10 min according to the equation shown, a Wittig-type reaction. The catalyst did not react with ethyl acetate or N,N-dimethylformamide for several weeks at room temperature; it does react with acetone but the products could not be traced down. Thus, even for the most resistant catalysts (such as Grubbs catalyst and **1**) it is worthwhile to protect the aldehydes as acetals [22]. Tolerance for carboxylic esters and amides was already known for the *in situ* prepared W and Re catalysts, as in the example mentioned above.

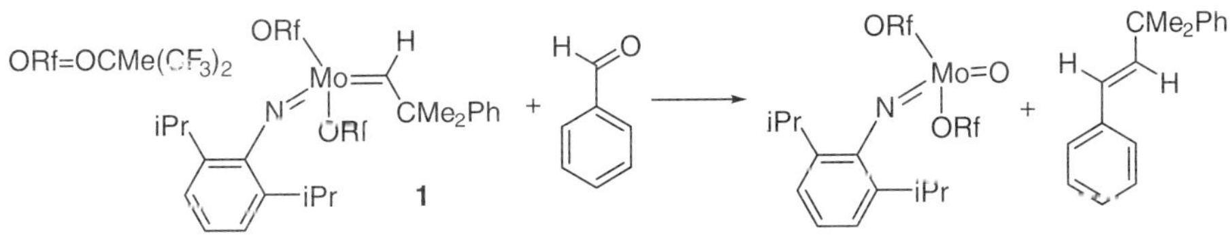

**Scheme 10.3** Reaction of molybdenum alkylidene with benzaldehyde.

An important feature of catalysts of type **1** is that they do not require ligand loss prior to being active as do the Ru-based Grubbs catalysts. It is a 14-electron species (assuming that Mo≡N is a triple bond) that can coordinate an alkene to form molybdenocyclobutane as the key catalytic intermediate. The bulky imido group prohibits dimerization, and the fluoroalkoxy species render electrophilicity to the metal to enhance the reactions with alkenes.

Simple alkenes can give turnover numbers of the order of several 100 000 with W- or Mo-based catalysts, including the *in situ* prepared catalysts (e.g., $WCl_6$, PhOH, $SnBu_4$), provided that the alkene has been thoroughly purified. A convenient purification method is percolation of the alkene over neutral alumina to remove hydroperoxides. Alumina converts the allylic hydroperoxides in allyl alcohols and enones and if these are not adsorbed they may still stop catalysis. Industrial practice is to purify alkenes over Na or Na/K dispersed on a heterogeneous support. This also removes alkynes and conjugated dienes. This procedure has also improved the turnover number of ruthenium catalysts for unsubstituted alkenes, showing that here the problem still exists, albeit much less pronounced (see Section 10.4).

The predecessors of the tetracoordinate Schrock Mo and W catalysts are the alkylidene W complexes reported by Osborn and coworkers in 1982, which were five-coordinate complexes $(tBuCH_2O)_2X_2W{=}CHtBu$, and required abstraction of the halide X to give extremely active catalysts for *cis*-2-pentene metathesis [23], thus precluding Schrock's rule that a metathesis alkylidene ETM catalyst should be four-coordinate. No quantitative data are available, but one must assume that these catalysts are more oxophilic than the Schrock catalysts. When $GaBr_3$ was used as the Lewis acid, TOFs at room temperature of the order of $300\,000\,h^{-1}$ were found for *cis*-2-pentene [24].

Hundreds of Schrock catalysts based on W, Mo, and Re have been made in the last 20 years, differing in their alkoxide, imido, and alkylidene substituents. Their activity, selectivity, and stability depend strongly on their substitution pattern [25]. The availability of such large numbers is of great value for the large variety of synthetic targets that may be addressed, because "It is increasingly unlikely that a small number of metathesis catalysts will carry out a wide variety of metathesis reactions" [26]. This also means that any potential neutral or anionic ligand that can replace one of the ligands in the catalyst, will have a dramatic effect on the catalyst's properties, and usually the change is for the worse! With the large amount of knowledge gathered one might be able to solve these problems by further modification or by catalyst screening. Protic groups are routinely protected and the nature of the protecting group changes the catalytic outcome, but for Ru-based catalysts this is not always needed.

Catalyst **1** is an extremely active 2-pentene metathesis catalyst [21]; in 1 min at 25 °C 250 turnovers were obtained, while the W analog [27] was even faster at $>1000\,min^{-1}$. The effect of the nature of the alkoxy group on Mo can be nicely illustrated as the relative rates for $OCMe_3$, $OCMe_2(CF_3)$, $OCMe(CF_3)_2$ are 1, $10^2$, and $10^5$. In this instance, for pure hydrocarbons, the catalyst mixtures *in situ* prepared from metal halide, phenols, and DEAC may form more active catalysts than the well characterized catalysts.

Tungsten catalyst **2** (Scheme 10.4) is also active as a metathesis catalyst for methyl oleate (*cis*-methyl-9-octadecenoate) [28]. At room temperature in 3 h between 200 and 300 turnovers were reached, after which the catalyst became inactive. The alkylidene signal had disappeared and only the W=O product was observed. It also reacts rapidly with acetone and ethyl acetate at room temperature in a Wittig fashion, as shown in Scheme 10.4. This demonstrates clearly that the Mo catalyst **1** is much more resistant to functional groups than the W analogue **2**.

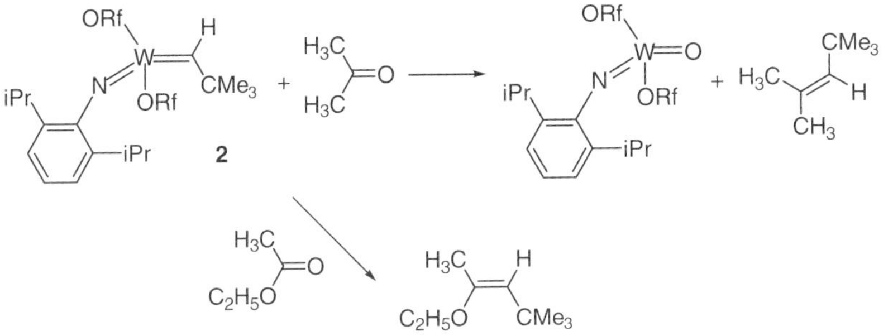

**Scheme 10.4**  Reactions of **2** with carbonyl functionalities.

Also, when reactive groups in the substrate or solvent are absent the catalysts decompose, as has been studied by Schrock and coworkers, both for W and for Mo. The decomposition reaction concerns a bimolecular or monomolecular decomposition giving alkene(s) and a reduced Mo(IV) or W(IV) species. Bimolecular decomposition of alkylidenes is slowest for neopentylidene or neophylidene complexes and most rapid for methylene complexes. At room temperature these reactions may sometimes take a day or more, and thus for fast catalytic conversions with high TONs they may not be important, apart from the fact that the catalyst cannot be stored in solution. In particular methylidene species are prone to dimeric coupling (Scheme 10.5, path A) and any β-hydrogen present in the metallacyclobutane may give β-hydrogen elimination and alkene formation (Scheme 10.5, path B). The resulting Mo dimer contains two tetrahedrally coordinated Mo atoms.

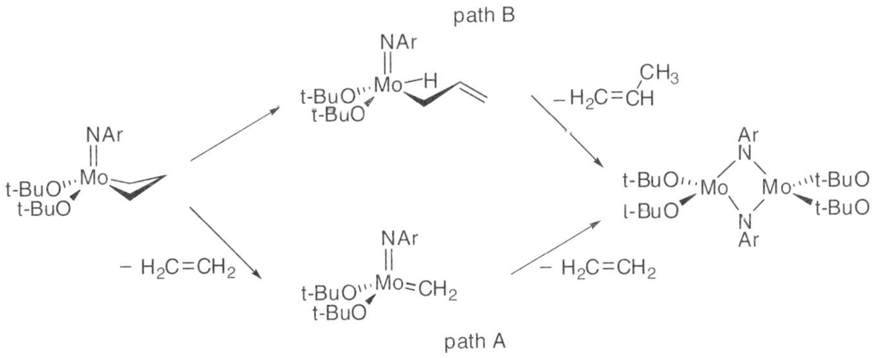

**Scheme 10.5**  Formation of reduced molybdenum dimers from metallacyclobutane.

The first example of a catalyst giving substantial amounts of propene via the path B mechanism of Scheme 10.5 is complex **3** containing an unsubstituted metallacyclobutane ring [29]. It was developed for asymmetric ring closing metathesis (ARCM) by Schrock and Hoveyda. RCM often produces ethene and this particular catalyst in the presence of ethene forms **3** as the resting state. The decomposition at room temperatures takes 10 days and thus for catalysis it may not be very relevant. ARCM was also slow, depending on the substrate, but the reaction could be accelerated by removing ethene from the solution. The organometallic products could not be identified; it could well be that the dimer formation is blocked by the bulky tris-isopropylphenyl groups.

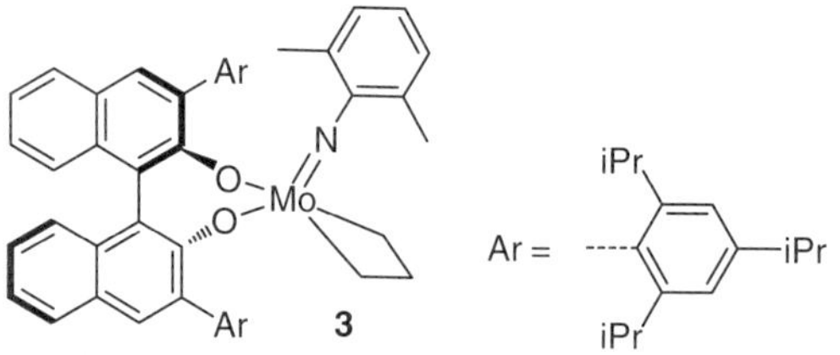

The products formed and the rate of alkylidene loss depend strongly on the ligands in the complex; one might obtain the impression that each "Schrock" compound has its unique reactivity, more so perhaps than the variety of phosphine complexes of late transition metals. We will restrict ourselves to giving a few examples and the interested reader should consult the reviews by Schrock for further reaction details [25, 26]. Tungsten complexes that contain a chiral bisphenol ligand reacted with ethylene to give tungstacyclobutane complexes, an ethylene complex, and a tungstacyclopentane complex. Interestingly, $^{13}$C NMR studies showed the formation of a heterochiral bimetallic methylene complex in which the methylenes are bridging asymmetrically (Scheme 10.6).

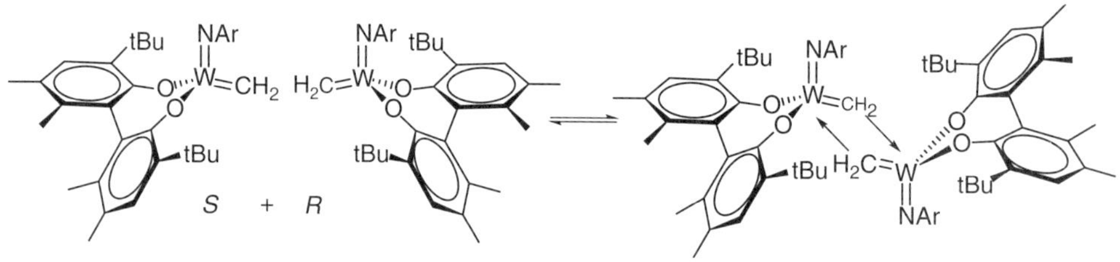

**Scheme 10.6** The formation of a heterochiral bimetallic methylene complex.

Moreover, the homochiral version of this dimer was not observed when the enantiomerically pure bisphenol ligand was employed. This is reminiscent of chiral poisoning and activation reviewed in Section 7.8 and the selective formation of the racemic complexes when $C_2$-symmetric $\eta^2$-anions are complexed to tetrahedrally surrounded metal atoms such as Zn [30] via a mechanism called self-discrimination [31], while for square planar complexes the reverse is observed, self-recognition [32]. Thus, alkylidenes in racemic versus enantiomerically pure systems could decompose at different rates and perhaps more bulky, remote substituents influencing dimer formation (e.g., o-aryl substituted binol ligands as in [33]) might even inhibit this type of decomposition. The monometallic release of alkenes from metallacyclobutanes via β-hydride elimination cannot be prevented in this way. In related studies of molybdenum systems dimers analogous to that shown in Scheme 10.6 were not observed.

Another reaction observed is the formation of dimers containing a W=W double bond after elimination of one molecule of alkene [34] (Scheme 10.7). The reaction takes place when the catalyst is treated with 2-pentene and the crystals of the dimer were obtained in low yield after 16 h at 25 °C.

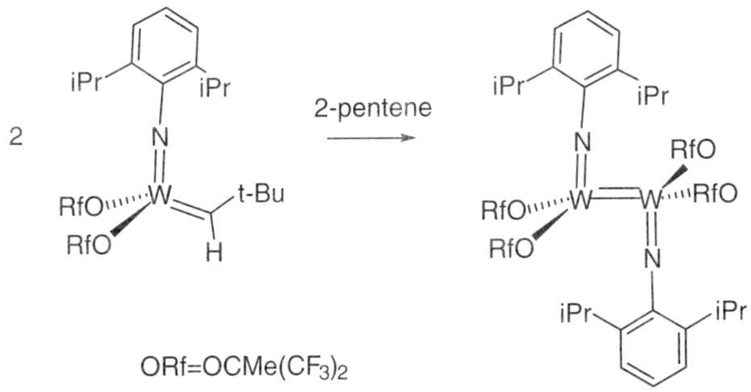

ORf=OCMe(CF3)2

**Scheme 10.7** Formation of a W=W double bond during metathesis.

The W=W distance in the X-ray structures is 2.47–2.49 Å. The W=W bonded species reported here are characterized by a 90° W–W–N angle and a 180° N–W–W–N dihedral angle. The W=W double bond is very stable and no exchange between unlike dimers was observed. One can imagine that the dimer forms from an intermediate similar to the dimers in Scheme 10.5, and that, in the absence of steric obstructions, the alkene can eliminate. In view of the stability of this dimer the reverse reaction, to regenerate an active catalyst, does not seem very likely, unless highly activated alkenes are used; on the other hand, such activated alkenes may provide catalytically inactive Fischer-type alkylidene complexes.

When W complexes with 2,6-dichlorophenylimido ligands are used the reactivity is different [35]. As is often observed for W, with ethene the stable product containing a WC₃ ring, a metallacyclobutane, is formed (Scheme 10.8) Decomposition gives the dimer with bridging imido groups and the production of both ethene and propene, showing that both reaction mechanisms are operative. Treatment of the t-Bu-alkylidene of W with 2-Me-butene-2 leads directly to the dimer, in which an interaction between the Cl atom and W is observed.

Replacement of RfO with pyrrolides in W complexes leads to tungsten imido alkylidene bispyrrolide complexes, which react with ethene to give stable alkylidene

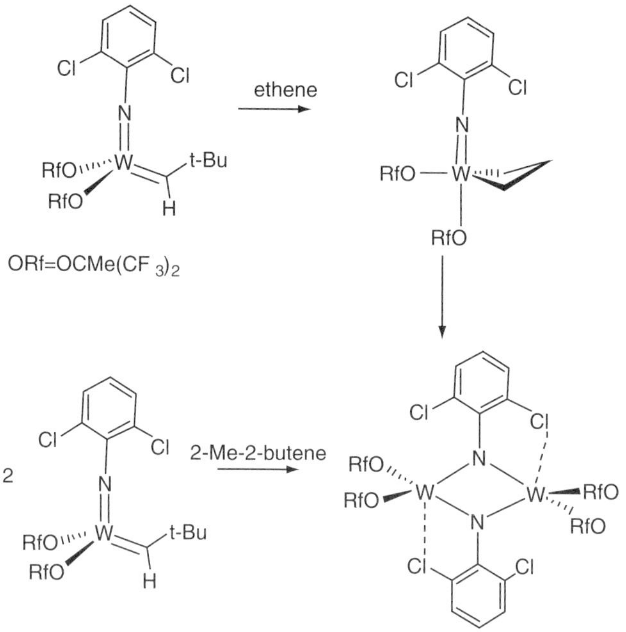

**Scheme 10.8** Formation of imido bridged dimers with Cl–W interaction.

complexes of W=CH$_2$, which are unusually stable, but their catalytic reactivity is low [36].

A substrate-induced stabilization for a Mo catalyst has been achieved by employing an ether modified substrate. The acyclic diene metathesis (ADMET) polycondensation of 1,4-diheptyloxy-2,5-divinylbenzene with the Schrock-type alkylidenes complex Mo(NAr)(CHCMe$_2$Ph)(ORf)$_2$ had to be carried out at somewhat higher temperatures (60 °C) than the alkyl-substituted divinylbenzenes. The reason was that ether linkages with Mo were formed, which have to be broken to obtain an active species. As a result the catalyst was also more stable and could be kept in solution for 24 days without decomposition [37] (Scheme 10.9). Both alkylidene and molybdenacyclobutane intermediates carrying the stabilizing ether linkage were detected.

## 10.2.2
### Regeneration of Active Alkylidenes Species

In a recent review Schrock discusses the issue of how to regenerate decomposition products into active alkylidene catalysts [26], which shows that in the last decade the interest in prolonging catalyst lifetime or increasing turnover numbers has received a lot more attention than in the first three decades of the development of homogeneous catalysts. Schrock focuses on the regeneration of reduced catalysts (Mo(IV) and W (IV), that is, the decomposition products obtained in the absence of decomposition routes initiated by reactions with substrates or impurities. In passing we have already

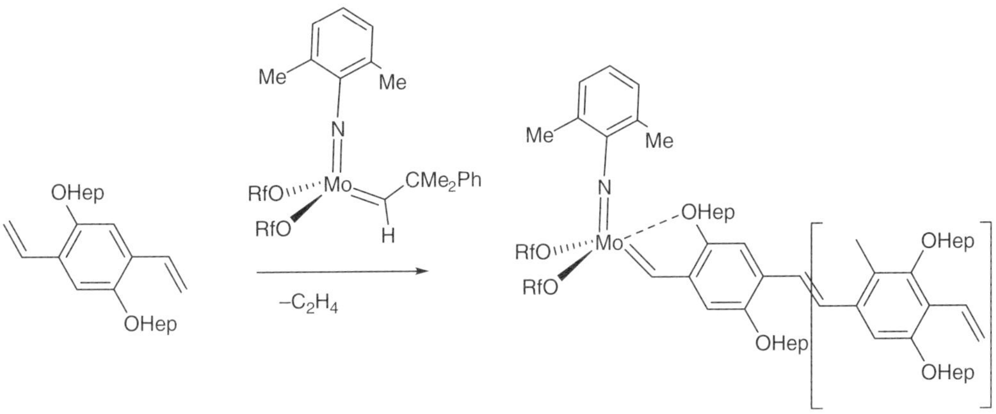

**Scheme 10.9** Ether coordination in Mo alkylidenes.

mentioned the methods available to generate metal alkylidene species, and we will return to these below. First, a few lines will be spent on deactivated complexes of high valence states. Most likely, as we have seen above, these will involve species that contain M=O units that have replaced the metal alkylidene M=C moieties (Schemes 10.3 and 10.4). It is unlikely, on thermodynamic grounds, that alkenes would induce the reverse reaction. In heterogeneous systems, operated as plug flow reactors, the continuous removal of traces of aldehyde and the high temperature applied may overcome the thermodynamic barrier. Furthermore, for tetraoxometallates the thermodynamics may be more favorable (see Section 10.3.2).

Double methylation followed by $\alpha$-elimination might be a feasible pathway, especially for M(VI) dihalides, the conventional catalysts. For M=O it seems less likely, although the activity of heterogeneous catalysts consisting of $Re_2O_7$ on an alumina support can be increased 10-fold or restored by treatment with $R_4Sn$ (R=alkyl) under mild conditions [38]. Addition of tetramethyltin, $(CH_3)_4Sn$, enhances also the activity of the $MoO_3/Al_2O_3$ catalyst. In propene metathesis at 303 K, the activity is about 20 times higher than that of a catalyst without a tetramethyltin promoter. The optimal Sn:Mo molar ratio of 0.05 for $MoO_3/Al_2O_3$ catalysts suggests that the amount of active sites generated by tetramethyltin on the $MoO_3/Al_2O_3$ surface is not more than 10% of the total Mo atoms in the catalyst [39]. When a $MoO_x/TiO_2$ catalyst was treated with $Me_4Sn$ the activity increased $10^3$ times [40].

The initial catalysts based on $WCl_6$ and DEAC showed increased lifetimes or activities when alcohols, carboxylic acids, and dioxygen were added. For the present Schrock catalysts this is not an option. Diols might lead to restoration of metal alkylidenes (see Section 10.3.2), which is only attractive for simple halide or oxide catalysts. The role of dioxygen in the old systems may well be oxidation of M(IV) to M(VI), followed by alkylation and $\alpha$-elimination to form metal alkylidenes.

Restoration of metathesis activity in $(RO)_2M=NAr$ (tetravalent metal) might be feasible with the common reagents for making metal alkylidenes: diazomethane

(Scheme 10.10, path A) [9], cyclopropene [41], transfer from another metal complex or phosphor and sulfur ylides [42], via a metallacyclopentane formed via cyclometallation of the metal with two alkene molecules, which can ring contract to a metallacyclobutane (Scheme 10.10, path B) [43].

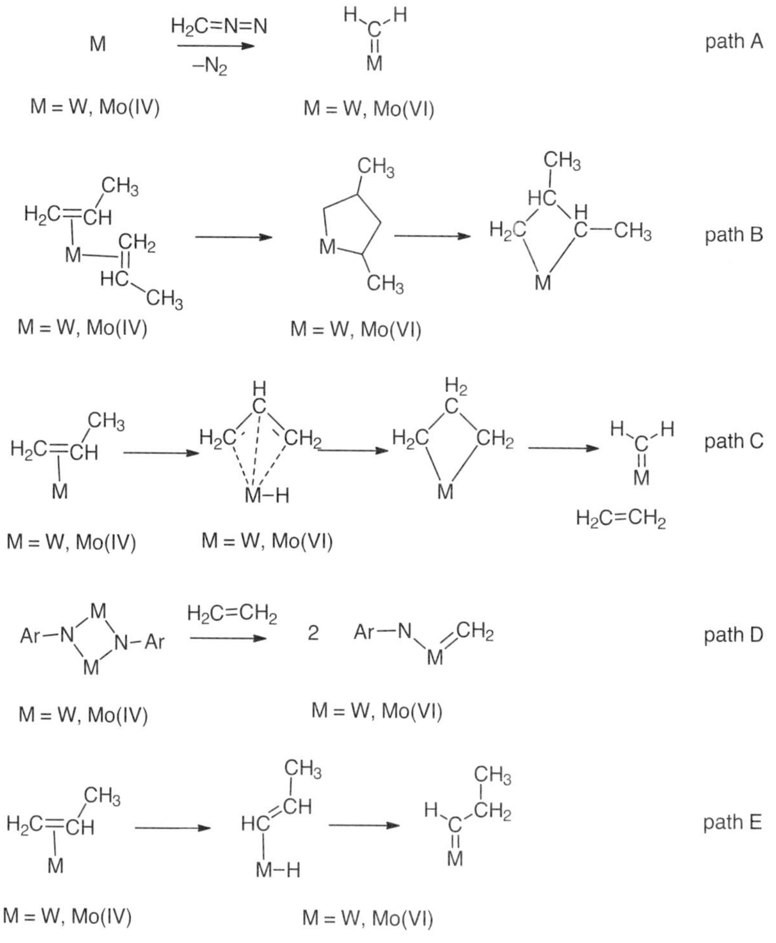

**Scheme 10.10** Metal alkylidenes formation as proposed for metal oxide catalysts.

Another possibility would be a reaction of the reduced dimeric species via the reverse reactions of Scheme 10.5, but the "oxidizing" alkene would have to be of a different nature than the one that was eliminated in the deactivation step (Scheme 10.10, paths C, D). In the heterogeneous catalysis literature one also finds speculation on oxidative addition of an alkene C−H bond to give a vinyl hydride, followed by rearrangement (path E, Scheme 10.10). There are examples of the reaction of Scheme 10.10 proposed for Mo and W oxides, for example, for molecular Ta compounds, but these compounds are not catalysts for metathesis [44].

Another reported activation is the reaction of the reduced metal with cyclopropane; the latter has been shown to be an activator in heterogeneous systems [3, 45, 46]. For instance, silica-molybdena reduced with CO under irradiation, reacts with cyclopropane at room temperature to give a 10-fold more active catalyst for propene metathesis. One molecule of surface $Mo=CH_2$ is formed and one Mo ethene complex. The Mo-methylidenes are stable up to 400 °C and react with water vapor at 120 °C (to give methane) according to IR studies [47]. There is consensus about heterogeneous Mo catalysts that the preferred precursor is Mo(IV), undergoing one of these oxidative reactions [48]; the intrinsic deactivation involves the formation of an alkene-Mo(IV) complex from the Mo(VI) metallacyclobutane.

The prerequisite of all reagents is of course that they should not interfere with the normal metathesis catalysis but only react with the $M=O$ (hexavalent) or dimeric reduced (tetravalent) species. From the reactivity of the metal alkylidenes and the low reactivities of their inactive states one would rather expect the opposite. Schrock concludes that, as yet, none of these approaches has been very successful [26] for the delicate, fine-tuned homogeneous catalysts. Perhaps the best solution is to continue fine-tuning for each substrate and reduce deactivation.

One might speculate on a mild version of the old reactivation methods for $WCl_6$ and $WOCl_4$ with dioxygen and DEAC. For instance, using a mild halogenating agent (organic or inorganic) and an unreactive alkylating agent such as $SnR_4$. More recent findings confirm the effect of $O_2$, for instance, the catalyst $W(O-2,6-C,H,Ph_2)_2Cl_4/SnBu_4$ is more active in the presence of dioxygen [49]. Halide-producing oxidizing agents may be more attractive for regeneration with metal alkyls. For chlorination or bromination a whole range of less to very reactive species is available (e.g., $\alpha,\alpha$-dichlorotoluene [50] would generate one dichlorometal species and one phenylmethylidene catalyst!). The alkylation will be much less trivial, as we know from Schrock's ETM work and his many colleagues now in the field that alkylation in high-valent ETM compounds is a very tricky business and most often the alkylidene is not the last ligand added to the complex.

## 10.2.3
### Decomposition Routes of Alkyne Metathesis Catalysts

Metathesis of alkynes has also been known since the early days of the first discoveries of heterogeneous metathesis catalysts [51]. Unlike the in situ prepared alkene metathesis catalysts that showed high activity, for pure hydrocarbons, comparable with that of today's well-defined catalysts, the alkyne catalysts prepared by in situ mixing showed a poor performance. Already in 1972 Mortreux proposed that breaking of the triple bond occurred and not breaking of a σ-bond and exchange of groups. Metathesis of $^{14}CH_3C \equiv CPr$ on $MoO_3$-$SiO_2$ at 300–400 °C gave $PrC\equiv CPr$ and $^{14}CH_3C \equiv C^{14}CH_3$ [52]. Alkyne metathesis has been enormously improved by replacing the in situ mix-and-stir catalysts by well-defined alkylidyne catalysts; it is an example of how mechanistic ideas can lead to far better catalysts.

The initial homogeneous catalyst was obtained by heating molybdenum carbonyl with resorcinol or chlorophenol at high temperatures (110–160 °C) and gave a few turnovers per hour for aryl substituted alkynes [53]. A considerable improvement was obtained by using $O_2Mo(acac)_2$ with phenols and $AlEt_3$ as the alkylating agent, a much more likely way to arrive at molybdenum alkylidyne species. Turnover frequencies as high as 17 000 per hour were now achieved at 110 °C [54]. Many tungsten alkylidyne complexes were tried, but many showed no reactivity. Moreover, the synthesis was often not straightforward and efficient and the efficient procedures had to be developed (Schrock and coworkers). Wengrovius and Schrock [55] found that $t\text{-}BuC\equiv W(O\text{-}t\text{-}Bu)_3$ was an extremely efficient catalyst for the metathesis of 3-heptyne with a turnover frequency of hundreds of thousands per hour at room temperature! Even for the fastest Mortreux catalyst obtained *in situ* this means that, given the temperature difference, it is a million times slower, and perhaps the concentration of the catalytic species amounts to only a few ppm of the total Mo (it should be noted that most Mo compounds are much less active than the W ones, unless RfO is used for Mo).

Depending on the size of the alkoxide groups used, the reaction has as a resting state the metallacyclobutadiene complex or the alkylidyne complex, as was concluded from the kinetics. The former reaction is zero order in alkyne concentration ("dissociative"), the latter is first order in alkyne concentration ("associative") [56], see Scheme 10.11.

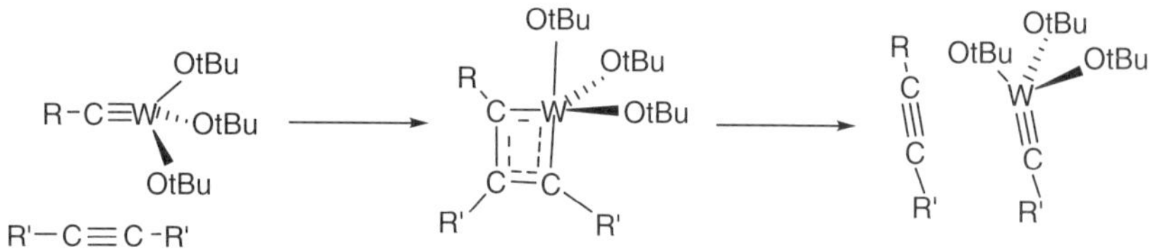

**Scheme 10.11** Alkyne metathesis with a tungsten alkylidyne complex.

1-Alkynes show very low conversion in a metathesis reaction and this was already explained by Schrock and coworkers in 1983 [57]. What happens is a transfer of the hydrogen atom (formally a β-hydride elimination) from the metallacyclobutadiene to the metal giving a deprotiotungstenacyclobutadiene complex followed by elimination of ROH (Scheme 10.12).

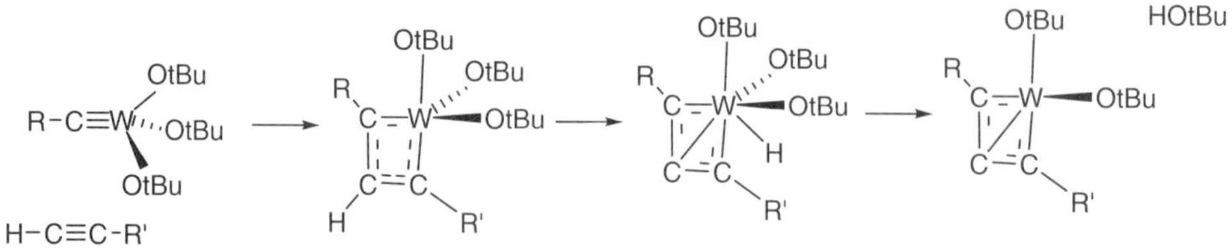

**Scheme 10.12** Decomposition of 1-alkyne metathesis catalyst.

Hydrochloric acid, phenols and carboxylic acids add to RC≡W(OR)₃ complexes to give alkylidene complexes, as shown in Scheme 10.12 [58]. It can be imagined that a phenol formed according to Scheme 10.12 reacts with the alkylidyne catalyst to give an alkylidene complex, Scheme 10.13. Metathesis of 1-alkynes, as reported by Mortreux using Schrock–Wengrovius type catalysts, gives, in addition to initial fast metathesis, polymerization of alkynes, the product being polyacetylenes [59], but less so in diethyl ether as the solvent. Polymerization of alkynes with the use of alkylidene complexes is well established [60]. Perhaps the combined decomposition reactions explain the polymer formation.

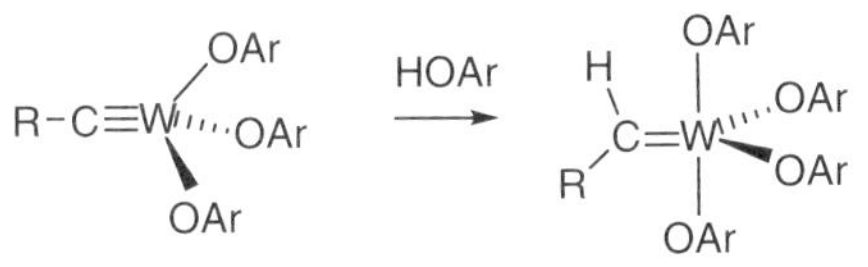

**Scheme 10.13** Formation of metal alkylidene complexes by addition of acids to alkylidyne complexes.

Several other reactions can be imagined that lead to metal alkylidenes, among them is the rearrangement of the deprotonated tungsten metallacyclobutadiene complex, although that would be a tetravalent W species, not an active catalyst (Scheme 10.14) [61].

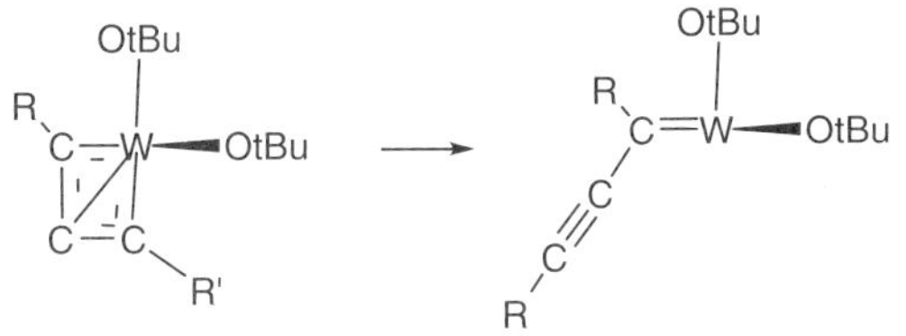

**Scheme 10.14** Alkylidene formation from deprotonated metallacyclobutadiene.

Mortreux and coworkers found that addition of quinuclidine to tBuC≡W(OtBu)₃ changed the selectivity of the reaction with 1-heptyne to almost exclusively metathesis, suppressing polymerization, albeit at very low rates (10 turnovers per hour at room temperature) [62].

A plethora of reactions has been described for Mo and W alkylidyne complexes with alkynes and many more other impurities may form that are the actual catalysts for alkene metathesis observed with alkylidyne complexes or alkyne polymerization. When 1-alkynes and alkenes were used with tBuC≡WCl₃(dme) [63] the authors proposed as an explanation the alkyne reactions found by Schrock (for Mo [64], for W [65]), depicted schematically in Scheme 10.15, but most likely the activity for alkene metathesis and 1-alkyne polymerization is due to traces of W(VI) alkylidenes formed via Scheme 10.13. Moreover, for 1-alkynes the system with Mo takes another course, Scheme 10.16, [64] where eventually a Mo(IV) alkylidene is formed. None of these reactions can be generalized as they depend fully on the substituents at the metal and the alkyne.

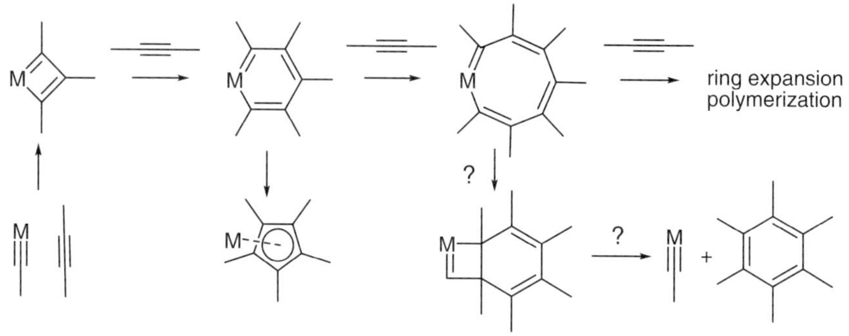

**Scheme 10.15** Formation of cyclopentadienyl moieties and benzene with alkylidyne M(VI) complexes.

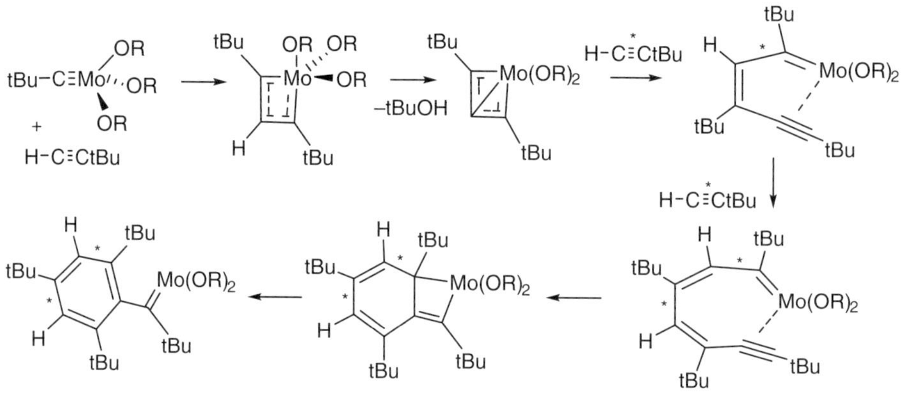

**Scheme 10.16** Mo(IV) alkylidene formation from alkylidyne. The asterisks indicate where the new alkyne has been incorporated via a metathesis-like reaction, that is, a 2 + 2 addition and rearrangement.

We have added to Scheme 10.15 the formation of benzene from the metallacyclooctatetraene; on heterogeneous $WO_3$ on silica this product has been reported by Moulijn and coworkers for both internal alkynes and terminal alkynes [66]. Other mechanisms cannot be ruled out, as low-valent Mo compounds will also trimerize alkyne carboxylates and the involvement of high-valent metal alkylidynes seems unlikely [67].

The immobilized, well-defined catalysts developed recently by Copéret and Schrock show a strong dependence on the alkoxy and imido substituent in terms of activity, TON and life-time. The Mo catalysts are very active for acyclic alkenes, but much less so in RCM catalysis. They present one of the best catalysts for oleate esters [17, 68]. Unfortunately, it is still too early to discuss their decomposition pathways (see Section 10.3.2 for related Re compounds). For tungsten the situation is the same [18, 69]. In this instance the well-defined immobilized catalysts are far less active than the *in situ* prepared homogeneous catalysts of unknown composition.

## 10.3
## Rhenium Catalysts

### 10.3.1
### Introduction

The history of rhenium as a homogeneous metathesis catalyst is relatively recent. Rhenium's major use as a catalyst is in the petroleum industry in the reforming process ("rheniforming", or more general platforming) that converts low octane number linear and cyclic alkanes (naphtha) into high octane number gasoline containing branched alkanes and aromatics, with coproduction of hydrogen. The catalyst is a combination of Pt and Re on silica or silica-alumina support. Heterogeneous Re catalysts have been used for fine chemical synthesis, on a small scale $(1400\,t\,a^{-1})$ only, for a number of years since 1986 by Shell Chimie in Berre, France [70]. Two target reactions were addressed: the ethenolysis of cyclooctene and cyclooctadiene to make $\alpha,\omega$-dienes, and isobutenolysis of cyclooctadiene, which was of interest for the synthesis of starting materials for terpene chemistry. For propene production from ethene and butene in a liquid phase a process has been developed (Meta-4, by IFP and CPC-Taiwan) using $Re_2O_7$ on alumina [71]. Alumina is the support material of choice and usually high loadings are applied [45].

In 1988 the development of homogeneous catalysts started, although prior to that several Re precursors in the presence of alkylating agents and Lewis acids were known to give some activity in metathesis. In the late 1980s Herrmann started his work on $MeReO_3$ (MTO) as a catalyst both in epoxidation and in metathesis. The synthesis of MTO had been reported before [72], but the chemistry and catalysis was explored and developed by Herrmann *et al.* [73]. MTO by itself is not active as a metathesis catalyst, but, like other Re compounds, needs the presence of alkylating agents and acidic cocatalysts [74]. Much higher activities were obtained when MTO was deposited on a support in a collaboration between the Basset and Herrmann groups. The support has a large influence on the activity and on $Nb_2O_5$ as a support for 2-pentene metathesis at room temperature, activities as high as 6000 turnovers per hour were achieved [75]. The effect was ascribed to the presence of both Lewis acidic and Brønsted acidic sites, as neither Lewis acidic $TiO_2$ nor weakly Brønsted acidic silica showed appreciable activity. One might have thought that the first alkylidene generated stems from the methyl group, but this is not the case, as was shown by Buffon and coworkers [76]. This was particularly tempting, because treatment of $Re_2O_7$ with tetraalkyltin activators lead to metathesis activity and one way to make MTO is reacting $Re_2O_7$ with $Me_4Sn$. A metathesis reaction of *trans*-2,5-$Me_2$-3-hexene with [13]C $Me_4Sn$, however, did not give the expected labeled 3-Me-1-butene, [13]C-1 and it was concluded that initiation took place via the allylic hydrogen activation mechanism (Scheme 10.10, path C). The latter gives indeed 3-Me-1-butene, but this does not contain [13]C. This reaction requires the presence of Re in a lower valence state (IV, or V), which could be identified. Moreover, it was shown that upon exposure to alkenes their concentration diminished.

In his attempts to arrive at well-defined molecular Re metathesis catalysts Schrock turned to a variety of neopentylidene Re imido-alkoxy complexes, inspired by the success of this approach in Mo and W chemistry [77], but Horton and Schrock did not find any activity for these complexes, not even for the alkylidene complexes of formula $(ArN)_2(RfO)Re=CH\text{-}tBu$. More recently tris(adamantylimido)MeRe was studied by Wang and Espenson [78]. No metathesis of alkenes was reported and we assume that this trisimido complex is not active either. It did react with benzaldehyde in a metathesis reaction, but ketones did not. At longer reaction times and with excess of aldehyde, MTO was obtained. Imines could be converted but catalytic metathesis of imines, as known for MTO, was not observed.

The first preliminary report on well-defined homogeneous Re catalysts dates from 1990, in which Toreki and Schrock described the synthesis and isolation of $tBuC\equiv Re$ $(=CH\text{-}tBu)(ORf)_2$ as an orange, unstable oil [79] which was active as a metathesis catalyst for terminal, internal, and functionalized alkenes. More metathesis data were reported in 1993 [80], but the catalyst remained orders of magnitude less active than the W and Mo isoelectronic counterparts.

Copéret, Basset, and coworkers [16] immobilized $(t\text{-}BuCH_2)_2Re(=CH\text{-}t\text{-}Bu)(\equiv C\text{-}t\text{-}Bu)$, which by itself is not a metathesis catalyst, on silica eliminating neopentane and obtained the most active rhenium catalyst so far, approaching the activity of Mo (Scheme 10.17). The compounds on the surface were completely characterized by high level NMR techniques. Catalyst lifetime is short, but the TONs obtained are higher than those of most other Re catalysts.

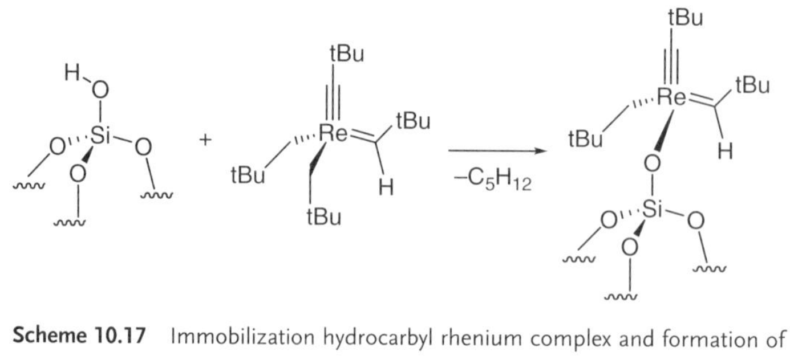

**Scheme 10.17** Immobilization hydrocarbyl rhenium complex and formation of active catalyst.

10.3.2
## Catalyst Initiation and Decomposition

Both initiation and catalyst decomposition are relatively unknown for rhenium catalysts, be they homogeneous or heterogeneous. Only for the well-defined Schrock catalysts is the initiating species the alkylidene complex itself, which does show incubation as we will see later. For its immobilized version a major, intrinsic decomposition pathway has been established.

In hindsight, the *in situ* prepared homogeneous catalysts prepared prior to 1990 were very poor catalysts. Several rhenium carbonyl species were used and together with alkylaluminum dichloride (and probably oxygen) they afforded some alkene metathesis activity, often at high temperatures and accompanied by many other reactions, as in the work by Farona [81]. He proposed an addition of $R-AlCl_2$ on coordinated CO to obtain the first (Fischer) alkylidene complex. The catalyst is long-lived (at 90 °C) and loses half of its activity in 5 days; most likely a new initiator is slowly formed while deactivation occurs, and only a tiny amount of the complex is active. Warwel reported that $MeRe(CO)_5$ in the presence of i-$BuAlCl_2$ gave activity at 70 °C [82].

The activity of heterogeneous $Re_2O_7/Al_2O_3$ depends strongly on the loading, and the activity per Re atom increases up to 18 wt% $Re_2O_7$ when a complete coverage by a monolayer is reached [83]. A rhenium coverage of one Re per 0.35 nm$^2$ is obtained and most likely it is present as mainly $Al-O-Re(=O)_3$ species. Heterogeneous $Re_2O_7/Al_2O_3$ catalysts were much more active, and when $R_4Sn$ alkylating agents were used they were active at room temperature [84], even for methyl oleate as a substrate. Co-metathesis of methyl oleate and *trans*-3-hexene was also successful. As was apparent from the reports of the Mol group, the interesting aspect of the heterogeneous catalyst is that, after treatment with $O_2$ at high temperature followed by the addition of $Me_4Sn$, the activity can be completely restored [83]. The alkyltin promoted catalysts are more active than the non-promoted ones, but they deactivate more rapidly. Moreover, oxidative restoration of the tin-containing catalyst in the longer term leads to coverage by tin oxides and low to nil activity [85].

Gates compared a number of Re heterogeneous catalysts and concluded that Re must be in a high oxidation state to show activity and that only a small portion of Re is in a high oxidation state [86]. He also adheres to the explanation that only a small fraction of Re takes part in catalysis and that a reservoir of precursors that is slowly oxidized ensures that the overall activity is retained. In Section 10.3.1 we already mentioned the work by Buffon *et al.* [76] in which it was shown that initiation requires the presence of Re in a lower valence state (IV, or V), which species were indeed identified. It was shown that upon exposure to alkenes these sites disappeared with concomitant start of metathesis catalysis.

The thermodynamic parameters for deactivation of $Re_2O_7$/alumina catalysts were studied by Spronk [87] for 1-octene and by Behr for 1-pentene [88]. In the absence of impurities the intrinsic decomposition was studied and the apparent activation barriers for metathesis and deactivation were determined. Spronk found that the apparent activation energy for metathesis was higher than that of the deactivation, but

Behr reported the reverse. Both modeled the reaction kinetics and proposed a gradual rise in temperature in order to maintain the same conversion at the same flow rate in a plug-flow reactor. Other solutions are possible, but from an industrial point of view raising the temperature is the most practical. Given the activation energies found, the required temperature increase for Spronk's catalyst system was smaller. The FEAST process was operated in this fashion; these authors ascribed deactivation for highly pure feeds mainly to pore blocking by fouling with heavy side-products [70].

Recent deactivation studies of 1-pentene metathesis over $Re_2O_7$/alumina show more details about poisons [89]. As this 1-pentene was prepared via Fischer–Tropsch it contained a variety of oxygenates which had to be carefully removed. Pent-2-en-1-ol was found to be a stronger poison than pent-2-en-1-al. The latter has about the same effect as the ketone tested, pent-1-en-3-one. A BTS catalyst (Cu on $M_gSiO_3$) could be used to lower the oxygenates content to 10 ppm. Treatment with alumina brought the level of oxygenates below the level of detection. Water is a stronger catalyst poison than pent-2-en-1-ol on a weight basis. The rate and mechanism of the intrinsic decomposition are independent of the flow rate of 1-pentene. The authors proposed path B of Scheme 10.10 and reductive elimination of cyclopropane as the mechanism (Scheme 10.18). After 30 h on stream a modest TON of a few thousand was found and the activity for the ultrapure feed had dropped to 60%, while in the reaction with a feed containing 50 ppm of pent-2-en-1-ol the activity had dropped to 52%, showing that the main cause is the intrinsic decomposition.

$$
\begin{array}{ccc}
& \underset{\overset{|}{C}}{H_2} & \\
H_2C \diagup \quad \diagdown CH_2 & \longrightarrow & H_2C \underset{\overset{|}{C}}{H_2} CH_2 \\
\diagdown \quad \diagup & & \\
M & & M
\end{array}
$$

M = W, Mo(VI),Re(VII)        M = W, Mo(IV),Re(V)

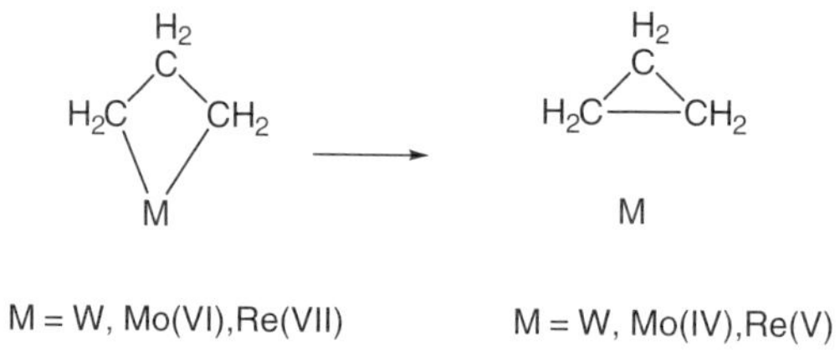

**Scheme 10.18**  Reductive elimination of cyclopropane.

Farona found that initiation does not take place with ethene, which rules out the metathesis of Re=O and ethene as the alkylidene-forming reaction [90]. Propene does give activity and thus the allylic hydride mechanism was proposed as the reaction that leads to the first alkylidenes (Scheme 10.10, path B). The mechanism points to Re(V) as the catalyst precursor species, similar to Mo(IV) and W(IV). Intrinsically the activity can be extremely high, but since only a small percentage reaches an active state the observed activity can vary by orders of magnitude.

In a more recent study, the inability of ethene to make the initial alkylidene catalyst was confirmed using deuterio-labeled ethene and non-labeled ethene [91]. However, cis-stilbene does lead to homo- and cross-metathesis with ethene, showing that another mechanism than the π-allylic one should be operative. The authors proposed that metal alkylidene formation takes place via Wittig-like cross metathesis of Re=O and alkenes. The aldehyde by-products could not be detected; only 2% of active Re was formed, and, besides, aldehydes are strongly adsorbed on alumina.

Also in the development of the Meta-4 process, many authors found that deactivation had two causes, one resulting from impurities in the feed and the other from an intrinsic deactivation mechanism, most likely the reverse of the reactions of Scheme 10.10 [92].

In an attempt to mimic heterogeneous catalysts Commereux [93] mixed $(ArO)_2Al$-i-Bu with $Re_2O_7$ (Ar = bulky phenoxide) to obtain a mixture of compounds with the initial formation of $(ArO)_2Al$–$OReO_3$ and i-BuReO$_3$. The precipitate formed was an active metathesis catalyst but given the low activity for 2-pentene (TOF = 60 h$^{-1}$) one must conclude that only an extremely small fraction is active. Methyl oleate is metathesized even more slowly. The TOF for norbornene ROMP was much higher at room temperature (3500 h$^{-1}$). Often norbornene is found to be effective in creating active sites, which explains the higher activity and indicates that upon mixing the aluminum and rhenium compounds only, not much catalyst is formed. Acetone and THF inhibit the reaction. In a later publication [94] catalyst activities were improved enormously by making a wide variety of aryloxy aluminum perrhenate adducts. For an *in situ* prepared catalyst, activities for 2-pentene metathesis as high as 6000 mol mol$^{-1}$h$^{-1}$ were measured. Dialkyl ether showed high activity. The complex or catalyst could not be isolated, but reaction with quinuclidine yielded complex **4**, enabling an approximate characterization of the highly labile catalyst or its precursor.

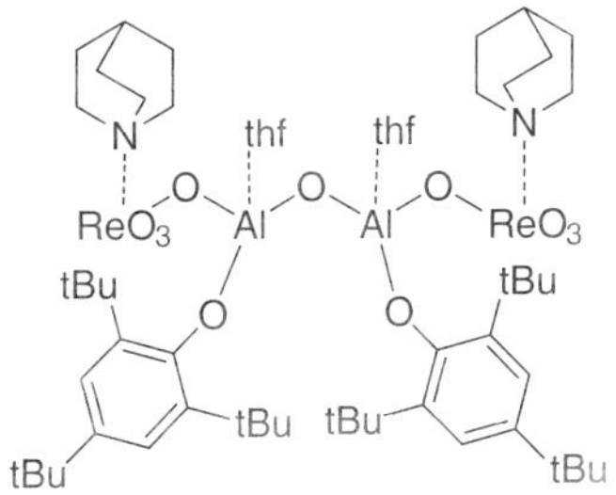

**4**

Grubbs reported an activation of Re(V) by 2-diphenylcyclopropene, equivalent to the activation of Ru(II) and W(IV) [95], Scheme 10.19. The resulting complex is not active as a metathesis catalyst, which is not surprising in view of the hexacoordinate character. After reaction with Lewis acids such as GaBr$_3$, only cyclic, strained alkenes could be subjected to a ROMP reaction. The Lewis acid probably removed the molecules of THF and one of the ORf groups. We have seen before that strained cyclic alkenes are more suited to form metal alkylidene catalysts than acyclic alkenes, which in this instance do not show metathesis activity.

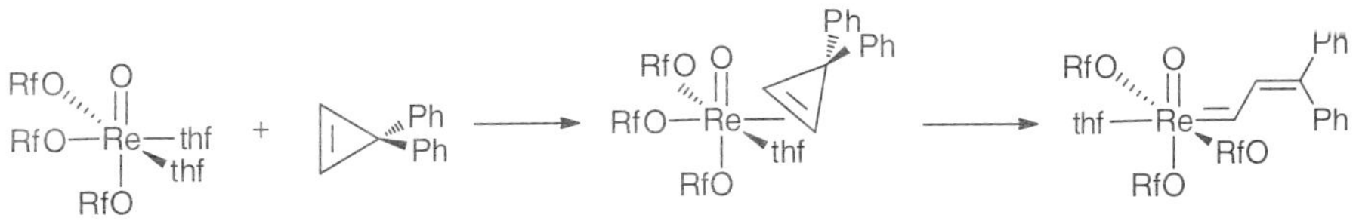

**Scheme 10.19** Grubbs Re(V) to Re(VII) conversion with 2-Ph$_2$-cyclopropene.

The search for a nitrosyl-containing rhenium catalyst was continued by Berke and coworkers [96]. The reaction of $[Re(H)(NO)_2(PR_3)_2]$ complexes with BArF acid etherate gave the corresponding cations $[Re(NO)_2(PR_3)_2][BArF_4]$. The addition of phenyldiazomethane to benzene solutions of these cations afforded the moderately stable cationic, five-coordinate rhenium benzylidene complexes. These complexes catalyze the ring-opening metathesis polymerization (ROMP) of highly strained nonfunctionalized cyclic alkenes such as norbornene and dicyclopentadiene. Turnover frequencies were only $650\,h^{-1}$. The initiation reaction was studied with ESI MS/MS and a unique reaction mechanism was found that led to the active catalyst (Scheme 10.20). Important steps are the nucleophilic attack of the phosphine at the coordinated norbornene, and attack of the phosphorus alkylidene on one of the nitrosyl groups forming an iminate. The final rhenium iminate is not necessarily the molecule that starts the catalysis, as two earlier alkylidenes in the sequence may also start the ROMP reaction. The molecular weights are high and, in the absence of chain transfer, at most 10% of the Re is active. As in heterogeneous systems, analysis of the bulk may be misleading. On the other hand, the example also shows us that much more complicated reactions may be going on than those shown in Scheme 10.10, both for initiation and for decomposition of the catalyst in systems like these, especially when only a small percentage of the metal present participates in the catalysis.

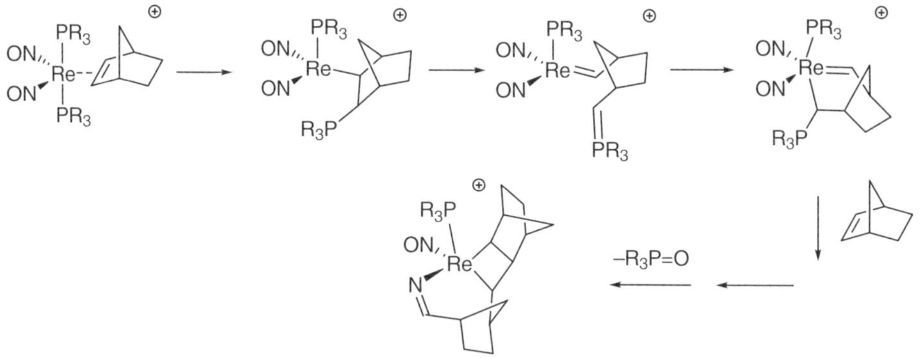

**Scheme 10.20** Intermediates observed in nitrosyl rhenium-catalyzed ROMP of norbornene.

$CH_3ReO_3$ on silica-alumina is a better catalyst than $NH_4ReO_4$ on the same support for 1-hexene metathesis at room temperature [97]. $CH_3ReO_3$ activated with $MeAlCl_2$ is a catalyst for norbornadiene ROMP with a TOF of $>1000\,h^{-1}$, but again the high MW $(5 \times 10^5)$ shows that only part of the Re present is active.

Basset and Copéret [98] showed that $MeReO_3$ on alumina gave, in part, a species containing an aluminum–rhenium bridged methylidene species, analogous to Tebbe's reagent (containing Ti as the early transition metal) [14]. They proposed that this is the precursor to the active catalyst. This is in disagreement with other studies that show that the $CH_3$ group of MTO is not involved in the formation of the first alkylidene species. For propene an initial TOF of $2600\,mol\,mol^{-1}\,h^{-1}$ was achieved, but rapid deactivation occurred. While ethene in other studies was found to be inactive as an initiator, Basset and coworkers suspect that ethene causes deactivation. The two things may be related to one another. Below we will see how ethene causes catalyst deactivation in a SOMC catalyst developed by Copéret.

Immobilized complexes originating from $Np_2Re(=CH\text{-}t\text{-}Bu)(\equiv C\text{-}t\text{-}Bu)$ form the most active Re catalysts known with an initial TOF of $7200\,h^{-1}$ [16]. They are more active than the homogeneous complexes $(RO)_2Re(=CH\text{-}t\text{-}Bu)(\equiv C\text{-}t\text{-}Bu)$ and all heterogeneous catalysts known so far. One explanation could be site isolation on the surface that prohibits any bimolecular mechanism. Copéret, Eisenstein, and coworkers gave an important explanation for the activity of immobilized well-defined Re catalysts [99], which is also valid for other $d^0$ catalysts (W, Mo) [100]. Their work concerns DFT studies of complexes of the formula $XYRe(=CHtBu)(\equiv CtBu)$ in which X and Y are monodentate anions, such as an alkoxy group, an amide group, or alkyl group. They found that coordination of ethene to the four-coordinate species forming a trigonal bipyramid is greatly enhanced when one of X,Y is a strong σ-donor (X in Scheme 10.21) and the other Y is a weak donor. The fastest reaction is found when X is C and Y is O, thus an alkyl group and a surface silanol group, for instance. Preparing the catalyst for ethene coordination implies a much higher barrier than expected, and this barrier may be higher than that of the formation of the metallacyclobutane. The latter is stabilized by poor X,Y σ-donors. Entropy plays an important role, especially in the rigid metallacycle, and no firm conclusion could be drawn yet as to which step is rate controlling.

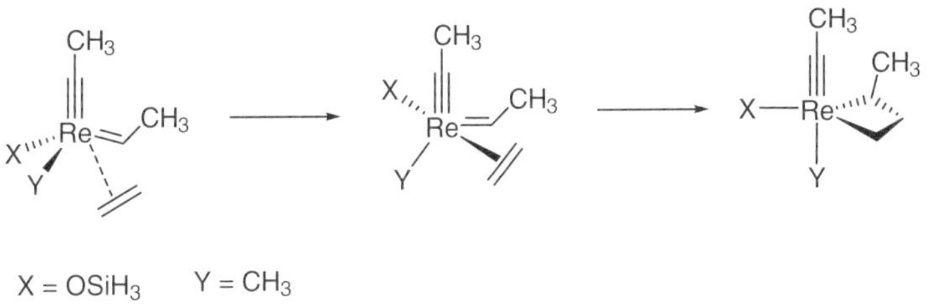

X = OSiH$_3$    Y = CH$_3$

**Scheme 10.21** Model compounds used in the calculation concerning the effect of the nature of X and Y.

The large percentage of active catalyst (70% of all rhenium present) and the short life-time made it possible to study the decomposition of the catalyst in propene metathesis with unprecedented detail [101]. Catalyst deactivation turned out to be first order in ethene and active sites. It produced 1-butene and pentenes in quantities greater than the number of sites, which means that its formation is neither associated with site activation nor with irreversible deactivation of the catalytic sites. Furthermore, in the liquid, leaching of a rhenium complex took place, also contributing to deactivation. DFT calculations and $^{13}C$ labeling studies led to the mechanism shown in Scheme 10.22. After β-hydride elimination from the metallacycle the reaction with the lowest barrier is the insertion of ethene. The allyl ethyl rhenium complex then undergoes α-elimination to produce 1-butene and the initial ethylidene catalyst; thus

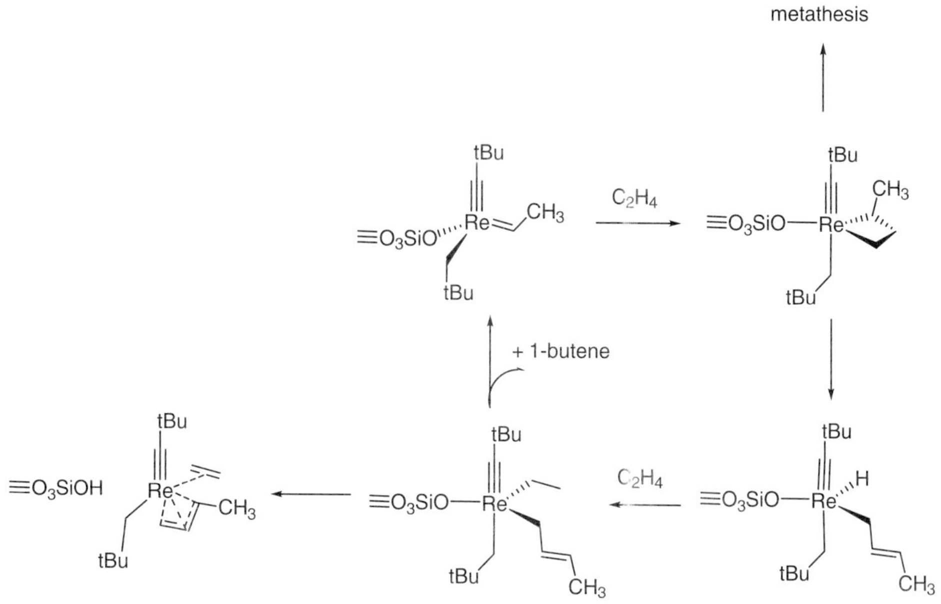

metathesis

**Scheme 10.22** Decomposition pathway of immobilized, highly active XYRe(=CHtBu)(≡CtBu) catalysts.

1-butene can form without catalyst decomposition. If elimination takes place from the hydride, as we have seen before, an inactive Re(V) species forms (cf. Scheme 10.5, upper path for Mo, or Scheme 10.10, the reverse of Path B). A slightly higher barrier gives β-hydride elimination with formation of a silanol and a Re(V) complex. This explains the role of ethene and the ethyl group as the neopentyl group present from the start does not contain β-hydrogens.

Several other reactions can be envisaged (some of which we have seen above), but the scheme shown contains the pathways with the lowest calculated barriers. Dimeric pathways were not considered as the average distance between Re atoms is >10 Å and the mobility was considered to be low on the calcined silica surface.

## 10.4
## Ruthenium Catalysts

### 10.4.1
### Introduction

In 1972 the metathesis nature of the ruthenium polymerization catalysts was already recognized [102]. Porri continued research along these lines and in 1974 he reported the ROMP of norbornene with ruthenium and iridium salts in protic media [103]. Incubation and reaction times reported were rather long. Ruthenium also showed activity for ROMP of cyclopentene, but only after the addition of dihydrogen. These were the first reports on catalysts made from elements of the right-hand side of the periodic table. This was very remarkable, because until then metathesis catalysts were incompatible with polar, and certainly protic, solvents. The work remained relatively unnoticed, although an industrial process was based on it to make a polymer named Norsorex in 1976 (CdF-Chimie, later Atofina, Arkema, and Astrotech). As in the Porri system, $RuCl_3$ is used in an alcohol as the solvent (*t*-BuOH) and *in situ* formation of the presumed ruthenium alkylidene took place. Norsorex is the trans- (90%) polymer of norbornene obtained via ROMP with a molecular weight of 2 M. The interest in ruthenium catalysts in protic media was "reactivated" by Novak and Grubbs in 1988 [104]. Rh(II) tosylate was used as a catalyst for the polymerization of 7-oxa-norbornene derivatives in water and an incubation time was noticed, necessary to form the initial ruthenium-carbene species. Water dramatically decreased the incubation time, but later we will see that water also leads to decomposition of the alkylidene catalyst!

10.4.2
**Initiation and Incubation Phenomena**

There are a few examples of *in situ* formation of the catalyst with the aid of initiating agents, which are a direct follow up of the Porri and Novak work involving *in situ*, uncontrolled formation of the catalyst, but these will be mentioned after the introduction of the Grubbs catalyst. These *in situ* methods make use of the knowledge that ruthenium is highly reactive towards alkynes and thus vinylidene species often result from the reaction, although earlier literature reported that bisphosphine ruthenium dichloride vinylidenes only show activity for strained cyclic alkene ROMP and that they are inactive for acyclic alkenes [105].

The occurrence of tungsten and molybdenum alkylidene complexes as active metathesis catalysts without further activation stimulated research in this direction for the ruthenium catalysts. Addition of ethyl diazoacetate as a carbene precursor (as Dolgoplosk did for tungsten [106]) to ruthenium hydrated salts gave active catalysts [107], as yet used *in situ*. The first successful synthesis of a ruthenium alkylidene was obtained with the use of diphenylcyclopropene as the carbene precursor [108], Scheme 10.23, and a common ruthenium phosphine dichloride precursor.

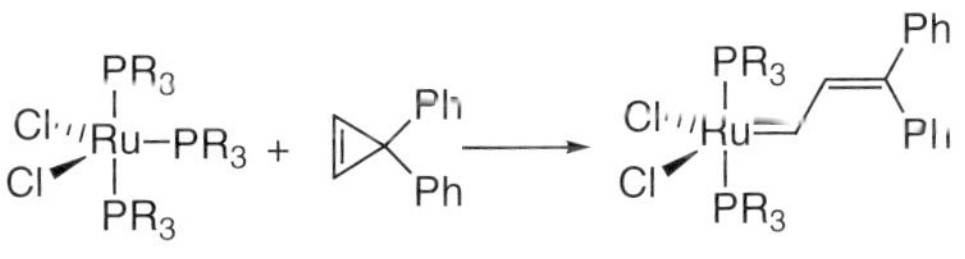

**Scheme 10.23**   The first isolated ruthenium carbene catalyst.

The latter method was utilized in the synthesis of ruthenium vinylalkylidene complexes of the type $RuCl_2(=CHCH=CPh_2)(PR_3)_2$ (R = Ph [109] and R=Cy [110]).

The phenyl carbene species (benzylidene) was isolated when diazomethylbenzene was used [111], Scheme 10.24.

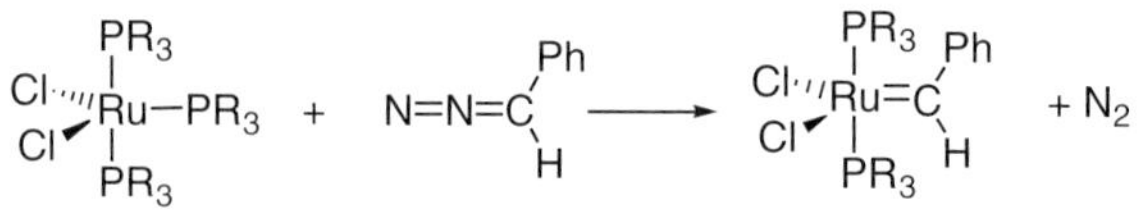

Scheme 10.24   Ruthenium benzylidene catalyst, now known as the Grubbs I catalyst.

While ruthenium dichlorides always showed long initiation times, these novel catalysts start the reaction with little delay. For compounds not containing functional groups their turnover frequencies can be several thousands per hour, but for polar molecules the total turnover may be as low as fifty, obtained in several hours. The Grubbs I catalyst – containing benzylidene as the alkylidene – has shorter incubation times than the first isolated active alkylidene ruthenium catalyst, diphenylvinyl alkylidene [112]. In ROMP, the latter shows initiation rates that are much lower than the rate of propagation, while for the former they are of comparable magnitude [111].

Initially, the presence of the two phosphine molecules in the complex was considered to be of crucial importance. Grubbs has shown that in fact the complex containing only *one* phosphine ligand is more active [113] and that dissociation of one phosphine was needed to arrive at an active catalyst. There is an equilibrium between the bis-phosphine complex and the mono-phosphine complex ( + free phosphine). Higher activities were obtained when CuCl was added to the reaction mixture; CuCl removes free phosphine from the solution, thus shifting the equilibrium to the side of the mono-substituted ruthenium complex. Ligands with lower $\chi$-values (e.g., tricyclohexylphosphine) give higher rates. Anions follow the reverse order: $Cl > Br > I$. This detailed study was conducted on the RCM reaction shown in Scheme 10.25, often chosen as the model test reaction after Grubbs had introduced his catalysts. In general the ruthenium catalysts are still not as fast as the tungsten- or molybdenum-based ones, but their flexibility and scope is enormous.

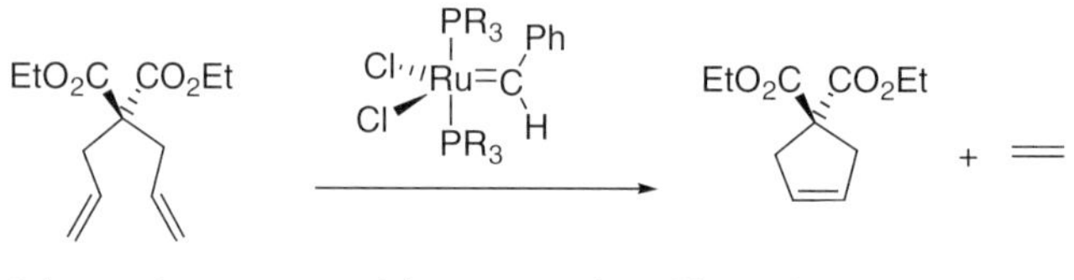

Scheme 10.25   RCM model reaction with Grubbs catalyst.

Another approach to remove one $Cy_3P$ ligand is shown in Scheme 10.26. Dias and Grubbs reported the synthesis of binuclear catalysts obtained from the reaction of the Grubbs I catalyst with Lewis acidic dimers (Scheme 10.26) such as cymene ruthenium dichloride dimer and $CpRhCl_2$ dimers [114]. The catalysts were used in ROMP of 1,5-cyclooctadiene. The new dimers were 10–90 times more active than Grubbs I catalyst (note that the order in the concentration of the latter is 0.5 when phosphine is

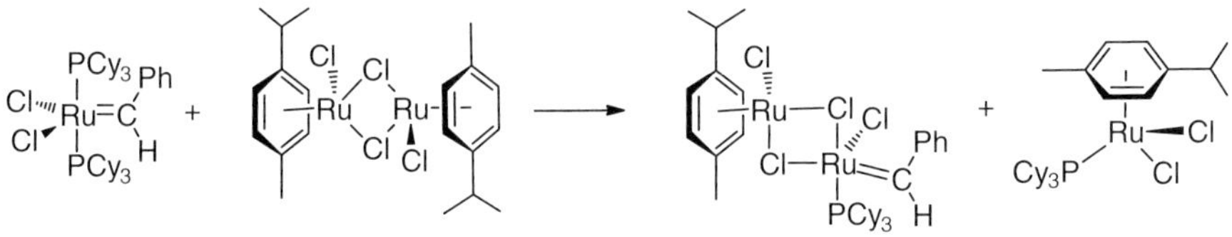

**Scheme 10.26** Lewis acids in combination with Grubbs I catalyst.

not removed from the solution and that comparison might be tricky). It was observed that the reaction of the dimers with alkenes was associative, unlike the reaction sequence of Grubbs I, which starts as a dissociative phosphine release. This was explained by the more electron-poor ruthenium center as a result of the Lewis acidic center forming the dimer.

In an attempt to enhance the dissociation of one phosphine, Werner and coworkers made the tricyclooctylphosphine analogue of the Grubbs I catalyst [115]. For reasons not known, the catalyst with the more bulky phosphines was less active than the one containing $Cy_3P$, and thus the order of activities found was $Cy_3P > Cyclooctyl_3P > iPr_3P$.

Phosphine dissociation as the initiation reaction and the rates of ROMP were extensively studied by Grubbs and coworkers for a variety of ligands, anions, and solvents [116]. Both bis-phosphine complexes and NHC-phosphine complexes (Grubbs II catalysts, to be discussed below) were studied. The rate of phosphine dissociation increases with increasing polarity (dielectric constant) of the solvent (pentane < toluene < diethyl ether < dichloromethane < tetrahydrofuran). Dissociation of $PCy_3$ is much slower when the para-ligand is an NHC ligand than when it is a phosphine. Why this is, is not well understood, but it probably has a steric background. Alkene binding to the unsaturated intermediate can be as much as four orders of magnitude faster for the NHC catalysts than for the $PCy_3$ catalysts. This is explained by the much stronger back-donation to the alkene in the NHC complexes, as the NHCs are much stronger donors than phosphines [117]. Within the group of NHCs the stronger donors give the fastest alkene coordination (and as a result also the fastest catalysts). In addition to stabilizing the alkene complex, electron donation from NHCs is expected to accelerate the oxidative addition required for metallacyclobutane formation. The NHC catalysts are much more thermally stable than the Grubbs I catalyst. As mentioned above, decomposition of alkylidene or benzylidene catalysts is a bimolecular process that takes place after phosphine dissociation, and thus, if phosphine dissociation is much slower and occurs to a lesser extent, bimolecular decomposition will also be much slower. In the presence of alkenes the rate of decomposition may even further decrease for the NHC catalysts as they react rapidly with alkenes and the equilibrium concentration may not be attained.

Another interesting precursor that leads to very fast catalysts was introduced by Castarlenas, Dixneuf and coworkers [118]. Scheme 10.27 shows how a cationic cymene-ruthenium allenylidene rearranges to an indenylidene complex under the influence of triflic acid. After the loss of cymene, this indenylidene species leads to an active species derived from a cationic Grubbs I catalyst as it contains $PCy_3$ as the

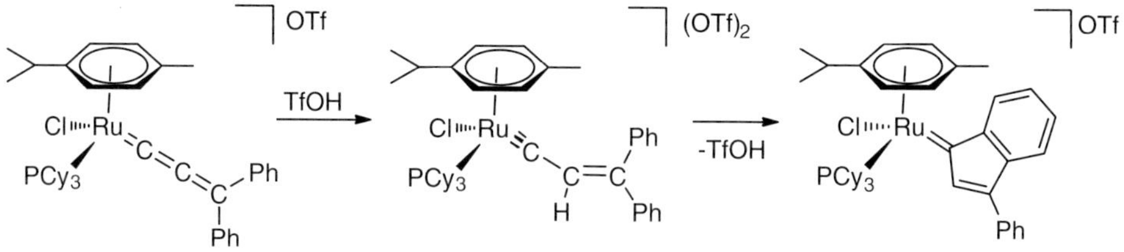

**Scheme 10.27** Rearrangement of allenylidene ruthenium to indenylidene ruthenium.

ligand that remains in the complex. Rates for ROMP of cyclopentene and cyclooctene are extremely high, almost $300\ mol\ s^{-1}$. Apparently cymene dissociation is much faster than phosphine dissociation and high concentrations of the active species are formed. At substrate to ruthenium ratios of 300 000 to 1 hardly any activity was observed due to impurities in the system.

Ruthenium complexes react with 1-alkynes to give, *inter alia*, vinylidene ruthenium complexes, of which many examples have been reported [119]. Several groups have studied the *in situ* formation of alkylidenes from alkynes to form the initial ruthenium vinylidene or alkylidene species. Nubel and Hunt from BP Amoco reported the successful use of butyne-1,4-diol diacetate as an initiating agent for ruthenium trihalides and phosphines in alcoholic media, which gave highly active catalysts in the presence of dihydrogen [120]. Turnover numbers as high as 110 000 were obtained for 1-octene at 90 °C in 2 h.

The synthesis of Grubbs-type catalysts from terminal alkynes was already reported by Werner, who focused on catalyst synthesis rather than catalysis [121]. A neat *in situ* method was developed by the same group a few years later, utilizing magnesium as the reducing agent for the hydrate of $RuCl_3$ in the presence of the desired phosphine and dihydrogen; after the reduction had taken place a terminal alkyne was added at low temperature (Scheme 10.28) [122]. In related recipes 2-butanol was used as the hydrogen source. Acetylene could also be used, to give ethylidene ruthenium species.

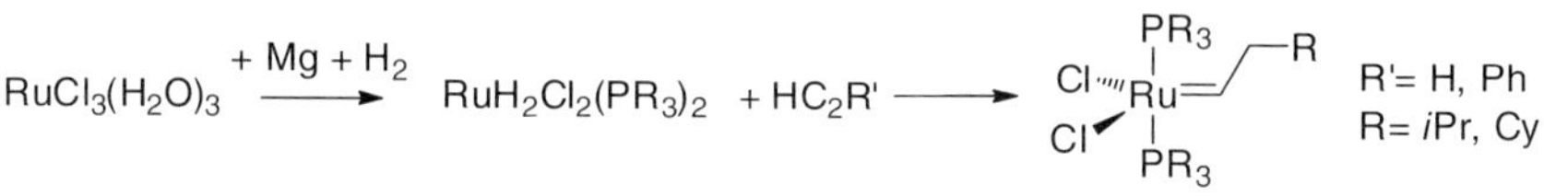

**Scheme 10.28** *In situ* generation of Grubbs I catalyst by Werner.

The fact that this method uses ruthenium hydrides is interesting, as these are known to form via decomposition of alkylidene catalysts with water and alcohols, and to give rise to isomerization of alkenes. Nubel and Hunt emphasized that their catalyst prepared *in situ* in this way gave very little isomerization. Thus, while the medium used – alcohols and water – gives rise to faster decomposition, the alkynes present convert the ruthenium hydrides, active as isomerization catalysts, rapidly into alkylidene species active in metathesis.

The above findings concerning *in situ* preparation of metathesis catalysts from ruthenium chloride, phosphine, hydrogen, and alkynes were applied by Vosloo and coworkers in the metathesis of 1-octene [123]. Hydrogen was used during the reaction to obtain the highest yield of primary metathesis products. Metathesis of 1-octene showed the best results at temperatures between 80 and 90 °C if acetic acid or ethanol was used as solvent with a $PCy_3/Ru$ molar ratio of 1. Both internal and terminal, alkyl and aryl alkynes could be used for the *in situ* activation.

Thus, alkynes have two functions, they convert inactive hydride species into active alkylidenes, and, in so doing, they remove ruthenium hydrides, which otherwise would cause deleterious isomerization of the alkene starting materials or products. In later sections we will come back to how these hydrides are formed. Grubbs and coworkers studied what other additives could be used to passivate the hydrides [124]. A wide variety of additives was tested and quinones were found to be very effective as hydride traps.

An important development has been the introduction of N-heterocyclic carbene ligands to replace one of the phosphine ligands. Since strong donor ligands such as $PCy_3$ ligands perform better than $PPh_3$ ligands, one might expect that carbene ligands, being stronger σ-donors, would be even better. The structure of the complex is the same, carbene and phosphine occupying trans-positions. The catalyst is known as the second generation Grubbs catalyst (Grubbs II), which in many applications is more active than the first generation Grubbs catalyst. As already mentioned the initiation of Grubbs II catalysts is slow, but dissociation of the phosphine could be accelerated by tuning of the phosphines without affecting the high rate of metathesis [125]. The rates of phosphine dissociation (initiation) from these complexes increase with decreasing phosphine donor strength. Complexes containing a triarylphosphine exhibit dramatically improved initiation relative to $(H_2IMes)(PCy_3)$ $(Cl)_2Ru=CHPh$. Increased phosphine dissociation leads to faster olefin metathesis reaction rates.

Various ways of phosphine scavenging for NHC-containing catalysts have been introduced by Grubbs and coworkers. Addition of ethereal HCl turned out to be an efficient way to remove $PCy_3$ as its phosphonium salt [126]. Herrmann reported on the use of cymene ruthenium dichloride scavengers, as shown in Scheme 10.26 for the Grubbs I catalyst [127].

Coordination of a weak ligand trans to the NHC ligand in order to accelerate initiation is a recurring theme. If a ligand is a too weak, however, the complex might decompose or dimerize, but one way to stabilize weak-ligand complexes somewhat is by incorporating the weak ligand as a bidentate in the initiating carbene or in an anion. The ether-carbene variant is known as the Hoveyda–Grubbs catalyst (Figure 10.1). When a carboxylate is used as one of the anions, the weak ligand can be connected to the anion. More examples are shown in Figure 10.1. This technique does not always result in high rates and most likely the increased stability makes only a small proportion of the applied catalyst active. The thio-ether analogue of the Grubbs–Hoveyda catalyst has also been reported [128].

As phosphine catalysts, NHC catalysts can also be generated conveniently *in situ* from (cymene)$RuCl_2$ dimer, *t*-Bu-ethyne, and an NHC precursor in the absence of hydrogen and phosphine [134]. The *in situ* formed catalyst is more active than the $PCy_3$-containing Grubbs II catalyst, because in this way the slow phosphine

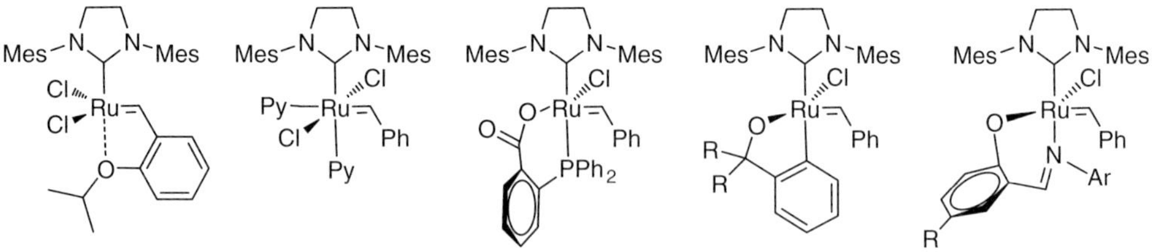

**Figure 10.1** The (bidentate) weak-ligand concept (respectively [129–133]).

dissociation step is circumvented. In the absence of *t*-Bu-ethyne the systems showed a much lower activity, proving that alkynes accelerate ruthenium alkylidene formation.

Another way to circumvent the slow initiation process of Grubbs's catalysts was reported by Piers and coworkers [135]. They took advantage of the stoichiometric metathesis reaction reported by Heppert and coworkers between a Grubbs catalyst precursor and a methylene cyclopropane known as Feist's ester [136]. Metathesis and elimination of diethyl fumarate provided unusual ruthenium carbides (Scheme 10.29). Protonation with an acid of a weakly coordinating anion $[H(OEt_2)_2]^+[B(C_6F_5)_4]^-$ resulted in the migration of a phosphine to the carbide carbon atom. The cationic four-coordinate, 14-electron phosphonium alkylidene complexes obtained were found to be the catalyst precursors with the fastest initial rates. The initiation with the first alkene molecule gives a vinyl phosphonium salt and the unsaturated reactive species.

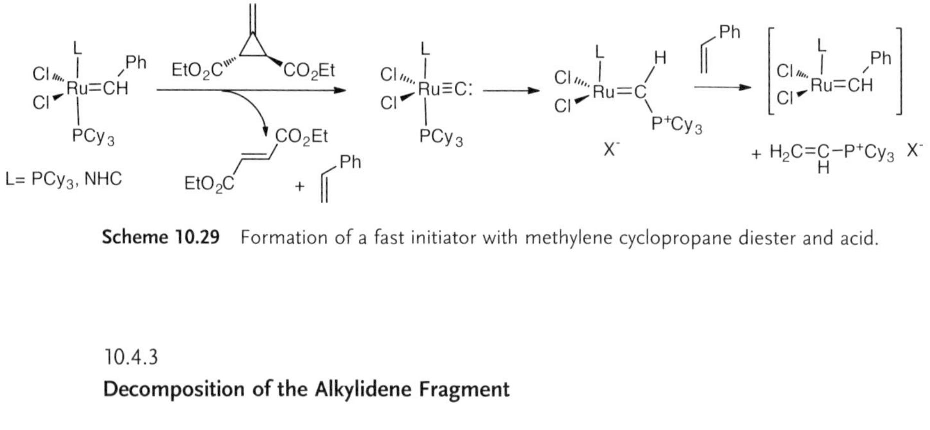

**Scheme 10.29** Formation of a fast initiator with methylene cyclopropane diester and acid.

### 10.4.3
### Decomposition of the Alkylidene Fragment

Ulman and Grubbs studied the decomposition mechanism of various Grubbs I catalysts in the RCM reaction [137]. Substituted alkylidenes were found to decompose through bimolecular pathways, while the unsubstituted methylidene was found to decompose unimolecularly, as indicated by the kinetics. The bimolecular reaction requires dissociation of a phosphine and the organic product is an alkene; the inorganic product could not be identified, but it included ruthenium hydrides. This decomposition pathway can be retarded by addition of phosphine, but this also slows the metathesis reaction. The methylidene ruthenium decomposition data fit a first-order kinetics plot, and the presence of excess free phosphine did not affect the rate of decomposition. No ethylene formation was observed by $^1$H NMR. The ruthenium complex could not be identified. The PCy$_3$-based initiator was significantly better for RCM than the P-$i$-Pr$_3$-based analogue.

The decomposition pathway of the most sensitive catalyst intermediate, the ruthenium methylidene species, was further studied by Grubbs and coworkers. It was found that the liberated PCy$_3$ during the initiation carries out a nucleophilic attack on the methylidene fragment [138]. The phosphorus ylid formed dissociates from the ruthenium complex and, subsequently, the highly basic phosphorus ylid abstracts a proton from another ruthenium methylidene and methyltricyclohexylphosphonium chloride is formed, one of the products observed previously. The resulting alkylidyne ruthenium complex reacts with the other, coordinatively unsaturated ruthenium species, it was proposed, with formation of the dimer shown in Scheme 10.30. The dimer was formed with a yield of 46%.

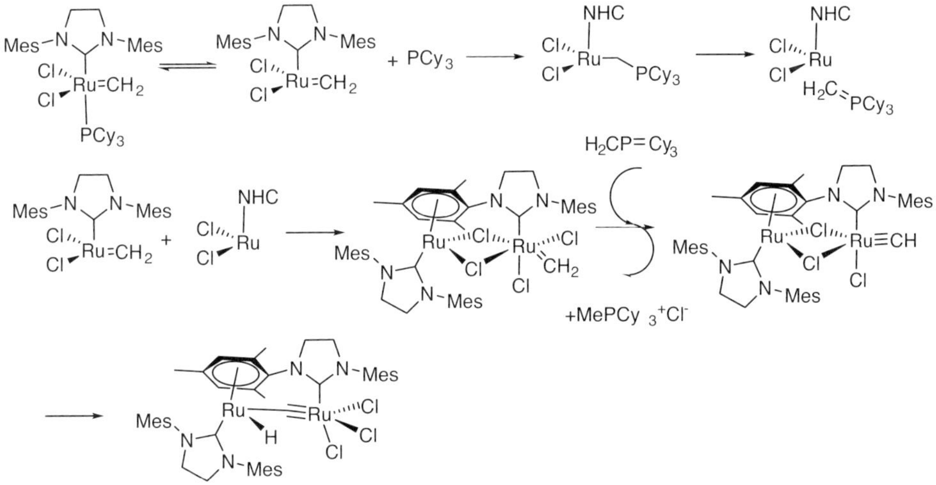

**Scheme 10.30** Decomposition of Ru=CH$_2$ in Grubbs II catalyst.

Before the above discovery, Hofmann and coworkers described the attack of electron-rich phosphines on the *benzylidene* fragment coordinated to a Grubbs I catalyst; in this instance the phosphine is a bidentate and the attack may be aided by the intramolecular character [139]. This would not be a common procedure to arrive at known active catalysts, which are all trans-diphosphine complexes, but they suggested that this reaction might be of importance in catalyst decomposition, as dissociation of phosphine is required from the ruthenium precursor to start catalysis. The reaction is shown in Scheme 10.31. After the formation of the ylid, triphenylphosphine dissociation takes place and the complex dimerizes. Remarkably, a catalyst based on ($t$Bu$_2$P)$_2$CH$_2$ only as the ligand – coordinating in a cis fashion! – was at the time the fastest catalyst for ROMP of cyclooctene, developed by Hofmann and coworkers [140].

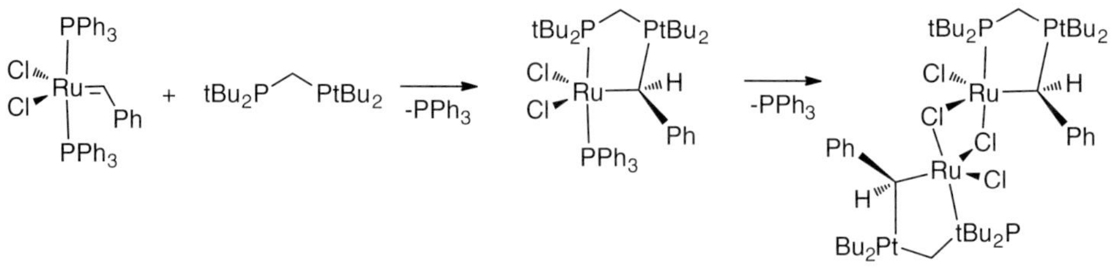

**Scheme 10.31** Nucleophilic attack of phosphine on carbene.

Maughon and coworkers conducted a kinetic study on the metathesis of methyl oleate and ethene with the Grubbs I catalyst [141]. At 25 °C and 4 bar of ethene, catalysis resulted in the selective formation of 1-decene and methyl 9-decenoate with very high initial rates. At higher conversions the rate dropped below commercially acceptable levels. They concluded that the terminal alkene products cause an increasing amount of non-productive metathesis, and the concentration of the methylidene ruthenium species increases, the main cause for decomposition. DFT calculations confirmed that ethene coordination to an alkylidene intermediate is highly favored with respect to oleate coordination.

The instability of Grubbs I methylidene catalyst was confirmed in a continuous flow reactor in the metathesis of ethene and 2-butene [142].

Delaude and coworkers (and others, see references therein) have shown that highly active catalysts can be prepared from the cymene ruthenium dimer in several steps via reaction with ethene and ligand (Cy$_3$P or NHC), see Scheme 10.32. The ethene complex was converted into catalysts by the procedure discussed above (see Section 10.4.2) involving reaction with derivatives of propargylic alcohols, obtaining, eventually, indenylidene precursors. In the presence of ethene or terminal alkenes they observed rapid deactivation of the highly active catalysts. A stoichiometric reaction of the indenylidene or benzylidene catalysts (both NHC- and Cy$_3$P-based) reverted the alkylidene precursors rapidly to the ethene-cymene precursor; in view of the high yield observed for this reaction and the absence of other products the authors proposed that this is a bimolecular process converting methylidene into ethene [143]. As the PCy$_3$ ligand is retained in the product the mechanism does not seem to involve the mechanism found by Grubbs, which starts with an attack of PCy$_3$ on the

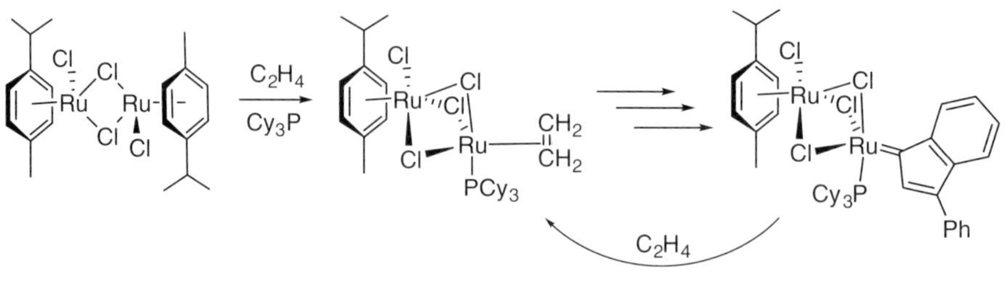

**Scheme 10.32** Formation of ruthenium-alkylidene Lewis acid adducts and their reversal to the ethene complex.

methylidene fragment, releasing the ylid (or phosphonium salt, see Scheme 10.30), and subsequent reaction with ethene giving an ethene complex [138a].

## 10.4.4
### Reactions Involving the NHC Ligand

Carbene-metal decomposition routes, be they ligand or coordinated substrate fragment, were presented in Section 1.7.3, and several of the reactions presented there are concerned with metathesis catalysis. The reactions discussed in this part concern mainly C—H activation reactions of the NHC ligand. Grubbs and coworkers described metallation of a methyl group of the $H_2IMes$ (NHC) ligand under "a moderately rigorous inert atmosphere" [144]. The overall reaction is shown in Scheme 10.33. A similar C–H activation for ruthenium hydrides was reported by Whittlesey [145], who also observed insertion of Ru into C—C bonds (methyl-phenyl), with elimination of methane. The complex resulting from C—H activation contains a carbonyl group, which usually evolves from an alcohol, but in this case details were missing.

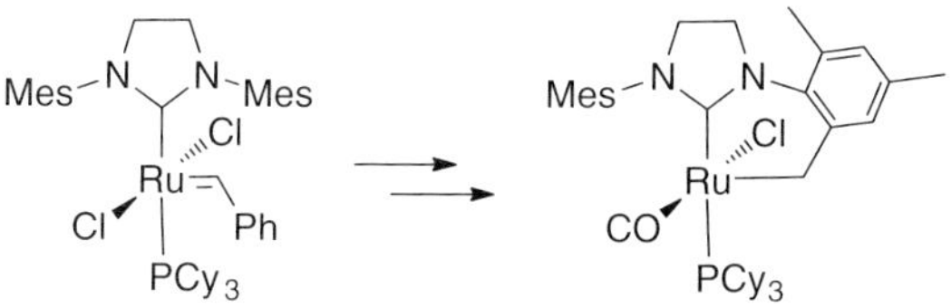

**Scheme 10.33**   The first C—H activation in Grubbs II catalyst.

Metallation of both phenyl rings of the NHC was reported by Grubbs and co-workers (see Scheme 10.34). After the first C—H activation most likely insertion of the benzylidene into Ru–H takes place, followed by reductive elimination of the phenyl and benzyl intermediate [146]. The second metallation is followed by reductive elimination of HCl and phosphonium salt formation. The reaction is much faster in dichloromethane than in benzene.

Blechert and coworkers found that the Hoveyda–Grubbs catalyst undergoes a reversible cyclization (formally a Cope rearrangement) rather than C—H activation, shown in Scheme 10.35, which, after oxidation of the ruthenacycle, renders the compound inactive. The oxidation with oxygen re-establishes the aromatic ring and another alkylidene–ruthenium bond, not suitable for catalysis, is formed [147]. The compound was obtained in low yield and most likely the reaction can be prevented by the appropriate substitution pattern on the aryl ring.

In the absence of oxygenates and oxygen as the cause of catalyst decomposition, C—H activation of the aryl ring on the NHC ligand is an important starting point for decomposition reactions. In recent years, many groups have focused on variation of the NHC ligand in order to obtain a more stable catalyst (see references in [148b]). As mentioned above, the rate of this reaction depends strongly on the particular catalyst

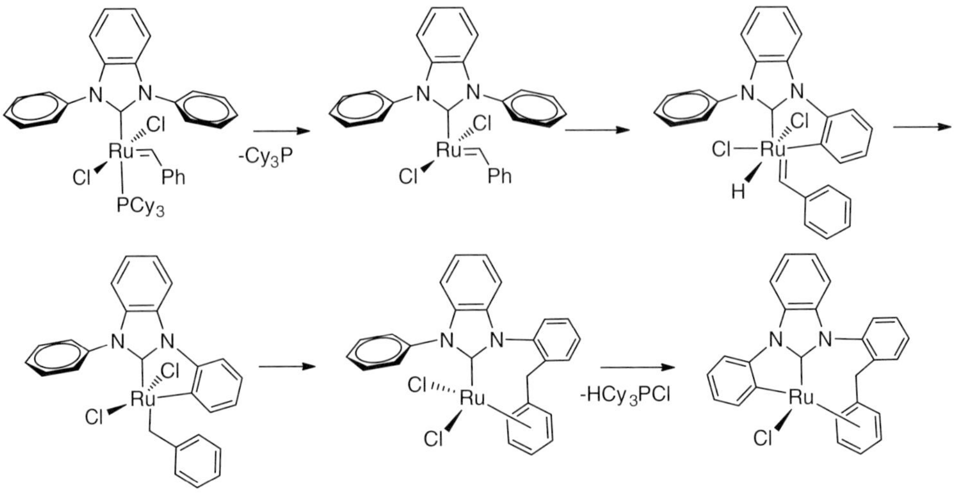

**Scheme 10.34** Double C—H activation in a Grubbs II catalyst.

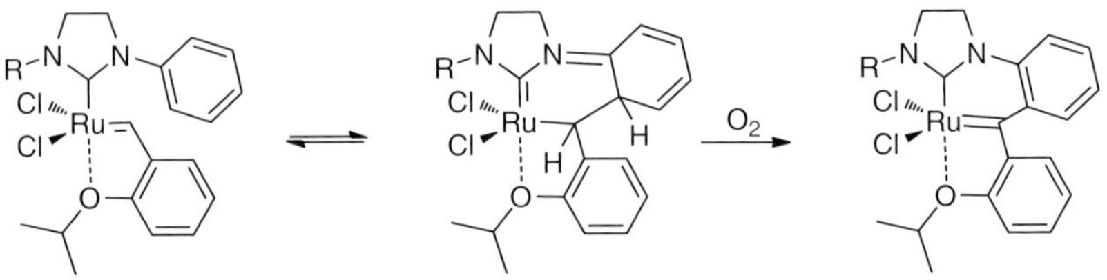

**Scheme 10.35** Reversible cyclization involving NHC ligand and alkylidene in Hoveyda–Grubbs catalyst.

structure and it can be reduced effectively by choosing the proper substitution pattern on the aromatic rings and the backbone of the NHC ligand, as was shown by Grubbs and coworkers [148]. They studied a whole range of NHC ligands in the RCM of diethyl diallylmalonate, often chosen as a benchmark reaction. They developed a very sensitive assay and were able to use as little as 25 ppm of catalyst in order to obtain data for the more stable catalysts. At these concentrations of catalyst, oxygen had to be excluded rigorously. It was found that the less bulky N-aryl on the NHC ligands led to increased activity, but, not unexpectedly, it also decreased the stability. Increased backbone substitution led to increased catalyst lifetimes and decreased reaction rates. The most stable catalyst for this reaction, used at 15–25 ppm catalyst level, is shown in Figure 10.2 under strictly oxygen-free conditions. Restricted rotation prevents close contact between Ru and the aryl group and thus a fast C—H activation reaction.

Dorta and coworkers studied substituted 1-naphthyl groups as the aryl groups at the NHC ligand [149] of a Grubbs II catalyst, that is, it contains PCy$_3$. Especially, 2-isopropyl-1-naphthyl (with another substituent at C-6 or C-7) gave stable and fast catalysts, but the study was done at a catalyst concentration of 0.1 mol%.

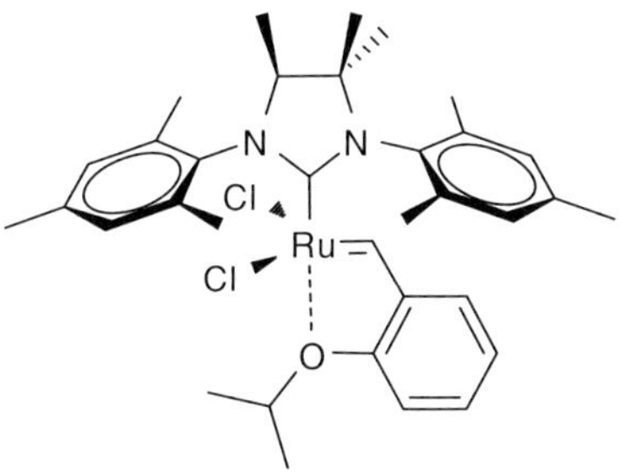

**Figure 10.2** The most stable and active catalyst for RCM of diethyl diallylmalonate [148].

Wagener and coworkers undertook a deuterium labeling study to determine the mechanism of olefin isomerization during the metathesis reactions catalyzed by a second-generation Grubbs catalyst [150]. Metathesis allyl-1,1-*d2* methyl ether gave evidence that a metal hydride addition–elimination mechanism was operating as deuterium was observed in vinylic positions. Subsequently, an NHC ligand was used bearing deuterated *o*-methyl groups on the aromatic rings. Its decomposition afforded several metal hydride species, but no deuteride complex was detected. Metathesis of 1-octene led to a variety of deuterated alkenes (see Scheme 10.36), indicating the presence of a deuteride complex, albeit not observable spectroscopically. They proposed that the H/D exchange is promoted by a ruthenium hydride intermediate, the formation of which is closely related to the decomposition of the methylidene catalyst (Scheme 10.30).

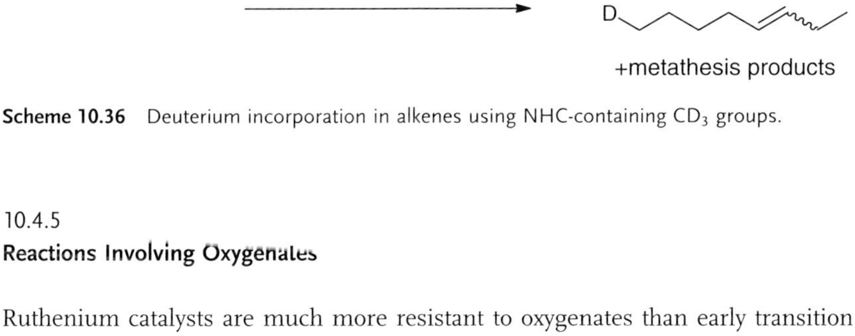

+metathesis products

**Scheme 10.36** Deuterium incorporation in alkenes using NHC-containing $CD_3$ groups.

## 10.4.5
## Reactions Involving Oxygenates

Ruthenium catalysts are much more resistant to oxygenates than early transition metal catalysts, as has been known since 1965 [4], but nevertheless, in the case of oxygenates in organic synthesis often 1% of catalyst is used. Before the discovery of the alkylidene and NHC decomposition reactions the focus of catalyst deactivation was on the removal of impurities in the feed. Turnover numbers for hydrocarbon alkene substrates may be as high as one million if hydroperoxides are removed down to the ppm level [120]. Oxygenates were generally considered as being intrinsically reactive toward metal alkylidenes, and thus limited turnover numbers were expected. Note, however, that tungsten catalysts (Me$_4$Sn activated) will also convert up to 500 mol of oleate substrates [151]. Mol and coworkers, using Grubbs I catalyst, achieved 2500 turnovers for methyl oleate [152]. This was improved to almost 20 000 by researchers from Sasol [153], also after careful purification of the substrate, with the use of a new ligand instead of PCy$_3$, namely 9-cyclohexyl-9-phospha-9$H$bicyclo-nonane (used as a 3:1 mixture of [3.3.1]- and [4.2.1]-bridged isomers). Turnover numbers remain lower than those of pure alkenes, but the intrinsic deactivation is lower than one might conclude from the amounts of catalyst used in organic syntheses.

Substrates containing aldehydes show low turnover numbers, and α-β unsaturated aldehydes show no activity at all. Protection of the aldehyde functionality as an acetal gave cross metathesis conversion, but still 2–5% catalyst was needed [154].

During the synthesis of Grubbs II catalysts Grubbs and coworkers found that the presence of primary alcohols led to hydrido carbonyl ruthenium chlorides [144]. Simultaneously, Dinger and Mol [155] identified the formation of such compounds via the deactivation of Grubbs I catalysts in the presence of alcohols. As the mechanism they proposed the scheme presented in Scheme 10.37. Similar results were observed by other groups [156].

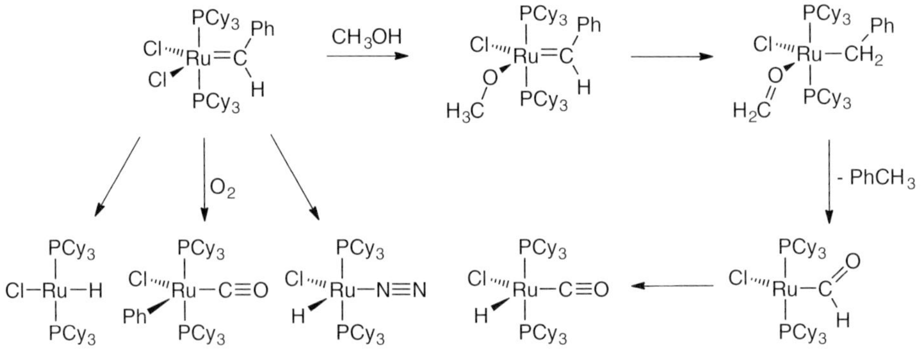

**Scheme 10.37** Grubbs I catalyst decomposition with alcohols and dioxygen.

The Grubbs II catalyst (Scheme 10.38) reacted faster with methanol under basic conditions than the Grubbs I catalyst and gave a mixture of products, including a ligand disproportionation reaction giving the same hydride that results from Grubbs I; the mechanism was proposed to be the same as that shown in Scheme 10.37 [157]. The disproportionation to bis-phosphine complexes is surprising, as NHCs usually bind much more strongly than phosphines. The fate of the NHC could not be established.

The hydrides formed are highly active isomerization catalysts, and thus their formation could drastically influence the selectivity of the metathesis reaction. The

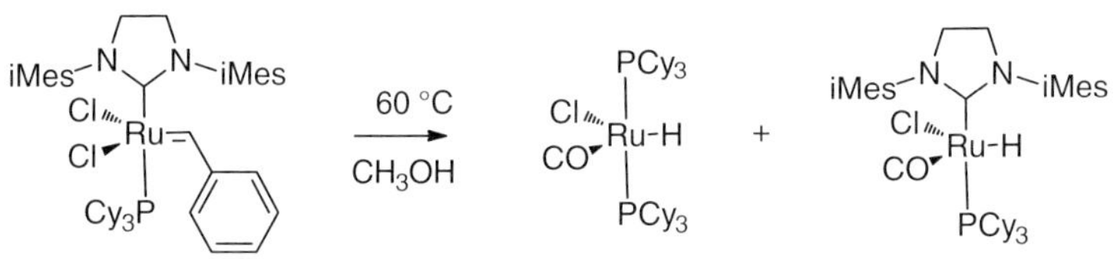

**Scheme 10.38** Grubbs II catalyst decomposition with alcohols and dioxygen.

metathesis reaction is still faster than the methanolysis, and thus carrying out the reaction in alcohols has only a small effect on the outcome. Dinger and Mol observed that at high substrate to catalyst ratio (100 000:1) in 1-octene metathesis, and thus with longer reaction times, considerable isomerization took place.

Also in 2003, Werner and coworkers reported on the catalyst decomposition of Grubbs I when allyl alcohol was used as the substrate [158]. The decomposition route starts with a normal metathesis giving styrene and a hydroxy-ethylidene ruthenium complex, but the latter decomposes giving a carbonyl ruthenenium dichloride (Scheme 10.39).

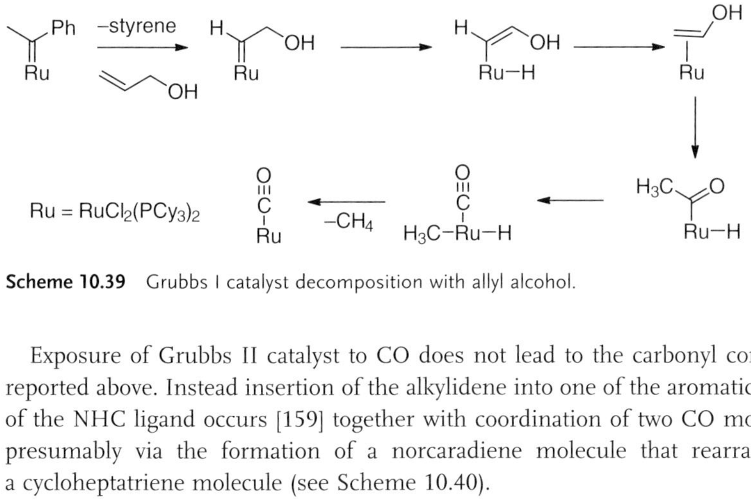

**Scheme 10.39** Grubbs I catalyst decomposition with allyl alcohol.

Exposure of Grubbs II catalyst to CO does not lead to the carbonyl complexes reported above. Instead insertion of the alkylidene into one of the aromatic groups of the NHC ligand occurs [159] together with coordination of two CO molecules, presumably via the formation of a norcaradiene molecule that rearranges to a cycloheptatriene molecule (see Scheme 10.40).

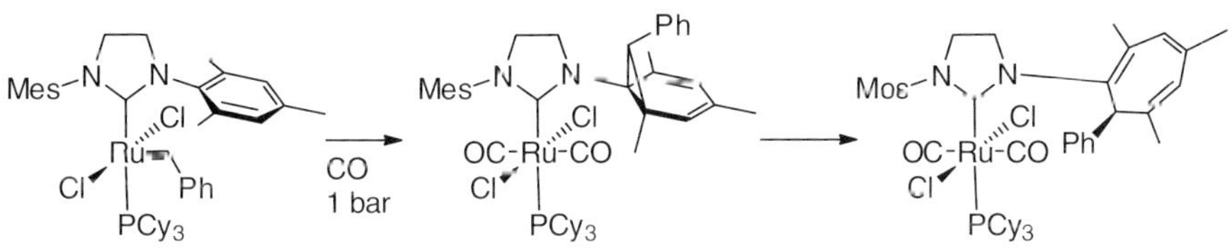

**Scheme 10.40** Benzylidene expulsion by CO, and norcaradiene, cycloheptatriene formation.

A ditriflate derivative of a Grubbs II catalyst containing benzylidene under slightly acidic conditions in water and acetonitrile reacts in 4 days at room temperature with water to give benzaldehyde quantitatively, according to the proposed mechanism shown in Scheme 10.41 as described by Lee and coworkers [160]. The reaction most likely starts with a nucleophilic attack of water on the carbene carbon atom coordinated to a cationic ruthenium ion. It might be compared with the formation of phosphine oxides from phosphines coordinated to divalent palladium and water (Sections 1.4.2 and 1.4.4). A deuterated benzylidene was used to show that there was no hydrogen exchange with water.

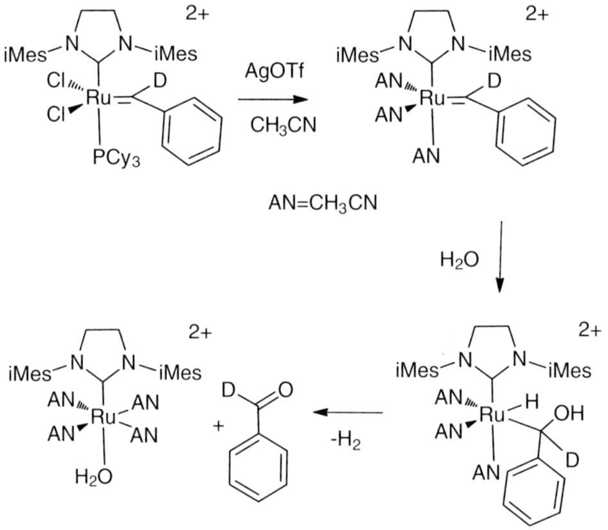

**Scheme 10.41** Reaction of a cationic NHC catalyst with water.

Using water-soluble catalysts, RCM in water hardly gives any metathesis, while ROMP may proceed well [161]. This is due to the instability in water of the methylidene intermediate, formed after the first cycle in which the benzylidene is replaced; methyl- or phenyl-substituted $\alpha,\omega$-dienes gave much better results in RCM than the unsubstituted ones [162]. Hong and Grubbs synthesized a water-soluble Hoveyda–Grubbs catalyst with a polyethyleneglycol attached to the saturated backbone of the NHC. This catalyst was used successfully in RCM and CM in water, although 5% catalyst was still needed.

Also in ionic liquids as the solvent for metathesis, water leads to catalyst decomposition. Stark and coworkers studied the effect of impurities on the activity of Grubbs I, Grubbs II and the Hoveyda–Grubbs catalyst [163] in the self-metathesis of 1-octene. The latter turned out to be less sensitive to the impurities usually present in ionic liquids, viz. water, halides, and 1-methylimidazole. Below the limit of detection of these impurities the Grubbs II and Hoveyda's catalysts perform equally well. On a molar basis 1-methylimidazole is the strongest inhibitor of the three poisons. At the concentrations used (at RT) 6 moles of 1-methylimidazole per ruthenium atom caused a decrease in activity by a factor of 11. Grubbs II catalyst is most sensitive to water and 1-methylimidazole, which might mean that the NHC ligand is the target of these impurities (as in Schemes 10.38–10.40). The intrinsic activity of the catalysts in ionic liquids seemed higher than that in pure 1-octene.

Vosloo and coworkers compared the influence of water, 1-butanol, and acetic acid on the self-metathesis of 1-octene with Grubbs I catalyst in various solvents [164]. The system was most active in chlorobenzene and the activity and selectivity showed the order water > 1-butanol > acetic acid. A Schrock catalyst $W(O-2,6-C_6H_3Ph_2)_2Cl_4/Bu_4Sn$ was half as active and more sensitive to the impurities studied.

Effective catalyst "poisons" that react rapidly with the metal alkylidene species can be used to our advantage as quenching reagents, for instance in polymerization reactions. In anionic living polymerizations the polymer chain can be functionalized either at the initiating side of the chain or at the end by an appropriate quenching agent. Kilbinger and coworkers introduced various reagents that can be used as terminators for a living ROMP reaction providing a functional group at the end of the terminated chain [165]. They used the monomer PNI, see Scheme 10.42, as a model cycloalkene (this one is already polar, and applications would concern rather norbornene or cyclooctene). The reagent shown is vinylene carbonate, which yields a polymer terminated with an aldehyde group. The electron-poor alkene vinylene carbonate reacts quantitatively with the growing chains, forming a Fischer carbene. This particular Fischer carbene eliminates a carboxylic acid leaving a ruthenium carbide behind [166], which is not an active metathesis catalyst unless it is reactivated. With the use of 3-Br-pyridine as the ligand in the complex (to reduce incubation times), the end-capping gave perfect results.

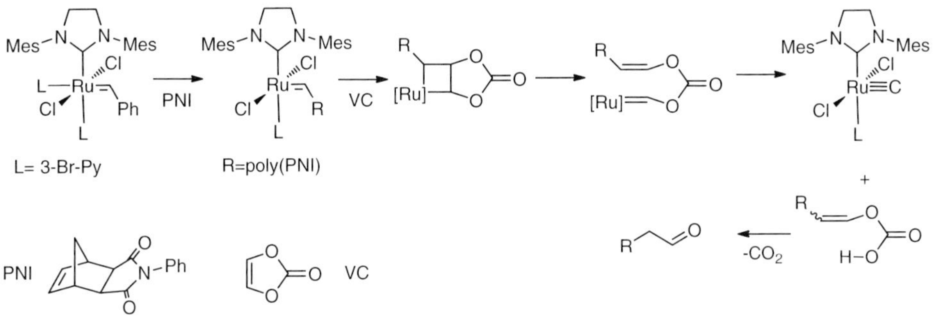

**Scheme 10.42** End-capping of living ROMP polymers with vinylene carbonate.

## 10.4.6
## Tandem Metathesis/Hydrogenation Reactions

The hydrides formed via any of the decomposition reactions reviewed so far are highly active hydrogenation catalysts and can be used as such. In addition, the alkylidene catalysts were transformed deliberately into a hydride species in order to carry out a tandem metathesis/hydrogenation reaction. The first application as a tandem catalyst is probably from McLain and coworkers in a brief summary of syntheses of polymers, in which they reported briefly the ROMP of the methyl ester cyclooctene-carboxylic acid, obtained from 1,5-cyclooctadiene and palladium catalyzed methoxycarbonylation [167]. After metathesis, the product was hydrogenated by the metathesis Grubbs I type catalyst by adding hydrogen at high pressure and temperature. The authors speculated that the metathesis alkylidene catalyst was converted to $(Cy_3P)_2RuCl_2(H_2)$, an active hydrogenation catalyst, under the influence of dihydrogen.

Prior to tandem reactions $(Cy_3P)_2RuClH(CO)$, the decomposition product of the Grubbs I catalysts with water and alcohols, was used as a hydrogenation catalyst in several instances. The use of $(Cy_3P)_2RuClH(CO)$ as a hydrogen transfer catalyst was reported by Graser and Steigerwald in 1980 and at the time it was the fastest phosphine ruthenium catalyst for this reaction [168]. $\alpha,\beta$-Unsaturated aldehydes were selectively hydrogenated in bulk to unsaturated alcohols with ruthenium complexes as homogeneous catalysts by Strohmeier [169]. Of the tested complexes, which did not include $(Cy_3P)_2RuClH(CO)$, $RuCl_2(CO)_2[PCy_3]_2$ was the most effective catalyst for this reaction. Sánchez-Delgado and coworkers found that, for the hydrogenation of ketones to alcohols, there was not that much difference in rate between $(Cy_3P)_2RuClH(CO)$ and its $PPh_3$ analogue [170]. Replacing the chloride ion by trifluoroacetate gave the fastest catalyst. Rempel and coworkers, also in 1997, reported on the hydrogenation of nitrile-butadiene rubber with the use of $(Cy_3P)_2RuClH(CO)$ [171].

As complexes of arenes and ruthenium dichloride are easily converted into active hydrogenation catalysts [172] with bases, Dias and Grubbs [114] added triethylamine at the end of the ROMP metathesis reaction carried out with the dimer shown in Scheme 10.26 and, after hydrogenation, indeed obtained saturated polymers (PE, in their case). In hindsight, the presence of the fragments of (cymene)$RuCl_2$ may not have been a prerequisite for the hydrogenation to take place.

Yi and coworkers studied the activity of $(Cy_3P)_2RuClH(CO)$ as a hydrogenation catalyst for alkenes [173], initially without reference to metathesis literature, where it was described as the decomposition product resulting from Grubbs I catalysts and water or primary alcohols, as reviewed above (Section 10.4.5). Hydrogenation of 1-hexene was achieved with a TOF of $12\,000\,h^{-1}$ at 4 bar and room temperature. The rate could be increased threefold by the addition of $HBF_4 \cdot OEt_2$, which led to the formation of a new hydride species, containing only one phosphine, the other phosphine molecule being removed as its phosphonium salt [174]. Yi and Nolan replaced one $Cy_3P$ ligand by an NHC ligand (IMes) and the rate of hydrogenation increased to $24\,000\,h^{-1}$ under the same conditions for 1-hexene [175].

Watson and Wagener reported on a tandem reaction which consisted of an ADMET or ROMP reaction, followed by a hydrogenation reaction using a Grubbs I catalyst [176]. After the metathesis reaction silica was added to the solution, at which the ruthenium carbene or its decomposition products precipitated and they were converted to a heterogeneous hydrogenation catalyst (8 bar, 90 °C). In this way for example, aliphatic polyesters were obtained (see Scheme 10.43).

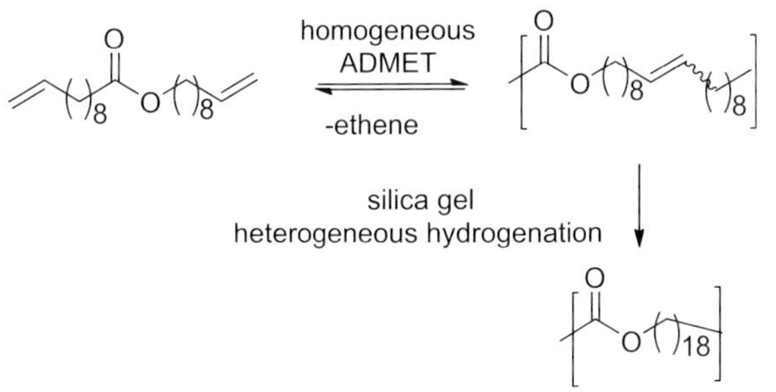

**Scheme 10.43** Homogeneous ADMET followed by heterogeneous hydrogenation.

Grubbs and coworkers studied a variety of substrates in metathesis reactions (RCM, cross metathesis) using Grubbs I and II catalysts followed by *in situ* hydrogenation [177]. The Grubbs I catalyst reacts with hydrogen in the presence of base to give $(Cy_3P)_2RuClH(H_2)$ [178]. Addition of base was not necessary for the hydrogenation of alkenes, but in order to obtain activity for ketone hydrogenation ethylenediamine, base and isopropanol were added, in analogy with the catalysts described above and reported before the metathesis era. Enones could be hydrogenated stepwise, first the alkene in the absence of the basic additives, and then the ketone function after adding the bases and alcohol, but the yield of the latter reaction was modest. Aromatics were not suitable as solvents for the hydrogenation as they gave only isomerization of the alkene metathesis products. A hydrogen transfer protocol could also be used for the hydrogenation of ketones with isopropanol as the hydrogen donor.

Fogg and coworkers reported how a catalyst could be used in a sequence of ROMP–hydrogenation–ROMP [179]. The Grubbs I catalyst was converted to a hydrogenation catalyst by addition of methanol and hydrogen. It was found that methanol functions as a powerful accelerating agent in the hydrogenation chemistry, and rates increased with increased MeOH concentration. We now know that methanol converts the metathesis catalyst into $(Cy_3P)_2RuClH(CO)$. $Et_3N$ was added to promote the formation of active hydrido species. Metathesis activity was restored by addition of 3-chloro-3-methyl-1-butyne [121b].

Dinger and Mol achieved TOFs as high as $160\,000\,\mathrm{mm^{-1}h^{-1}}$ for 1-octene hydrogenation at $100\,^\circ C$ and 4 bar of hydrogen using $(NHC)PCy_3Ru(CO)HCl$ which was obtained by reaction of Grubbs II catalysts with methanol and $Et_3N$ [180].

Schmidt and Pohler added sodium hydride to the solution after RCM using Grubbs II catalyst to convert the ruthenium species into active hydrogenation catalysts [181]. Hydrogen can also be generated *in situ* by adding protic solvents to an excess of NaH present.

Another tandem reaction is isomerization, often an undesired side-reaction of metathesis. Snapper and coworkers used it to their advantage [182]. Grubbs II and Hoveyda–Grubbs' ruthenium alkylidenes were shown to be effective catalysts for cross-metatheses of allylic alcohols with cyclic and acyclic olefins, as well as for isomerization of the resulting allylic alcohols to alkyl ketones. The net result of this

new tandem methodology is a single-flask process that provides highly functionalized, ketone-containing products from simple allylic alcohol precursors. Tandem
catalysis of metathesis and isomerization was reviewed by Schmidt [183]. Subsequently, Schmidt showed that RCM could be continued by several reactions,
depending on the conditions applied to the ruthenium catalyst, very much in line
with that described above (see Scheme 10.44) [184].

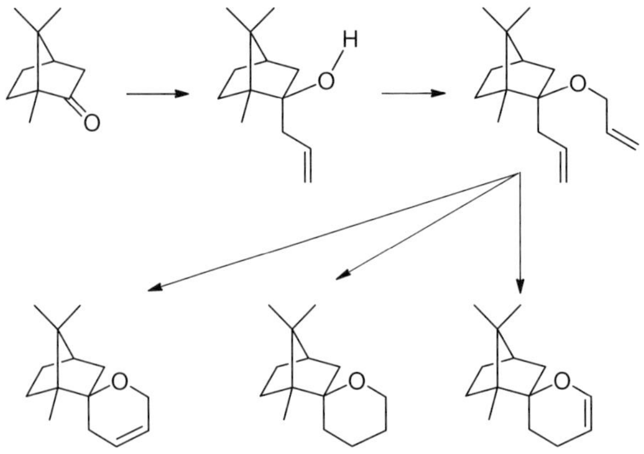

**Scheme 10.44** Combination of RCM, hydrogenation and isomerization.

Tandem reactions of metathesis and other reactions have been reviewed by
Dragutan and Dragutan [185]. Background is provided for most applications encountered to date where fundamental non-metathetical synthetic transformations (hydrogenation, oxidation, isomerization, allylation, cyclopropanation, and so on) and a
variety of name reactions (Diels–Alder, Claisen, Heck, Ugi, Pauson–Khand, Kharasch
addition, and so on) are occurring in tandem, as concurrent or sequential processes,
with every known type of metathetical reactions catalyzed by ruthenium complexes.
The large number of applications stresses the fertility of ruthenium catalysts.

## 10.5
## Conclusions

In the last 15 years metathesis of functionalized alkenes has given access to a huge
number of new compounds, many of which are extremely difficult to make via other
routes. Metathesis has enriched chemical syntheses enormously. Its popularity needs
no further explanation [186]. In spite of the astounding progress in this area there is
still a lot to gain. Several substrates and catalysts show TOFs higher than $100\,000\,h^{-1}$
and TONs of half a million. The substrates in these reactions are unfunctionalized
alkenes, used in CM, ROMP, and RCM. Potentially, TONs may be even higher, as
poisoning experiments in heterogeneous catalysis show that sometimes only 1% of
the metal used is active [187]. For more complex molecules often 1–5% of catalyst is still

used, and, in exceptional cases, even 30–50%. If in the latter instance all catalyst complex were active, this would even be deleterious for the yield in a CM reaction as the alkylidene in the starting catalyst ends up as a fragment of one of the products. Thus there is room for enormous improvement in TOF and TON for functionalized substrates. If catalysts for simple alkenes could be tuned to even higher TONs, neither recycling nor separation of the catalyst from the product may be needed. As yet there seem to be no large-scale applications of homogeneous ruthenium catalysts for simple alkenes, because the amount of catalyst needed is too high. As mentioned above (see Section 10.4.2), the *in situ* regeneration of (part of) the decomposed catalyst (ruthenium) brings it almost within reach. Even for high value-added products the use of 5% of catalyst may be an obstacle for industrial application, although in the laboratory, in view of the extraordinary syntheses that can be achieved, this is not a limitation.

Therefore, it is not surprising that a large research effort is being devoted to deactivation mechanisms and how to avoid these, also by the champions of the area. Apart from polymerization by organometallic complexes there is no area where such a large proportion of research is devoted to catalyst stability, decomposition mechanisms, and regeneration (not counting the many reports on applications).

Heterogeneous catalysts may not be as active as homogeneous catalysts on an average molar basis, but they can be generated by burning off the residues and returning the metal to the desired valence state by the proper treatment. Alkyl-metal activated ones, suitable for metathesis of functionalized alkenes, pose an extra difficulty here as the metal oxides cannot be removed easily. For simple alkenes they have been used in large-scale operations for decades (SHOP, Phillips Triolefin process, OCT).

In the homogeneous catalysis area efforts continue to find a universal catalyst, or recipe, although Schrock stated that it is more likely that for each process a new, optimized catalyst has to be developed [26].

In large-scale applications the use of ionic liquids to aid the separation of catalyst and products may be an option, also because the intrinsic activity in ILs seems somewhat higher than that in pure alkenes (or other solvents) [163, 188]. For fine chemicals the removal of the catalyst could be done by adding silica to the solution to precipitate the catalyst on it; this solution would be acceptable if sufficiently high turnover numbers were achieved, which in the case of low-priced fine chemicals would still amount to 0.5–1 million turnovers per mole of catalyst.

Decomposition of the alkylidene metal catalysts can take place via numerous reactions. For all metals (Mo, W, Re, Ru) the methylidene complexes are the least stable ones. One can distinguish intrinsic instability and reactions with impurities or functionalized substrates. Molybdenacyclobutane decomposes to give propene. Mo and W methylidenes can decompose giving dimers (dimetallacyclobutanes), or produce ethene and M—M double bonds, or nitrogen-bridged dimers when imido ligands are present. More bulky ligands can be applied to prevent dimerization reactions. Both Mo and W alkylidenes were shown to react with aldehydes in a Wittig-type reaction giving metal oxides and alkenes, thus showing the oxophilic character of these metals. By proper substitution this reactivity can be diminished. Many protic compounds lead to decomposition reactions, but a wide range of catalysts has been developed that will resist polar, functional groups in the substrates.

*In situ* prepared Mo and W catalysts are often more active for non-functionalized hydrocarbons than well-defined catalysts. An upper limit for TON seems to be of the order of half a million for thoroughly purified alkenes (over Na/K); in this way not only oxygenates are removed, but also alkynes and dienes. As these purification methods cannot be applied to functionalized alkenes, the turnover numbers of the latter will not reach easily such high values. On the other hand the intermediates obtained by synthesis may not contain the typical refinery impurities, alkynes and dienes!

Internal alkynes can be rapidly metathesized with well-defined alkylidyne precursors, but 1-alkynes cannot, as they give alkylidenes after β-hydride elimination in the metallacyclobutadiene intermediate giving a deprotiotungstenacyclobutadiene complex followed by elimination of ROH. The equilibrium can be restored by adding more alcohol. Unfortunately, addition of ROH to an alkylidyne complex also takes place, which leads to an alkylidene complex, which instead of alkyne metathesis will lead to alkyne polymerization.

Rhenium metallacyclobutanes show β-hydride elimination followed by insertion of ethene into the rhenium hydride bond and further reactions.

A common decomposition reaction of ruthenium catalysts is the one with protic solvents or impurities giving ruthenium hydride species. Hydrides are a cause of isomerization of the substrates or products and, in most cases, it is desirable to avoid this. Ruthenium hydrides can be removed effectively by quinones or alkyne derivatives. The latter regenerate alkylidene ruthenium species which are active catalysts again. Alternatively, ruthenium hydrides can be generated deliberately after completion of the metathesis reaction, to continue the synthesis with hydrogenation of alkenes or other unsaturated intermediates (tandem synthesis).

Ruthenium methylidene is the species with the lowest intrinsic stability, which may end up as part of methylphosphonium salts and/or carbide species. NHC ligands undergo a variety of reactions, such as metallation via C–H activation, or ring expansion with the alkylidene.

Several reactions that destroy the active alkylidene catalyst have been used as quenchers in ROMP reactions, thus putting a molecule with desired properties at the end of the polymer chain.

In conclusion, catalyst decomposition mechanisms have been studied in detail in the metathesis field, including effective use of these reactions, and regeneration reactions. In view of the importance of metathesis for organic synthesis and polymer synthesis many more interesting phenomena will be discovered.

## References

1 (a) Eleuterio, H.S. (1960) US Patent
    3,074,918; *Chem., Abstr* (1961) **55**, 16005;
    (b) Eleuterio, H.S. (1991) *J. Mol. Catal.*,
    **65**, 55.
2 van Leeuwen, P.W.N.M. (2004)
    *Homogeneous Catalysis; Understanding
    the Art*, Ch. 9, Kluwer Academic
    Publishers, Dordrecht, the Netherlands
    (now Springer).
3 Ivin, K.J. and Mol, J.C. (1997)
    *Olefin Metathesis and Metathesis
    Polymerization*, Academic
    Press, San Diego, USA,
    London, UK.

4 Natta, G., Dall'Asta, G., and Porri, L. (1965) *Makromol. Chem.*, **81**, 253.

5 (a) Michelotti, F.W. and Keaveney, W.P. (1965) *J. Polym. Sci.*, **A3**, 895; (b) Rinehart, R.E. and Smith, H.P. (1965) *Polym. Lett.*, **3**, 1049.

6 Novak, B.M. and Grubbs, R.H. (1988) *J. Am. Chem. Soc.*, **110**, 960 and 7542.

7 Hérisson, J.-L. and Chauvin, Y. (1971) *Makromol. Chem.*, **141**, 161.

8 Dolgoplosk, B.A., Makovetskii, K.L., and Tinyakova, E.I. (1972) *Dokl. Akad. Nauk SSSR*, **202**, 871 (Engl. transl. p. 95–97).

9 Dolgoplosk, B.A., Golenko, T.G., Makovetskii, K.L., Oreshkin, I.A., and Tinyakova, E.I. (1974) *Dokl. Akad. Nauk SSSR*, **216**, 807.

10 (a) Sharp, P.R. and Schrock, R.R. (1979) *J. Organomet. Chem.*, **171**, 43; (b) Johnson, L.K., Frey, M., Ulibarri, T.A., Virgil, S.C., Grubbs, R.H., and Ziller, J.W. (1993) *J. Am. Chem. Soc.*, **115**, 8167.

11 (a) Binger, P., Müller, P., Benn, R., and Mynott, R. (1989) *Angew. Chem., Int. Ed. Engl.*, **28**, 610; (b) Johnson, L.K., Grubbs, R.H., and Ziller, J.W. (1993) *J. Am. Chem. Soc.*, **115**, 8130.

12 Schrock, R.R. (1974) *J. Am. Chem. Soc.*, **96**, 6796.

13 (a) Howard, T.R., Lee, J.B., and Grubbs, R.H. (1980) *J. Am. Chem. Soc.*, **102**, 6876; (b) Lee, J.B., Ott, K.C., and Grubbs, R.H. (1982) *J. Am. Chem. Soc.*, **104**, 7491; (c) Straus, D.A. and Grubbs, R.H. (1985) *J. Mol. Catal.*, **28**, 9.

14 Tebbe, F.N., Parshall, G.W., and Reddy, G.S. (1978) *J. Am. Chem. Soc.*, **100**, 3611–3613.

15 Rendón, N., Blanc, F., and Copéret, C. (2009) *Coord. Chem. Rev.*, **253**, 2015–2020.

16 Chabanas, M., Baudouin, A., Copéret, C., and Basset, J.-M. (2001) *J. Am. Chem. Soc.*, **123**, 2062–2063.

17 Rendón, N., Berthoud, R., Blanc, F., Gajan, D., Maishal, T., Basset, J.–M., Copéret, C., Lesage, A., Emsley, L., Marinescu, S.C., Singh, R., and Schrock, R.R. (2009) *Chem. Eur. J.*, **15**, 5083–5089.

18 Blanc, F., Berthoud, R., Copéret, C., Lesage, A., Emsley, L., Singh, R., Kreickmann, T., and Schrock, R.R. (2008) *Proc. Nat. Acad. Sci.*, **105**, 12123–12127.

19 (a) Van Dam, P.B., Mittelmeijer, M.C., and Boelhouwer, C. (1972) *J. Chem. Soc., Chem. Commun.*, 1221; (b) Verkuylen, E. and Boelhouwer, C. (1974) *J. Chem. Soc., Chem. Commun.*, 793; (c) Verkuylen, E., Kapteijn, F., Mol, J.C., and Boelhouwer, C. (1977) *J. Chem. Soc., Chem. Commun.*, 198.

20 Grubbs, R.H. and Hoppin, C.R. (1977) *J. Chem. Soc., Chem. Commun.*, 634–635.

21 Schrock, R.R., Murdzek, J.S., Bazan, G.C., Robbins, J., DiMare, M., and O'Regan, M. (1990) *J. Am. Chem. Soc.*, **112**, 3875.

22 O'Leary, D.J., Blackwell, H.E., Washenfelder, R.A., Miura, K., and Grubbs, R.H. (1999) *Tetrahedron Lett.*, **40**, 1091–1094.

23 Kresss, J., Wesolek, M., and Osborn, J.A. (1982) *J. Chem. Soc., Chem. Commun.*, 514–516.

24 Kresss, J. and Osborn, J.A. (1983) *J. Am. Chem. Soc.*, **105**, 6346–6347.

25 Schrock, R.R. (2002) *Chem. Rev.*, **102**, 145.

26 Schrock, R.R. (2009) *Chem. Rev.*, **109**, 3211–3226.

27 Schrock, R.R., DePue, R., Feldman, J., Schaverien, C.J., Dewan, J.C., and Liu, A.H. (1988) *J. Am. Chem. Soc.*, **110**, 1423.

28 Schaverien, C.J., Dewan, J.C., and Schrock, R.R. (1986) *J. Am. Chem. Soc.*, **108**, 2771–2773.

29 Tsang, W.C.P., Schrock, R.R., and Hoveyda, A.H. (2001) *Organometallics*, **20**, 5658–5669.

30 Takacs, J.M., Reddy, D.S., Moteki, S.A., Wu, D., and Palencia, H. (2004) *J. Am. Chem. Soc.*, **126**, 4494–4495.

31 (a) Kim, H.-J., Moon, D., Lah, M.S., and Hong, J.-I. (2002) *Angew. Chem., Int. Ed.*, **41**, 3174–3177; (b) Portada, T., Roje, M., Hamersak, Z., and Zinic, M. (2005) *Tetrahedron Lett.*, **46**, 5957–5959.

32 (a) Enemark, E.J. and Stack, T.D.P. (1998) *Angew. Chem., Int. Ed.*, **37**, 932–935; (b) Rowland, J.M., Olmstead, M.M., and Mascharak, P.K. (2002) *Inorg. Chem.*, **41**, 1545–1549.

33 Zhu, S.S., Cefalo, D.R., La, D.S., Jamieson, J.Y., Davis, W.M., Hoveyda, A.H., and Schrock, R.R. (1999) *J. Am. Chem. Soc.*, **121**, 8251–8259.

**34** Lopez, L.P.H., Schrock, R.R., and Müller, P. (2006) *Organometallics*, **25**, 1978–1986.

**35** Arndt, S., Schrock, R.R., and Müller, P. (2007) *Organometallics*, **26**, 1279–1290.

**36** Kreickmann, T., Arndt, S., Schrock, R.R., and Müller, P. (2007) *Organometallics*, **26**, 5702.

**37** Peetz, R.M., Sinnwell, V., and Thorn-Csanyi, E. (2006) *J. Mol. Catal. A: Chem.*, **254**, 165–173.

**38** Tarasov, A.L., Shelimov, B.N., Kazansky, V.B., and Mol, J.C. (1997) *J. Mol. Catal.*, **115**, 219–228.

**39** Handzlik, J. and Ogonowski, J. (2002) *Catal. Lett.*, **83**, 287–290.

**40** Tanaka, K. and Tanaka, K.-I. (1984) *J. Chem. Soc., Chem. Commun.*, 748–749.

**41** (a) Binger, P., Mueller, P., Benn, R., and Mynott, R. (1989) *Angew. Chem., Int.Ed. Engl.*, **28**, 610; (b) de la Mata, F.J. and Grubbs, R.H. (1996) *Organometallics*, **15**, 577; (c) Johnson, L.K., Grubbs, R.H., and Ziller, J.W. (1993) *J. Am. Chem. Soc.*, **115**, 8130.

**42** Tonzetich, Z.J., Schrock, R.R., and Mueller, P. (2006) *Organometallics*, **25**, 4301.

**43** Yang, G.K. and Bergman, R.G. (1985) *Organometallics*, **4**, 129.

**44** Freundlich, J.S., Schrock, R.R., and Davis, W.M. (1996) *J. Am. Chem. Soc.*, **118**, 3643.

**45** Mol, J.C. and van Leeuwen, P.W.N.M. (2008) Metathesis of alkenes, in *Handbook of Heterogeneous Catalysis*, 2nd edn, vol. 7 (eds G. Ertl, H. Knözinger, F. Schüth, and J. Weitkamp), Wiley-VCH Verlag GmbH, Weinheim, pp. 3240–3256.

**46** Elev, I.V., Shelimov, B.N., and Kazansky, V.B. (1989) *Kinet. Katal.*, **30**, 895–900.

**47** Vikulov, K.A., Elev, I.V., Shelimov, B.N., and Kazansky, V.B. (1989) *J. Mol. Catal.*, **55**, 126–145.

**48** Vikulov, K.A., Shelimov, B.N., Kazansky, V.B., and Mol, J.C. (1994) *J.Mol. Catal.*, **90**, 61–67.

**49** Vosloo, H.C.M., Dickinson, A.J., and du Plessis, J.A.K. (1997) *J. Mol. Catal. A: Chem.*, **115**, 199–205.

**50** Maynard, H.D. and Grubbs, R.H. (1999) *Macromolecules*, **32**, 6917–6924.

**51** Pennella, F., Banks, R.L., and Bailey, G.C. (1968) *Chem. Commun.*, 1548.

**52** Mortreux, A. and Blanchard, M. (1972) *Bull. Soc. Chim. France*, 1641.

**53** Mortreux, A. and Blanchard, M. (1974) *J. Chem. Soc. Chem. Commun.*, 786.

**54** Bencheick, A., Petit, M., Mortreux, A., and Petit, F. (1982) *J. Mol. Catal.*, **15**, 93.

**55** Wengrovius, J.H., Sancho, J., and Schrock, R.R. (1981) *J. Am. Chem. Soc.*, **103**, 3932.

**56** Freudenberger, J.H., Schrock, S.R.R., Churchill, M.R., Rheingold, A.L., and Ziller, J.W. (1984) *Organometallics*, **3**, 1563.

**57** (a) McCullough, L.G., Listermann, M.L., Schrock, R.R., Churchill, M.R., and Ziller, J.W. (1983) *J. Am. Chem. Soc.*, **105**, 6729–6730; (b) Freudenberger, I.H. and Schrock, R.R. (1986) *Organometallics*, **5**, 1411.

**58** Freudenberger, J.H. and Schrock, R.R. (1985) *Organometallics*, **4**, 1937.

**59** Mortreux, A., Petit, F., Petit, M., and Szymanska-Buzar, T. (1995) *J. Mol. Catal. A: Chem.*, **96**, 95–105.

**60** Schrock, R.R., Luo, S., Lee, J.C. Jr., Zanetti, N.C., and Davis, W.M. (1996) *J. Am. Chem. Soc.*, **118**, 3883.

**61** Bray, A., Mortreux, A., Petit, F., Petit, M., and Szymanska-Buzar, T. (1993) *J. Chem. Soc., Chem. Commun.*, 197–199.

**62** Coutelier, O. and Mortreux, A. (2006) *Adv. Synth. Catal.*, **348**, 2038–2042.

**63** Weiss, K., Goller, R., and Loessel, G. (1988) *J. Mol. Catal.*, **46**, 267–275.

**64** Strutz, H., Dewan, J.C., and Schrock, R.R. (1985) *J. Am. Chem. Soc.*, **107**, 5999–6005.

**65** Pedersen, S.F., Schrock, R.R., Churchill, M.R., and Wasserman, H.J. (1982) *J. Am. Chem. Soc.*, **104**, 6808–6809.

**66** Moulijn, J.A., Reitsma, H.J., and Boelhouwer, C. (1972) *J. Catal.*, **25**, 434–459.

**67** Ardizzoia, G.A., Brenna, S., LaMonica, G., Maspero, A., and Masciocchi, N. (2002) *J. Organomet. Chem.*, **649**, 173–180.

**68** Blanc, F., Berthoud, R., Salameh, A., Basset, J.-M., Copéret, C., Singh, R., and Schrock, R.R. (2007) *J. Am. Chem. Soc.*, **129**, 8434–8435.

69 Rhers, B., Salameh, A., Baudouin, A., Quadrelli, E.A., Taoufik, M., Copéret, C., Lefebvre, F., Basset, J.-M., Solans-Monfort, X., Eisenstein, O., Lukens, W.W., Lopez, L.P.H., Sinha, A., and Schrock, R.R. (2006) *Organometallics*, **25**, 3554–3557.

70 Chaumont, P. and John, C.S. (1988) *J. Mol. Catal.*, **46**, 317–328.

71 Cosyns, J., Chodorge, J., Commereuc, D., and Torck, B. (1998) *Hydrocarbon Process.*, March **77**, 61–66.

72 Beattie, I.R. and Jones, P.J. (1979) *Inorg. Chem.*, **18**, 2318–2319.

73 Herrmann, W.A. (1990) *J. Organomet. Chem.*, **382**, 1–18.

74 Herrmann, W.A., Kuchler, J., Felixberger, J.K., Herdtweck, E., and Wagner, W. (1988) *Angew. Chem.*, **100**, 420–422.

75 Buffon, R., Auroux, A., Lefebvre, F., Leconte, M., Choplin, A., Basset, J.-M., and Herrmann, W.A. (1992) *J. Mol. Catal.*, **76**, 287–295.

76 Buffon, R., Choplin, A., Leconte, M., Basset, J.-M., Touroude, R., and Herrmann, W.A. (1992) *J. Mol. Catal.*, **72**, L7–L10.

77 Horton, A.D. and Schrock, R.R. (1988) *Polyhedron*, **7**, 1841–1853.

78 Wang, W.-D. and Espenson, J.H. (1999) *Organometallics*, **18**, 5170–5175.

79 Toreki, R. and Schrock, R.R. (1990) *J. Am. Chem. Soc.*, **112**, 2448–2449.

80 Toreki, R., Vaughn, G.A., Schrock, R.R., and Davis, W.M. (1993) *J. Am. Chem. Soc.*, **115**, 127–137.

81 Greenlee, W.S. and Farona, M.F. (1976) *Inorg. Chem.*, **15**, 2129–2134.

82 Warwel, S. and Siekermann, V. (1983) *Makromol. Chem. Rapid Commun.*, **4**, 423–427.

83 Mol, J.C. (1999) *Catal. Today*, **51**, 289–299.

84 Bosma, R.H.A., Van den Aardweg, G.C.N., and Mol, J.C. (1983) *J. Organomet. Chem.*, **255**, 159–171.

85 Rodella, C.B., Cavalcante, J.M., and Buffon, R. (2001) *Appl. Catal. A: Gen.*, **274**, 213–217.

86 Kirlin, P.S. and Gates, B.C. (1985) *Inorg. Chem.*, **24**, 3914–3920.

87 Spronk, R., Dekker, F.H.M., and Mol, J.C. (1992) *Appl. Catal. A: Gen.*, **83**, 213–233.

88 Behr, A. and Schüller, U. (2009) *Chem. Ing. Technol.*, **81**, 429–439.

89 Behr, A., Schueller, U., Bauer, K., Maschmeyer, D., Wiese, K.-D., and Nierlich, F. (2009) *Appl. Catal. A: Gen.*, **357**, 34–41.

90 McCoy, J.R. and Farona, M.F. (1991) *J. Mol. Catal.*, **66**, 51–58.

91 Salameh, A., Copéret, C., Basset, J.-M., Böhm, V.P.W., and Röper, M. (2007) *Adv. Synth. Catal.*, **349**, 238–242.

92 Amigues, P., Chauvin, Y., Commereuc, D., Hong, C.T., Lai, C.C., and Liu, Y.H. (1991) *J. Mol. Catal.*, **65**, 39–50.

93 Commereuc, D. (1995) *J. Chem. Soc., Chem. Commun.*, 791–792.

94 Doledec, G. and Commereuc, D. (2000) *J. Mol. Catal. A: Chem.*, **161**, 125–140.

95 Flatt, B.T., Grubbs, R.H., Blanski, R.L., Calabrese, J.C., and Feldman, J. (1994) *Organometallics*, **13**, 2728–2732.

96 Frech, C.M., Blacque, O., Schmalle, H.W., Berke, H., Adlhart, C., and Chen, P. (2006) *Chem. Eur. J.*, **12**, 3325–3338.

97 Herrmann, W.A., Wagner, W., Flessner, U.N., Vokhardt, U., and Komber, H. (1991) *Angew. Chem. Int. Ed. Engl.*, **30**, 1636–1638.

98 Salameh, A., Baudouin, A., Soulivong, D., Boehm, V., Roeper, M., Basset, J.-M., and Copéret, C. (2008) *J. Catal.*, **253**, 180–190.

99 Solans-Monfort, X., Clot, E., Copéret, C., and Eisenstein, O. (2005) *J. Am. Chem. Soc.*, **127**, 14015–14025.

100 Poater, A., Solans-Monfort, X., Clot, E., Copéret, C., and Eisenstein, O. (2007) *J. Am. Chem. Soc.*, **129**, 8207–8216.

101 Leduc, A.-M., Salameh, A., Soulivong, D., Chabanas, M., Basset, J.-M., Copéret, C., Solans-Monfort, X., Clot, E., Eisenstein, O., Böhm, V.P.W., and Röper, M. (2008) *J. Am. Chem. Soc.*, **130**, 6288–6297.

102 Porri, L., Rossi, R., Diversi, P., and Lucherini, A. (1972) *Polym. Prepr. (Am Chem. Soc. Div. Polym. Chem.)*, **13**, 897.

103 Porri, L., Rossi, R., Diversi, P., and Lucherini, A. (1974) *A. Makromol. Chem.*, **175**, 3097.

104 Novak, B.M. and Grubbs, R.H. (1988) *J. Am. Chem. Soc.*, **110**, 7542–7543.

105 (a) Bruneau, C. and Dixneuf, P.H. (1999) *Acc. Chem. Res.*, **32**, 311–323; (b)

Katayama, H. and Ozawa, F. (1998) *Chem. Lett.*, 67–68; (c) Schwab, P., Grubbs, R.H., and Ziller, J.W. (1996) *J. Am. Chem. Soc.*, **118**, 100–110.

**106** Dolgoplosk, B.A., Golenko, T.G., Makovetskii, K.L., Oreshkin, I.A., and Tinyakova, E.I. (1974) *Dokl. Akad. Nauk SSSR*, **216**, 807.

**107** France, M.B., Paciello, R.A., and Grubbs, R.H. (1993) *Macromolecules*, **26**, 4739.

**108** Johnson, L.K., Grubbs, R.H., and Ziller, J.W. (1993) *J. Am. Chem. Soc.*, **115**, 8130; For this method, see also: Binger, P., Müller, P., Benn, R., and Mynott, R. (1989) *Angew. Chem., Int. Ed. Engl.*, **28**, 610.

**109** Nguyen, S.T., Johnson, L.K., Grubbs, R.H., and Ziller, J.W. (1992) *J. Am. Chem. Soc.*, **114**, 3974.

**110** Nguyen, S.T., Grubbs, R.H., and Ziller, J.W. (1993) *J. Am. Chem. Soc.*, **115**, 9858.

**111** Schwab, P., Grubbs, R.H., and Ziller, J.W. (1996) *J. Am. Chem. Soc.*, **118**, 100–110.

**112** Nguyen, S.T., Johnson, L.K., and Grubbs, R.H. (1992) *J. Am. Chem. Soc.*, **114**, 3974–3975.

**113** Dias, E.L., Nguyen, S.T., and Grubbs, R.H. (1997) *J. Am. Chem. Soc.*, **119**, 3887.

**114** Dias, E.L. and Grubbs, R.H. (1998) *Organometallics*, **17**, 2758–2767.

**115** Stuer, W., Wolf, J., and Werner, H. (2002) *J. Organomet. Chem.*, **641**, 203–207.

**116** Sanford, M.S., Love, J.A., and Grubbs, R.H. (2001) *J. Am. Chem. Soc.*, **123**, 6543–6554.

**117** McGuinness, D.S., Cavell, K.J., Skeleton, B.W., and White, A.H. (1999) *Organometallics*, **18**, 1596.

**118** Castarlenas, R., Vovard, C., Fischmeister, C., and Dixneuf, P.H. (2006) *J. Am. Chem. Soc.*, **128**, 4079–4089.

**119** Katayama, H. and Ozawa, F. (2004) *Coord. Chem. Rev.*, **248**, 1703.

**120** Nubel, P.O. and Hunt, C.L. (1999) *J. Mol. Catal. A*, **145**, 323.

**121** (a) Grünwald, C., Gevert, O., Wolf, J., Gonzalez-Herrero, P., and Werner, H. (1996) *Organometallics*, **15**, 1960–1962; (b) For propargyl chloride, see: Wilhelm, T.E., Belderrain, T.R., Brown, S.N., and Grubbs, R.H. (1997) *Organometallics*, **16**, 3867–3869.

**122** Wolf, J., Stüer, W., Grünwald, C., Werner, H., Schwab, P., and Schulz, M. (1998) *Angew. Chem. Int. Ed.*, **37**, 1124–1126.

**123** van Schalkwyk, C., Vosloo, H.C.M., and Botha, J.M. (2002) *J. Mol. Catal. A: Chem.*, **190** 185–195.

**124** Hong, S.H., Sanders, D.P., Lee, C.W., and Grubbs, R.H. (2005) *J. Am. Chem. Soc.*, **127**, 17160–17161.

**125** Love, J.A., Sanford, M.S., Day, M.W., and Grubbs, R.H. (2003) *J. Am. Chem. Soc.*, **125**, 10103–10109.

**126** Morgan, J.P. and Grubbs, R.H. (2000) *Org. Lett.*, **2**, 3153–3155.

**127** Weskamp, T., Kohl, F.J., Hieringer, W., Gleich, D., and Herrmann, W.A. (1999) *Angew. Chem. Int. Ed.*, **38**, 2416–2419.

**128** Ben-Asuly, A., Tzur, E., Diesendruck, C.E., Sigalov, M., Goldberg, I., and Lemcoff, N.G. (2008) *Organometallics*, **27**, 811–813.

**129** Garber, S.B., Kingsbury, J.S., Gray, B.L., and Hoveyda, A.H. (2000) *J. Am. Chem. Soc.*, **122**, 8168–8179.

**130** Gessler, S., Randl, S., and Blechert, S. (2000) *Tetrahedron Lett.*, **41**, 9973–9976.

**131** Samec, J.S.M. and Grubbs, R.H. (2008) *Chem. Eur. J.*, **14**, 2686–2692.

**132** Denk, K., Fridgen, J., and Herrmann, W. (2002) *Adv. Synth. Catal.*, **344**, 666–670.

**133** Allaert, B., Dieltiens, N., Ledoux, N., Vercaemst, C., Van Der Voort, P., Stevens, C.V., Linden, A., and Verpoort, F. (2006) *J. Mol. Catal. A*, **260**, 5482–5486.

**134** Louie, J. and Grubbs, R.H. (2001) *Angew. Chem. Int. Ed.*, **40**, 247–249.

**135** Romero, P.E., Piers, W.E., and McDonald, R. (2004) *Angew. Chem. Int. Ed.*, **43**, 6161–6165.

**136** Carlson, R.G., Gile, M.A., Heppert, J.A., Mason, M.H., Powell, D.R., Vander Velde, D., and Vilain, J.M. (2002) *J. Am. Chem. Soc.*, **124**, 1580.

**137** Ulman, M. and Grubbs, R.H. (1999) *J. Org. Chem.*, **64**, 7202–7207.

**138** (a) Hong, S.H., Wenzel, A.G., Salguero, T.T., Day, M.W., and Grubbs, R.H. (2007) *J. Am. Chem. Soc.*, **129**, 7961–7968; (b) Hong, S.H., Day, M.W., and Grubbs, R.H. (2004) *J. Am. Chem. Soc.*, **126**, 7414–7415.

**139** Hansen, S.M., Rominger, F., Metz, M., and Hofmann, P. (1999) *Chem. Eur. J.*, **5**, 557–566.

**140** Hansen, S.M., Volland, M.A.O., Rominger, F., Eisenträger, F., and Hofmann, P. (1999) *Angew. Chem. Int. Ed.*, **38**, 1273–1276.

**141** Burdett, K.A., Harris, L.D., Margl, P., Maughon, B.R., Mokhtar-Zadeh, T., Saucier, P.C., and Wasserman, E.P. (2004) *Organometallics*, **23**, 2027–2047.

**142** Lysenko, Z., Maughon, B.R., Mokhtar-Zadeh, T., and Tulchinsky, M.L. (2006) *J. Organomet. Chem.*, **691**, 5197–5203.

**143** Sauvage, X., Borguet, Y., Zaragoza, G., Demonceau, A., and Delaude, L. (2009) *Adv. Synth. Catal.*, **351**, 441–455.

**144** Trnka, T.M., Morgan, J.P., Sanford, M.S., Wilhelm, T.E., Scholl, M., Choi, T.-L., Ding, S., Day, M.W., and Grubbs, R.H. (2003) *J. Am. Chem. Soc.*, **125**, 2546–2558.

**145** Jazzar, R.F.R., Macgregor, S.A., Mahon, M.F., Richards, S.P., and Whittlesey, M.K. (2002) *J. Am. Chem. Soc.*, **124**, 4944–4945.

**146** Hong, S.H., Chlenov, A., Day, M.W., and Grubbs, R.H. (2007) *Angew. Chem. Int. Ed.*, **46**, 5148–5151.

**147** Vehlow, K., Gessler, S., and Blechert, S. (2007) *Angew. Chem. Int. Ed.*, **46**, 8082–8085.

**148** (a) Chung, C.K. and Grubbs, R.H. (2008) *Org. Lett.*, **10**, 2693–2696; (b) Kuhn, K.M., Bourg, J.-B., Chung, C.K., Virgil, S.C., and Grubbs, R.H. (2009) *J. Am. Chem. Soc.*, **131**, 5313–5320.

**149** Vieille-Petit, L., Luan, X., Gatti, M., Blumentritt, S., Linden, A., Clavier, H., Nolan, S.P., and Dorta, R. (2009) *Chem. Commun.*, 3783–3785.

**150** Courchay, F.C., Sworen, J.C., Ghiviriga, I., Abboud, K.A., and Wagener, K.B. (2006) *Organometallics*, **25**, 6074–6086.

**151** (a) Baker, R. and Crimmin, M.J. (1977) *Tetrahedron Lett.*, 441; (b) Verkuijlen, E. and Boelhouwer, C. (1974) *J. Chem. Soc., Chem. Commun.*, 793–794.

**152** Buchowicz, W. and Mol, J.C. (1999) *J. Mol. Catal. A: Chem.*, **148**, 97–103.

**153** Forman, G.S., McConnell, A.E., Hanton, M.J., Slawin, A.M.Z., Tooze, R.P., Janse van Rensburg, W., Meyer, W.H., Dwyer, C., Kirk, M.M., and Serfontein, D.W. (2004) *Organometallics*, **23**, 4824–4827.

**154** O'Leary, D.J., Blackwell, H.E., Washenfelder, R.A., Miura, K.,

and Grubbs, R.H. (1999) *Tetrahedron Lett.*, **40**, 1091–1094.

**155** Dinger, M.B. and Mol, J.C. (2003) *Organometallics*, **22**, 1089–1095.

**156** (a) Fürstner, A., Ackermann, L., Gabor, B., Goddard, R., Lehmann, C.W., Mynott, R., Stelzer, F., and Thiel, O.R. (2001) *Chem. Eur. J.*, **7**, 3236; (b) Fogg, D.E., Amoroso, D., Drouin, S.D., Snelgrove, J., Conrad, J., and Zamanian, F. (2002) *J. Mol. Catal. A: Chem.*, **190**, 177–184; (c) Drouin, S.D., Zamanian, F., and Fogg, D.E. (2001) *Organometallics*, **20**, 5495–5497; (d) Werner, H., Grünwald, C., Stüer, W., and Wolf, J. (2003) *Organometallics*, **22**, 1558–1560.

**157** (a) Dinger, M.B. and Mol, J.C. (2003) *Eur. J. Inorg. Chem.*, 2827; (b) Banti, D. and Mol, J.C. (2004) *J. Organomet. Chem.*, **689**, 3113–3116.

**158** Werner, H., Grünwald, C., Stüer, W., and Wolf, J. (2003) *Organometallics*, **22**, 1558–1560.

**159** Galan, B.R., Gembicky, M., Dominiak, P.M., Keister, J.B., and Diver, S.T. (2005) *J. Am. Chem. Soc.*, **127**, 15702–15703.

**160** Kim, M., Eum, M.-S., Jin, M.Y., Jun, K.-W., Lee, C.W., Kuen, K.A., Kim, C.H., and Chin, C.S. (2004) *J. Organomet. Chem.*, **689**, 3535–3540.

**161** Hong, S.H. and Grubbs, R.H. (2006) *J. Am. Chem. Soc.*, **128**, 3508–3509.

**162** Kirkland, T.A., Lynn, D.M., and Grubbs, R.H. (1998) *J. Org.Chem.*, **63**, 9904–9909.

**163** Stark, A., Ajam, M., Green, M., Raubenheimer, H.G., Ranwell, A., and Ondruschka, B. (2006) *Adv. Synth. Catal.*, **348**, 1934–1941.

**164** van Schalkwyk, C., Vosloo, H.C.M., and du Plessis, J.A.K. (2002) *Adv. Synth. Catal.*, **344**, 781–788.

**165** Hilf, S., Grubbs, R.H., and Kilbinger, A.F.M. (2008) *J. Am. Chem. Soc.*, **130**, 11040–11048.

**166** Caskey, S.R., Stewart, M.H., Johnson, M.J.A., and Kampf, J.W. (2006) *Angew. Chem., Int. Ed.*, **45**, 7422–7424.

**167** McLain, S.J., McCord, E.F., Arthur, S.D., Hauptman, E., Feldman, J., Nugent, W.A., Johnson, L.K., Mecking, S., and Brookhart, M. (1997) *Polym. Mater. Sci. Eng.*, **76**, 246–247.

**168** Graser, B. and Steigerwald, H. (1980) *J. Organomet. Chem.*, **193**, C67–C70.

**169** Strohmeier, W. and Holke, K. (1980) *J. Organomet. Chem.*, **193**, C63–C66.

**170** Sanchez-Delgado, R.A., Valencia, N., Marquez-Silva, R.-L., Andriollo, A., and Medina, M. (1986) *Inorg. Chem.*, **25**, 1106–1111.

**171** Martin, P., McManus, N.T., and Rempel, G.L. (1997) *J. Mol. Catal. A: Chem.*, **126**, 115–131.

**172** Hinze, A.G. (1973) *Recl. Trav. Chim. Pays-Bas*, **92**, 542–552.

**173** Yi, C.S. and Lee, D.W. (1999) *Organometallics*, **18**, 5152–5156.

**174** Yi, C.S., Lee, D.W., He, Z., Rheingold, A.L., Lam, K.-C., and Concolino, T.E. (2000) *Organometallics*, **19**, 2909–2915.

**175** Lee, H.M., Smith, D.C. Jr., He, Z., Stevens, E.D., Yi, C.S., and Nolan, S.P. (2001) *Organometallics*, **20**, 794–797.

**176** Watson, M.D. and Wagener, K.B. (2000) *Macromolecules*, **33**, 3196–3201.

**177** Louie, J., Bielawski, C.W., and Grubbs, R.H. (2001) *J. Am. Chem. Soc.*, **123**, 11312–11313.

**178** Drouin, S.D., Yap, G.P.A., and Fogg, D.E. (2000) *Inorg. Chem.*, **39**, 5412–5414.

**179** (a) Drouin, S.D., Zamanian, F., and Fogg, D.E. (2001) *Organometallics*, **20**, 5495–5497; (b) Fogg, D.E., Amoroso, D., Drouin, S.D., Snelgrove, J., Conrad, J., and Zamanian, F. (2002) *J. Mol. Catal. A: Chem.*, **190**, 177–184.

**180** Dinger, M.B. and Mol, J.C. (2003) *Eur. J. Inorg. Chem.*, 2827–2833.

**181** Schmidt, B. and Pohler, M. (2003) *Org. Biomol. Chem.*, **1**, 2512–2517.

**182** Finnegan, D., Seigal, B.A., and Snapper, M.L. (2006) *Org. Lett.*, **8**, 2603–2606.

**183** Schmidt, B. (2004) *Eur. J. Org. Chem.*, 1865–1880.

**184** Schmidt, B. and Staude, L. (2006) *J. Organomet. Chem.*, **691**, 5218–5221.

**185** Dragutan, V. and Dragutan, I. (2006) *J. Organomet. Chem.*, **691**, 5129–5147.

**186** Hoveyda, A.H. and Zhugralin, A.R. (2007) *Nature*, **450**, 243–251.

**187** Chauvin, Y. and Commereuc, D. (1992) *J. Chem. Soc., Chem. Commun.*, 462–464.

**188** Williams, D.B.G., Ajam, M., and Ranwell, A. (2006) *Organometallics*, **25**, 3088–3090.

# Index

*Homogeneous Catalysts: Activity – Stability – Deactivation*, First Edition. Piet W.N.M. van Leeuwen
and John C. Chadwick.
© 2011 Wiley-VCH Verlag GmbH & Co. KGaA. Published 2011 by Wiley-VCH Verlag GmbH & Co. KGaA.